W9-CBK-639

CONSTRUCTION PROJECT ADMINISTRATION

EIGHTH EDITION

Edward R. Fisk, PE
Wayne D. Reynolds, PE

Upper Saddle River, New Jersey
Columbus, Ohio

Library of Congress Cataloging-in-Publication Data

Fisk, Edward R.,
Construction project administration / Edward R. Fisk, Wayne Reynolds.
p. cm.
Includes bibliographical references and index.
ISBN 0-13-099305-0
1. Construction industry—Management. 2. Building—Superintendence. I. Reynolds, Wayne, II. Title.
TH438.F57 2006
624′.068—dc22

2005005093

Assistant Editor: Linda Cupp
Production Coordination: Carlisle Publishers Services
Production Editor: Holly Shufeldt
Design Coordinator: Diane Ernsberger
Cover Designer: Thomas Borah
Cover Art: Getty One
Production Manager: Deidra Schwartz
Marketing Manager: Mark Marsden

This book was set in Usherwood Book by Carlisle Communications, Ltd. It was printed and bound by R. R. Donnelley & Sons Company. The cover was printed by Phoenix Color Corp.

Pearson Education Ltd.
Pearson Education Singapore Pte. Ltd.
Pearson Education Canada, Ltd.
Pearson Education—Japan
Pearson Education Australia Pty. Limited
Pearson Education North Asia Ltd.
Pearson Educación de Mexico, S.A. de C.V.
Pearson Education Malaysia Pte. Ltd.

PEARSON
Prentice Hall

10 9 8 7 6 5 4 3 2 1
ISBN 0-13-099305-0

PREFACE

The principal objective of this book is to provide those of us who are active in the construction industry with a single source of information that will help address the responsibilities and risks that we are likely to encounter. The book not only introduces students, design professionals, project managers, and owners to the special problems of construction, but also serves as a ready reference to experienced contract administrators and construction engineers as well.

The first edition was addressed to students of construction management, on-site representatives, engineers, and inspectors to provide them with a ready source of information in preparing for the responsibilities they could expect to confront on modern construction projects.

However, during the many seminars held by the authors throughout the United States, Guam, Canada, Jamaica, and Mexico, and in the courses they have taught for the University of California, Berkeley, Institute of Transportation Studies; University of Washington, Seattle, Engineering Professional Programs; Eastern Kentucky University; and the American Society of Civil Engineers, it became evident that the project managers, contract administrators, and other management personnel who worked with or exercised control over the on-site project representatives had special problems that also needed to be addressed if the project team concept was to be realized. Thus, the concept for the second edition was born: to bring together the office and field personnel and present them with a workable system for operating as an effective construction team.

The third edition continued the concept of developing the project team approach, with the added consideration of claims-avoidance methods to reduce claims losses. Each member of the project team needed to become intimately familiar with the principles of construction project administration. It was toward this end that the author strove to meet the particular needs of the project team in today's changing construction environment. A considerable amount of new material was added, and some of the chapters were reorganized for a more logical flow of information.

Later editions provided the updating necessary to remain current with state-of-the-art techniques in construction and to add new material, including references to AIA, EJCDC, and FIDIC documents, so that the book can literally become a single source for most construction-phase activities.

As a part of the continuing effort to stay abreast of the state of the art of the construction industry, and in recognition of the federal declaration to make the metric system (SI) the basic system of measurement in the United States and that federal agencies be required to use it exclusively, the fifth edition was updated to emphasize its use and included supplementary information to assist civil and construction engineers in utilizing metric (SI) civil engineering units in construction. In addition, all of the original material was reviewed and updated, the subject of partnering was addressed, and the index was made more user-friendly.

The author is grateful to the many contributions made through the years since this book was first published. Contributors to previous editions included Julius (Jim) Calhoun, Esq., Asst. General Counsel for Montgomery-Watson in Pasadena, CA (ret.); Gary L. McFarland, PE, and Charles H. Lawrance, PE, President and Vice-President, respectively, of Lawrance, Fisk, & McFarland, Inc., of Santa Barbara, CA; Wendell Rigby, PE, former Senior Civil Engineer of the City of Thousand Oaks, CA; Harold Good, CPPO, Director of Procurement and Contracting of the City of Palm Springs, CA; Albert Rodriguez, CPCU, ARM, President, Rodriguez Consulting Group, Inc., Jacksonville, FL; Robert Rubin, Esq., PE, of Postner & Rubin, Attorneys-at-Law, New York, NY; Joseph Litvin, Esq., PE, Attorney-at-Law, Dayton, OH; Arthur Schwartz, Esq., General Counsel for the National Society of Professional Engineers, Alexandria, VA; Robert Smith, Esq., PE, of Wickwire Gavin, PC of Madison, WI, General Counsel for the Engineers Joint Contract Documents Committee (EJCDC); and the members of the EJCDC whose contributions to the tools of the contract administrator are without equal.

The author extends his particular thanks and appreciation to Donald Scarborough, President of Forward Associates, Ltd., of Novato, CA, for his valuable contributions to the updated chapters on CPM scheduling; to William W. Gurry, President of Wm. Gurry & Associates, Atlanta, GA, for his contributions on design–build contracts; and to the Associated General Contractors of America for its input on the concept of partnering.

Special thanks and appreciation is offered to my daughter, Jacqueline, and to her son, John Stamp, PhD, who did most of the indexing for the sixth edition. Thanks also to my son, Edward, who provided all of the computer expertise, both editorially and in a support capacity, for the last three editions of this book.

PREFACE TO THE EIGHTH EDITION

Some notable changes have taken place in the eighth edition. To produce this and all subsequent editions, author Ed Fisk has joined with Wayne Reynolds, Associate Professor at Eastern Kentucky University, who has a long record of experience in heavy construction. Together, the authors have updated all of the chapters to twenty-first century technology.

For the seventh edition, author Ed Fisk added a new chapter, "Electronic Project Administration" (Chapter 5), and welcomed two new contributors to the chapter who are experts in their own right on computer application as applied to procurement and project administration. They were Mr. Harold Good, CPPO, formerly Director of Procurement and Contracting for the City of Palm Springs, a leader in his field, and W. Gary Craig, PE, President of ProjectEDGE.

For this eighth edition, the authors have revised much of Chapters 13 and 14 on construction scheduling, and Chapter 16 on value engineering. Numerous updates have been made to other chapters as well.

To access supplementary materials online, instructors need to request an instructor access code. Go to **www.prenhall.com**, click the **Instructor Resource Center** link, and then click **Register Today** for an instructor access code. Within 48 hours after registering you will receive a confirming e-mail including an instructor access code. Once you have received your code, go to the site and log on for full instructions on downloading the materials you wish to use.

My special thanks to each of you for your generous contributions of time and effort to make the eighth edition a continued success.

CONTENTS

1

THE PROJECT DELIVERY SYSTEM

Throughout the ages, human beings have been building to meet the needs of their habitation on this earth. Then, just as now, the planning and building of each such project involved the collective efforts of many workers, all with different skills and types of specialized knowledge. At first the methods were primitive but effective. As the products of modern technology replaced the older, outdated tools of these early builders, the methods of construction and the types of skills and specialized knowledge required to complete a construction project had to change to keep pace. Now, in the twenty-first century, we are again experiencing change as the computer has revolutionized the way that projects can be administered, both on the Web and in extranet applications.

Project Participants

Whether the project is a building, bridge, dam, pipeline, sewage treatment plant, water supply system, or any one of numerous other types of projects, it requires the skills and services of a project team comprised of three principal participants, or only two participants if we consider the concept of a design–build contract.

The owner
The designer
The builder
The design-builder

In practice, the owner usually enters into a contract with an architect/engineer or a design–build contractor to plan and design a project to satisfy the owner's particular needs. The owner participates during the design period to set criteria for design, cost, and time limits for completion and to provide decision-making inputs to the architect/engineer or design–build contractor.

Under conventional contracts, upon completion of the planning and design process the project is ready for construction, and the advertising or selection process to obtain one or more qualified construction contractors begins.

After selection of one or more qualified construction contractors, or, as in the case of public works projects, selection of the lowest qualified bidders, the owner enters into a contract directly with each prime contractor, who will then be fully responsible directly to the owner or the owner's designated representative for building the project in accordance with the plans, specifications, and local laws. The contractor has the further responsibility for the integrity of the new structure that has been built—in effect, the contractor must guarantee the work. Although the architect/engineer may be obligated to make field visitations to the construction site during the progress of the work, such periodic visits are for the purpose of observing materials and completed work to evaluate their *general* compliance with plans, specifications, and design and planning concepts only. Such basic services should not be interpreted as including full-time inspection for quality control and assurance.

Thus, on the typical project, there are usually only two prime contracts with the owner: one with the architect/engineer for the design and planning of the project, the other with a single construction contractor or occasionally several prime construction contractors to build the project.

As is frequently the case on a modern, complex project, numerous special types of construction are involved, and the contractor who entered into an agreement with the owner to build a project finds that the work can better be accomplished by subcontracting with a specialty contractor to do this portion of the work. Such subcontracts are agreements between the prime or "general" contractor and the subcontractor only, and involve no contractual relationship between any subcontractor and the owner. Under the owner's contract for construction, the general contractor is fully responsible for the entire work, whether or not subcontractors have been utilized to accomplish any portion of it. The traditional contractual arrangement is illustrated in Figure 1.1.

Figure 1.1 does not take into account the relationship between an owner with its own in-house engineering staff (such as many public agencies and utility companies) and the construction contractor. However, by combining the functions of *owner* and *architect/engineer,* as in the diagram, the relationships would be similar.

Construction Administration

Construction administration and contract administration are terms easily confused. As used in this book, the term *contract administration* means the management or handling of the business relations between the parties to a contract, which is popularly thought of as being limited to the administrative paperwork or electronic project management applications. In this book the term *construction administration* is used to refer to the much broader responsibility of relating to *all* project-related functions between the parties to a contract—not only the traditional contract administration duties, but also the conduct of the parties, relations with the contractor,

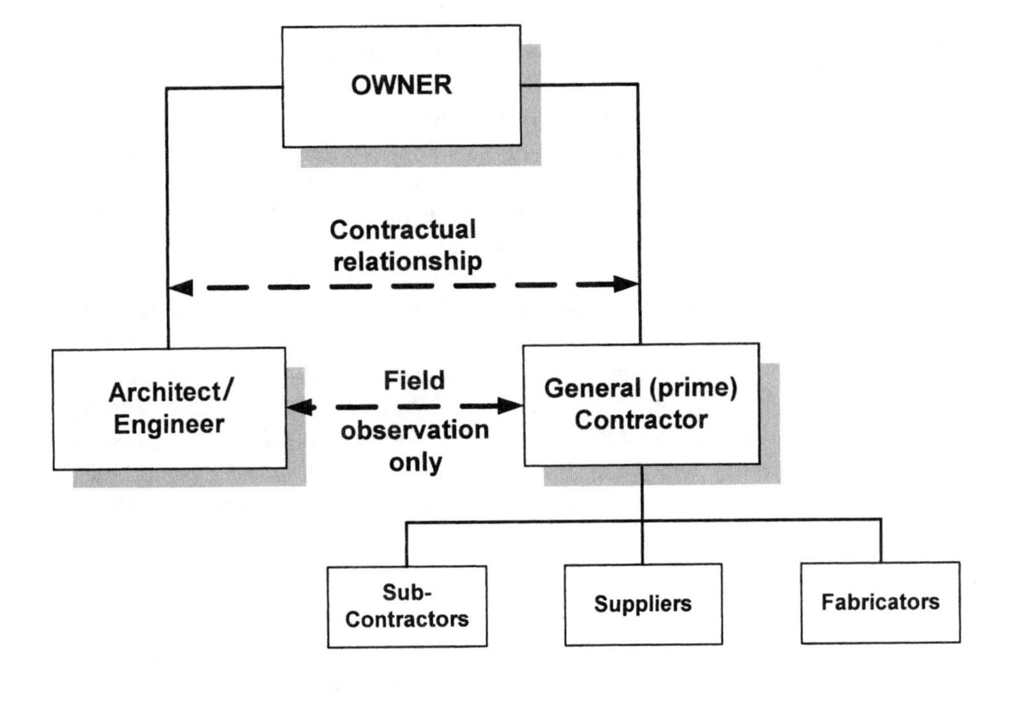

FIGURE 1.1 Traditional Construction Contract Relationships.

communications, business systems, procedures, responsibility, authority, and duties of all of the parties, documentation requirements, construction operations, planning and scheduling, coordination, materials control, payment administration, change orders and extra work, dispute and claim handling, negotiations, and all project closeout functions, including punch list inspections, final cleanup, and administrative closeout. Thus, as used in this book, contract administration, whether electronic or traditional paperwork, is just a part of construction project administration.

It is not uncommon for the architect/engineer's or owner's Project Manager to function as the contract administrator, working out of the home office, who jealously guards the control of the job by reserving all meaningful project administration duties to him- or herself, while the authority of the Resident Project Representative at the project site is often limited to inspection and routine clerical duties. However, it is organizational hierarchies such as that which are the root cause of numerous construction claim losses to the owner or architect/engineer.

The mark of a good manager is the ability to select and hire qualified people and then to be willing to delegate as much authority as possible to such people. As long as a manager refuses to delegate and reserves all or most of the contract administration tasks to him- or herself, the capabilities of that manager will be severely inhibited, and furthermore, the Project Manager will be incurring considerable risk of loss to the parent organization through potential delay-claim losses. A manager's authority is in no way diminished through delegation, but rather is strengthened.

As a means of implementing such a sound relationship between a Project Manager and a Resident Project Representative, the organizational chart shown in Figure 1.2 suggests a division of responsibility between field and office management personnel. A Project Manager can efficiently handle several projects without needlessly delaying any one of them by delegating the authority to make decisions on matters that should be decided at the Resident Project Representative's level. This contributes to the smooth and efficient operation of the construction activities and lessens the risk of contractor delay claims that would normally follow the delays

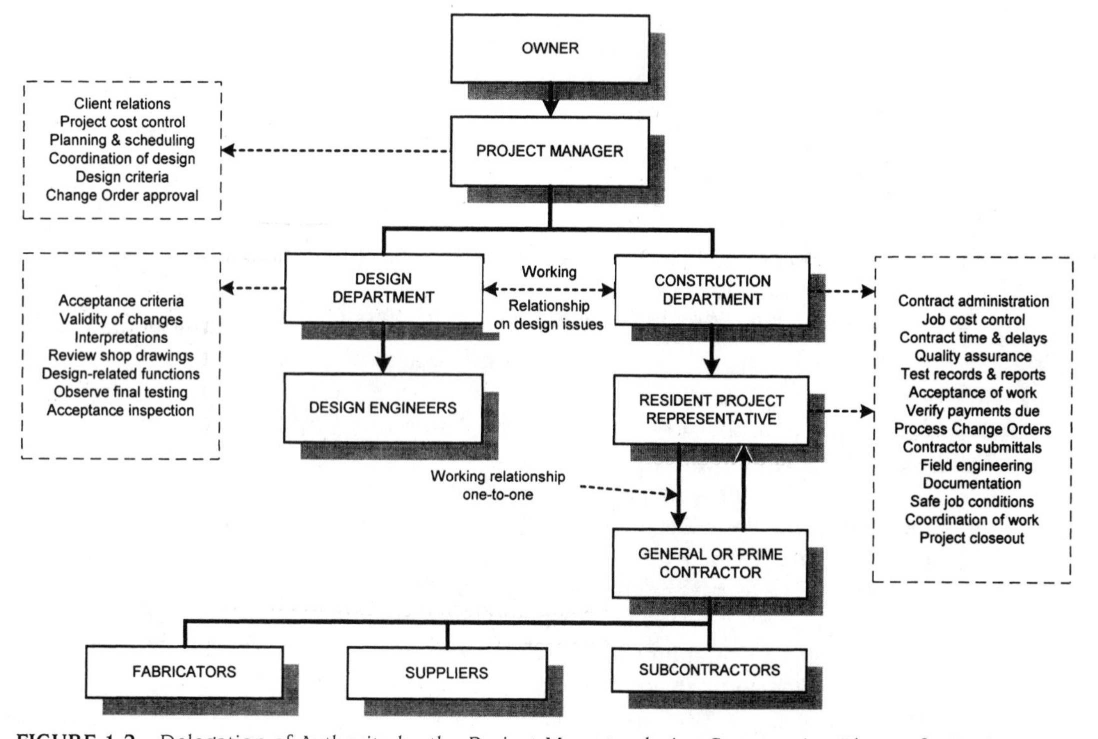

FIGURE 1.2 Delegation of Authority by the Project Manager during Construction Phase of a Project.

caused by routing routine matters through the home office as a prerequisite to obtaining permission to act.

The Resident Project Representative is under obligation to keep the Project Manager informed every step of the way. Where the Project Manager's decision is required, the Resident Project Representative should obtain a decision by telephone, fax, or e-mail or through the use of extranet or Internet applications, prior to issuing a consent order to a contractor to proceed with some particular work or corrective action. Then, of course, proper administrative paperwork or electronic documentation must be completed to confirm the actions taken. Failure to expedite decisions often results in otherwise preventable claims that have a way of escalating into major claims the longer they take to be resolved.

The most effective Project Manager, or Contracting Officer (as he or she is known on U.S. federal projects), is the person who is willing to delegate as many contract administration functions as possible to the Resident Project Representative in the on-site field office (the one-to-one concept). If unwilling to delegate, the Project Manager's only alternative to save the job is to relocate the Project Manager's office to the job site and run the project from there. On matters affecting time or money, however, only the Project Manager or Contracting Officer is empowered to execute contract modifications or Change Orders.

The One-to-One Concept

One of the single most important philosophies in construction project administration, the *one-to-one concept,* is a vital administrative procedure that can eliminate much conflict, reduce exposure to claims-producing problems, and result in greater efficiency for all parties to the contract.

Under this concept, the owner or architect/engineer designates a single individual, preferably located at the project site, to be the sole spokesperson representing the owner's interests. This person should be the Resident Project Representative, sometimes simply referred to as the "Project Representative." Under this arrangement, *all* orders issued to the contractor must be issued through the Resident Project Representative, and no one in either the owner's or the architect/engineer's or construction manager's office should be permitted to make any commitments to or issue orders or instructions directly to the contractor or any of its subcontractors, except by communicating such orders to the Resident Project Representative for issuing to the contractor. Failure to follow this procedure may place the owner and the contractor in a difficult contractual position. Under the contract law principle of implied authority, it is generally held that the contractor may receive orders from any individual whom it has reason to believe has the authority to issue such orders on behalf of the owner (see "Apparent Authority" in Chapter 2). Thus, the project manager, department heads, vice presidents, city or county engineers, or other persons of authority might otherwise visit the site and make statements that result in the creation of constructive changes (see Chapter 19) and not only bind the owner, but also lay the foundation for a contractor claim.

One situation where the author visited a construction site with a principal of an engineering firm was a classic example of what *not* to do. Upon arrival at the site, the principal went to the field office to confer with the Resident Project Representative. Then the principal toured the project site (one of his monthly site visits) with the Resident Project Representative and the contractor's representative. Up to this point, everything was done "by the book." However, from this point on, the principal's actions became a classic example of what not to do.

The principal listened to the contractor's side of the difficulties experienced during the previous month, including failure to achieve certain high standards of quality and workmanship in certain areas. The principal listened, then unbelievably made commitments to the contractor by accepting such nonconforming work without ever talking it over with the Resident Project Representative. In short, he gave away the store!

To complicate matters further, the principal's actions totally stripped the Resident Project Representative of his authority and ability to deal effectively with the contractor, as after that the contractor realized that all that would be necessary to avoid unpopular decisions made in the field would be to do an end run around the Resident Project Representative and go directly to the principal to obtain concessions. Thus the principal's workload is increased, the effectiveness of the on-site inspection forces is diminished, and the risk of claims is greatly increased.

What should have been done would be for the principal to listen to the contractor's comments about the project without offering comment at that time, then go back to the field office with the Resident Project Representative and, behind closed doors, discuss the events and issue orders to the Resident Project Representative as to the acceptability or nonacceptability of the contractor's work. This would have placed the Resident Project Representative in a position of receiving backing from the home office, and the contractor would have realized that in the end, all orders will be received only from the Resident Project Representative.

The principal is still the only person with the authority to make the final determination, but is advised to issue those orders only through the Resident Project Representative to preserve the one-to-one relationship. One of the greatest difficulties, where a project is being administered by an architect or engineer on behalf of an owner, is to keep the owner from violating this vital management concept.

As a part of the one-to-one concept, the contractor, too, must organize so that a single management person located at the project site is designated as the contractor's sole agent. This is best set up as a provision of the specifications. Then, it should be arranged during the preconstruction conference that the contractor's agent should be designated in writing, and that no substitutions are permitted under the contract without the written authority of the corporate office. The designated person should be capable of speaking officially for the contractor, although it is certainly acceptable to use an on-site superintendent or project manager as the agent of the contractor, just as the engineer or architect uses his or her Resident Project Representative. The military has a word for this. It is called "chain-of-command."

The 5-Step Process of Initiating a Project

An important part of organizing a project so as to avoid later difficulties, which could include award disputes, charges of preference, loss of money due to bidder default, and

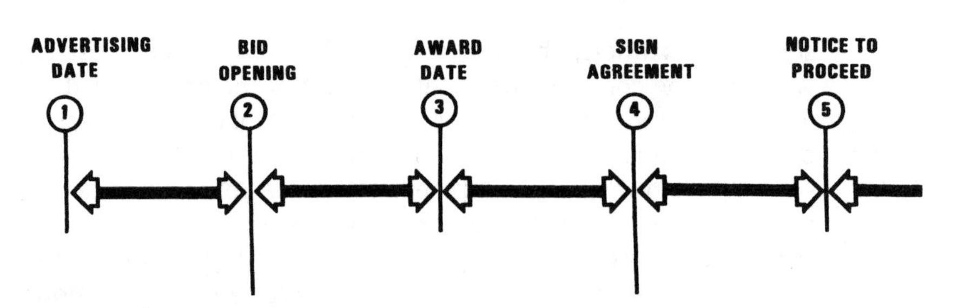

FIGURE 1.3 The 5-Step Process.

later disputes over lost time and delays in the work, is the initiation of the project according to an orderly administrative procedure. This process is what the author calls the "5-step project initiation process" (Figure 1.3). It holds that there are five vital steps that must be followed when initiating a project, especially in public works projects:

United States	International (FIDIC)
1. Advertise for bids.	Solicit tender.
2. Open bids.	Open tender.
3. Award contract.	Issue letter of acceptance.
4. Sign agreement or "contract."	Execute contract agreement.
5. Issue Notice to Proceed.	Set commencement date.

While it is common for many owners and architects or engineers to follow most of these steps, items 4 and 5 are, unfortunately, often combined, sometimes even with the uninformed blessing of the owner's attorney, as well.

An important point should be noted here. The procedures for the signing of the agreement often take time. If a cautious contractor chooses to wait until it actually "sees" the agreement with the owner's signature, and if the project time is stated in the agreement as beginning as of the date of signing the agreement, there is the possibility of a valid delay claim against the owner even before the project begins. The contractor can rightfully claim that (1) it could not start until a signed contract was received, (2) the time lost in receiving the signed document was part of its construction time, and (3) it should be compensated with an extension of project time to cover the days lost while waiting for a signed agreement.

Throughout the book references will be made to the 5-step process and partial diagrams will retain the identification numbers in the foregoing list to identify any of the five tasks listed.

Control of Quality in Construction

Without definition, the term *quality control* in construction can have several meanings. To be sure, the actual quality of construction depends largely upon the control of the construction itself, thus involving the contractor to a great extent. What constitutes quality control and quality assurance appears to be the subject of dispute by some. For example, checking the placement of reinforcing steel in concrete formwork may be considered as quality control if the contractor does it, and as quality assurance if the owner observes or verifies that it has been done; yet the physical act of checking this work is exactly the same in either case.

Whether the subject is called quality control or quality assurance, the function performed is essentially that which has been recognized over the years as being construction inspection and testing of materials and workmanship to see that the work meets the requirements of the drawings and specifications. Inspection takes many forms, and its responsibilities vary somewhat depending upon the intended inspection objective. As an example, an inspector in the employ of the local building official is principally concerned with the safety and integrity of the structure being built, and whether it meets the local building code requirements. Quality of workmanship or aesthetics is largely beyond the code inspector's responsibility and, because his or her salary is paid by the public, quality of workmanship is, to a great extent, left to the owner to control, using the owner's, contractor's, or designer's personnel. However, inspection by the owner's representative is intended to include concern not only for the structural integrity and safety of the structure, but also for the quality of workmanship, selection of materials being used, aesthetic values, and similar matters involving compliance with the provisions of the contract plans and specifications.

Organizational Structure of a Construction Project

There is no single organization chart that will remotely approximate the organizational structure of the field forces of the owner, the design organization, or the contractor on all projects. Before the internal structure of any of the principals to a construction contract can be examined, some understanding of the several basic types of contractual relationships must be gained. Of the several types of contractual relationships frequently encountered in construction, four of the principal types are:

1. Traditional architect/engineer (A/E) contract
2. Design/construction manager (D/CM) contract
3. Professional construction manager (PCM) contract
4. Design–build contract (similar to turnkey construction)

Under the provisions of the *traditional architect/engineer* contract illustrated in Figure 1.1, the owner usually engages the services of an architect/engineer to perform planning and design services, including preparation of plans, specifications, and estimates. Professional services of the architect/engineer during the construction are generally limited to performance of intermittent field visitations and certain contract administration functions such as review of the contractor's payment requests, review of shop drawings, evaluation of contractor claims, interpretation of plans and specifications during construction, change order requests, and final inspection.

A *design/construction manager* contract, illustrated in Figure 1.4, is quite similar to the traditional A/E contract with the exception that the architect/engineer's project

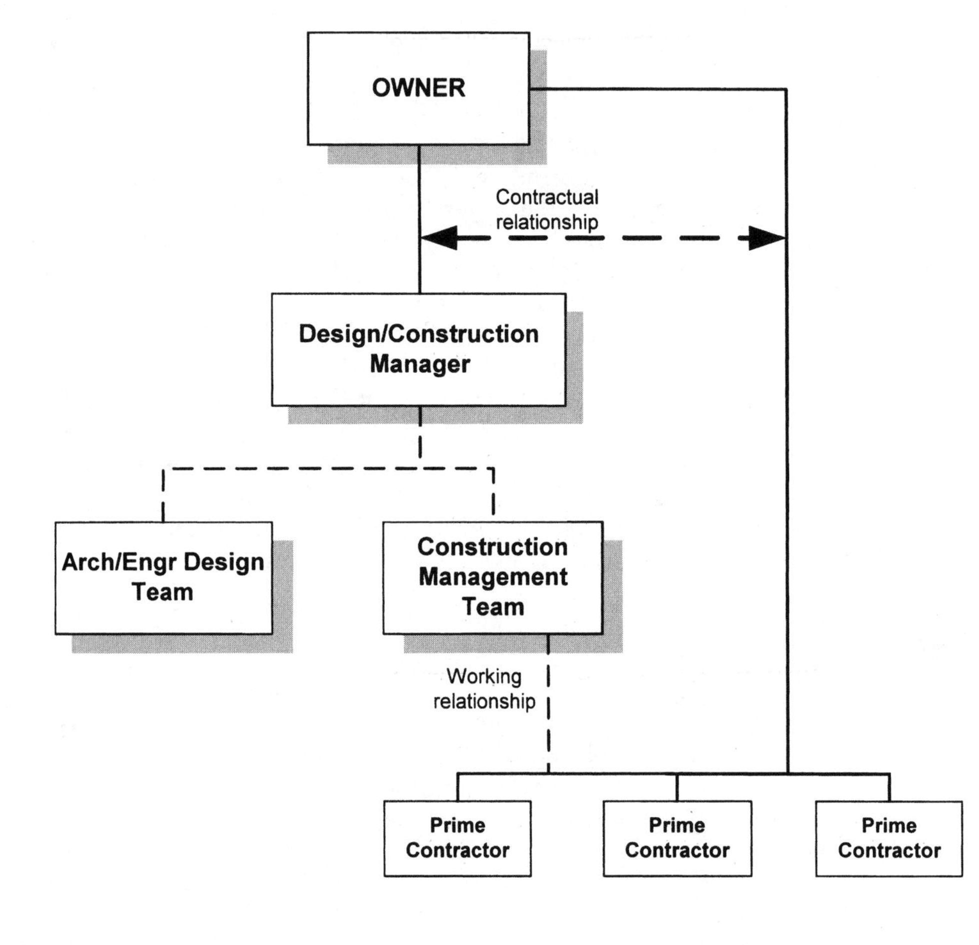

FIGURE 1.4 Contractual Relationships under a Design/Construction Manager-Type Contract.

manager is fully responsible to the owner during both the design and planning phases as well as the entire construction phase to provide for all project needs. This includes all scheduling, cost control, quality control, long-lead purchasing, letting of single or multiple contracts, and coordination of the work. The design/CM responsibilities do not terminate until final acceptance of the completed project by the owner.

These responsibilities include the examination of cost-saving alternatives during both the design and construction phases of the project, and the authority to require the design or construction changes necessary to accomplish the owner's objectives.

A *professional construction management (PCM)* contract is based upon a concept pioneered several years ago by the General Services Administration of the federal government, and for a time was used extensively by that agency for the construction of public buildings. Although the functions performed by the professional construction manager may be no different than those of a design firm doing construction management, the responsibilities and contractual status are significantly different. Under the professional construction manager (PCM) concept, illustrated in Figure 1.5, the owner engages a construction management firm under a separate contract in addition to a conventional architect/engineer and construction contractor contract. Thus, instead of only two contracts for a project, the owner has actually executed three. In keeping with the principles of this concept, the professional construction management firm performs no design or construction with its own forces, but acts solely in the capacity of an owner's representative during the life of the project. In many cases, the PCM is responsible for reviewing the architect/engineer's payment requests in addition to those of the contractor. In any case, the PCM is responsible for total project time and cost control and coordination as well as quality control and, as such, provides supervision and control over those functions of the architect/engineer and the contractor that relate to these important subject areas.

One important distinction is that a "construction manager" under this concept is an organization, not a single individual. Thus the construction management firm may provide a staff of both field and office personnel, including a project manager, estimators, schedulers, accountants, construction coordinators, field engineers, quality control personnel, and others.

A *design–build* contract, illustrated in Figure 1.6, sometimes called *turnkey construction,* is based upon the owner entering into an agreement with a single firm to produce all planning, design, and construction with its own in-house capabilities. Some organizations recognize a further distinction between design–build and turnkey construction in that both provide both design and construction by a single organization, or joint venture, but the turnkey contractor also assembles the financing package. Such design–build firms are generally licensed as both architect/engineers *and* as general construction contractors in those states that require it, and offer a complete package deal to the owner. Its principal advantages, where its use is permitted, are the elimination of contractor claims against the owner resulting from errors in the plans or specifications and the ability to begin construction on each separate phase of a project as it is completed, without waiting for overall project design completion—the "fast-track" concept. It is in the design—build industry that fast-track construction was born.

There is one disadvantage in the system when public funds are involved in construction. Under the laws of many states, a construction contractor must be obtained through a competitive bidding process, and the lowest bidder gets the job. Usually, design firms and construction management organizations are selected on the basis

PROFESSIONAL CONSTRUCTION MANAGER
(AGENCY CONTRACT)

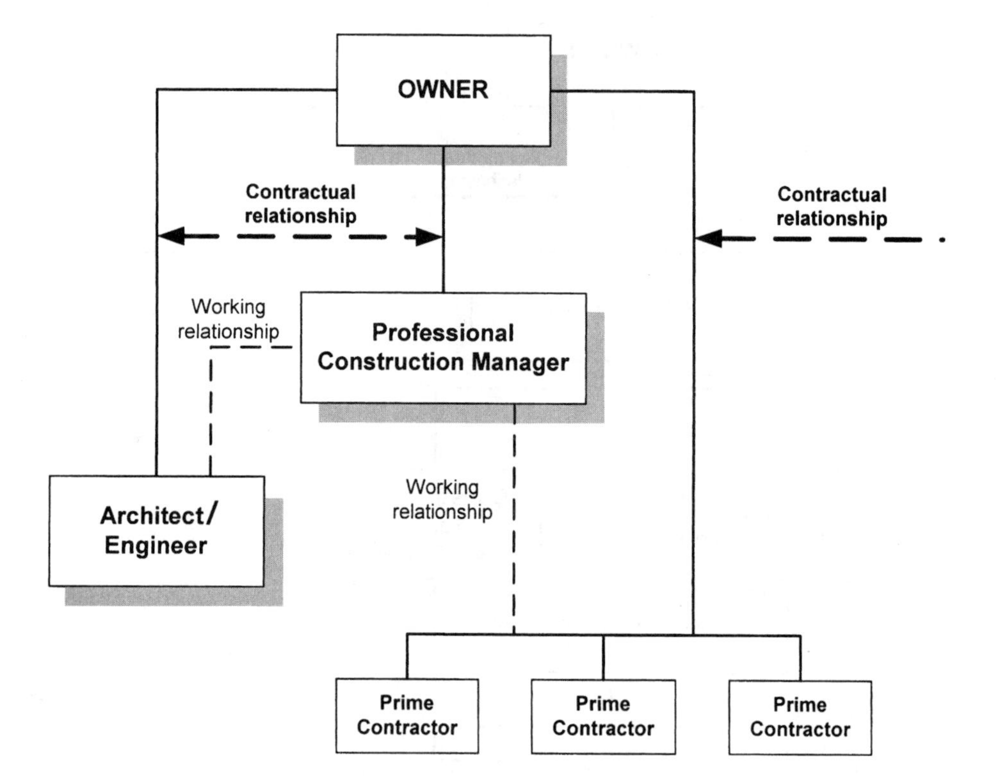

- Three-party team of owner, designer, and construction manager
- Fixed price or negotiated separate prime contracts directly with the owner
- Construction manager may act as owner's agent to extent delegated
- Negotiated professional fee for construction management services
- Negotiated professional fee for design services

FIGURE 1.5 Contractual Relationships under a Professional Construction Manager Contract.

of their individual expertise and previous experience in the type of work to be designed. Under this concept it is felt that the greatest savings and cost benefits to the owner will be obtained by careful planning during the design stage, and that the occasional cost savings that might result from competitively bidding the design responsibilities would be more than lost in the resultant higher construction cost that all too often follows a set of plans and specifications that had to be prepared in a hurry without checking.

DESIGN/BUILDER

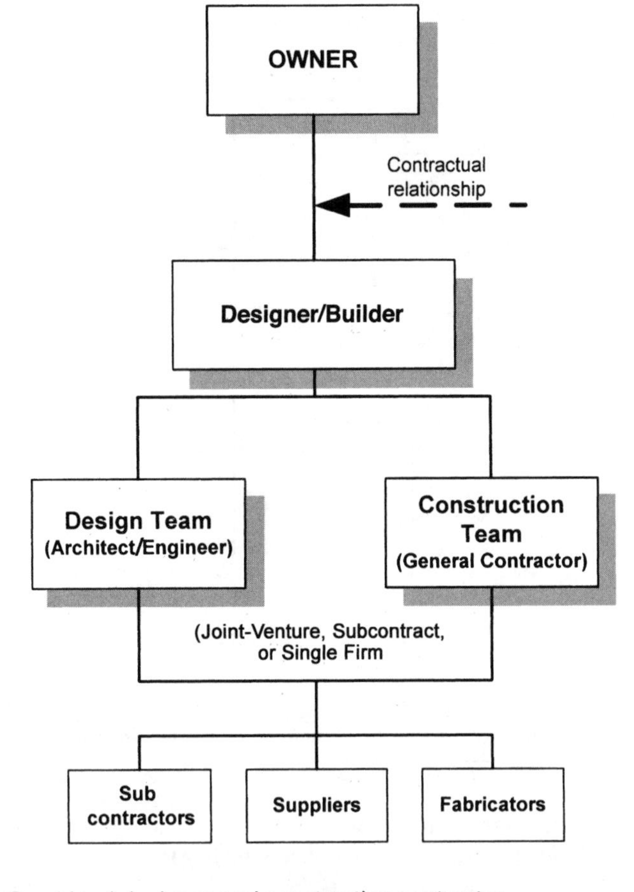

- Combined designer and construction contractor
- Numerous subcontractors
- Fixed price, guaranteed maximum price, or cost-plus-fixed-fee for design/build contract
- No separate fee for design services

FIGURE 1.6 Design-Build Contract Relationships (Similar to Turnkey Construction).

Staff Assignments for Construction Quality Assurance/Control

The staff requirements for the construction management and quality assurance/control activities of a construction project vary from job to job and from one employer to another. Although there seems to be a lack of uniformity in the structuring of many owners' or architect/engineers' field forces during construction, the average contractor organization seems to be extremely well organized in this area. This is probably to be expected, as the contractor organization is performing its primary function at the site, whereas the owner or architect/engineer is often on less familiar ground during the construction phase, even though the contract may call for the architect/engineer performance of some construction management functions.

In an attempt to compare job assignments and titles of positions of comparable authority from one organization to the next, the numerous titles of the same job emphasize the difficulty of determining position by title alone. Figure 1.7 is a chart of the normal functional relationships under a design/construction management contract, which will be used to illustrate the problem.

An example of supervisory job titles of comparable authority is shown in the following table, which is based upon actual job titles used by some contractor and architect/engineer offices to designate the various levels of supervisory and management personnel utilized during the construction phase of a project. The levels indicated are those used in Figure 1.7.

All of these levels share in the responsibility of administering various provisions of the construction contract for their respective employers. In addition to the foregoing list of full-time personnel on the project site, numerous tasks remain to be performed by specialty inspectors and representatives of the various local government agencies having jurisdiction over the project. These include the following public and private specialty and code enforcement inspectors:

1. Local building department (code enforcement)
2. Soils inspectors

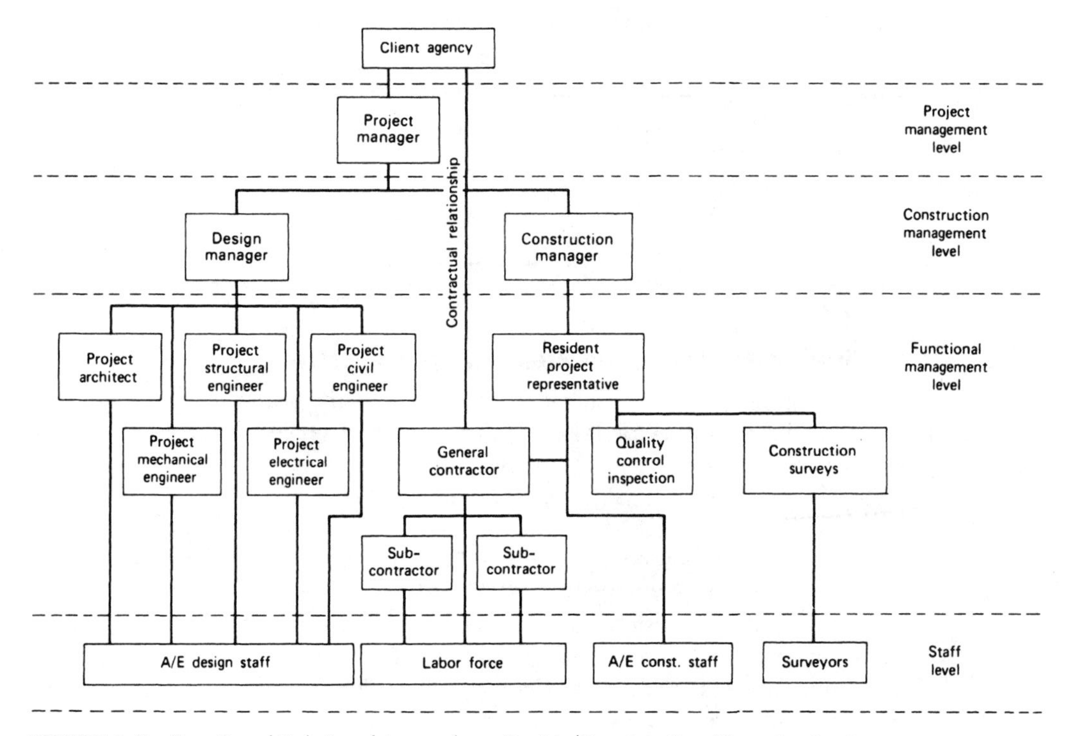

FIGURE 1.7 Functional Relationships under a Design/Construction Manager Contract.

3. Inspectors of other agencies whose facilities are involved
4. Utility company inspectors
5. Specialty inspectors (concrete, masonry, welding, electrical, etc.)
6. Manufacturers' representatives (special equipment or materials)
7. OSHA safety inspectors

Each of the specialty and code enforcement inspectors is responsible only for its particular specialty task; thus, the overall responsibility for project administration and quality control falls on the shoulders of the Resident Project Representative of the owner or design firm or the contractor's quality control (CQC) representative.

Level	Owner or Architect/Engineer	Contractor
Project Management level	Project Manager	Project Manager
	Project Engineer	
	Project Architect	General Superintendent
	Project Director	Construction Manager
	Contracting Officer	
Construction Management level	Construction Manager	Construction Manager
	Resident Engineer	Project Engineer
	Resident Architect	Superintendent
	Construction Coordinator	
	Resident Manager	
Functional Management level	Resident Project Representative	Project Engineer
	Project Representative	Superintendent
	Resident Engineer	Foreman
	Resident Inspector	CQC Representative
	Inspector	
	Quality Control Supervisor	

Full-Time versus Part-Time Project Representative

Not all construction projects subject to inspection by the owner or the design firm will require a full-time inspector. It is not infrequent for a single construction inspector to be assigned the responsibility of the Resident Project Representative for several projects at the same time. Usually, this is a method used to provide quality assurance inspections on smaller projects whose budgets or complexity of construction do not justify the financial burden of a full-time Resident Project Representative. The difference in responsibility is slight, as the administrative responsibilities are identical to those of the full-time Resident Project Representative on a single large project. As for the inspections performed, the inspector merely schedules field visits so as to be at each of the projects at key times during construction. Generally, an inspector should be on call and be able to respond to a specific field problem on short notice. Thus, in this book, the responsibilities and duties of the full-time Resident Project Representative should be understood to apply equally to a part-time project representative working directly out of the home office.

Professional Construction Management

The CM Controversy

CM by now has become a byword in the construction industry. But do most people actually know what it means? Even some industry giants seem to be confused.

The term *construction manager* is one of the most misunderstood titles of modern-day construction and defies an accurate definition that is acceptable to everyone. Definitions have ranged from applying the term to the Resident Project Representative, to the other extreme of being a third prime contract with the owner (the other two prime contracts being those of the architect/engineer and the general construction contractor). In the latter case, the construction manager is the owner's agent and the duties of the position require supervision over some of the functions of both the design firm and the contractor. The American Society of Civil Engineers refers to this function as *professional construction management* to distinguish it from the type of construction management practiced by the design/construction management firms; others call it *third-party CM* or a *CM agency contract*. Both the American Institute of Architects (AIA) and the Associated General Contractors of America (AGC) simply refer to this type of contract as *construction management*.

Each of the major professional and technical organizations seems to agree in principle on the concept that the construction manager should be a firm that has no direct connection with either the architect/engineer firm that designs the project or the general contracting firm that constructs it. Even when a general contractor acts as construction manager, the AGC recommends that it enter into contract under a professional services agreement, and that the firm does not use any of its own construction forces to build the project. The principal difference between the professional services agreement proposed by the AGC and, for example, that of the AIA is that the AGC contract provides for quoting a guaranteed maximum project cost after the construction management team has developed the drawings and specifications to a point where the scope of the project is clearly defined. Under the AIA contract form, no price guarantees are made.

Yet, where many organizations seem to miss the point is that CM involves participation by the Construction Manager from the very conception of the project, through the investigation and design process, selection of feasible separate bid packages, value engineering, constructability analysis, bidability analysis, preparation of input into the specifications and other front-end documents, assistance in examining bids and awarding the contract, and finally, participation in the construction phase in the form of scheduling, cost control, coordination, contract administration, and final closeout of a project. Yet many in the construction industry are still confused by the rather ill-chosen name *Construction Management* and still only recognize the construction-phase tasks as being construction management. This is easily understandable, as the term in itself does suggest that its duties should occur during the construction phase only. This is precisely why, a number of years ago, the American Consulting Engineers Council (ACEC) began to refer to third-party CM as the *ACEC Project Management System.* Strictly speaking, it is just that—a project management system. However, the term *CM* sticks to this day, so the industry needs to understand it better.

Unfortunately, the problem is further complicated by the fact that even in the prestigious *Engineering News-Record (ENR),* in its annual list of top-rated CM firms, some of them are known by the author to never have done a single project that met the definition of third-party CM. The services they actually performed were better described as "services during construction." The firms in question simply performed contract administration and inspection and believed it to be CM. Other firms in the list performed the full range of services, beginning with the conceptual phase and continuing through the design phase before finally entering the construction phase. *ENR* has no way of determining the difference in interpretation by the various companies, as its reporting is based upon reported CM revenues claimed by the listed companies.

Engineering Definition of Professional Construction Management

The Construction Management Committee of the American Society of Civil Engineers originally defined a professional *construction manager* as a firm or an organization specializing in the practice of construction management or practicing it on a particular project as a part of a project management team consisting of the owner, a design organization, and the construction manager (usually referred to as CM). As the construction professional on the project management team, the CM provides the following services or portions of such services, as appropriate:

1. Works with the owner and design organization from the beginning of design through completion of construction; provides leadership to construction team on all matters that relate to construction; makes recommendations on construction technology, schedules, and construction economies.
2. Proposes construction alternatives to be studied by the project management team during the planning phase, and predicts the effect of these alternatives on the project cost and schedule; once the project budget, schedule, and quality requirements have been established, the CM monitors subsequent development of the project to see that those targets are not exceeded without the knowledge of the owner.
3. Advises on and coordinates procurement of material and equipment and the work of all the construction contractors; monitors and inspects for conformity to design requirements; provides current cost and progress information as the work proceeds; and performs other construction-related services as required by the owner.
4. In keeping with the non-adversary relationship of the team members, the CM does not normally perform significant design or construction work with

its own forces. (In a more recent action, the ASCE Committee on Construction Management also accepts the concept of CM being performed by the firm responsible for design.)

The typical functional relationships associated with a professional construction management contract are best shown in Figure 1.8, which was prepared by the General Services Administration, Public Buildings Service of the U.S. federal government.

Fast-Track Construction

Frequently, a construction management contract is encountered that requires the letting and administering of multiple construction contracts for the same project, with each let at different times during the life of the project. Such staggered letting of construction contracts on the same project is referred to as *fast-track* contracting, and its principal objective is to shorten construction time for the overall project by starting some portions of the work as soon as it has been designed (Chapter 13), even though other portions of the project have not yet been designed. It is a risky process, depending heavily upon the careful selection of the various separate bid packages and the ability to schedule and control the design effort. Without this, fast-track can become the most costly method ever designed for completing a project late. Many times, such contracts also require purchase of special equipment or materials long before a contract has been let to install them. This is referred to as *long-lead procurement,* and such early purchases, along with the accompanying expediting and scheduling, is one of the functions required to be performed by the CM team. The Resident Project Representative is a vital link in the successful operation of a construction management contract and will be called upon to assist in many of the tasks described. When utilizing the fast-track process, the services of a construction management firm are essential, as the skills and experience required to complete a fast-track project successfully are seldom possessed by the design firm. On the other hand, if a single prime (general) contractor is used to build a project, the use of a construction management firm may actually be redundant, as many of the tasks that the CM is being paid for will ordinarily be done by the general contractor.

Design–Build Contracts

The design–build concept, as originally conceived, was based on the concept that a single firm had the in-house staff and expertise to perform all planning, design, and construction tasks. Later, increased interest in the concept had engineers, architects, and conventional contractors seeking to compete with the original design–build firms to meet the growing interest by owners in the project delivery process.

Under the current approach, instead of limiting design–build to firms with inhouse capability in both areas, the field has now been opened up to permit contracts with engineers who subcontract the construction portion to a contracting firm, with construction contractors that subcontract design services to an engineer or architect, and with engineers and architects in joint venture with contractor firms.

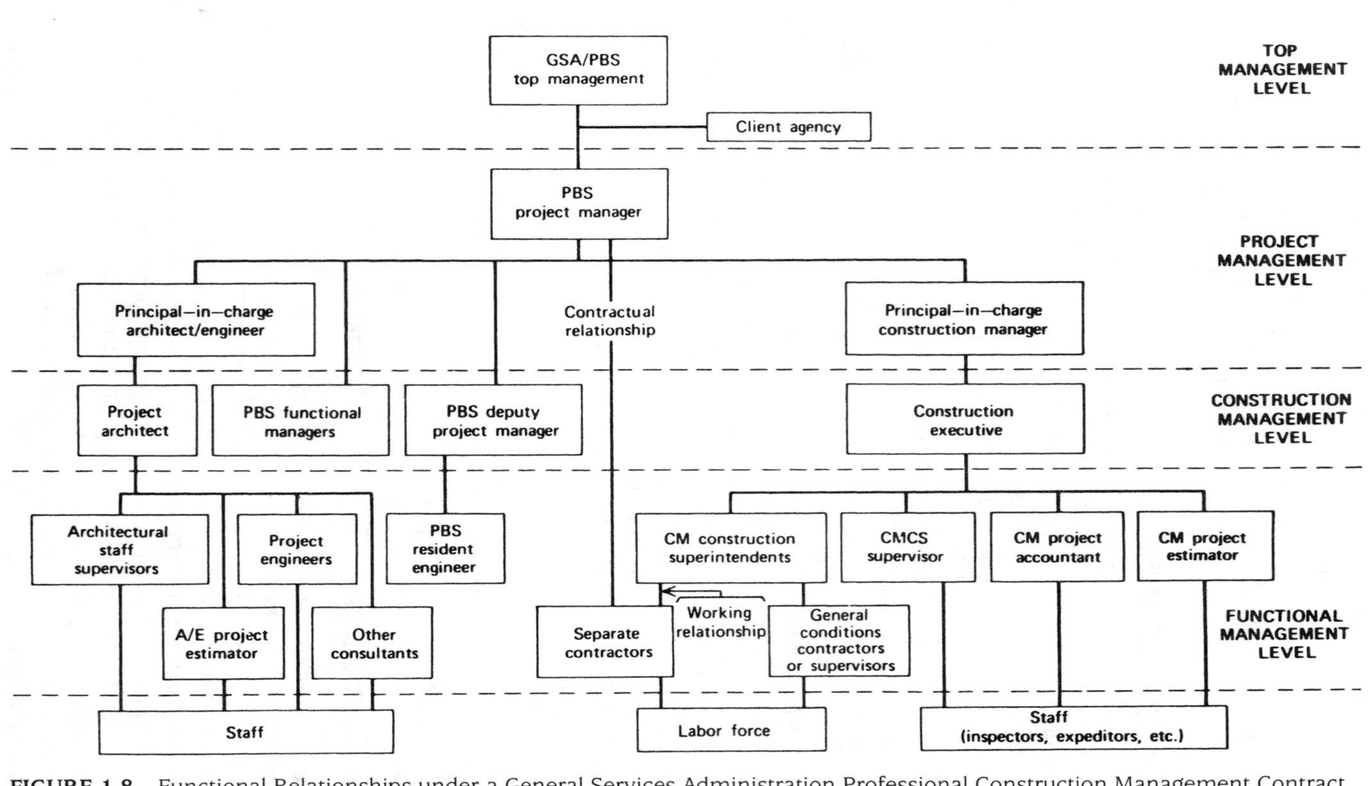

FIGURE 1.8 Functional Relationships under a General Services Administration Professional Construction Management Contract.

There are basically three types of design–build firms today; contractor-led, designer-led, and single firm. Contractor-led firms tend to dominate due to their experience in estimating, purchasing, cost control, and construction supervision, not to mention the contractor's better financial backing and ability to manage risk.

Design–Build Documents

In recognition of these relationships, a special group of standard contract documents for design–build projects was prepared by the Engineers Joint Contract Documents Committee (EJCDC).[1] They are available from ASCE, NSPE, or ACEC.[2] The documents, and an accompanying guide on how to use them, are probably the most comprehensive set of standard documents now available to those who work with design–build projects.

The special design–build documents cover a variety of situations-for example, the relationship between owner and design-builder, which includes both general conditions and two agreements (one for a stipulated sum contract, another for a cost-plus contract that has a provision for guaranteed maximum price). For design-builders without a professional in-house staff to perform design services, EJCDC has prepared a subagreement to be used by both design-builder and engineer. To cover the relationship between design-builders and construction subcontractors, another document covers general conditions and two subagreements (one for stipulated price, the other for cost-plus).

Among the highlights of the new documents are descriptions of the selection process; scope definition; differentiation of design and construction work in several areas (such as providing for a general warranty and guarantee for construction work as well as a standard of care for design work); a dispute resolution process; responsibilities for subsurface conditions; remedies in the event an engineer is asked by design-builders to compromise an engineer's legal and professional responsibilities; and communications among engineer, design-builder, and owner.

An integral part of the new group of documents is a separate guide describing their use. The guide includes a commentary, a guide for preparing requests for proposals, a suggested proposal form, and a "how-to" guide on the preparation of supplementary conditions.

Because many owners don't have the in-house expertise to prepare proposal documents (or review design-builders' design submittals, observe the quality of construction, and so forth), EJCDC is now preparing another document for use by the owner (and an owner's consultant) to cover such services.

Design–Build for Public Projects

While quite popular in the private sector, some public agencies find that the process still presents legal hurdles resulting from alleged conflict between the design professional selection process under the Brooks law and similar local regulations and the

---

[1]EJCDC is a multidisciplinary group made up of the American Society of Civil Engineers (ASCE), the National Society of Professional Engineers (NSPE), and the American Consulting Engineers Council (ACEC).

[2]ACEC referred to here is the American Consulting Engineers Council, not to be confused with the Association of Consulting Engineers of Canada, which shares the same acronym.

competitive bid process traditionally used for contractor award. At the federal level, however, legislation enacted in 1997 makes it possible to award design–build contracts under specified conditions.

The principal barrier to the use of the design–build process in the public sector lies in the difference between the procurement laws governing the selection process for architects and engineers versus that for construction contractors. Most state laws require the selection of architect/engineers on the basis of the most qualified, with the price set by negotiation. Construction contractors, on the other hand, must be selected on the basis of sealed bids, with award going to the lowest responsible bidder. As it can be seen, there is an immediate conflict when attempts are made to contract for both design and construction under the same contract. To further complicate matters, many states require bidders to list subcontractors in their bids, a near impossible task in a design–build contract, as the project has not yet been designed.

Federal Design–Build Contracts[3]

Effective October 10, 1997, the federal government was authorized to enter into design–build contracts using a two-phase design–build selection process authorized by 10 U.S.C. 2305a and 41 U.S.C. 253m. Under those provisions, a two-phase design–build selection process may be used when a contracting officer determines that the method is appropriate based upon the following considerations:

- At least three, but not more than five offers are considered
- Design work must be performed *before* developing price or cost proposals, and offerors will incur a considerable expense in preparing offers.

Proposals must be evaluated in phase one to determine who may submit proposals for phase two. The phase one contract must be awarded using competitive negotiation. After evaluating phase one proposals, the contracting officer must select the most highly qualified offerors, and only those offerors may submit a phase two proposal.

Phase two involves submittal of a cost proposal which is subject to consideration for technical evaluation factors, including design concepts, management approach, key personnel, and proposed technical solutions. It may take some time before the lawyers work out all of the wrinkles in this latest adaptation of the original two-step process, but it appears to offer considerable flexibility in combining the concept of competitive price bidding along with an equitable architect-engineer selection process that appears to be acceptable to the design community.

Legal Barriers for State and Local Public Projects

Due to the growing popularity of the design–build concept, the potential conflict between laws of the various states governing selection of design professionals as opposed to the competitive bidding process used in public works for award of a construction contract is

[3]c.f. 48 CFR Ch. 1 (10-1-97 edition) Subpart 36.3—Two-Phase Design–Build Selection Procedure.

being reexamined in many states. Many states are revising their statutes to permit the award of a design–build contract without violating the law.

The following summary was abstracted from an American Bar Association review of design–build contracts under the various state and local procurement laws as they existed in 1996.[4] State procurement laws, with respect to design–build, can be grouped into four categories. In some cases state laws are not the same for all agencies within the state, thus some states will fall into more than one of the following categories:

1. Laws that expressly prohibit design–build by any public agency (4 states)
2. Laws that pose obstacles to design–build by some public agencies (26 states)
3. Laws that pose no obstacles to design–build, even though design–build is not expressly permissible (22 states)
4. Laws that expressly allow design–build for all or some types of public projects (26 states)

Very few state statutes expressly prohibit the use of design–build, such as by requiring that a project be split into separate design and construction phases, and requiring the preparation of plans and specifications before bids are solicited. Other barriers to using the design–build method are laws that prohibit the award of a single construction contract to a general contractor by requiring multi-prime contracts, a practice widely used on public projects in the Northeast. In that instance, laws require the preparation of separate plans and specifications to allow for separate award of contracts for any number of trades. If a state agency cannot award a contract to a single general contractor for construction, it may also have trouble awarding a single contract for design and construction.

Definitions of Individual Construction Responsibilities

Local building codes often require intermittent and sometimes "continuous" inspection on certain critical types of construction, such as structural concrete, structural masonry, prestressed concrete, structural welding, high-strength bolting, and similar work to be performed by special inspectors. The word *continuous* in this context is sometimes confusing because there are cases in which "continuous" is not synonymous with "constant" without a reasonable interpretation. Work on structural masonry or concrete work that can be inspected only as it is being placed requires the *constant* presence of the inspector. Placement of forms for concrete or reinforcing for concrete can fairly be interpreted as requiring continuous monitoring of the work so as to miss nothing, yet would hardly be interpreted as requiring

[4] "Design–Build Contracts under State and Local Procurement Laws" by Kenneth M. Roberts, partner in the Construction Group at Schiff Hardin & Waite in Chicago, and Nancy C. Smith, partner at Nossaman, Guthner, Knox & Elliot in Los Angeles. *American Bar Association, Public Contract Law Journal,* Vol. 25, No. 4, Sept. 1996.

the inspector's presence during the entire time that the steel is being placed. Generally, any special inspector coming onto the job will be under the authority of the Resident Project Representative and should be advised to follow his or her instructions.

Project Manager

Every participating organization on a project has its *project manager.* Although the person is sometimes known by different names, the duties remain the same. Whether in the direct employ of the owner, the design firm, or the contractor, the project manager (PM) is usually the person responsible for the management of all phases of the project for his or her organization. For a design firm, the project manager controls the scheduling, budgeting, cost control coordination of design and construction, letting of contracts for the owner, and is normally the sole contact with the client as a representative of the design firm.

For the owner, a project manager is similarly responsible for all phases of a project, but may participate in architect/engineer selection and is the representative of the owner in connection with any business concerning the project. Where an architect/engineer firm has been engaged for design services only, the owner's project manager will provide construction contract administration and may employ a Resident Project Representative or other on-site quality control personnel to work under his or her supervision. Wherever this is the case, the architect/engineer may still be called upon to review shop drawings for the owner, and wherever a proposed design change is contemplated, the architect/engineer of record should always be consulted.

In a contractor's organization, "project manager" is also a frequently used title, although many very large firms still use the title "superintendent" for this function. As the title implies, the contractor's project manager is in complete charge of his or her project for the general contractor. This particular project manager or superintendent's responsibilities include coordination of subcontractors, scheduling, cost control, labor relations, billing, purchasing, expediting, and numerous other functions related to the project. To the owner, the PM or superintendent is the general contractor. Whether referred to as project manager or as superintendent, his or her duties are the same in many contractor organizations.

Professional Construction Manager

The services performed by the professional construction manager cover a broad range of activities and, to some extent, overlap those traditionally performed by both the architect/engineer and the construction contractor, involving both the design and construction phases of a project. A comprehensive construction management contract may easily cover any if not all of the tasks included within the following six categories:

1. Participation in determining the bidding strategy involved in a fast-track, multiple-prime-contract project so as to avoid conflicts during the letting of the separate contracts.

2. Design phase review, including review of formal design submittals, review of contract documents, and overall constructability analysis.
3. Cost management, including estimates of construction cost and development of the project budget.
4. Scheduling for all phases of a project, generally incorporating critical path techniques.
5. Bid opening and evaluation, and assistance in contractor selection.
6. On-site, construction-phase management to provide contract administration, inspection, coordination, and field management.

On-site, construction-phase management may include coordination of separate contracts, phased construction (fast-track) contracts, monitoring of individual phases of the work, adjustment of the work to accommodate changed conditions or unanticipated interferences, determination of whether materials and workmanship are in conformance with the approved contract drawings and specifications, arrangements for the performance of necessary field and laboratory tests where required, preparation of change orders and change proposals, and review of progress payments and recommendations to the owner for payments to the contractors.

The construction manager (CM) may also, in some contracts, provide certain services that would normally have been provided by a general contractor, had there been one. These might well include establishment, maintenance, and operation of temporary field construction facilities, provisions for site security, cleanup, temporary utilities, and similar General Conditions items of work. Such items of work are generally paid for under this type of contract on a reimbursable basis and are not a part of the construction manager's professional fee for services. In certain respects, the professional construction manager's responsibilities may overlap or even preempt those of the contractor and the architect/engineer under some contracts.

Quality Control Representative

Under the provisions of the construction contracts of numerous federal agencies, in particular the Corps of Engineers, Naval Facilities Engineering Command, National Aeronautics and Space Administration, and others, an inspection concept known as *contractor quality control* (CQC) is often implemented. Though not favored by much of the construction industry, under this system the contractor must organize and maintain an inspection system within the organization to assure that the work performed by the contractor and subcontractor forces conforms to contract requirements, and to make available to the government adequate records of such inspections. Under the majority of such plans, as implemented by their respective agencies, a government representative is stationed on-site to provide quality assurance inspections. In some cases, the design firm may be engaged to provide quality assurance inspection in addition to contractor quality control. In such cases, the design firm may have a full-time Resident Project Representative and supporting staff on-site to perform this function.

Resident Project Representative; Resident Engineer; Resident Inspector; Resident Manager; Project Representative

These titles usually refer to an on-site full-time project representative to whom has been delegated the authority and responsibility of administering the field operations of a construction project as the representative of the owner or the design firm. On some occasions, the inspection needs of a particular project may require that the Resident Project Representative be a qualified, registered professional engineer; in other cases, a highly qualified nonregistered engineer may be desired. Wherever a nonregistered engineer is permissible, it is often equally acceptable to use an experienced construction inspector for this purpose. The American Institute of Architects, in its documents, describes this individual as the "full-time project representative," whereas the Engineers Joint Contract Documents Committee[5] uses the term "Resident Project Representative." In this book the term *Resident Project Representative* is used to stand collectively for resident engineer, resident inspector, full-time project representative, resident manager, and project representative.

Inspector; Field Engineer; Quality Assurance Supervisor

These titles usually refer to a staff-level, on-site representative of the owner, design firm, or contractor who has the responsibility of observing the work being performed and of reporting any variations from the plans and specifications or other contract documents. In addition, the inspector should call to the attention of the quality control supervisor or Resident Project Representative any unforeseen field conditions in time for remedial measures to be taken without creating delays in the work or changes in existing work to correct a problem. The inspector is the on-site eyes and ears of his or her employer, and although not empowered to make field changes that depart from the plans and specifications, the inspector should be capable of evaluating field problems and submitting competent recommendations to his or her supervisor. On projects using a Resident Project Representative, the inspector will normally work under that person's direct supervision.

Except for the responsibility for construction field administration, which is one of the principal functions of the Resident Project Representative, the inspector's job is identical in all respects to that of the Resident Project Representative. Inspection of construction is an occupation that requires a highly qualified person with a good working knowledge of construction practices, construction materials, specifications, and construction contract provisions. It is not in itself a job title, as the inspector may be a registered professional engineer or architect, a field engineer, a quality control specialist, or any of a host of other classifications. In this book the term *inspector* will be used to stand collectively for field engineer, inspector, quality control supervisor,

[5]Comprised of American Consulting Engineers Council, American Society of Civil Engineers, National Society of Professional Engineers, and Construction Specifications Institute. Endorsed by Associated General Contractors of America.

or in some cases, the Resident Project Representative where the duties referred to apply to all field representatives on-site, whether full- or part-time.

Contractor's Engineering Section

The contractor's engineering section is an important tool for assisting the contractor's management forces in analyzing construction and engineering problems. Not all construction companies are set up the same; however, engineering functions may be set up to operate at three levels:

1. Project level, where responsibility is usually limited to a single project
2. Area or regional level, where responsibility includes several projects
3. Main office level

Interaction with the owner's or architect/engineer's Resident Project Representative at the above level is necessary under a one-to-one relationship or a partnering agreement.

Representative tasks that may be performed by the contractor's engineering section may include the following:

1. Estimate and prepare bid proposals
2. Plan and schedule
3. Stay alert to future work prospects
4. Maintain preconstruction liaison with joint venture partners
5. Review subcontract proposals
6. Prepare budget control estimates
7. Develop construction methods
8. Maintain contract relations
9. Lay out the work
10. Handle requisitioning, scheduling, and expediting
11. Document all work performed
12. Prepare and check payment estimates
13. Document costs and extra work
14. Document subcontract performance and payment estimates
15. Prepare reports of costs and construction activities
16. Issue progress reports: daily; weekly; monthly
17. Draft contract modifications, change orders, and claims
18. Provide assistance in contract settlements

However lengthy, the list is not exhaustive, and is subject to wide variation depending on the size and structure of the contractor's organization. A representative organization of an engineering department of a small contractor is illustrated in Figure 1.9.

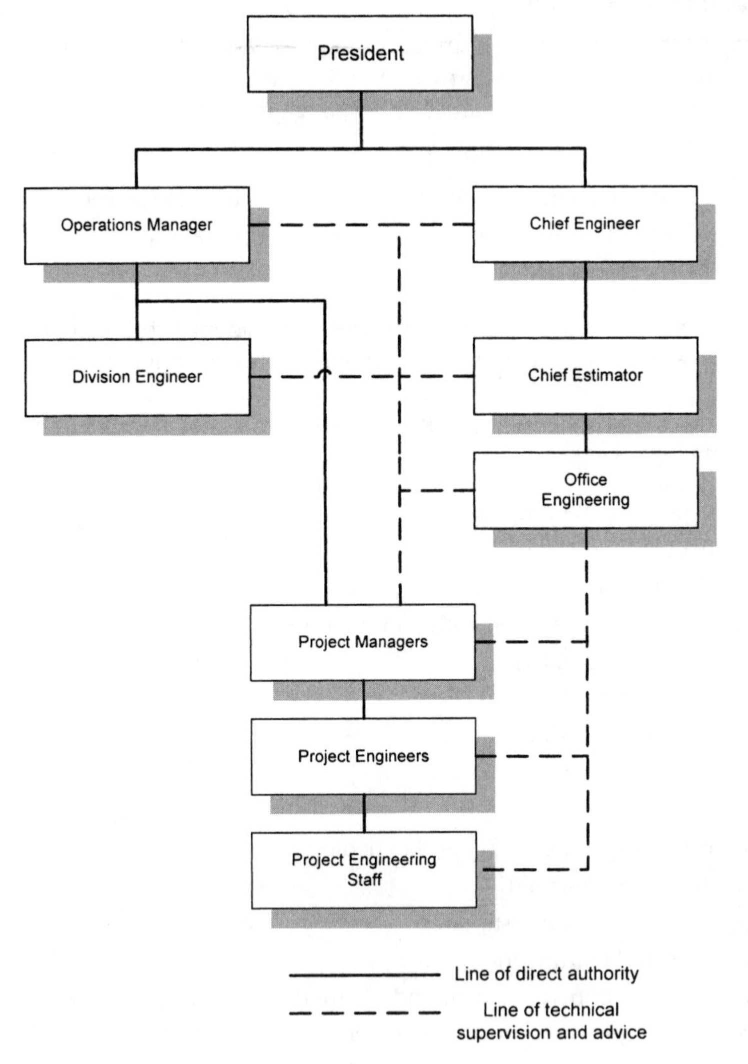

FIGURE 1.9 Typical Organization of an Engineering Department of a Small-Size Contractor.

DEFINING SCOPE OF WORK IN A CM CONTRACT

It is important, because of the differences in definition of the term *CM,* to clearly define the scope of the services to be performed in the CM contract with the owner. Without such definition, there are often claims and disputes between the architect/engineer as CM and the owner, due to a failure of the parties to achieve a meeting of the minds, as one party assumes that the term includes services that the other party never planned to provide for the basic fee agreed upon.

Construction Administration Task List

It is recommended that all anticipated tasks be listed and included in any CM contract executed with the owner. A checklist of some of the tasks that may be involved follows:

Conceptual Phase

Develop conceptual estimates
Develop conceptual schedules
Provide input to program risk analysis

Program Planning Phase

Provide constructability analysis
Identify potential major construction problems
Develop project resource requirements
Inventory available area resources
Assist in development of capital budgets
Assist in development of cash flow projections
Develop parametric estimates and cost budgets
Update preliminary schedule
Develop preliminary project control system
Develop preliminary project management information system
Develop project safety program
Develop project labor relations program
Assist in development of insurance program
Administer electronic data processing (EDP) services

Design Phase

Oversee overall project planning
Assist in development of project life-cycle costs
Evaluate cost trade-offs
Provide value engineering function
Qualify potential bidders
Procure long-lead-time items
Finalize bid work packages
Finalize prequalified contractor lists
Finalize project schedules
Finalize physical layout of construction areas
Finalize project control systems and management information systems

Assist in obtaining required permits and licenses
Provide input and review of contract documents

Construction Phase

Develop and administer area transportation system
Administer project EEO program
Enforce project safety program
Coordinate labor relations
Receive and evaluate bids and award prime contracts
Manage and perform general conditions tasks
Implement time- and cost-control systems
Manage daily construction activities of the owner or architect/engineer
Administer prime contracts
Receive, review, and approve contractor's requests for progress payments
Administer contract changes and claims
Quality assurance and inspection
Interpret contract documents

Closeout and Startup Phase

Oversee project closeout
Oversee systems validation, testing, and startup

Responsibility for Coordination of the Trades

Construction projects of any size usually require the efforts of several specialty contractors. Usually, of course, such work is performed under subcontract with a general contractor who is responsible for the coordination of all of the subcontractors.

A large number of claims involving changes, delays, and even site conditions result from coordination problems. As a result, the courts have been asked to determine who is responsible, as well as the scope of that responsibility. Changes in the methods of construction contracting, such as design–build and multiple-prime contracts, have made the problem worse and increased the volume of litigation.

Owner's Responsibility

Traditionally, the owner has had no responsibility for coordination of subcontractors. Scheduling and coordination of the trades has traditionally been the exclusive responsibility of the general contractor, and most standard construction contracts are clear on this issue. The owner's obligation is simply to avoid any active interference with the work of the various contractors.

As long as a single general contractor is used to construct a project, this will still hold true. However, if the owner awards more than one prime contract or elects to perform some of the work with its own forces, the owner inherits coordination responsibilities.

On a multiple-prime project, the owner is responsible for coordinating the various trade contractors with whom it has contracted in much the same way as a general contractor must coordinate its subcontractors. Few owners are equipped to do this, which has created the need for the professional construction manager.

Less apparent to owners is the fact that if they perform even a small portion of the work themselves, they have an implied obligation to coordinate the work in order to avoid economic harm to the prime contractors. Failure to meet this obligation can be expensive for the owners, as payment of delay damages will quickly offset the savings realized from performing the work with one's own forces.

On traditional projects, the project architect/engineer is not responsible for coordinating the various trades. The architect/engineer may be called upon to interpret the requirements of the contract documents to resolve scope-of-work problems, but will not become involved in the actual scheduling or coordinating of the work itself.

On a multiple-prime project, however, the owner's representative (usually called the construction manager) plays a very different role. In the absence of a general contractor, the owner must rely on the construction manager to provide overall scheduling and coordination of the trade contractors. In such a case, if improper coordination delays harm one of the contractors, the construction manager may be held liable.

The construction manager may be held directly liable to the contractor on a theory of negligence. More commonly, the contractor will sue the owner with whom it has contracted. The owner may then seek indemnification from the construction manager, arguing that any liability to the contractor is the result of the construction manager's failure to perform its duties properly, as established in the owner-construction manager agreement. This is part of what the owner pays for in awarding a lump sum contract to a single general contractor.

THE PARTNERING CONCEPT[6]

The Partnering concept is not a new way of doing business; some have always conducted themselves in this manner. It is going back to the way people used to do business when their word was their bond and people accepted responsibility. Partnering is not a contract, but a recognition that every contract includes an implied covenant of good faith.

While the contract establishes the legal relationships, the Partnering process is designed to establish working relationships among the parties through a mutually developed, formal strategy of commitment and communication. It attempts to create an environment where trust and teamwork prevent disputes, foster a cooperative bond to everyone's benefit, and facilitate the completion of a successful project.

[6] *Partnering, a Concept for Success,* © Copyright AGC of America, 1991, The Associated General Contractors of America, 1957 E. Street NW, Washington, DC 20006.

For the most effective results, parties involved in the contract should conduct a Partnering workshop, ideally at the early stages of the contract. The sole agenda of the workshop is to establish and begin implementing the partnering process. This forum produces the opportunity to initiate the key elements of Partnering.

The key elements of Partnering include the following:

- *Commitment*. Commitment to Partnering must come from top management. The jointly developed Partnering charter is not a contract, but a symbol of commitment.
- *Equity*. All of the parties' interests need to be considered in creating mutual goals, and there must be a commitment to satisfying each of the parties' requirements for a successful project by utilizing win/win thinking.
- *Trust*. Teamwork is not possible where there is cynicism about others' motives. Through the development of personal relationships and communication about each of the parties' risks and goals, there will be better understanding. Along with understanding comes trust, and with trust comes the possibility for a cooperative relationship.
- *Development of Mutual Goals/Objectives*. At a Partnering workshop the parties should identify all respective goals for the project in which their interests overlap. These jointly developed and mutually agreed upon goals may include achieving value engineering savings, meeting the financial goals of each party, limiting cost growth, limiting review periods for contract submittals, early completion, no lost time because of injuries, minimizing paperwork generated for the purpose of case building or posturing, no litigation, or other goals specific to the nature of the project.
- *Implementation*. The parties jointly should develop strategies for implementing their mutual goals and the mechanisms for solving problems.
- *Continuous Evaluation*. In order to assure effective implementation, the parties need to agree to a plan for periodic joint evaluation based on the mutually agreed upon goals to assure that the plan is proceeding as intended and that all of the parties are carrying their share of the load.
- *Timely Responsiveness*. Timely communication and decision making not only save money, but also can keep a problem from growing into a dispute. In the Partnering workshop the parties must develop mechanisms for encouraging rapid issue resolution, including the escalation of unresolved issues to the next level of management.

Partnering Benefits

For all of the parties involved in a project, Partnering is a high-leveraged effort. It may require increased staff and management time up front, but the benefits accrue in a more harmonious, less confrontational process, and at completion a successful project can be realized without litigation and claims. The Partnering process empowers the project personnel of each of the parties with the freedom and authority to accept responsibility—to do their jobs by encouraging decision making and problem solving at the lowest possible level of authority. It encourages everyone to take pride in their efforts and tells them it's OK to get along with each other.

Partnering is also an opportunity for public sector contracting, where the open competitive-bid process ordinarily keeps the parties at arm's length prior to award, to achieve some of the benefits of closer personal contact that are possible in negotiated or design–build contracts.

Potential Problems

Partnering requires that all parties to a contract commit themselves to the concept. The Partnering concept is endangered if there is not true commitment.

Those people who have been conditioned in an adversarial environment may be uncomfortable with the perceived risk in trusting. Some may even find it impossible to adapt their thinking to a relationship based on trust. Giving only lip service to the term or treating the concept as a fad is not true commitment.

For some, changing the myopic attitude that leads to thinking that it is necessary to win every battle, every day, at the other parties' expense will be very difficult. Win/win thinking is an essential element for success in this process. It must be remembered that the win/win concept, after all, is based upon the acceptance of the concept of compromise between the parties.

The Partnering Process

In the words of the Associated General Contractors of America, "The partnering process involves the empowerment of the project personnel of all participating organizations with the freedom and authority to accept responsibility—to do their jobs by encouraging decision making and problem solving at the lowest possible level of authority." This, of course, involves a major commitment by top management of all participating organizations to delegate such authority to the lowest possible management level at the site, such as the owner/engineer's Resident Project Representative and the contractor's project manager or superintendent.

The Partnering process emphasizes the need to educate your organization to the facts about the concept. It is important to make Partnering intentions clear from the beginning and obtain a commitment from top management from the start.

The Partnering workshop must involve the following:

1. Creation of a Partnering charter
2. Development of an issue-resolution process
3. Development of a joint evaluation process
4. Discussion of individual roles and concerns
5. Presentation of a facilitated workshop

For success of the Partnering concept, it is necessary to provide for periodic evaluation. Occasional escalation of an issue may surface, of course, but the process is structured to accommodate such problems. In the end, final evaluations must be made, and refinements may be considered to improve future Partnering contracts.

Closing

Many of us have witnessed the construction industry evolve into an adversarial, confrontational business where the parties' energies are misdirected away from the ultimate goal of constructing a quality product, on time, and within budget. Partnering changes mind-sets. It helps everyone in the construction process to redirect his or her energies and to focus on the real issues associated with achieving our ultimate objective.

Specifying Partnering

In the specifications, the owner needs to state its intention to encourage the formation of a cohesive Partnering relationship with the contractor and its subcontractors. This Partnering relationship must be structured to draw on the strengths of each organization to identify and achieve reciprocal goals. The objectives are effective and efficient contract performance, intended to achieve completion within budget, on schedule, and in accordance with plans and specifications.

An example of a provision for partnering in a DOT specification is illustrated in Figure 1.10.

The Partnering relationship must be bilateral in makeup, and participation must be totally voluntary. Any cost associated with effecting this partnership should be agreed to by both parties and should be shared equally with no change in contract price. To implement this Partnering initiative, it is anticipated that within 60 days of the Notice to Proceed, the contractor's on-site project manager and the owner's on-site representative should attend a Partnering development seminar followed by a team-building workshop to be attended by both the contractor's key on-site staff and

5-1.16 PARTNERING

The State will promote the formation of a "Partnering" relationship with the Contractor in order to effectively complete the contract to the benefit of both parties. The purpose of this relationship will be to maintain cooperative communication and mutually resolve conflicts at the lowest possible management level.

The Contractor may request the formation of such a "Partnering" relationship by submitting a request in writing to the Engineer after approval of the contract. If the Contractor's request for "Partnering" is approved by the Engineer, scheduling of a "Partnering" workshop, selecting the "Partnering" facilitator and workshop site, and other administrative details shall be as agreed to by both parties.

The costs involved in providing a facilitator and a workshop site will be borne equally by the State and the Contractor. The Contractor shall, pay all compensation for the wages and expenses of the facilitator, and of the expenses for obtaining the workshop site. The State's share of such costs will be reimbursed to the Contractor in a change order written by the Engineer. Markups will not be added. All other costs associated with the "Partnering" relationship will be borne separately by the party incurring the costs.

The establishment of a "Partnering" relationship will not change or modify the terms and conditions of the contract and will not relieve either party of the legal requirements of the contract.

FIGURE 1.10 Example of Specifications Provision for Partnering from a DOT Specification.

the owner's personnel. Follow-up workshops should be held periodically throughout the duration of the contract as agreed to by the contractor and owner.

An integral aspect of Partnering is the resolution of disputes in a timely, professional, and nonadversarial manner. Alternative dispute resolution (ADR) methodologies need to be encouraged in place of the more formal dispute resolution procedures. ADR will assist in promoting and maintaining an amicable working relationship to preserve the partnership. ADR in this context is intended to be a voluntary, nonbinding procedure available for use by the parties to this contract to resolve any dispute that may arise during performance.

Contracting for Public Works Projects

A few words of wisdom are offered for the contractor's benefit. The words are not about how a contractor should manage its construction projects, however. That task is up to the contractor. These are guidelines that may help a contractor sidestep some of the usual pitfalls when expanding its business from private contracts to public works contracting. The author makes no pretense of offering management methods that will revolutionize a contractor's business. Only that, if you follow some basic guidelines, you will be less likely to be caught unaware when you suddenly realize that you failed to include enough money in your public works bid to cover the many unforeseen costs that are commonplace in public works construction contracts.

If you have been a successful small contractor in the private sector and wish to cash in on that pot of gold at the end of the public works rainbow, listen up!

> If you are low bidder on your first public project, you probably made a serious mistake in your bid.

The newcomer to the business of public works contracting may be in for a shocking surprise if the work is judged by private work of a similar nature. Those appealing advertisements in the daily construction newspapers and the local newspapers seem to suggest that there is a virtual pot of gold at the end of the construction rainbow, and that all you need do to reach it is to underbid the competition, and the rest should be all downhill. If anything is true, it is the opposite. That is precisely where the contractor's trouble begins. The methods used in managing the construction projects that worked so well in the private sector cannot be readily applied to the much more regimented and documented area of public works contracting. The old concept of simply providing the agency with a "good job" is no longer sufficient. The new game must be played "by the book."

It all begins with the Notice Inviting Bids or another similarly named document. The bidders should be prepared to follow all instructions to the letter. The often-followed practice among contractors in the private sector of submitting a "qualified" bid by marking over portions of the bidding documents serves little in public works contracting, except to have the entire bid rejected as being "nonresponsive" without any opportunity for the bidder to negotiate or make corrections. Then all of the time and money spent by the contractor to prepare the bid is needlessly wasted. There are no second chances. The bid must be right the first time, as there will be no opportunity to talk about it later on.

In the business of public works contracting, the bidder had better be prepared to read all of the specifications, including the front-end documents, before submitting its bid; otherwise, later surprises may be sufficient to force a small contractor into bankruptcy. Some contractors refer to public bidding as "a method of finding a contractor willing to work at a loss." Interpretations of apparent ambiguities in the contract documents can be the source of serious problems.

If the new entrant into the field of public works contracting assumes that the contractor's mission is "just to build a project as it appears on the plans and specifications," then it is time to review the facts. The project that is supposed to be built may, after a while, seem to be treated as a secondary by-product by the agency, and the contractor may soon feel that the true objective is to generate reams of paperwork and make work for local politicians. The paperwork is, indeed, overwhelming at times, but it is important. In fact, it may even be well for the contractor to initiate a few in-house paperwork systems of its own. Otherwise, when you want to file claims for recovery of job costs, your case may be lost even before it is ready to present.

Many newcomers to the contracting business feel that they are fairly shrewd people and may feel that they know a few tricks of their own. In all too many cases, the opportunity to use some of these innovative methods may never arrive, as the die may have been cast the moment those voluminous contract documents were signed, without your taking the time to study the terms as carefully as you should have. Generally, those contract documents were compiled by experts whose only objective (it sometimes seems) was to place all of the risk on the contractor's shoulders and none on the public agency.

Remember:

> The Contract Documents were written for a reason, and it was probably not to help you.

It should be remembered that the architect/engineer and the public agency are not generally out to "get" the contractor, but are just trying to protect their own positions (though sometimes somewhat excessively). They are usually just following the numerous laws governing the administration of public works contracts at all levels. A public agency may not be aware of a newcomer's unfamiliarity with the special circumstances of public works contracting and may merely treat each contractor in the same manner as it would any other of the more experienced public works contractors. The agency is required by law to administer its contracts in a fashion that will provide maximum protection to the public, not the contractor.

The subject of this book is intended to serve owners, architect/engineers, and contractors of both the public and the private sectors. In addition, it's going to tell the contractor what the owner and the architect/engineer are up to. A contractor can increase its profitability if it is aware of all that is involved.

The best solution is to not try to fight city hall by attempting to bypass the rules. That approach will only cost both ill will and your money. To be profitable, a contractor needs to learn to respond by playing the game by the owner's rules and winning.

PROBLEMS

1. What are the three principal concerns of a Resident Project Representative or an inspector with respect to his or her project during the construction phase?
2. Ideally, a Resident Project Representative or an inspector should be under the direct management control of whom? (Indicate job title.)
3. Who should have the authority to make design changes in the field during construction: the resident engineer, the owner's representative, the design engineer, or the inspector?
4. What type of organizational structure is best suited to fast-track construction?
5. Under a construction management (CM) contract, should the CM firm's responsibilities normally begin at the construction phase, the design phase, the planning phase, or the conceptual phase?
6. A design–build firm may be (check all that apply): (a) a single firm with both design and construction capability; (b) a joint venture contract between a contractor and a design firm; (c) a design firm that subcontracts the construction portion to a construction contractor; or (d) a construction contracting firm that subcontracts the design work to an architect or engineer firm.
7. True or false? Under the Partnering concept, all differences are intended to be resolved at the lowest management level—preferably at the project site.
8. What type of relationships are intended to be established under a Partnering agreement?
9. What are the five steps necessary to set up a Partnering workshop?
10. Is Partnering a totally informal process, or should it be addressed in the specifications?
11. True or false? Under the CQC process, inspection is provided by the contractor.
12. What is the 5-step process of initiating a project? Name each of the steps in proper order.
13. Explain the one-to-one concept.
14. What is the principal barrier to the use of design–build contracts on public works projects?

2

RESPONSIBILITY AND AUTHORITY

The Resident Project Representative and Inspectors as Members of the Construction Team

The effective management of a construction operation can only be achieved through a well-coordinated team effort. The Resident Project Representative and the inspectors are vital members of that team, for without them there would normally be no direct involvement in the construction of the project by the architect/engineer or the owner. The Resident Project Representative and the inspectors are the eyes and the ears of the architect or engineer and the owner. Their authority and responsibility on the project is largely based upon that concept. To be sure, one architect or engineer may delegate more or less authority to the Resident Project Representative or inspector than another architect or engineer; however, that is a matter of employer–employee relations between the Resident Project Representative or inspector and his or her employer. In addition, many federal construction contracts require a resident inspector in the employ of the contractor under the principle of *contractor quality control* (CQC). Generally, in the absence of specific instructions to the contrary, the guidelines in this chapter are considered by many members of the construction industry to be the normal standards.

Lines of Authority on Construction Projects

Almost all construction projects involve organizations. The participants—owner, contractor, subcontractors, and suppliers—are not individuals. They are corporations, partnerships, or other forms of business associations.

Organizational decision making can be ponderous. Yet decisions and approvals are required on a daily basis during the performance of a construction contract. It is obvious that each organization must establish lines of authority by designating the individuals who are authorized to make necessary decisions (see "The One-to-One Concept" in Chapter 1). If this is done in a careful, thoughtful manner, the project will benefit. If it is done in a haphazard manner, the likelihood of misunderstandings and disputes will increase.

Agency Relationship

An agency relationship is established when one party, such as the owner or architect/engineer, designates another party, such as a Resident Project Representative or inspector (the agent), to act on its behalf. Once the agency relationship is established, the owner or architect/ engineer is bound by the acts and omissions of its Resident Project Representative or inspector, as its agent. If on-site representatives are not designated and given the authority that they need, communication will break down and confusion will result. There will also be some unanticipated legal ramifications.

Actual Authority

The most common way for an agency relationship to be established is by an express grant of actual authority. The contract documents simply state that a particular individual shall have the authority to make certain decisions. On public construction projects, the existence of actual authority is important because statutes protect public agencies against acts or decisions of unauthorized individuals.

Apparent Authority

An agency relationship can also be created when the owner or architect/engineer allows another party to operate with the *appearance* of authority to act on behalf of the owner or architect/engineer. The contractors who reasonably believe that they are dealing with a duly authorized agent will be able to hold the owner or architect/engineer accountable for the acts or knowledge of that *apparent* "agent."

The doctrine of apparent authority does not apply to public agency owners, as statutes shield them from liability for the acts of anyone other than the actual designated agent. Private organizations can protect themselves by contractually designating an individual with authority to act for them and disclaiming responsibility for the acts of anyone else.

This is not totally foolproof, however, as the contractual disclaimer can be constructively waived by a persistent pattern of conduct that is inconsistent with the terms of the contract. A similar result may occur through *ratification*. An owner or a contractor may, by its conduct, ratify the unauthorized actions of a person at the job site. The most common form of ratification is silence or acquiescence. In other words, if an owner or a contractor sits silently by while one of its people takes charge at the job site, it will be difficult later to argue that the individual lacked authority to make a particular decision or issue a particular directive.

Delegation of Authority

Once actual authority has been conferred on an agent, such as a Resident Project Representative, it is possible to allow that agent to delegate his or her authority, say, to one of the inspectors. It is sometimes common to grant certain authority to a specific individual "or his or her designated representative."

The phrase "or designated representative" may create problems, however. Other companies may not be able to determine whether there has been an effective delegation of authority. The option of delegating authority also increases the likelihood of creating the appearance of authority. It is then inconsistent with any disclaimer of "apparent authority."

Actual authority may be delegated even by a public agency. As an example, an excavation contractor was obligated to follow the directives of a resident inspector because the government contracting officer had delegated on-site authorities to the inspector (*Appeal of Stannard Construction Co., Inc.,* ENG BCA No. 4767 [May 17, 1984])[1]

Limited Authority

It is also possible to grant limited authority to an agent. This can be done in order to give the Resident Project Representative certain responsibilities and authority without allowing that representative to make fundamental contract changes, changes in time or cost, or to waive legal rights. This arrangement is preferable to naming an authorized individual in the home office "or his or her designated representative."

To be effective, a grant of limited authority must include clear, objective contractual limitations on the scope of the agent's authority. For instance, a Resident Project Representative might be given the authority to approve changed or extra workup to a stated dollar amount.

Summary

The foregoing provisions can best be summarized in the following four guidelines:

1. Establish actual authority in the contract with a specific individual. Do not designate a position, job title, or group of individuals.
2. Disclaim any apparent authority and establish administrative procedures consistent with the disclaimer language.
3. Do not give designated representatives the right to further delegate their authority.
4. When granting limited authority, state objective guidelines to delineate the scope of that authority.

---

[1]See *Construction Claims Monthly,* Sept. 1984, p.5, Business Publishers Inc., 8737 Colesville Road, Suite 1100, Silver Spring, MD 20910-3928.

WHY HAVE AN INSPECTOR?

It is not enough to leave the assurance of quality workmanship and materials entirely in the hands of the contractor. The architect/engineer or other design firm that was responsible for the determination of site conditions and for the preparation of the plans and specifications should be retained during the construction phase to provide field administration and quality assurance for the owner, the safety of the public, and the professional reputation of both the design firm and the contractor. Many architects and engineers seem to believe that if their design is adequate and the plans and specifications are carefully prepared, the field construction will take care of itself. Experience has proven this to be far from true.

Although a design firm acting as the agent of the owner during the construction phase of a project does not guarantee the work of the contractor, nor does such agent in any way relieve the contractor of any responsibilities under the terms of the construction contract, the design firm, through its field inspection forces, must endeavor to guard the owner against defects and deficiencies in the work. When, in the judgment of the designer, the plans and specifications are not being followed properly, and the design firm has been unable to obtain compliance by the contractor, the owner should be notified so that appropriate measures can be taken. Inspection should be performed during the progress of the work; inspection after completion defeats the purpose of providing quality control and assurance on the job, as many potential difficulties must be detected during construction; otherwise, they may be permanently covered. The result would be a latent defect that may not be discovered for years; then, when it is discovered, it may be too late, as it may have been instrumental in contributing to a structural failure or other disaster.

Often, the word *supervision* has been used in the past in connection with field inspection. It is a legally risky term, and as such, its use has long been discouraged by all technical and professional societies of the construction industry. In the case of *U.S. Home Corp. v. George W. Kennedy Construction Co.* [610 F. Supp. 759 (D.C.Ill.) 1985] a federal court ruled that if a project owner agrees to "oversee" or "supervise" a contractor's performance of the Work, the owner may share responsibility for defective work performed by the contractor. The bottom line is this: If the owner simply "inspects" the Work for compliance with the plans and specifications, the owner does not assume responsibility for the sufficiency of the contractor's work. If, on the other hand, the owner agrees to "supervise" the contractor's means and methods of performance, it gives rise to the issue presented in the *U.S. Home Corp. v. George W. Kennedy Construction Co.* case.

In the 1996 edition of the Standard General Conditions of the Construction Contract of the Engineer's Joint Contract Documents Committee (EJCDC), published jointly by the American Consulting Engineers Council, the American Society of Civil Engineers, the Construction Specifications Institute, and the National Society of Professional Engineers, as approved and endorsed by the Associated General Contractors of America, the general contractor is allowed greater management control, whereas the engineer's involvement during the construction phase is lessened. The 1996 edition of the EJCDC General Conditions (Article 9.10.B) states that the engineer "will not supervise, direct, control, or have authority over or be responsible for Contractor's means, methods, techniques, sequences, or procedures of construction, or the safety precautions and programs incidental thereto." The AIA General Conditions makes similar reservations. Therefore, the Resident Project Representative for the design firm and the owner should not direct, supervise, or assume control over the means, methods, techniques, sequences, or procedures of construction except as specifically called for in the project specifications. Instead, the Resident Project Representative should exercise authority on behalf of the owner so that such activities will result in a project substantially in accordance with the requirements of the contract documents. The FIDIC documents are silent on this issue, stating only that "the Engineer shall have no authority to relieve the Contractor of any of his obligations under the Contract."

Construction administration and construction quality control by the contractor or quality assurance by the design firm or owner should include continuous on-site inspection during all structural construction of a building by one or more competent, technically qualified, and experienced inspectors. If employed directly by the owner or by the design firm, all such inspectors should be under the architect/engineer's supervision and direction and should report any discrepancies directly to the architect/engineer. If there is a staff of several inspectors on a project, all other inspectors should be under the direct supervision of the Resident Project Representative, and all communications between them and the owner or design firm should be through the Resident Project Representative. It is the responsibility of these other inspectors to see that all details of the engineer's or architect's design drawings, shop drawings, barplacing drawings, and similar documents that have been approved by the engineer or architect are constructed in strict accordance with their respective requirements. In addition, each inspector must see that all the requirements of the specifications have been met and that all workmanship and construction practices are equal to or in excess of the standards called for in the construction contract documents. The inspector should also make certain, as the job progresses, that the mechanical and electrical installations are constructed in accordance with the approved drawings, and that nonstructural items do not adversely interfere with structural elements.

Supervision by the architect or engineer of record and by the Resident Project Representative or inspector are really quite different, particularly in one very important respect. Neither the Resident Project Representative nor the inspector should have the authority to change plans or specifications, or to make his or her own interpretations, even though he or she may be a qualified engineer with both design and construction experience. If any question of interpretation arises, or if there is a disagreement on a technical matter between the inspector and the contractor, or if there appears to be any possibility of error or deviation from good construction practice that is noticed by the inspector, it should immediately be brought to the attention of the architect/engineer's project manager for decision. Inspection by the contractor under a CQC contract should follow similar procedures. However, supervision by the architect/engineer or its designated project manager, as distinct from that of the Resident Project Representative (or resident inspector), may include the authority to modify the plans and specifications consistent with the contract provisions between the owner and the design firm and between the owner and the contractor, if job conditions indicate that a change would be in the interest of improvement of the structure or if such a change were otherwise justified and consistent with sound design principles followed in the original design. Any such change should be supported by a formal change order, however.

Authority and Responsibility of the Resident Project Representative

The Resident Project Representative can function effectively only when given certain specific authority. In addition, the contractor should be made fully aware of the authority of the Resident Project Representative. Although a number of documents have been published that set forth a recommended scope of an inspector's authority and responsibility, the following is recommended because it represents a set of standards that have been established as the result of evaluating the answers to numerous pertinent questions contained in questionnaires that have been circulated by the Task Committee on Inspection of the Construction Division of the American Society of Civil Engineers to a nationwide cross section of contractors; owners and owner representatives; engineers; federal, state, and local governmental agencies; independent inspection agencies; and others.

Responsibility[2]

As the Resident Project Representative for a design firm or the owner, the resident engineer or inspector is responsible for seeing that the work being inspected is constructed in accordance with the requirements of the plans and specifications. This, however, does not confer the right to unnecessarily or willfully disrupt the operations of the contractor. In the performance of assigned duties, the Resident Project Representative, who is referred to in the following guidelines simply as the "inspector," would normally assume the following responsibilities:

1. The inspector must become thoroughly familiar with the plans and specifications as they apply to the work to be inspected and should review them frequently. The inspector must be capable of immediately recognizing if the work being inspected conforms to the contract requirements.
2. If any material or portion of the work does not conform to the requirements, the inspector should so notify the contractor, explain why it does not conform, and record it in the daily diary. Should the contractor ignore the notice and continue the operation, the inspector should promptly advise the architect/engineer or the owner.

[2]Committee on Inspection of the Construction Division, "Recommended Standards for the Responsibility, Authority and Behavior of the Inspector," *Journal of the Construction Division*, Vol. 101, No. C02, June 1975. Proceedings of the American Society of Civil Engineers, pp. 360–363, inclusive. Reproduced by permission.

3. As a member of the construction team, the inspector must perform all duties in a manner that will promote the progress of the work. The inspector should be familiar with the construction schedule and should know how the work that is being inspected fits into the overall schedule. Completion of the work within the contract time is also of importance to the owner.
4. The inspector must studiously avoid any inspection, testing, or other activity that could be construed as a responsibility of the contractor; otherwise, the owner's position may be prejudiced in the event of a dispute or claim. This applies particularly to the contractor's quality control program for testing and inspecting the contractor's materials and workmanship, as a part of his or her contractual responsibility.
5. When the inspector is assigned to any operation, it should be covered as long as the work is proceeding or see to it that another inspector takes over should the original inspector have to leave. This applies particularly to work that will not be viewed again, such as driving piles, laying pipe in a trench, and placing concrete.

READ THE CONTRACT...

SOMEBODY PREPARED IT FOR A REASON, AND IT WAS PROBABLY NOT TO HELP YOU

6. The inspector's daily report and diary should include a recording of the day's happenings, the contractor's activity on the work being inspected, instructions given the contractor, and any agreements made. The inspector must remember that in the event of contract disputes, the daily reports and diary may assume legal importance.
7. In the matter of on-site testing, tests should be performed expeditiously and carefully, test samples must be carefully handled and protected, and test failures must be reported to the contractor without delay. It is a needless waste of time and money when a contractor is informed of an unsatisfactory result of a test that was performed two or three days previously.
8. Inspections and tests should be made promptly and timely:
   (a) Materials should be checked as soon after they are delivered as possible. An inspector who rejects materials after they have been placed in their permanent position is not working in the best interest of the owner.
   (b) Preparatory work such as cleanup inside the forms, fine grading of footing areas, winter protection for concrete, and so on, should be checked promptly to minimize delay to subsequent operations.
   (c) Work should be inspected as it progresses. For example, postponing the inspection of the placing of reinforcing steel and other embedded items until they are 100 percent complete does nothing but delay progress.
   (d) An inspector has the responsibility to be available at all times to provide prompt inspection and a decision on acceptance when required. A contractor should not be required to delay his or her work while the inspector is locating the architect/engineer's or owner's project manager to make a decision. Of course, by the same token, the contractor is expected to give adequate notice to the inspector when the Work will be ready for inspection on an operation.
9. If any specific tolerance governing the contractor's work is found to be unrealistic, it is the responsibility of the inspector to so report it to the architect/engineer or owner.
10. Too literal an interpretation of the specifications can cause problems if it is not applicable to the particular situation. In such an instance, the inspector must review the conditions and seek guidance from the project manager, if necessary.
11. Whenever possible, problems should be anticipated in advance of their occurrence. The contractor's superintendent or foreman may seem to be unaware of a sleeve or other embedded item that must be set in the forms. It is incumbent upon the inspector to point this out to the superintendent or foreman. By this advance notice, the inspector contributes to maintaining the progress of the work.

12. Unacceptable work should be recognized in its early stages and reported to the contractor before it develops into an expensive and time-consuming operation. The notification should be confirmed in writing where necessary. For example, if the contractor is using the wrong form lining, stockpiling unacceptable backfill material, or placing undersized riprap material, the contractor should be informed of this at the first opportunity. An inspector who is thoroughly familiar with the contract requirements can recognize these situations almost immediately.
13. Occasionally, a problem may arise that the inspector is unable to handle alone. This should be reported to the architect/engineer or owner for prompt action. Unresolved problems can sometimes develop into critical situations and claims.
14. Rather than make a hasty decision, the inspector should thoroughly investigate the situation and its possible consequences. Many embarrassing situations develop from decisions made prematurely. For example, a request by the contractor to be permitted to begin placing concrete at one end of a long footing while the crew is completing the reinforcing at the far end should be given consideration and not be automatically denied. If necessary, the inspector should seek advice from the architect/engineer or owner's engineering staff.
15. When work is to be corrected by the contractor, the inspector should follow it up daily. Otherwise, corrections may be forgotten or the work soon covered over.
16. The inspector should stand behind any decisions made on issues concerning the contractor's work. An untrue denial by the inspector can cause immeasurable damage to the relations between the contractor and inspection personnel.
17. In the course of his or her work, the inspector must be capable of differentiating between those items that are essential and those that are not, as defined by the architect/engineer or owner's engineering staff.
18. The inspector should be safety minded. If a dangerous condition is observed on the job, there is a responsibility to call it to the attention of the contractor and then note it in his or her daily diary. The mere physical presence of the owner's representative on the site creates a responsibility to report a recognizably unsafe condition.
19. The inspector has a responsibility to be alert and observant. Any situation that threatens to cause a delay in the completion of the project should be reported to the architect/engineer or owner.

Authority[3]

The Resident Project Representative, referred to herein as the inspector, must be delegated certain authority to perform the required duties properly. The close working relations with the contractor demand it. The inspector should use the given authority when the situation demands it and should not abuse it. In addition, the contractor is entitled to know what the specific authority of the inspector is, and when the work is not proceeding in an acceptable manner.

1. The inspector should have the authority to approve materials and workmanship that meet the contract requirements and should give approvals promptly, where necessary.
2. The inspector should *not* be given the authority to order the contractor to stop operations (c.f. "Disapproving or Stopping the Work" in Chapter 7). When a contractor is ordered to halt an active operation immediately, it becomes a costly item, particularly if expensive equipment and material such as concrete are involved. If the stop order is not justifiable by the terms of the contract, the contractor has just cause to demand reimbursement for the damage that has been suffered. Because of the nature of the inspector's duties, the inspector cannot be familiar with all the details of the contract nor with all the other contractual relationships. Authority for the issuance of a stop order should be left to the judgment of the architect/engineer's or owner's project manager (see Chapter 7 for possible exceptions to the rule). Furthermore, the power to "stop the work" is often interpreted by the courts as *control of the work*. This may expose the architect or engineer to tort liability for any accidents or injuries on the job.
3. The inspector should not have the authority to approve deviations from the contract requirements. This can be accomplished properly only with a change order.
4. The inspector should not require the contractor to furnish more than that required by the plans and specifications.
5. Under no circumstances should the inspector attempt to direct the contractor's work; otherwise, the contracting firm may be relieved of its responsibility under the contract.
6. Instructions should be given to the contractor's superintendent or foreman, not to workers or to subcontractors.

[3]Committee on Inspection of the Construction Division, "Recommended Standards for the Responsibility, Authority and Behavior of the Inspector," *Journal of the Construction Division*, Vol. 101, No. C02, June 1975. Proceedings of the American Society of Civil Engineers, pp. 360–363, inclusive. Reproduced by permission.

The following is quoted from a news report that exemplifies one of the difficulties that can be encountered by attempting to direct the contractor's work or methods of construction. The event was reported in *ENR* (*Engineering News Record*) on 5 May 1997 under the heading of "Direction of the Work by the Inspector—Boring Machine Wanders Off Course by 18 Feet."

> Seattle Public Utilities engineers are trying to figure out options after a misdirected 3.5-ft-dia micro-tunneling machine left the intended route of a new underground storm sewer and *ended up instead beneath the basement of a downtown store.* The 18-ft-long, $600,000 Soltau mole, operated by tunneling subcontractor Northwest Boring Co., Inc., Woodinville, Wash., was driven 18 ft out of the right-of-way, *apparently as a result of a city inspector's error in plotting the mole's course.* City Project Manager Pamela Miller says the error occurred after workers moved a planned manhole from the store front. A new directional mark that was 1 ft out line was apparently placed on the pavement some 25 ft from the mole's starting point, leading operators to drive the machine off course. Engineers discovered the error when the mole did not appear in the receiving pit after 460 ft of pipe had been inserted into a 475 section of tunnel. The glitch could add up to two months to the overall 2-mile, $10.2-million project. It was scheduled to finish in July. The city has not decided if it will continue the present course by negotiating a permanent easement with property owners or remove the mole and fill in the tunnel. Northwest President Don Gonzales was unavailable for comment.

The Seattle Post-Intelligencer reported the event on 2 May 1997 under the headline "City Shoots for 2nd Try with Tunneler—$600,000 Funneled to Fix Eastlake Error," which is quoted in part as follows:

> A miscalculation by a city inspector caused the micro-tunneling machine to aim slightly to the left of where it was supposed to go and to bore under the restaurant supply store.

Although most documents that define the inspector's responsibility and authority are the result of studies and recommendations by professional societies such as the American Consulting Engineers Council (ACEC), American Society of Civil Engineers (ASCE), The Construction Specifications Institute, Inc. (CSI), the National Society of Professional Engineers (NSPE), and the American Institute of Architects (AIA), in at least one state such requirements have been incorporated into law for certain types of work.

When the Resident Project Representative is a licensed engineer and may properly be termed a *Resident Engineer,* it may be desirable to delegate greater authority and discretion to that position. The document illustrated in Figure 2.1 presumes that the Resident Project Representative is not a licensed engineer, and if more authority is to be delegated, some changes must be made in these documents. This document has important legal consequences, and consultation with an attorney is encouraged if questions arise.

SUGGESTED FORMAT
(for use with No. 1910-1, 1996 Edition)

This is **EXHIBIT D**, consisting of _____ pages, referred to in and part of the **Agreement between OWNER and ENGINEER for Professional Services** dated ____________________, ________.

Initial:
OWNER ______
ENGINEER ____

Duties, Responsibilities, and Limitations of Authority of Resident Project Representative

Paragraph 1.01C of the Agreement is amended and supplemented to include the following agreement of the parties:

D6.02 *Resident Project Representative*

A. ENGINEER shall furnish a Resident Project Representative ("RPR"), assistants, and other field staff to assist ENGINEER in observing progress and quality of the Work. The RPR, assistants, and other field staff under this Exhibit D may provide full time representation or may provide representation to a lesser degree.

B. Through such additional observations of Contractor's work in progress and field checks of materials and equipment by the RPR and assistants, ENGINEER shall endeavor to provide further protection for OWNER against defects and deficiencies in the Work. However, ENGINEER shall not, during such visits or as a result of such observations of Contractor's work in progress, supervise, direct, or have control over the Contractor's Work nor shall ENGINEER have authority over or responsibility for the means, methods, techniques, sequences, or procedures selected by Contractor, for safety precautions and programs incident to the Contractor's work in progress, for any failure of Contractor to comply with Laws and Regulations applicable to Contractor's performing and furnishing the Work, or responsibility of construction for Contractor's failure to furnish and perform the Work in accordance with the Contract Documents. In addition, the specific limitations set forth in section A.1.05 of Exhibit A of the Agreement are applicable.

C. The duties and responsibilities of the RPR are limited to those of ENGINEER in the Agreement with the OWNER and in the Contract Documents, and are further limited and described as follows:

1. *General:* RPR is ENGINEER's agent at the Site, will act as directed by and under the supervision of ENGINEER, and will confer with ENGINEER regarding RPR's actions. RPR's dealings in matters pertaining to the Contractor's work in progress shall in general be with ENGINEER and Contractor, keeping OWNER advised as necessary. RPR's dealings with subcontractors shall only be through or with the full knowledge and approval of Contractor. RPR shall generally communicate with OWNER with the knowledge of and under the direction of ENGINEER.

2. *Schedules:* Review the progress schedule, schedule of Shop Drawing and Sample submittals, and schedule of values prepared by Contractor and consult with ENGINEER concerning acceptability.

3. *Conferences and Meetings:* Attend meetings with Contractor, such as preconstruction conferences, progress meetings, job conferences and other project-related meetings, and prepare and circulate copies of minutes thereof.

4. *Liaison:*
   a. Serve as ENGINEER's liaison with Contractor, working principally through Contractor's superintendent and assist in understanding the intent of the Contract Documents.

Page 1 of ___ Pages
(**Exhibit D** - Resident Project Representative)

FIGURE 2.1 Engineer's Joint Contract Documents Committee suggested *Duties, Responsibilities, and Limitations of Authority of Resident Project Representative. (Copyright © 1996 by National Society of Professional Engineers.)*

b. Assist ENGINEER in serving as OWNER's liaison with Contractor when Contractor's operations affect OWNER's on-Site operations.

c. Assist in obtaining from OWNER additional details or information, when required for proper execution of the Work.

5. *Interpretation of Contract Documents:* Report to ENGINEER when clarifications and interpretations of the Contract Documents are needed and transmit to Contractor clarifications and interpretations as issued by ENGINEER.

6. *Shop Drawings and Samples:*
   a. Record date of receipt of Samples and approved Shop Drawings.

   b. Receive Samples which are furnished at the Site by Contractor, and notify ENGINEER of availability of Samples for examination.

   c. Advise ENGINEER and Contractor of the commencement of any portion of the Work requiring a Shop Drawing or Sample submittal for which RPR believes that the submittal has not been approved by ENGINEER.

7. *Modifications:* Consider and evaluate Contractor's suggestions for modifications in Drawings or Specifications and report with RPR's recommendations to ENGINEER. Transmit to Contractor in writing decisions as issued by ENGINEER.

8. *Review of Work and Rejection of Defective Work:*
   a. Conduct on-Site observations of Contractor's work in progress to assist ENGINEER in determining if the Work is in general proceeding in accordance with the Contract Documents.

   b. Report to ENGINEER whenever RPR believes that any part of Contractor's work in progress will not produce a completed Project that conforms generally to the Contract Documents or will prejudice the integrity of the design concept of the completed Project as a functioning whole as indicated in the Contract Documents, or has been damaged, or does not meet the requirements of any inspection, test or approval required to be made; and advise ENGINEER of that part of work in progress that RPR believes should be corrected or rejected or should be uncovered for observation, or requires special testing, inspection or approval.

9. *Inspections, Tests, and System Startups:*
   a. Consult with ENGINEER in advance of scheduled major inspections, tests, and systems startups of important phases of the Work.

   b. Verify that tests, equipment, and systems start-ups and operating and maintenance training are conducted in the presence of appropriate OWNER's personnel, and that Contractor maintains adequate records thereof.

   c. Observe, record, and report to ENGINEER appropriate details relative to the test procedures and systems startups.

   d. Accompany visiting inspectors representing public or other agencies having jurisdiction over the Project, record the results of these inspections, and report to ENGINEER.

Page 2 of ___ Pages
(**Exhibit D** - Resident Project Representative)

FIGURE 2.1 (continued)

10. *Records:*
   a. Maintain at the Site orderly files for correspondence, reports of job conferences, reproductions of original Contract Documents including all Change Orders, Field Orders, Work Change Directives, Addenda, additional Drawings issued subsequent to the execution of the Contract, ENGINEER's clarifications and interpretations of the Contract Documents, progress reports, Shop Drawing and Sample submittals received from and delivered to Contractor, and other Project related documents.

   b. Prepare a daily report or keep a diary or log book, recording Contractor's hours on the Site, weather conditions, data relative to questions of Change Orders, Field Orders, Work Change Directives, or changed conditions, Site visitors, daily activities, decisions, observations in general, and specific observations in more detail as in the case of observing test procedures; and send copies to ENGINEER.

   c. Record names, addresses and telephone numbers of all Contractors, subcontractors, and major suppliers of materials and equipment.

   d. Maintain records for use in preparing Project documentation.

   e. Upon completion of the Work, furnish original set of all RPR Project documentation to ENGINEER.

11. *Reports:*
   a. Furnish to ENGINEER periodic reports as required of progress of the Work and of Contractor's compliance with the progress schedule and schedule of Shop Drawing and Sample submittals.

   b. Draft and recommend to ENGINEER proposed Change Orders, Work Change Directives, and Field Orders. Obtain backup material from Contractor.

   c. Furnish to ENGINEER and OWNER copies of all inspection, test, and system startup reports.

   d. Report immediately to ENGINEER the occurrence of any Site accidents, any Hazardous Environmental Conditions, emergencies, or acts of God endangering the Work, and property damaged by fire or other causes.

12. *Payment Requests:* Review Applications for Payment with Contractor for compliance with the established procedure for their submission and forward with recommendations to ENGINEER, noting particularly the relationship of the payment requested to the schedule of values, Work completed, and materials and equipment delivered at the Site but not incorporated in the Work.

13. *Certificates, Operation and Maintenance Manuals:* During the course of the Work, verify that materials and equipment certificates, operation and maintenance manuals and other data required by the Specifications to be assembled and furnished by Contractor are applicable to the items actually installed and in accordance with the Contract Documents, and have these documents delivered to ENGINEER for review and forwarding to OWNER prior to payment for that part of the Work.

14. *Completion:*
   a. Before ENGINEER issues a Certificate of Substantial Completion, submit to Contractor a list of observed items requiring completion or correction.

   b. Observe whether Contractor has arranged for inspections required by Laws and Regulations, including but not limited to those to be performed by public agencies having jurisdiction over the Work.

FIGURE 2.1 (continued)

c. Participate in a final inspection in the company of ENGINEER, OWNER, and Contractor and prepare a final list of items to be completed or corrected.

d. Observe whether all items on final list have been completed or corrected and make recommendations to ENGINEER concerning acceptance and issuance of the Notice of Acceptability of the Work.

D. **Resident Project Representative shall not:**

1. Authorize any deviation from the Contract Documents or substitution of materials or equipment (including "or-equal" items).

2. Exceed limitations of ENGINEER's authority as set forth in the Agreement or the Contract Documents.

3. Undertake any of the responsibilities of Contractor, subcontractors, suppliers, or Contractor's superintendent.

4. Advise on, issue directions relative to or assume control over any aspect of the means, methods, techniques, sequences or procedures of Contractor's work unless such advice or directions are specifically required by the Contract Documents.

5. Advise on, issue directions regarding, or assume control over safety precautions and programs in connection with the activities or operations of OWNER or Contractor.

6. Participate in specialized field or laboratory tests or inspections conducted off-site by others except as specifically authorized by ENGINEER.

7. Accept Shop Drawing or Sample submittals from anyone other than Contractor.

8. Authorize OWNER to occupy the Project in whole or in part.

Page 4 of ___ Pages
(**Exhibit D** - Resident Project Representative)

FIGURE 2.1 (continued)

Problems

1. What is apparent authority?
2. Give an example of delegated authority.
3. What are the dangers of allowing the inspector to supervise the work?
4. When should materials be checked by the inspector?
5. How much authority should be allowed the Resident Project Representative or inspector—should he or she be permitted to approve Change Orders affecting time or cost?
6. Is it permissible for the owner's representative to issue instructions or orders to a subcontractor as long as they are directed to a management representative of that subcontractor?
7. Who should have the sole authority to stop the work?
8. Which of the following is an acceptable practice by the engineer, the Resident Project Representative, or the inspector?

    a. To control the work
    b. To supervise the work
    c. To have the right to stop the work

9. True or false? If requested by the contractor, the Resident Project Representative or inspector may offer suggestions as to how to do the work.
10. True or false? If an inspector discovers that the work of a subcontractor is deficient, he or she should direct the subcontractor to make the necessary corrections, then tell the General Contractor what was done.
11. What are the three types of authority exercised in a construction contract?
12. It is a well-known principle that the engineer should not tell the contractor how to do the work; but, if the engineer observes that the contractor's construction methods appear to be unable to meet the specifications, what should the engineer do?

3

RESIDENT PROJECT REPRESENTATIVE OFFICE RESPONSIBILITIES

SETTING UP A FIELD OFFICE

The basic requirements concerning the setting up of a field office for the Resident Project Representative (resident engineer or inspector) will have been established in the contract documents long before either of them reports to the project site. It is normally the project specifications that call for a field office for the Resident Project Representative, and these same provisions often include requirements for field office furnishings, utilities, janitorial services, sanitary facilities, and telephone. However, if the specifier fails to call for field office facilities in the specifications for a project that requires the services of a full-time Resident Project Representative, the contractor that provided such extras at no extra cost would soon be out of business; after all, if the contractor was aware of the requirement for such facilities at the time the job was bid, money would have been allowed in the bid for this purpose—if not, it would be unfair to expect it.

Many agencies that are involved in frequent construction contracts already own their own trailer-type offices (Figure 3.1) and will normally have one moved to the construction site at the beginning of the work. Thus, the contractor has no reason even to suspect that the omission of field office facilities from the specifications was an oversight.

What is to be done if a field office is needed but was not specified? Perhaps the Resident Project Representative can prevail on the contractor to provide a prefabricated structure for the inspector's use, but the inspector and employer should be aware that this often takes the form of a tool shed structure that uses flap covers over side openings instead of windows. Under the circumstances, the field office probably would not be provided with power or lights either, so if it rains the resident engineer or inspector might be faced with three choices: leave the flaps open and get the

FIGURE 3.1 Part of a Large Agency-Owned Construction Field Office Complex, Aguamilpa Hydroelectric Project, Nayarit, Mexico. *(Acting agency in the above photograph: Comisión Federal de Electriciodaol.)*

plans wet; close the flaps and sit in the dark waiting for the rain to stop; or close the shed and stay outside in the rain. Thus, if the resident engineer or inspector has been hired prior to advertising the job for bids, it might be wise to ask the design firm or the owner to assure that the proper facilities will be provided for the inspector in the specifications. It is not too late even if the job is already out for bids, for as long as the bids have not been opened, an addendum to the specifications may still be issued that can provide for these facilities. One other alternative presents itself. Generally, a contractor will provide a field office trailer for his or her own use (Figure 3.2). It is usually not too difficult to obtain permission to set up a corner for the Resident Project Representative's own trailer—it sure beats sitting in the rain.

Almost all construction contractors today are using the trailer-type offices that are either available for purchase or can be rented on a month-to-month basis. Most of the larger trailers have inside toilets, are air conditioned and heated, and are fairly easy to keep clean; some are even carpeted.

Where the specifications do call for a Resident Project Representative's field office, it is not uncommon for a contractor to offer to "share" the field office with the Resident Project Representative by partitioning it across the middle, setting aside one end for the contractor and the other end for the architect/engineer. If the specifications will support the Resident Project Representative in a request for a separate trailer for an office, there should be no compromise or settlement for a shared unit, as it is not a preferable arrangement. There can be numerous reasons in support of the need for a separate trailer and some of them are mentioned later in the book. One of the principal reasons is security for both architect/engineer or owner and the contractor and the respective records and property of each and the ability to hold

FIGURE 3.2 Typical Contractor's Field Office Trailer.

confidential meetings in a separate unit. In a joint-use trailer office, there are no secrets—when someone wants a private conversation, whether it be the contractor or the architect/engineer, you will notice a desire to go for a walk around the site.

Familiarization with Construction Documents

Upon being engaged to provide resident inspection on a project, the Resident Project Representative or inspector should obtain a complete set of all contract documents, including all contract drawings, standard drawings, specifications and specification addenda, and copies of all reference specifications, standards, or test requirements cited. Sufficient time should be spent carefully studying all of these documents, until the inspector is thoroughly familiar with all phases and details of the project as shown on the plans and specifications. This type of review should ideally take place in the office of the design organization so that the inspector can obtain first-hand responses to questions from the design staff. This enables a better understanding of the project by the inspector, and as a result, a more smoothly run project. The inspector should mark all areas containing key provisions and each area where special care must be taken. Cross-references should be carefully noted so that the affected sections in the specifications can be flagged to indicate locations where the different trades must interface or the work of different contractors must be coordinated. It is also wise to mark those areas where special tests or inspections are specified and where samples are required.

At the same time, the Resident Project Representative must study the contract General Conditions ("boilerplate") of the specifications, as these provisions set the stage for almost all of the construction administration functions that will need to be performed. It is an especially good idea to set up a chart showing all the tests required, each type of test, the reference standard, the frequency of testing, and the specification section reference where these are called for (look ahead to Figure 3.13). This should be done prior to starting the work at the site.

Early in the project, a complete list should be compiled in chronological order that shows the date that every milestone event on the project is to take place (Figure 3.3). This should include meetings, submittals, tests, delivery dates for equipment, contractor partial payment requests, scheduling of surveys, final date for submittal of "or equal" items, and all other "milestone" events

Equipping the Field Office

Field office equipment and supply requirements will vary from one job to the next, but on many public agency projects or larger private ventures such as high-rise buildings, hospitals, schools, and similar projects where the field management of the construction is more organized and formal, the Resident Project Representative's field office facilities may include any or all of the following:

1. Several desks, chairs, and a conference table
2. One or more plan tables and drafting stools
3. A plan rack or "stick file"
4. A four- or five-drawer filing cabinet
5. Telephone service (not coin-operated)
6. Bookcase
7. Inside toilet and lavatory (or adjacent portable unit)
8. Water, power, and lights
9. Heating and cooling facilities, as required
10. Janitorial services to clean the facility

In addition to furnishings, as listed above, it is not uncommon to require the following equipment for the owner's or architect/engineer's use during the life of the project:

1. Electrostatic copy machine
2. Fax machine with dedicated telephone line
3. Radio pagers
4. PC computer, modem, printer, and word processor and construction administration and scheduling software with dedicated phone line or DSL (Digital Subscriber Line) if Internet access is to be provided

PROJECT No. 1553-04
SPECIFIED SUBMITTALS & EVENTS: CHRONOLOGICAL

Submittal or Event Item	By	Reference
PRE BID		
1. Bidders Qualifications	NASA	IB
2. Pre Work Conference	NASA	SP
BID OPENING + 10 DAYS		
1. Pre Award Survey	NASA	IB
WITHIN 5 DAYS AFTER NOTICE TO PROCEED		
1. Commence Work (Also Resident Engr on site)	Contr/A&E	IB 22
WITHIN 14 DAYS AFTER NOTICE TO PROCEED		
1. Working Schedule	Contr	IB 29; SP 5.1
WITHIN 15 DAYS AFTER NOTICE TO PROCEED		
1. CPM Job Schedule	Contr	IB 26
2. Shop Drawing & Equipment List Schedule	Contr	IB 23
3. Shop Drawing & Equipment List—1st Submittal	Contr	IB 33
4. Quality Control Program—1st Submittal	Contr	IB 70
5. Notify Contr Officer of Struct Steel Deliv Locn	Contr	SP 13.1
6. AC Water Pipe Materials List	Contr	TS 579
WITHIN 30 DAYS AFTER NOTICE TO PROCEED		
1. Receive & Review Schedule of Submittals	NASA/A&E	IB 23
2. Report Status of Subcontr & Purchase Orders	Contr	IB 56
3. Schedule of Submittals (Final)	Contr	SP 9.2
4. Electrical Shop Dwgs & Lists	Contr	TS 587
WITHIN 45 DAYS AFTER NOTICE TO PROCEED		
1. Quality Control Program Submittal (Final)	Contr/A&E	SP 4.5
WITHIN 60 DAYS AFTER NOTICE TO PROCEED		
1. Plumbing: Proposed Matls List	Contr	TS 407
2. Electrical: Proposed Matls List	Contr	TS 470
3. Low Press Comp Air: Proposed Matls List	Contr	TS 480
4. Fire Protec Equipt: Prop. Matls List & Shop Dwgs	Contr	TS 507
5. Ventil. System: Proposed Matls List	Contr	TS 554
30 DAYS BEFORE COMPLETION		
1. Equipment Manuals	Contr	IB 24; SP 9.5
BEFORE COMPLETION (NO SPECIFIED TIMETABLE)		
1. Dwg. Change Incorporation (As-Built Dwgs)	Contr	IB 33; SP 1.5
NO SPECIFIED TIME FROM NOTICE OR AWARD		
1. Report of Subcontractors over $10,000 Update annually	Contr	IB 54
2. 5 Days after award of subcontract: Statement to NASA of Subcontr and data	Contr	IB 53
3. 30 Days before starting specific phase of work: Detailed quality control plan	Contr	IB 71
4. Prior to fabrication (General items): Shop Dwg Approvals	Contr	SP 1.1
5. Return of Contr Submittals (Appr/Not Appr)	A&E	SP 1.6
6. Plan for tracking & processing contr Submittals	A&E	A&E Contract
7. Concrete affidavits from Weighmaster Before Concrete Placement	Contr	TS 176; 102
8. Conc Matls Test Reports: As Work Progresses	Contr	TS 182; 185
9. Design Mix for Concrete: 15 Days before Start of Specified Work	Contr	TS 183
10. Hardware Samples	Contr	TS 344
Contract Cost Breakdown & Payroll Reports		IB 78
PERIODIC		
1. Monthly: Reports on Tests and Inspections	Contr	SP 4.7
2. Monthly: Progress Payment Certification	A&E	A&E Contract
3. Monthly: Quality Control Meetings	Contr/A&E	A&E Contract
4. Twice Monthly: Const Schedule & Progress Mtg	Contr/A&E	A&E Contract
5. Weekly: Construction Status Reports	A&E	A&E Contract

FIGURE 3.3 List of Project Milestones.

On larger projects, owners and architect/engineers often require the contractor to furnish a PC computer and printer along with a copy of their CPM scheduling program.

Most of these items might normally be specified to be furnished by the contractor and thus would be included in his or her bid price. In addition to the items provided by the contractor, the Resident Project Representative or inspector's employer would be expected to provide such additional items as a typewriter, an adding machine (tape type), a small calculating machine, a postal scale, reproduction equipment, and all expendable office supplies.

On smaller projects, the facilities provided would be scaled down accordingly. The items just listed might reasonably be expected to be provided on a project involving a construction cost of somewhere over $1 million and involving over a full year of construction time.

Ordering Supplies and Equipment

It is generally too late if the Resident Project Representative reports to the field office empty-handed, only to find that the wheels of the construction process are already in motion. Before leaving the home office to take up residence in the new field office, the Resident Project Representative would draw as many office supplies and equipment as one might expect to need. The full list might conceivably include many, if not all, of the following items:

1. Report forms
2. Field books or "record books" for diaries
3. Stationery
4. Transmittal forms
5. Envelopes (all sizes)
6. Blank bond typing paper
7. Columnar pads (for estimating)
8. Loose-leaf notebooks (8 1/2" × 11" three-ring type)
9. Pens, pencils, felt-tip pens, highlighter pens
10. Rejection or nonconforming tags
11. Minimum of two weeks' supply of film or electronic media for all cameras
12. Floppy disks

In addition to the other supplies and equipment previously listed, the resident engineer or inspector should acquire all of the normal personal protective equipment required under OSHA for the types of work and environmental field conditions that are likely to be encountered.

Establishment of Communications

Although this is one area in which the contractor almost always excels with regard to maintaining contact with members of its own field staff, some design firms and owners are beginning to realize the value to them, in both time and money, in investing in some means of direct communication with the Resident Project Representative at all times. The types of personal communications devices that offer the most value to the architect/engineer, owner, or contractor for maintaining contact with the field office include the following:

1. Field office telephones (not coin-operated)
2. Cellular telephones
3. Two-way radios (or combination radio/cell phones)
4. Voice or display pagers
5. Cordless land-line telephones
6. PDAs (Personal Digital Assistant) with wireless Internet access, such as Palm VIIs

Each of these pieces of communications equipment has its own particular best application and limitation, and it is seldom that all of them will be used on a single project.

Field Office Telephone

The field office telephone is by far the most common device, and regardless of whether or not any other communications devices are provided, every field office should be provided with a telephone (not a pay phone) for voice communications. In addition, a second telephone line should be provided for a fax machine. The design firm or owner should be careful in specifying its telephone needs, as it could find itself in the rather unique position of being furnished with a telephone that only connects to a private phone system. It may be necessary in some cases to specify that the contractor's field office must have a telephone and the Resident Project Representative's field office must also be provided with a telephone on a separate line connected to an established telephone exchange. If the contractor is connected to a private phone system, the Resident Project Representative's telephone should also be capable of being connected to the contractor's private phone system in addition to the established telephone system. The contractor should allow the design firm, the owner, or their authorized representatives or employees free and unlimited use of the field office telephone for all calls that do not involve published toll charges. Any toll charges received should be billed to the owner by the contractor at the rates actually charged the contractor by the telephone company.

Cellular Telephone

A cellular telephone is potentially valuable to a resident engineer or inspector, but it represents a cost that may be difficult to justify. Contrast this with the use of a pager, which can let the inspector know immediately to call the office, regardless of where the inspector may be at the time. There are other situations, however, where a cellular radio-telephone can be of great service in construction. For a roving engineer or inspector who must serve the needs of numerous project sites, and spends a great deal of time on the road within range of a cellular receiving station, or for the inspector of

FIGURE 3.4 Typical Cellular Phone.

a project such as a long pipeline, highway, or canal, a cellular radio-telephone used in a cellular service area can be an extremely useful tool, provided that uninterrupted service is available in the construction area.

The Resident Project Representative should not consider CB radio for this purpose, as it is not sufficiently reliable unless served by repeater stations at strategic locations—and that *really* gets costly.

Two-Way Radios

A two-way commerial radio (not CB radio) is a device best suited to on-site communications. Projects that involve a crew of several persons at remote locations within a large construction site, and operations that involve the issuance of instructions over relatively short distances, will find that two-way radios are great time-savers. They can eliminate the sometimes misunderstood hand signals, and a clear line of sight need not be maintained to communicate within reasonable distances.

Some variations include a combination two-way radio and cellular telephone (Figure 3.5). Typically, these have the full capability of a two-way radio and a limited capacity of a cellular telephone. In the cell-phone mode, they can be programmed to provide full service on both incoming and outgoing calls, or they can be restricted to allow the user to initiate calls, but not to receive calls.

Voice Pagers

Voice pagers are quite useful in some situations; however, their limitations should be evaluated before considering replacing all beepers with voice pagers or display pagers. To be sure, a central message center is not needed, and a short voice message such as a telephone number can be transmitted directly. However, with voice pagers if the construction area is noisy, the message cannot be clearly heard, and without the convenience of a central message center the call could be lost—or at

FIGURE 3.5 Two-Way Radio with a Cellular Phone Key Pad.

least difficult to confirm. Also, there is no privacy, and all messages received can be heard by anyone standing close to the receiver. If turned low, the inspector may not hear the message. Of course, most voice pagers can be switched over to a tone signal if message security is wanted at a special time. Yet unless the caller and the receiver knew that the audio was going to be turned off at any given time, how would either party know enough to call a central message center? In any case, to prevent misuse and abuse it would be in the best interests of the inspector and the employer to have all messages sent through a message center via a tone pager to preserve the one-to-one relationship so vital to effective job management.

Cordless Land-Line Telephone

For the small construction site, the cordless telephone appears to be a convenience well-suited to the one-person field office. Many different cordless land-line telephones are on today's market, and no specific recommendations are offered, but a reliable instrument can be the answer to the age-old problem of being able to reach the Resident Project Representative when away from the field office, but still on the site. A cordless telephone is *not* a substitute for a regular telephone in the field office, however, as the field office telephone is necessary to make the system work.

There are some limitations of use, however, that should be considered before investing in the system. Basically, it must be understood that all cordless telephones are in reality small handheld radio transmitter-receivers with very short transmission ranges. The system is only suited to small jobs where the user is never very far from the location of the field office where the base radio set is installed. Thus, for a pipeline or highway job, it appears to be quite unsuitable. One must also understand its physical limitations. For example, there are sometimes radio-shielded areas on a construction site where transmission and reception is limited. This condition is remedied merely by moving the position of the cordless unit. However, if a call were to have been received while the instrument was in a shielded area, the user would have no knowledge of it. If the caller to the site is aware that a cordless telephone is being used, the appropriate thing is simply to redial the call a few minutes later when it can be presumed that the Resident Project Representative or inspector is out of the shielded area.

A cordless land-line telephone set consists of two units. One is a small battery-powered hand unit that can be carried on the user's belt to anywhere on the site. The second is a combination land-line telephone and radio receiver that can be plugged into the standard telephone outlet wherever the field office telephone has been installed. The incoming calls are first picked up by the base unit and relayed by radio to the handheld unit being carried by the Resident Project Representative or inspector. For outgoing calls, you simply reverse the procedure.

Except for the first cost of acquiring the instrument itself, the operating cost is the same as an ordinary land-line telephone.

PDAs

A PDA (Personal Digital Assistant), such as Palm VII, can be a handy tool for real-time communications on a project. However, PDA applications to electronic project management program applications are generally limited to those that do not require a lot of input on the PDA screen. An added capability can be realized with a Palm i705, which provides wireless Internet capability. For example, updating a punch list is very efficient on a PDA because you can have a sort by room and item number, and the user is merely required to indicate whether to accept the item or not. Other uses include the ability to create safety or correction notices from the field. The key to effective use of the PDA is to make extensive use of pick lists instead of data entry by stylus.

In an ideal situation, a field person would have access to a computer in the field office that has a cradle attached to it and the PDA. Whenever the person is in the field office, he or she can put the PDA in the cradle and synchronize information with the PDA and the database. Then anything started in the field on the PDA can be completed in the database or just sent directly from there. Anything done by anyone else in the organization can be retrieved back to the PDA if that application resides there. As an example, new Punch List items could appear in the list after having been entered by the architect/engineer at some other location.

Handling Job-Related Information

The establishment of communications on a project does not stop with the procurement of communications hardware. *Field communications* is a term that must also be applied to the procedures for handling and transmitting job-related information from one party to the other, the determination of who is authorized to receive and give project information, and the routing instructions for the transmittal of all communications, records, and submittals. At the beginning of a job, one of the first and most important things to be determined is the establishment of the authorized line of communications and authority, and the method of handling such information (Figure 3.6). Such communications can be effectively handled electronically (see Chapter 5), or by the traditional hand methods.

Generally, it has been found preferable to establish the Resident Project Representative as the *only direct link* between the contractor and the design organization even though the matters being communicated may be intended for the project manager of either the owner or a separate firm. In this way, all transmittals will be received first by the Resident Project Representative, who will log them into the field office record book, and only then transmit them to the project manager. In the office of the design organization, a similar procedure is followed. All transmittals at that end should be received only by the project manager; if they are intended for other members of the design staff, they should be distributed through the project manager. In this way, there is always a single point of communication at each end: the Resident Project Representative in the field and the project manager at the design office.

Similarly, for the contractor, it is desirable to submit all data through the field office. No submittals should be permitted directly to the architect/engineer by any subcontractor or supplier at any time (see "Contractor Submittals" in Chapter 4).

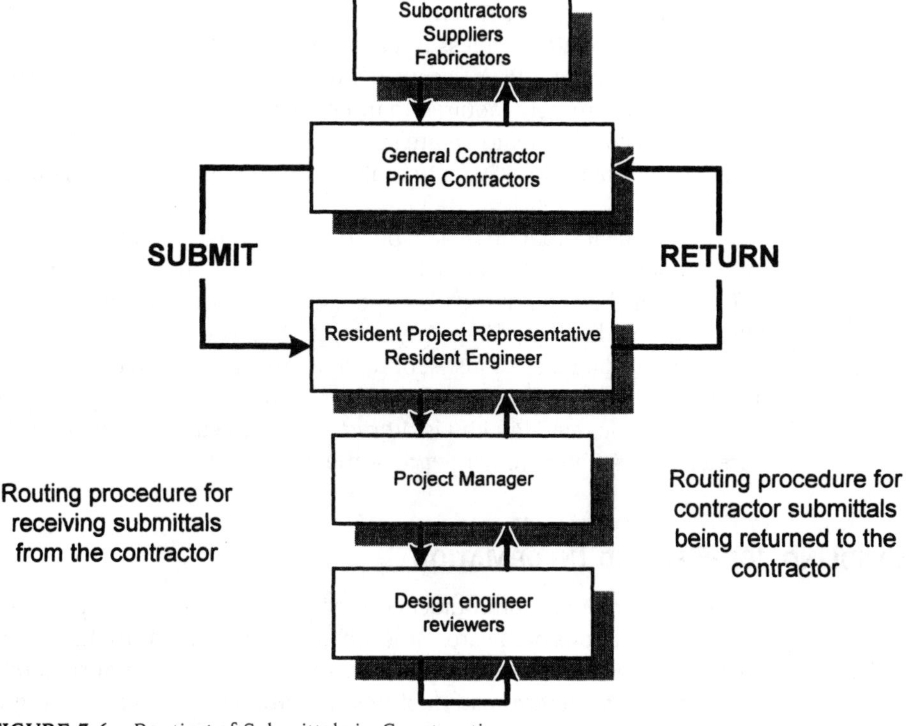

FIGURE 3.6 Routing of Submittals in Construction.

All requests for deviations, Change Order claims, shop drawing submittals, and similar transmittals should be refused by the project manager if they are not transmitted to him or her through the Resident Project Representative. Otherwise, the records would be incomplete, and, often, coordination is lacking because the information may have bypassed the project manager or Resident Project Representative, resulting in conflict or confusion in the field administration of the work. Similarly, submittals from subcontractors or materials suppliers should never be received by the Resident Project Representative directly, except through the hands of the general contractor. After all, it is the contractor's obligation to receive *and approve* all such transmittals before being considered for acceptance by the design firm. Officially, the subcontractor does not even exist, as far as the owner is concerned, because the owner entered into a contract with the general contractor only, and thus no contractual relationship exists between the owner and any subcontractors or material suppliers.

Staffing Responsibilities

Staffing a field inspection office is usually not the responsibility of the Resident Project Representative, but is normally done by the design firm, the owner, or in the case of a CQC contract, the contractor. However, the Resident Project Representative should understand the types of persons required on some representative types of projects, because the Resident Project Representative will usually have the responsibility of supervising their activities.

Generally, on a project large enough to support a full-time Resident Project Representative, the personnel needs of the project may vary from a single field representative of the owner or design firm under the responsible charge of a professional engineer or architect and backed up by occasional temporary special inspectors to ensure building code compliance, to a moderate-sized staff of three to five persons on a slightly larger project. Included as a part of this larger staff would be the Resident Project Representative, a full-time field inspector, possibly a field office assistant with estimating background, and a clerk-typist.

Staffing Level of Field Office

On an exceptionally large number of projects where there were major claims, it was found that there was a failure by the architect or engineer and the owner to provide an adequate field force at the site during construction. It seems to be a popular myth that a Resident Project Representative is all that is necessary and that assigning more persons to the site is merely gilding the lily. Nothing could be further from the facts.

Generally, on small projects of $1 million or less a single on-site representative may be all that is necessary. On certain types of larger projects, however, consideration must be given to employing two, three, or more persons at the site, full-time. Although there is no known study of all types of projects, some investigation was made on the construction of treatment plants, and some guidelines were formulated by the State of California Division of Water Quality as administrators for the EPA (Environmental Protection Agency) Clean Water Grant Program in that state.

Project Size	Inspection Person-Years/Year
Less than $1 million per year	1.0
$1 to $5 million per year	2.5
Over $5 million per year	2.5 plus 2.0 for each additional $5 million or portion thereof

FIGURE 3.7 EPA Field Office Staffing Guidelines for Wastewater Treatment Plant Construction.

Under their approach, only the *number* of on-site employees was addressed. In their analysis no distinction was made as to the types of persons at the site, or whether clerical help was necessary. The California DWQ guidelines are summarized in the table shown in Figure 3.7.

Subsequent to the development of these data, the author has done further study and determined qualitative guidelines as well. In the typical failure cases investigated, on-site representation was by a single but highly qualified registered professional engineer with significant field experience. The breakdown was traced to the demands upon the Resident Project Representative's time by both the administrative requirements and the on-site technical inspection requirements.

If you assume a situation involving one Resident Project Representative on a $6 million federally funded project, it is safe to say that over 80 to 95 percent of this person's time must be spent on administrative matters that preclude being in the field during many periods of critical need. Thus, serious problems can occur on site without the Resident Project Representative's timely knowledge. Unfortunately, the largest portion of these administrative matters involve tasks that do not require any knowledge or skill above the level of an ordinary clerk.

Types of Personnel Assigned to Field Office

The usual method of solving the understaffing problem in the field is to assign another person to the site to help. But in a classic case of misguided good intentions, the second person is usually an inspector, because in the architect/engineer's or owner's mind, that is where the help is needed. This person also has a heavy load of paperwork to do in support of his or her own activities on site. As if to complicate the problem further, the Resident Project Representative now must add the burden of supervision and personnel administration to the already heavy administrative load. Thus, the Resident Project Representative, the highest-paid on-site person, is 95 percent occupied with administrative tasks, of which probably 70 percent could be performed more economically by a clerk. Meanwhile, an inspector is on the site who can devote perhaps no more than 75 percent of chargeable time to on-site inspection.

The answer by now should be obvious. In assignment of personnel to a job that justifies more than one person, based upon the EPA table shown in Figure 3.7 and the breakdown of that data shown in Figure 3.8, the first assignment is the Resident Project Representative. The second assignment should be a field clerk, not an inspector. The field clerk will relieve the Resident Project Representative of up to

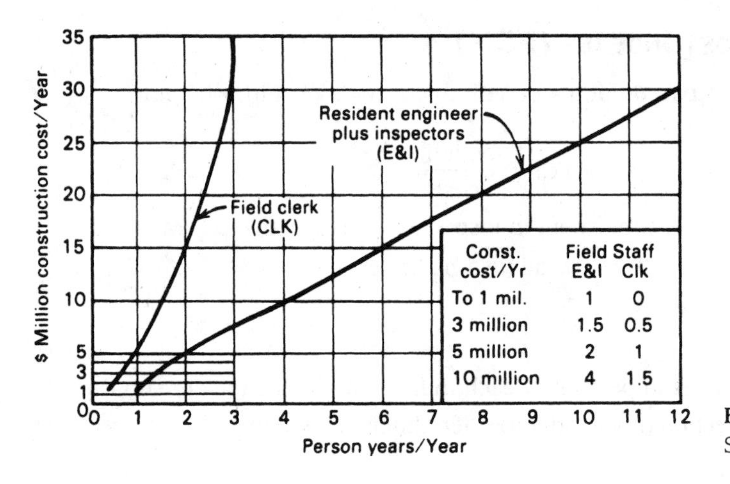

FIGURE 3.8 Breakdown of Field Office Staffing Guidelines.

70 percent of the routine paperwork burden, leaving the most qualified and highest-paid on-site person available to perform on-site inspections and troubleshoot to prevent field problems. This will leave the project with more time chargeable to inspection than the previous example utilizing an inspector as the second person, and at the same time it will substantially cut salary costs. As the work increases in complexity, inspectors can then be added as needed.

Percent of Time Expended by Each Classification toward Various Tasks

The table in Figure 3.9 illustrates the cost benefits of utilizing a clerk in the field.

Case Type	Classification	Administrative/ Management	Clerical Work	Inspection	Field Cost Index* (FCI)
1	Res. Engineer	25%	55%	20%	0.20
2 (not advised)	Res. Engineer Inspector	30% 5%	65% 20%	5% 75%	0.28
3 (preferred)	Res. Engineer Field Clerk	30% none	none 100%	70% none	0.53
4	Res. Engineer Field Clerk Inspector	30% none 5%	none 100% none	65% none 95%	0.59

* Field cost index (FCI) represents the value of inspections generated by each dollar spent on field staffing costs

FIGURE 3.9 Comparison of Field Cost Benefits by Effective Utilization of a Field Clerk.

Derivation of the Field Cost Indexes (FCIs)

The value of the FCI is based upon a benefit-to-cost relationship, wherein

$$FCI = \frac{\text{benefits}}{\text{costs}}$$

$$FCI = \frac{\text{percent of time spent in inspection} \times \text{respective pay rates}}{\text{all field pay rates}}$$

where

I = inspection time in percent of each individual's total daily time on the job

I_e and I_i = inspection time by the Resident Engineer and inspector, respectively

and

R = pay rate for each respective field classification

$R_e R_c$ and R_i = pay rates for the Resident Engineer, clerk, and inspector, respectively

The FCIs in the following examples were based upon the assumption of the billing rates listed below for the Resident Engineer, clerk, and inspector. The index may vary slightly depending upon the relative costs in other firms. The higher the FCI value, the greater the savings.

To apply this principle to a numerical example, assume the following field billing rates for the Resident Engineer, inspector, and clerk, respectively:

$$R_e = 58.00 \text{ per hour}$$
$$R_i = 29.00 \text{ per hour}$$
$$R_c = 18.00 \text{ per hour}$$

Then, based upon the work distribution percentages indicated in Figure 3.9, we can calculate the following cases.

Case 1: One Resident Engineer on-site only

$$FCI = \frac{I_e R_e}{R_e} = \frac{0.20 \times 58.0}{58.0} = 0.20$$

Case 2: One Resident Engineer plus one inspector at the site

$$FCI = \frac{I_e R_e + I_i R_i}{R_e + R_i} = \frac{(0.50 \times 58.00) + (0.75 \times 29.00)}{58.00 + 29.00} = 0.28 \text{ (low)}$$

Case 3: One Resident Engineer plus one clerk at the site

$$FCI = \frac{I_e R_e}{R_e + R_c} = \frac{0.70 \times 58.00}{58.00 + 18.00} = 0.53 \text{ (good)}$$

Case 4: One Resident Engineer, clerk, and inspector at the site

$$FCI = \frac{I_eR_e + I_iR_i}{R_e + R_i + R_c} = \frac{(0.65 \times 58.00) + (0.95 \times 29.00)}{58.00 + 29.00 + 18.00} = 0.59 \text{ (good)}$$

Then, by comparing the FCI values, it can be seen that if two persons were assigned to a field office, there would be an approximate 30 percent cost savings by utilizing a clerk as the second member of the field staff instead of an inspector, while providing approximately 88 percent of the total inspection hours that would have been available with an inspector.

As evidenced by the increase in the FCI through the use of a clerk at the project site, a significant cost savings can be achieved, while at the same time more technical expertise in construction is made available on site without added cost. A valuable benefit is the fact that not only is field morale increased, but also project documentation is usually better and more consistent. Thus, in case of claims, the owner and architect/engineer are better protected.

The chart in Figure 3.8 provides a general guideline for the assignment of field personnel to a project similar in complexity to a wastewater treatment plant or plant addition. It must be emphasized, however, that project cost alone is not an indication of the level of field staffing required, and that each case must be examined upon its merits. Staffing requirements for construction field offices will vary significantly depending upon the type of project, the number of areas in which the contractor will be working, the number of separate activities being concurrently pursued by the contractor, the type of construction contract (multiple prime, fast-track, etc.), and the complexity of the project. The average distribution of field costs under different staffing arrangements can be compared graphically in the chart in Figure 3.10.

The Resident Project Representative is the highest-paid member of the field team of the owner or architect/engineer, and thus it is incumbent upon the project manager to provide for the most effective utilization of his or her services. In Figure 3.11 two of the cases from Figure 3.10 are portrayed again graphically. It can be seen that the utilization of a field clerk in lieu of an inspector as the second assignment to a project (Case 3 in preference to Case 2) can improve project cost efficiency considerably.

Selection of Trailer-Type Field Offices

When specifying the field office requirements in the project specifications, consideration must be given at that time as to the staffing requirements of the project to assure adequate working space for all persons assigned to the project. Selection of office trailer size should be based upon the maximum ultimate occupancy load. The following guideline is suggested as a minimum based upon the use of trailer-type field offices:

Field Office (Or Primary Field Office, If More Than One)

Resident Engineer only	16 m^2 (2.4 m by 6.4 m body) 168 ft^2 (8 by 21 ft)
Resident Engineer and a clerk	19.3 m^2 (2.4 m by 7.9 m body) 208 ft^2 (8 by 26 ft)

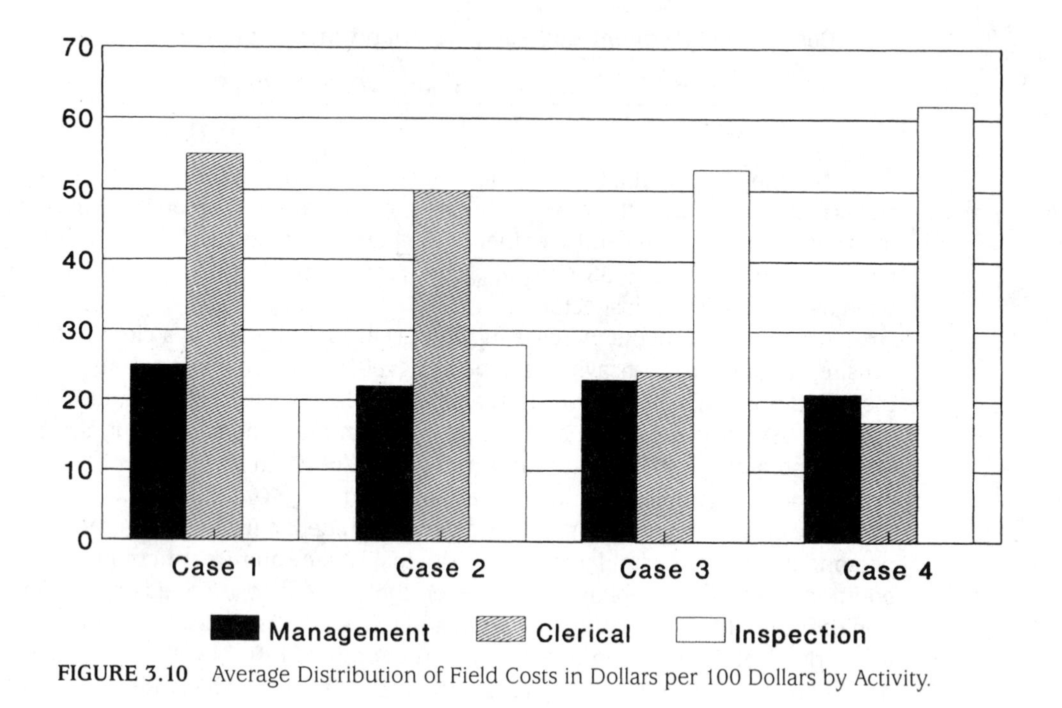

FIGURE 3.10 Average Distribution of Field Costs in Dollars per 100 Dollars by Activity.

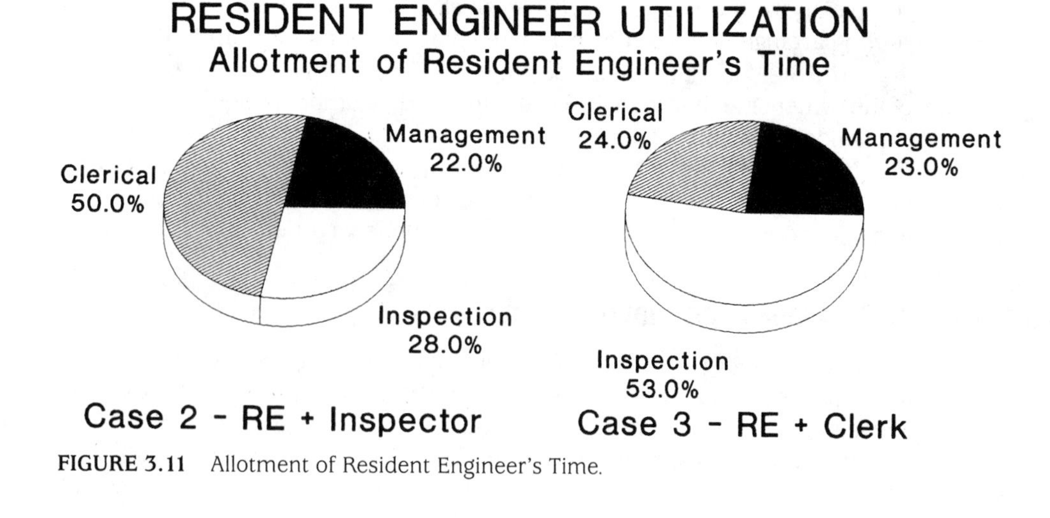

FIGURE 3.11 Allotment of Resident Engineer's Time.

Resident Engineer, one inspector, and a clerk	24 m^2 (2.4 m by 9.7 m body) 256 ft^2 (8 by 32 ft)
Resident Engineer plus two inspectors and a clerk	39 m^2 (3.0 m by 12.8 m body) 420 ft^2 (10 by 42 ft)
Secondary Field Office	
One inspector or field engineer	12 m^2 (2.4 m by 4.9 m body) 128 ft^2 (8 by 16 ft)

Figure 3.12 illustrates the typical floor plans and layout of field office trailers available from one major national organization dealing in sales, lease-purchase, and rental of such equipment.

Construction Safety

Although it is clearly understood that the Resident Project Representative's involvement in construction safety is limited (see Chapter 9), and that the general contractor has the principal responsibility for all construction safety compliance, there are certain considerations that should be kept in mind by each inspector on the site. The degree of the Resident Project Representative or inspector's involvement may vary somewhat . . . depending upon the terms of the specific contract provisions and the actual circumstances surrounding each case of a potential hazard.

Occasionally, a contract for inspection may require the inspector to monitor the contractor's safety program to assure that an effective safety program is being provided (not uncommon in federal contracts). The inspector may be required to be involved in meetings with the contractor to discuss safety measures, and where project safety has been made one of the inspector's duties to observe, personal assurance should be obtained that safe practices are being followed. Some of the matters of concern in this type of responsibility are:

1. Review of the contractor's accident prevention program that is required by the local OSHA compliance agency.
2. A code of safe practices developed by the contractor and checked by the inspector for each project.
3. Various permits that may be required prior to starting specific work items, such as excavation, trench shoring, falsework, scaffolding, crane certifications, and similar requirements to be verified before allowing the contractor to begin.
4. Other safety items that may be pertinent to the contract, such as blasting operations, personal protective gear required, backup alarms for equipment, rollover protection guards on equipment, traffic control, and similar protective requirements to be confirmed.
5. The reporting of fatal accidents or disabling accidents to the local safety compliance agency as required.

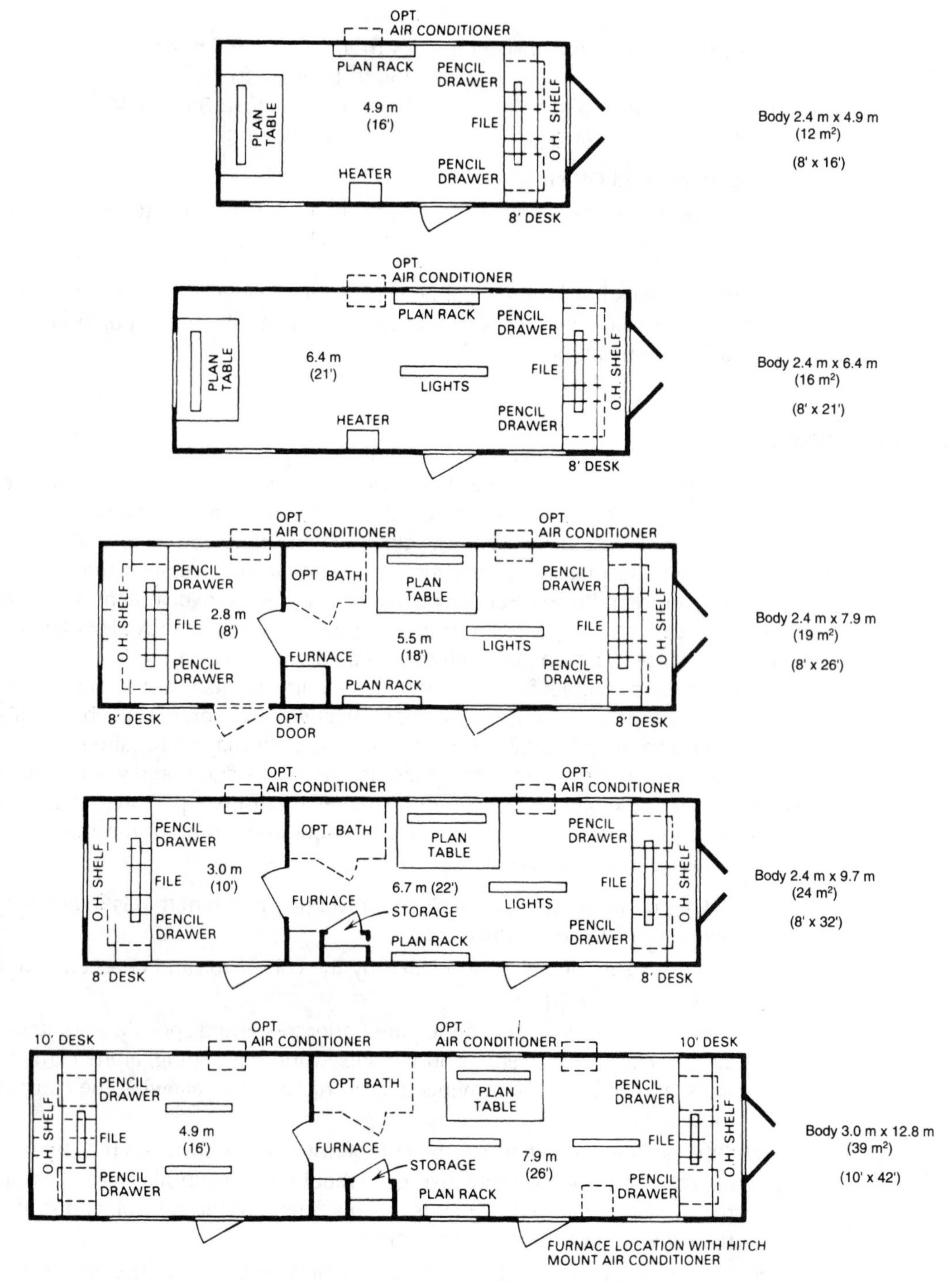

FIGURE 3.12 Typical Floor Layout of Available Construction Field Office Trailers.

The inspector's responsibilities with relation to the handling of hazardous conditions and the effect of various construction contract provisions on the administration of safety requirements are explored in greater detail in Chapter 9.

Development of an Inspection Plan

It is desirable for each inspector to take a systematic approach to the quality control or quality assurance functions that are required for the project. Even if no formal plan is required to be submitted to the owner as a part of the quality control provisions of the contract, such as in a federal CQC (contractor quality control) operation, such a plan is an excellent inspection tool, and the Resident Project Representative is urged to plan ahead by developing an outline of all the inspections that must be made, a checklist of points to look for, and a list of the types and frequencies of all tests that are required (see page 309).

An inspection plan for submittal to the owner agency might reasonably be expected to cover all or some of the following items:

1. Establishment of detailed inspection procedures.
2. Outline of acceptance/rejection procedures.
3. Preparation of a chart showing all tests required, when they are needed, the frequency of sampling and testing, the material being tested, and who is obligated to perform the tests (Figure 3.13).
4. Establishment of who will be responsible for calling the laboratory for pickup of samples for testing, who will call for special inspectors when needed, and to whom such outside people will be directly responsible on the project.
5. Identification of who must physically prepare samples for testing, the contractor or the inspector; determination of whether the contractor will provide a laborer to assist the inspector in obtaining samples and transporting samples for testing.
6. Establishment of ground rules for acceptable timing of work operations after sampling and testing; mandatory scheduling must be provided to assure not only time to make samples and tests, but also time to make corrections needed before work may be allowed to continue.

Often, on a federal project involving a CQC operation, if an architect/engineer firm is selected to provide quality assurance, it is required to submit a formal *construction surveillance and inspection plan* prior to beginning the work. Under such a program, the architect/engineer must provide a Resident Project Representative and an appropriate field staff at the construction site, plus home office support as required. Implementation of such a program requires the architect/engineer, the construction manager, and the Resident Project Representative to meet the contractor prior to beginning each phase of the work and to define the responsibilities of each party under the contract. At that time, the contractor should be asked to submit and

TESTING PLAN

S= Sampled
T= Tested

SHEET NO.

SPEC SECTION & PARAGRAPH	ITEM OF WORK	TESTS REQUIRED (ASTM,ACI, COMM'L STDS, ETC)	SAMPLED BY	TESTED BY	ON SITE	OFF SITE	FREQUENCY OF TESTING
02200	EARTHWORK						
2.04B	Sieve Analysis	ASTM C-36	ABC Lab	ABC Lab		S T	One passing test per source
2.0	Standard Moisture Density	ASTM D-1557 (Mod.)	ABC Lab	ABC Lab	S	S T	One - each source
2.06E	In-place Density	ASTM D-1556	ABC Lab	ABC Lab	S T		Min. 20 - (Random)
03300	CONCRETE						
1.04	Mix Design	ASTM C-192; C-39	ABC Lab	ABC Lab		S T	One per each req'd Strength - Aggregate size
1.05	Test cylinders	ASTM C-31	Inspector	ABC Lab	S	T	See Note #1
2.03	Slump	ASTM C-143	Inspector	Inspector	S T		One per batch
2.01	Air	ASTM C-231	ABC Lab	ABC Lab	S T		One set of cylinders
04200	MASONRY						
2.03	Efflorescence	See Spec. Par. 2.03A		ABC Lab		S T	One passing test
2.05	Mortar and Grout	See Spec. Par. 2.05	Inspector	ABC Lab	S	T	See Note #2
2.07	Drying Shrinkage	ASTM C-426 (See Spec.)	ABC Lab	ABC Lab		S T	One Passing
2.08	Air dry Condition	ASTM C-427	ABC Lab	ABC Lab	S T		One test per delivery

Data used are fictitious for illustration only

EXPLANATORY NOTES

Note (1) 6 cyl per day per 40 m^3 Test 3 AT 7 days; 3 AT 28 days

Note (2) One each design mix Min. 3 ea day Test one at 7 days; 2 ea at 28 days

FISK FORM 6-3

FIGURE 3.13 Example of a Testing Plan.

explain his or her established systems for quality control and his or her accident prevention program. These programs should then be reviewed by the architect/engineer or owner and the Resident Project Representative, and the programs compared with the specific requirements of the contract. Suggestions, if warranted, should be made by the Resident Project Representative at that time.

OTHER JOB RESPONSIBILITIES (Resident Proj. Rep)

In addition to the items covered previously, there are numerous other field responsibilities that the Resident Project Representative must expect to cope with. These responsibilities in connection with the tasks described are only highlighted in this chapter, as the technical details of administering each of the listed subjects is covered more thoroughly in the chapters that follow. Naturally, much of the Resident Project Representative's work will have to be done in the field office. Unfortunately, in today's complex society, the resultant paperwork required means that a good portion of the Resident Project Representative's time must be spent writing and recording data. Administration of a construction project is no longer as simple as it was in days past, when all that was needed was a thorough knowledge of construction—today, the paperwork is just as vital as the inspections themselves.

Construction Planning and Scheduling

A Resident Project Representative on a sizable project will most certainly become involved in construction scheduling, or at least in an evaluation of the construction schedules prepared by the contractor. An inspector who does not have a basic understanding of the principles involved will be hard-pressed to fulfill all administrative responsibilities. The two principal types of schedules that the inspector will be most likely to encounter are bar charts and network diagrams.

In a network diagram, the Resident Project Representative should be capable of recognizing logical and illogical construction of a schedule. In addition, realistic scheduling times, compatible delivery dates for equipment and materials, critical path operations, float (slack) times, and other related items of a network schedule must be checked. Also, frequent comparisons to see if the actual construction events are following the network diagram must be made, and the inspector should check to see that updated network data are provided as required.

Contractor's Plant and Equipment

Although it is rare that the resident engineer or inspector will be required to check the contractor's equipment, it is an occasional requirement under the provisions of some public agency contracts. If equipment inspection is required, all that is normally expected of the inspector is a check of each piece of major equipment and a determination of whether it has the necessary safety devices and that such devices are all in good working order. This includes devices such as safety cages, backup bells, guards over moving parts, and similar items. In addition, the equipment check should include an evaluation of whether the exhaust emissions are excessive, and that there are no cracked windshields or bad tires on automotive or other heavy motorized equipment.

Measurements for Progress Payments

One of the traditional responsibilities of the Resident Project Representative is the review of the contractor's monthly partial payment request to see that the quantities of materials and equipment delivered to the site or used in the work agree with the quantities for which the contractor has requested payment. It should be kept in mind that on a lump-sum project, the Resident Project Representative must still make monthly pay estimates of the contractor's work. However, the accuracy of the measurements are not as critical as they would be on a unit-price contract, because such differences will be made up in the end. This is not meant to suggest that the inspector can afford to be careless about quantities on a lump-sum job, as it is highly undesirable to allow the owner to pay for more work than has actually been accomplished as of the date of the payment request. Such overruns would completely defeat the purpose of the normal 5 to 10 percent retention that is usually held by the owner until the end of the project. If no retention money is held, even on lump-sum projects, the cumulative monthly progress payments must be carefully controlled to avoid overpayment.

On a unit-price contract, a precise measurement must be made of all construction quantities, and generally a system of measurement is spelled out in the specifications under the heading "Measurement and Payment," so that there can be no argument as to where, how, and when the measurements for pay purposes must be made. Errors in measurement or overlooked items under a unit-price contract can cost the owner an immense sum of money. Because of this fact, many measurements for such contracts are performed by the design firm utilizing the services of a survey crew to determine pay quantities of pipe; to cross-section a borrow area to determine the exact quantities of earth excavated; and to make similar measurements.

The subject of measurement and payment during construction is covered in detail in Chapter 17, where numerous methods are described and some of the typical pitfalls are discussed.

Filing of Notices and Certificates

Although the filing of legal notices is primarily the responsibility of the owner, that job may be delegated to the Resident Project Representative or project manager as an owner's representative. The notices and certificates themselves will normally have been prepared when the Resident Project Representative receives them; however, some knowledge of the process involved should be understood.

Such notices include the filing of the Notice to Proceed as well as the Certificate of Completion or Certificate of Substantial Completion at the close of construction. In addition, the inspector may be involved with the serving or filing of other forms of construction documents such as the Field Order and Deficiency Notice as well as special notices calling attention to imminent safety hazards that require immediate correction to remove a hazard to life or health.

Evaluation of Construction Materials and Methods

The Resident Project Representative is frequently called upon to evaluate construction materials and methods. The responsibilities should be clear, however, and although the expertise of the Resident Project Representative is necessary for the good of the project, any such recommendations should be made *only* to the responsible architect or engineer, in writing, who will be the final authority as to what action to take. It is to be hoped that the architect or engineer will also recognize the value of following a formal transmittal procedure and, after reaching a decision, will submit all instruction or responses directly through the Resident Project Representative instead of mailing or delivering them directly to the contractor.

The work on a project may also involve the preparation of work statements, estimates, and data to contractor-requested changes. The procedures are similar to those specified for evaluations of materials and methods. A fuller discussion of the handling of contractor submittals is covered under "Contractor Submittals" in Chapter 4.

RFIs (Requests for Information)

One of the most frequent contributors to construction delay claims is the handling (or mishandling) of contractor-generated requests for information or clarification. If you have a construction project under way and an RFI is received from the on-site contractor, my advice is start responding to the problem *NOW.* Don't wait until it is convenient, because it will never be convenient. As an immediate solution to the problem or inquiry, at the very least, send an e-mail to the contractor to tell it that you are working on it.

If you are operating on an electronic project management program, RFIs can be sent through the database on a workflow route determined by the owner or architect/engineer (see Figure 5.12). Routes can be set up for each design discipline and for combinations of several disciplines. The route participants as well as all authorized interested parties are notified of the pending and completed actions. RFIs are *action* items and can be coded for the decision log.

Record Drawings

A large portion of the work on many projects includes the posting of "as-built" information on a set of prints at the construction site. This is sometimes followed by a requirement that all such records of changes be drafted onto a set of reproducibles of the contract drawings. This is intended to provide the owner with a permanent record of each feature of a project as it was actually constructed. The normal construction contract usually calls for the *contractor* to make a set of *record drawings* by marking a set of prints with all changes from the original drawings as bid, including all Change Orders, alignment changes, depth changes of underground pipes and utilities, and all other items that are not the same as they were originally drawn.

Where computers are being used, record drawings, as well as contract drawings, can be attached to electronic mail for distribution and viewing, as well as nondestructive markup using a browser. A current drawing list can be accessed readily by any authorized user of construction management software that has been designed for contract administration by the owner or architect/engineer.

The term *as-built drawings* should be discouraged because of the legal implications involved when the architect or engineer signs a certificate that says that everything shown on the drawings is exactly as constructed—this act could haunt the architect or engineer for years. The drafting of such data onto a set of transparencies is often required, but the resident engineer or inspector should be particularly careful when making commitments that the condition of the contract that requires the preparation of record drawings does not of itself mean that they will be drafted on reproducibles. It merely means that all changes will be marked (usually with colored pencil) on a set of record prints at the site. If work on the transparencies is required, it will be performed as a separate contract item by copying from the record drawings.

Many field people have been very lax in assuring that all record drawings are kept up to date, and unfortunately, the oversight may not be discovered for several years if no further work is constructed in the same area. It is a vital concern and should not be overlooked. The most common procedure is for the contractor to prepare the record drawings as the project progresses, by clearly and legibly marking a set of prints that at the end of the project are turned over to the architect/engineer or to the owner's engineers for checking. After approval by the architect/engineer or owner, these record data are normally turned over to the owner, or if the contract calls for it, drafted on a set of tracings and then turned over to the owner.

PROBLEMS

1. Evaluation of the contractor's schedule submittal should be limited to what three principal concerns?
2. What risk is the owner or the engineer exposed to if he or she tells the contractor how to construct the Work?
3. Why is it desirable to prepare an inspection plan prior to construction?
4. What basic equipment (not furnishings) should be required in a construction field office?
5. A project has field trailers for the Resident Project Representative at the construction site and an off-site home office for the project manager. It is a traditional contract with an outside engineer/architect and a general contractor with three subcontractors constructing the work. From the following list specify, in numerical order, the routing of submittals from a subcontractor to the engineer/architect.

   a. ______ A/E project manager
   b. ______ Design reviewers
   c. ______ General contractor
   d. ______ Resident Project Representative
   e. ______ Subcontractors

6. You have been asked to plan the staffing of the Resident Project Representative's field office for a wastewater treatment plant that is going to be constructed. The estimated construction cost is $36,000,000.00 for the project to be constructed over a two-year period. What is the probable total number of field personnel recommended? Based upon the chart in Figure 3.8, list their classifications and the number of personnel in each field classification recommended for the project.
7. For a project that will be staffed by a Resident Project Representative, two inspectors, and a field clerk, what is the minimum size recommended for the field office trailer?
8. What is the correct terminology for the final set of project drawings that have had all of the latest changes and field conditions marked on the drawing set?
9. What is an RFI and how is it used?

4

DOCUMENTATION: RECORDS AND REPORTS

In the earlier years of construction, all that seemed to be needed to assure quality construction was the assignment of a full-time Resident Project Representative who possessed many years of experience and a track record of successful projects. The philosophy was based upon the premise that the Resident Project Representative would ensure that the owner received his or her money's worth by applying the knowledge learned through the years of construction experience. No detailed records were kept; in fact, many decisions were made in the field that should have been made by the architect or engineer, and many "deals" were made involving construction trade-offs, without any documentation.

Unfortunately, too many of the old-time inspectors are still operating in this manner. In a recent case involving the installation of an underground pipeline with a special joint detail that was causing some trouble, an inspector "solved" the field problems (or thought he did) and, upon completing the project, proudly moved on to another project where the virtues of his technique of handling the previous job were extolled. Shortly after completion, unknown to the inspector, most of the joints were found to leak. Unfortunately, no daily reports were filed nor did the inspector maintain a daily diary. A couple of years later a lawsuit followed in which the contractor wanted to recover the additional costs, claiming that the engineer's design was wrong, that the manufacturer's product was deficient, and that changes had been made at the direction of the inspector that cost the contractor additional money without solving the problem. The engineer, without the benefit of any documentation to support a counter position, was forced to settle at a considerable disadvantage.

The engineer of record was placed in a very vulnerable position as a direct result of the failure of the inspector to maintain adequate records. If daily reports had been made out and submitted to the engineer regularly, it is quite probable that the engineer may have had the opportunity to review the problem and to take corrective action at an early date, possibly preventing the occurrence of the problem.

It was vital to the engineer's and owner's defense that they be capable of documenting the day-by-day events that led up to the problem, as well as the substance of conversations that took place between the inspector and the contractor, and what commitments, if any, were made by either party. Of prime importance was the issue of whether the inspector had actually warned the contractor of the possibility of leakage in the joint. This inspector had previous experience in the installation of the same kind of pipe and pipe joints on a previous job, and had noted its tendency to leak if installed in a certain manner.

The inspector was located and interviewed by the author for the engineers who had originally designed the job more than two years before. Upon careful questioning, the inspector admitted that he had recorded nothing, but claimed he could remember each incident fairly well. Subsequent questioning disproved this, and the engineer's office was left with little defense except the memory of a witness who could easily be discredited.

An interesting fact should be recognized by all inspectors. Any project could become involved in litigation, and it could be several years after the incident before testimony of the inspector as a witness is requested. Any record that the inspector makes in writing, which is recorded in a form that will retain its credibility, may be referred to by the inspector while on the witness stand. This is an allowable method of refreshing a witness's memory. There are some limitations, however, and one of them is that the notes recorded by the inspector must be made on the same day that the incident or conversation took place. It is *not* acceptable to write notes on scratch paper, then at a later date transcribe them into the inspector's diary or log book. Such personal records of the inspector, because they are not a part of the agency's regular business record-keeping system, may not be considered as evidence in themselves, except for records made by certain types of public officials in the course of their official duties, and, possibly, except for records that are ruled admissible under the principle of "normal records kept in the regular course of business." Thus, these notes may not be entered as evidence, but can only be referred to in court by the party who wrote them—and then only as a memory-refreshing device.

It cannot be emphasized too strongly that the modern construction job is beset with numerous potential disputes or legal problems. *Any inspector who fails to keep adequate records is not performing a competent job and should be replaced.* Instead of providing the services to the owner that the latter is paying for, such an inspector is simply adding to the overhead cost of the project, or worse, because the owner is lulled into the feeling that with an inspector on the job, its interests are going to be adequately protected. Had the owner known in time, corrective action could have been taken.

Project Documentation as Evidence in Claims

Evidence of a written document made as a record of an act, a condition, or an event is normally admissible when offered to prove the act, condition, or event if:

1. The written document was made in the regular course of business (this may be applied to field diaries only if the owner or architect/engineer normally required that daily diaries be kept on all projects as a normal business record).

2. The document was written at or near the time of the act, condition, or event.
3. The possessor of the record or other qualified witness testifies to its identity and the mode of its preparation.
4. The sources of information and the method and time of preparation were sufficient to indicate its trustworthiness.

FILES AND RECORDS

It is preferable to establish a filing system for an entire company or agency rather than to depend upon record-keeping systems designed for each individual project. Unfortunately, many organizations never have established a filing system broad enough to include the special problems of the construction phase of a project. Any effort to utilize a design-oriented filing system for construction is doomed from the start.

Elsewhere in this chapter, emphasis is placed upon the Resident Project Representative as the developer of a competent construction field office record-keeping system, because in the absence of an established company system of records filing, the Resident Project Representative may be the only person to whom we can turn to develop and maintain competent project records.

Construction Filing System of a Major Engineering Firm

A sample of the last six categories of the filing system of a major engineering firm in the United States is shown in the following example. Each of these categories relates exclusively to construction phase activities.

8.0 BID PHASE ACTIVITIES
8.1 Advertisement for Bids
8.2 Bidder List (Documents Issued)
8.3 Bid Opening Reports
8.4 Summary and Evaluation of Bids
8.5 Preaward Submittals
9.0 PRECONSTRUCTION PHASE
9.1 Inspection and Testing Manual
9.2 R/W, Easement, and Permit Documents
9.3 Preconstruction Conference
9.4 Contractor Submittals
  9.4.1 Bonds and Insurance
  9.4.2 Bid Breakdown (Schedule of Values)
  9.4.3 Preliminary Schedule (CPM, etc.)
9.5 Notices to Contractor
  9.5.1 Award
  9.5.2 Proceed
10.0 CONSTRUCTION PHASE

10.1 Inspection Records and Reports
- **10.1.1** Daily Construction Reports
- **10.1.2** Field Diaries
- **10.1.3** Certificates and Delivery Tickets
- **10.1.4** Nonconformance Reports
- **10.1.5** Batch Plant Records
- **10.1.6** Special Inspection Reports

10.2 Quality/Materials Testing
- **10.2.1** Pipe
- **10.2.2** Concrete
- **10.2.3** Soils
- **10.2.4** Asphalt Products
- **10.2.5** Welding
- **10.2.6** Other Materials

10.3 Changes and Extra Work
- **10.3.1** Change Orders
- **10.3.2** Work Directive Changes
- **10.3.3** Field Orders
- **10.3.4** Estimates of Change Order Costs
- **10.3.5** Requests for Proposals
- **10.3.6** Extra Work Reports
- **10.3.7** Change Order Log
- **10.3.8** Deviation Requests

10.4 Payment for Work or Materials
- **10.4.1** Progress Payment Estimates
- **10.4.2** Contractor's Pay Requests
- **10.4.3** Materials Delivered (Not Yet Used)

10.5 Progress of the Work
- **10.5.1** Contractor's Work Schedules (Diagrams)
- **10.5.2** Schedule Updates (Computer Printouts)
- **10.5.3** Monthly Progress Reports and Job Status

10.6 Time of Work
- **10.6.1** Delays in the Work
- **10.6.2** Time Extensions
- **10.6.3** Suspension of Work

10.7 Contractor Submittals
- **10.7.1** Shop Drawings
- **10.7.2** Samples
- **10.7.3** Certificates
- **10.7.4** Mix Designs
- **10.7.5** Sheeting, Shoring, and Bracing Plans

10.8 Record Drawings
- **10.8.1** Updates During Construction
- **10.8.2** Final Record Drawings

10.9 Photographic Records
- **10.9.1** Progress Photos
- **10.9.2** Claims Photos
- **10.9.3** Safety Hazard Photos
- **10.9.4** Accident Photos
- **10.9.5** Public Relations Photos

10.10 Disputes, Protests, and Claims
- **10.10.1** Contractor-Initiated Actions
- **10.10.2** Owner/Engineer Documentation

10.11 Safety and Health (OSHA)

10.12 Beneficial Use/Partial Utilization

10.13 Maps

10.14 Outside Services
- **10.14.1** Surveys
- **10.14.2** Testing Laboratories
- **10.14.3** Special Inspections
- **10.14.4** Consultants

11.0 PROJECT CLOSEOUT

11.1 Operational Testing and Evaluation

11.2 Punch Lists

11.3 Final Submittals from Contractor
- **11.3.1** Record Drawings
- **11.3.2** Keying Schedule
- **11.3.3** Spare Parts
- **11.3.4** Tools

11.4 Notice of Completion

11.5 Final Progress Payment

11.6 Release of Retainage and Withholding

12.0 O&M AND PROJECT STARTUP

12.1 Correspondence with Contractors and Manufacturers

12.2 Training
- **12.2.1** Manufacturer's Training
- **12.2.2** Training Manual Draft
- **12.2.3** Operator Certification Material
- **12.2.4** Audiovisual Aids and Materials

12.3 O&M Manual
- **12.3.1** Draft O&M Manual
- **12.3.2** Review Comments from Client, EPA, etc.
- **12.3.3** Staff Review/Technical Manual Summaries
- **12.3.4** Graphic Materials; Photos

12.4 Startup
- **12.4.1** Equip Inspection/Review Report
- **12.4.2** Troubleshooting/Process Problems
- **12.4.3** Scheduling (Plan of Operation)

- **12.4.4** Startup Meeting Summary
- **12.4.5** Equipment Warranties/Plant Acceptance
- **13.0** PROJECT FOLLOW-UP
- **13.1** Site Visit Notes and Memos
- **13.2** Photos
- **13.3** Final Project Accounting

Individual Project Records

It is often the Resident Project Representative's responsibility to determine what the specific needs of the employer are with regard to the types of construction records that must be established and maintained for a specific project. One principal exception is the conduct and administration of federal agency projects in which the government agency often provides a very specific list of all types of records, reports, and other documentation that is required, plus some specific requirements concerning the form in which such records must be maintained. Often the printed forms themselves are provided.

Many local public agencies, as well as some of the larger private architect/engineer firms, have preprinted forms to assist the inspector in the recording of pertinent job information, and many have procedures established for the handling, distribution, and storage of job records as well.

Without regard for whether an architect/engineer, public agency, owner, or other interested party has established such record keeping as a policy matter, *each* inspector, not just the resident engineer or inspector, should *always* maintain a daily diary in which notes and records of daily activities and conversations are kept. Such a diary should contain abstracts of all oral commitments made to or by the contractor, field problems encountered during construction, how such problems were resolved, notices issued to the contractor, and similar information. It should be remembered, however, that the daily construction diary is *not* a substitute for the Daily Construction Report, which describes the construction progress and normally receives wider distribution. The information recorded in the inspector's diary is generally of a private nature and is intended for the use of the inspector and his or her employer only.

Construction Records

The following is a list of the principal types of construction records that the Resident Project Representative should maintain on every project:

1. *Progress of the work*. Maintain a Daily Construction Report (Figure 4.1) containing a description of the work commenced, new work started, status of work in progress, labor and equipment at the site, weather, and visitors to the site. If no work was performed at all, a daily report should still be filed, stating "no work." On projects where several inspectors are involved, this report is compiled from each inspector's Daily Record of Work Progress (Figure 4.2).

DAILY CONSTRUCTION REPORT

DATE 31 JAN 2000

DAY: S M(X) T W TH F S

PROJECT RESERVOIR I-D AND PUMP STATION
JOB NO. 00-03
CLIENT CITY OF FULLERTON
CONTRACTOR H & H CONSTRUCTORS, INC.
PROJECT MANAGER E. R. FISK

WEATHER	Brite Sun	Clear	Overcast (X)	Rain	Snow
TEMP.	To 32	32-50	50-70 (X)	70-85	85 up
WIND	Still (X)	Moder	High	Report No 197	
HUMIDITY	Dry	Moder (X)	Humid		

AVERAGE FIELD FORCE

Name of Contractor	Non-manual	Manual	Remarks
H&H CONSTRUCTORS, INC.	1	12	PRIME
S&M CONSTRUCTION CO.	1	2	EARTHWORK SUB.
GLADMORE ENGRG.	0	2	TEST LAB

VISITORS

Time	Representing	Representing	Remarks
14:15	RICHARD CLEMENT	ARMCO	

EQUIPMENT AT THE SITE CAT 623B ELEV. SCRAPER; (2) CAT D-9 DOZERS (ONE IDLE/UNDER REPAIR) CAT 140G GRADER; HITACHI EX60URG EXCAVATOR; CAT 970F LOADER

CONSTRUCTION ACTIVITIES SECOND STAGE EXCAVATION OF ACCESS RAMP WAS CONT'D WITH THE CAT D-9 DOZER. EXCESS MATL IS BEING PICKED UP WITH THE CAT 623B SCRAPER AND DISTRIBUTED ON SITE TO BRING TO FIN. GRADE. ROUGH SLOPE TRIMMING IS CONTINUING WITH THE DOZER.
PIPE LAYING: LAYING OF 400mm RCCP FOR THE BY-PASS LINE HAS PROGRESSED TO NW COR. OF SITE. AS OF 08:30 HRS CORNER BENDS WERE BEING LAID. SAND BACKFILL IS BEING PLACED IN THE PIPE ZONE AND CONSOLIDATED BY FLOODING.
BASIN AREA SLOPES ARE BEING HAND-TRIMMED TO FIN. GRADES AND THICKENED EDGE SECTIONS ARE BEING CUT TO TEMPLATE. BOTTOM OF BASIN IS BEING TRIMMED TO GRADE WITH THE CAT 970 LOADER.
TRENCH BACKFILL: MUD THAT WAS BEING WASHED INTO THE ID-2 PRESSURE LINE TRENCH DURING PREVIOUS RAIN IS BEING REMOVED AND REPLACED WITH SAND.
THRUST BLOCK CONCRETE WAS PLACED IN THE TWO 45-DEG BENDS IN THE 400mm BY-PASS LINE AND CONC. ENCASEMENT OF THE CROSSOVER RISER WAS CAST USING CLASS "C" CONCRETE PER SPEC.

Data used are fictitious for illustration only

DISTRIBUTION: 1. Proj. Mgr. 2. Field Office 3. File 4. Client

PAGE 1 OF 1 PAGES

BY R. E. Barnes TITLE RES. ENGR.

Fisk Form 8-1

FIGURE 4.1 Daily Construction Report. (*Fisk, Edward R.,* Construction Engineer's Complete Handbook of Forms, *1st Edition, © 1993. Reprinted by permission of Pearson Education, Inc., Upper Saddle River, NJ.*)

INSPECTOR'S DAILY RECORD OF WORK PROGRESS

DAY: S M T W TH F S (W marked X)

For submittal to Resident Project Rep. to compile Daily Construction Report

Project Title: APOLLO RESERVOIR Project No.: 00-03
Feature: EMBANKMENT CONTROL Date: 2 FEB 2000
Contractor: SLM CONST (SUB) Type of Work: EARTHWORK

WEATHER	Brite Sun	Clear	Overcast ✓	Rain	Snow
TEMP.	To 32	32-50	50-70 ✓	70-85	85 up
WIND	Still ✓	Moder	High	Report No.	
HUMIDITY	Dry	Moder ✓	Humid	207	

CONTRACTOR'S WORK FORCE (Indicate classifications, including Subcontractor personnel)
ONE FOREMAN; 2 EQUIP OPERATORS

EQUIPMENT IN USE OR IDLED (Identify which)
ONE CAT 623B ELEV. SCRAPER; (2) CAT D-9 DOZERS (ONE DOWN FOR REPAIR)
CAT 140 G GRADER; HITACHI EX 60 URG EXCAVATOR; CAT 970F LOADER

MATERIALS OR EQUIPMENT DELIVERED VCP PIPE DELIVERED TO SITE

NON-CONFORMING MATERIALS OR WORK (Describe reason for non-conformance)
VCP DELIVERY TODAY INCL. 14 LENGTHS W/REJECTABLE DEFECTS.

FIELD PROBLEMS (which could result in delay or claim) DISCOVERY OF EXPANSIVE CLAY IN AREA NOT SHOWN IN SOILS REPORTS. REPORT TO ENGR. FOR INSTRUCTIONS.

QUANTITIES OF PAY ITEMS PLACED 95m PIPE TRENCH EXCAVATION; 46m OF PIPE BEDDED, LAID, JOINTED. 27m FULLY BACKFILLED AND COMPACTED. 12m PIPE ZONE BACKFILLED W/SAND ONLY. NO PAVING OVER TRENCHES YET.

SUMMARY OF CONSTRUCTION ACTIVITIES
EXCAVATION OF BASIN AREA CONTINUING.
TRENCH EXCAVATION & BACKFILL FOR INLET/OUTLET PIPELINE
DELIVERY OF PIPE MATERIALS TO SITE
CONSTRUCTION FENCING CONTINUING

Data used are fictitious for illustration only

FOLLOWUP INSPECTIONS OF PREVIOUSLY REPORTED DEFICIENCIES
RE-CHECKED EXCAVATION & RECOMPACTION OF TRENCH BACKFILL OVER 600mm RCCP INLET/OUTLET LINE TO ASSURE 90% DENSITY. IT IS OK NOW. CHECKS OUT AT 92 & 93% AT ALL LEVELS.

DISTRIBUTION 1. Field Office
2. Inspector

Inspector [signature]

Fisk Form 8-5

FIGURE 4.2 Inspector's Daily Record of Work Progress. (*Fisk, Edward R.,* Construction Engineer's Complete Handbook of Forms, *1st Edition, © 1993. Reprinted by permission of Pearson Education, Inc., Upper Saddle River, NJ.*)

2. *Telephone calls.* All telephone calls made or received should be logged and a note made indicating the identities of the parties as well as a brief phrase indicating the nature or purpose of the call.
3. *Tests of materials.* A record should be kept of all material samples sent out to the laboratory for testing (Figure 4.3) as well as those tests performed at the site. The report should include space for later inclusion of the test results, as well as the location in the structure where the particular material was to be installed.
4. *Diary.* A daily diary should be maintained by each member of the field staff. This book may end up in court, so it should be neatly and accurately recorded. An entry should be made every day, whether or not work was performed. The detailed contents and form of the diary will be described later in this chapter (see Figure 4.7).
5. *Log of submittals.* All material being transmitted to the architect/engineer through the Resident Project Representative should be logged in and out on a submittal log such as the one shown in Figure 4.4.

Construction Field Office Files

All field office files should be kept up to date and should be maintained for ready reference at the job site during the entire construction phase of the project. Upon completion of the work, the files should be turned over to the architect/engineer, who will retain some and forward others to the owner for retention. The field office files should include the following categories:

1. *Correspondence.* Copies of all correspondence concerning the project that have been sent to the Resident Project Representative should be maintained and filed by date.
2. *Job drawings.* Drawings of clarification or change or drawings that contain supplemental information should be filed at the field office, in addition to a complete set of all contract drawings as bid.
3. *Shop drawing submittals.* The Resident Project Representative should maintain a drawing log and should maintain a shop-drawing file of submittals that have received final review and approval (Figure 4.4).
4. *Requisitions.* Copies of all approved requisitions for payment should be kept at the site for field reference and as a guide for initial review of the next month's partial pay requisition from the contractor.
5. *Reports.* Copies of all reports of all types should be filed by date.
6. *Samples.* All approved samples showing material and/or workmanship should be kept at the job site as a basis of comparison, and should be appropriately tagged and logged.
7. *Operating tests.* The Resident Project Representative is responsible for seeing that all required tests are performed at the proper time. The files should include the results of all such testing.
8. *Deviation requests.* Whenever a request for deviation is received, a copy should be maintained with the disposition of the request.

ENGINEERING INSPECTION TESTING

Twining Laboratories *of Southern California, Inc.*

3310 Airport Way / Mailing Address P.O. Box 47 / Long Beach, CA 90801 / (213) 426-3355 / (213) 636-2386 / (714) 828-6432
BRANCH OFFICE: 1514-D North Susan Street/ Santa Ana, CA 92703 / (714) 554-2645

EXAMINATION NO. 87-19685

COMPRESSION TESTS ON CONCRETE CYLINDERS

E.R. Fisk & Associates
1224 East Katella
Suite 105
Orange, CA 92667

PROJECT	Galaxy Office Tower		
ADDRESS	12891 Galaxy Way, El Segundo, CA		
ARCHITECT	J.K. Jones Association	CONTRACTOR	Smith Construction
ENGINEER	Martin & Brayton	SUBCONTRACTOR	Blayman Concrete
SAMPLE FROM	Second floor slab, A-C to 7-9		
MIX	LB 87-1061	SPECIFIED STRENGTH	4000 @ 28 days
DATE CAST	1-28-87	SLUMP $4\frac{1}{2}$"	SPECIMEN SIZE 6" x 12"
DATE TESTED	2-4-87	2-25-87	2-25-87
SPECIMEN NO.	2A	2B	2C
COMPRESSION	7 days	28 days	28 days
TOTAL LOAD LBS.	86,000	122,500	125,000
POUNDS PER SQ. IN.	3,040	4,330	4,420
AREA. SQ. IN.	28.27		
	28 day average = 4,375		
COMPLIANCE	28 day test complies with specified strength		

	MADE BY	DELIVERED BY	RECEIVED ON
SPECIMENS	Harold Johnson	TLSC	1-29-87

DISTRIBUTION
E.R. Fisk & Assoc. - 2
Smith Construction -1
Blayman Concrete - 1
El Segundo Building Dept. - 1
C & C Ready Mix - 1

FIGURE 4.3 Laboratory Test Report.

CONTRACTOR SUBMITTAL LOG

Project: 5 MGD Treatment Plant Addition   Project Mgr.: R. E. Barnes

Job No.: 00-042   Contractor: ABC Constructors Inc.

Date Rec'd.	Trans-mittal No.	Description	Subcontractor / Ref. Spec. Section	Con-tractor Trans. No.	No. Copies Rec'd.	Action: No Exceptions Taken	Action: Make Correc-tions Noted	Action: Revise & Resubmit	Action: Rejected	Date Ret'd.	No. Copies Ret'd.	Remarks
6-11-99	78	Plan View--Aeration Basin Secondary Sed. Tanks Dissolved Air Flot. Thickener	01300		6			X		6-21-99	5	
6-12-99	79	Electrical materials list	A. J. Peterson 16050	76	6	X				7-17-99	5	
6-12-99	80	Warranty covering Acme Kitchen Unit	11460	3A	6		X			6-14-99	5	
6-12-99	81	Metal Compartment Dwgs---Color Card	10150	68	6		X			7-19-99	5	
6-12-99	8	Cell-tite Resin system	09800	69	6			X		6-18-99	5	
6-12-99	83	Pumps, Flow Detector, Motors, Frequency Drive	Harris & Foote 02660		6					7-19-99	5	See Change Order No. CO-003
6-12-99	84	Chlorination Equipment	Wallace & Tiernan 11200	55	6				X	6-18-99	5	
6-12-99	85	Curing Compound	03370	74	6	X				6-18-99	5	Approved "or equal" items
6-18-99	86	Wall spool Blower Solids Bldg	15050	80	6	X				6-21-99	5	
6-21-99	87	S. Primary Sludge Digester Slab	11300	77	6	X				6-22-99	5	

Contractor Submittal Log provides a permanent record of all submittals by the contractor of shop drawings, samples, and other requested data received during construction.

FIGURE 4.4 Contractor Submittal Log. (*Fisk, Edward R.,* Construction Engineer's Complete Handbook of Forms, *1st Edition, © 1993. Reprinted by permission of Pearson Education, Inc., Upper Saddle River, NJ.*)

Construction Progress Record

The most commonly accepted form of construction progress record is in the form of a Daily Construction Report, which is filled in by the Resident Project Representative or, if applicable, by the contractor's CQC representative on a daily basis even if no work was performed at the site that day. Usually, such reports are executed in carbon copies or by the use of forms printed on NCR paper that will provide the necessary number of copies.

The daily report is highly necessary as a progress record, and the use of this report in combination with an inspector's daily diary allows two types of information to be recorded in separate documents. In this manner, the more privileged type of information can be restricted to recording in the diary, and the true work progress can be recorded on the Daily Construction Report, where it will receive wider distribution.

Electronic Record Keeping

Widespread use is being made of PC computers for construction record keeping and contract administration activities. The concept has merit, but a note of caution is offered here regarding some precautions that should be taken to provide for the security and integrity of key records, such as the Daily Construction Report and the Construction Diary.

In an article published in *Smart Computing* magazine, a compilation was made of the life expectancy of various storage media. It appears evident from that study that electronic media showed very little merit as an archival storage tool. Estimated useful life of electronic storage media that exceeds the number of years that the media has been in existence must be considered as speculative, at best.

Remember, too, that regardless of the potential life span of the storage media involved, all electronic storage is subject to the risks of computer memory loss, equipment breakdown, corrupted files, and, last but not to be ignored, the ease of altering records without detection by someone out to falsify the records or cover up errors. Thus, the Daily Construction Report is wisely generated and updated in the computer, as electronic storage affords easy access and search capability. However, in addition to computer storage, it would be wise to print a hard copy of each daily report at the end of each business day and have it signed by the Resident Project Representative and placed in the file. Thus, if litigation later occurs, an original hard copy is available from the files, which can be compared with the electronic storage version.

The Daily Construction Diary is another matter. This document should be kept in a stitch-bound book, with all entries handwritten contemporaneously. Don't listen to the advocates of computerizing everything, as the credibility of the diary itself is a product of the manner in which it was prepared and updated, and electronic storage of the diary effectively removes all of the safeguards otherwise provided. Computer records are excellent in their place, but the diary is not one of them. See Chapter 5 for more detail on electronic contract administration and record keeping.

Construction Reports

Daily Construction Reports

The content of a Daily Construction Report should include the following information (however, it should be remembered that as long as a separate daily diary is being kept, the Daily Construction Report should contain items relating to work progress, not to conversations or to other transactions):

1. Project name and job number.
2. Client's name (name of project owner).
3. Contractor's name (general contractor only).
4. Name of the Project Manager for the design organization.
5. Report number and date of report (use consecutive numbering).
6. Day of the week.
7. Weather conditions (wind, humidity, temperature, sun, clouds, etc.).
8. Average field force, both supervisory and nonsupervisory.
   (a) Name of each contractor or subcontractor on the job that day.
   (b) Number of manual workers (journeymen and apprentices) at the site.
   (c) Number of nonmanual workers (superintendents and foremen) at the site.
9. Visitors at the site; include names, employers, and time in and out.
10. List identity, size, and type of all major pieces of construction equipment at the site each day. Indicate if idle, and reason, if applicable.
11. Log all work commenced; status of all work in progress; and all new work started. Identify location of the work as well as its description, and which contractor or subcontractor is performing it.
12. Sign the daily report with your full name, title, and date.

On large projects, items 1 through 4 are often preprinted on the Daily Construction Report form to avoid needless duplication of effort by field personnel. Where electronic reporting is used, a companion input form can be provided for gathering input data from the field. The Daily Report form provided under an electronic construction management program, such as described in Chapter 5, includes weather information, what the contractors are doing, and any significant events, including owner, architect/engineer, or Resident Project Representative directions. The program also can instruct the user not only to record items like weather, but also to comment on whether the weather had any observed impact on construction activities. An electronic reporting system also can provide the means of capturing daily inspection reports (usually scanned images).

Monthly Reports

In addition to Daily Construction Reports, it is not uncommon to require monthly reports as well. Generally, such reports are grouped in two categories: Monthly Report of Contract Performance, as in Figure 4.5; and a General Project Status Report, as illustrated in Figure 4.6.

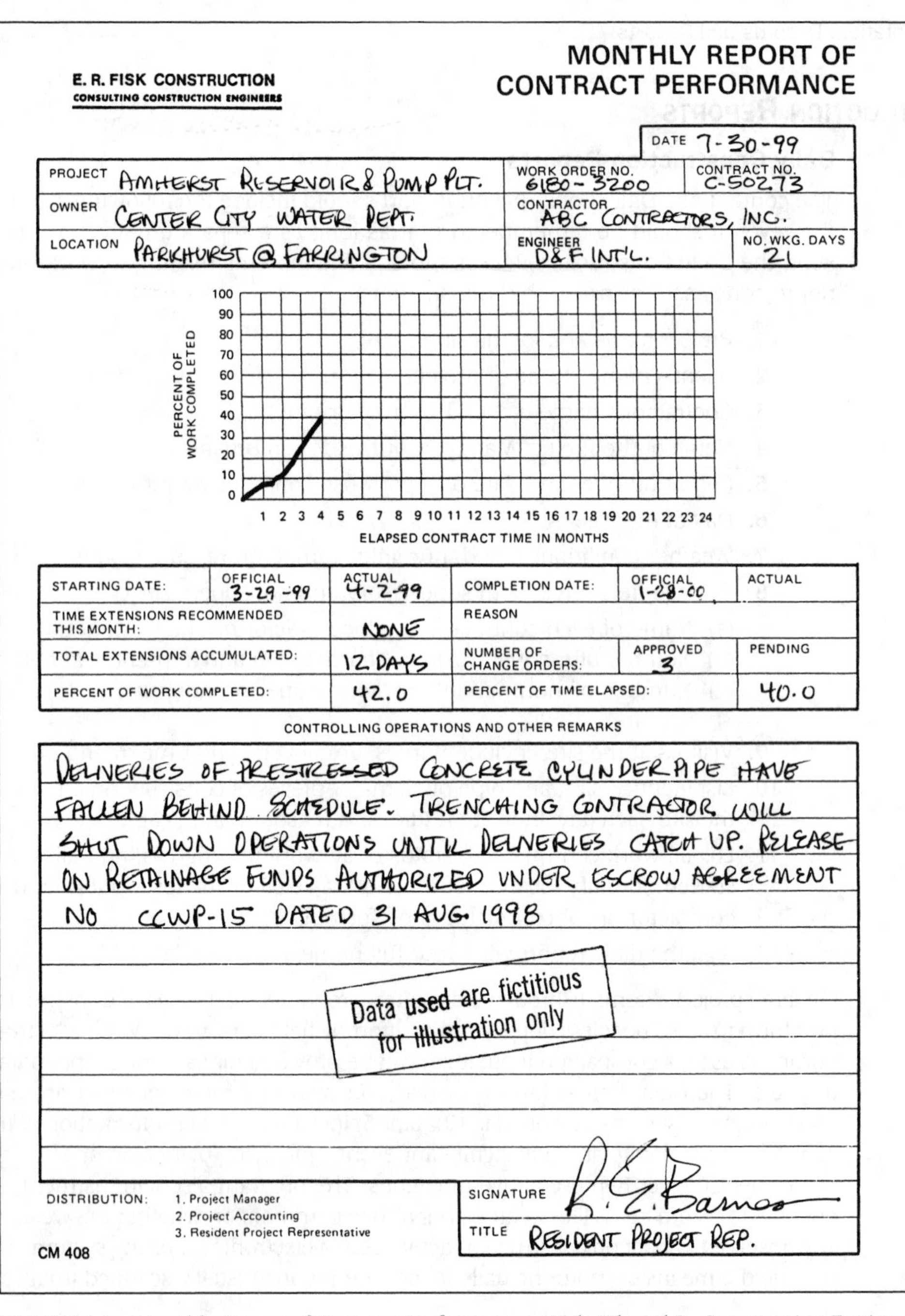

E. R. FISK CONSTRUCTION
CONSULTING CONSTRUCTION ENGINEERS

MONTHLY REPORT OF CONTRACT PERFORMANCE

DATE 7-30-99

PROJECT AMHERST RESERVOIR & PUMP PLT.	WORK ORDER NO. 6180-3200	CONTRACT NO. C-50273
OWNER CENTER CITY WATER DEPT.	CONTRACTOR ABC CONTRACTORS, INC.	
LOCATION PARKHURST @ FARRINGTON	ENGINEER D&F INT'L.	NO. WKG. DAYS 21

STARTING DATE:	OFFICIAL 3-29-99	ACTUAL 4-2-99	COMPLETION DATE:	OFFICIAL 1-28-00	ACTUAL
TIME EXTENSIONS RECOMMENDED THIS MONTH:		NONE	REASON		
TOTAL EXTENSIONS ACCUMULATED:		12 DAYS	NUMBER OF CHANGE ORDERS:	APPROVED 3	PENDING
PERCENT OF WORK COMPLETED:		42.0	PERCENT OF TIME ELAPSED:		40.0

CONTROLLING OPERATIONS AND OTHER REMARKS

DELIVERIES OF PRESTRESSED CONCRETE CYLINDER PIPE HAVE FALLEN BEHIND SCHEDULE. TRENCHING CONTRACTOR WILL SHUT DOWN OPERATIONS UNTIL DELIVERIES CATCH UP. RELEASE ON RETAINAGE FUNDS AUTHORIZED UNDER ESCROW AGREEMENT NO CCWP-15 DATED 31 AUG. 1998

Data used are fictitious for illustration only

DISTRIBUTION: 1. Project Manager
2. Project Accounting
3. Resident Project Representative

SIGNATURE R. E. Barnes

TITLE RESIDENT PROJECT REP.

CM 408

FIGURE 4.5 Monthly Report of Contract Performance. (*Fisk, Edward R.,* Construction Engineer's Complete Handbook of Forms, *1st Edition, © 1993. Reprinted by permission of Pearson Education, Inc., Upper Saddle River, NJ.*)

GENERAL PROJECT STATUS REPORT

PROJECT Hydroelectric Project
OWNER City of Thousand Oaks, CA — PROJECT NO. 89/90-00
CONTRACTOR XYZ Constructors, Inc. — CONTRACT NO. 22 Dec 1988
PROJECT MANAGER R. E. Barnes — DATE 1 Mar 1990

Amount of Original Contract		$	65,123,405.00
Approved Change Orders to Date			62,037.00
Anticipated Over-run or (Under-run) in Uncompleted Work			25,622.00
Actual Over-run or (Under-run) in Completed Work			----
Rental Revenue from Contract Owner not included above			----
Estimated Total Amount of Principal Contract		$	65,211,164.00
Other Contract Work not Included in Principal Contract			----
Total Estimated Contract Volume		$	65,211,164.00
Contract Revenue to Date	$ 27,176,039.00		
Less: Contract Advances for Materials on Hand	(-2,657,123.00)		
Contract Advances for Plant and Move-in	(-1,058,416.00)		
Revenue Reduction for Uncompleted Work	(-806,014.00)		
Other	----		
Total Amount of Work Completed to Date	22,584,486.00	$	22,584,486.00
Uncompleted Contract Volume		$	42,646,678.00

Percent Complete Based on Original Contract	34.66	%
Percent Complete Based on Total Estimated Contract	34.03	%
Time Allotted by Original Contract	1,400	Days
Extension of Contract Time	10	Days
Total Contract Time	1410	Days
Contract Time Elapsed	434	Days
Percent of Original Contract Time Elapsed	31.00	%
Percent of Total Time, Including Extensions, Elapsed	30.78	%
Date Contract was Physically Completed—If Completed	----	
Expected Date of Physical Completion—If Not Complete	2 Nov 1992	

Wiley-Fisk Form 8-7

FIGURE 4.6 General Project Status Report. (*Fisk, Edward R.,* Construction Engineer's Complete Handbook of Forms, *1st Edition, © 1993. Reprinted by permission of Pearson Education, Inc., Upper Saddle River, NJ.)*

The *Monthly Report of Contract Performance* is intended to supplement the Daily Construction Reports, and as such summarizes the work progress for the immediately preceding month both graphically (S-curve diagram) and verbally, through a summary of contract time and change orders, and in a brief narrative report summarizing the work progress during the past month.

The *General Project Status Report* is a monthly statement of time and cost and should be of special interest to the project manager. The document not only places all time and cost figures in one handy document, but also records all changes that accrued since the beginning of the project. This is a strongly recommended document for whoever is responsible for management of the contract on behalf of the owner or architect/engineer.

Construction Diary

Often called by different names, the construction diary (Figure 4.7) is an important document. The requirements for maintaining an unimpeachable legal record in the form of a daily diary are indicated in the list that follows shortly. Although variations may occur without destroying the credibility of the document, the recommendations provided here should assure the greatest degree of reliability.

It should be remembered that it is frequently necessary to consult a field diary to give testimony during a court trial. In some cases, the book itself may not be generally admissible as evidence but could be used as a memory refresher by the person who made the original entries while giving testimony on the witness stand. It is because of this provision that certain basic record-keeping rules are considered mandatory to preserve the integrity of the record.

Mention is sometimes made of the "privileged" nature of some of the contents of the construction diary. This is *not* meant to imply that it is a private document to be seen and possessed by the inspector alone. On the contrary, the record normally belongs to the design firm or owner, and although it is wise for the inspector to retain a copy of its contents, the diary must normally be turned in with the job records when it is full or at the end of the construction project. During the progress of the work, it may be advisable to submit the daily diary at regular intervals to the Project Manager of the design organization or owner to allow inspection of its contents. In this manner the Project Manager can be advised of all the transactions that have been taking place in the field. Normally, the Project Manager may want to make copies of its pages at that time; however, that does not preclude the requirement that the filled books be turned in to the Project Manager at the end of the job, at which time they are often stored in the vault with other permanent job records.

The diary requirements can be grouped into two significant categories: *format* and *content*. Each is equally important in its own way.

Format of the Construction Diary

1. Use only a hardcover, stitched-binding field book such as used by surveyors for their note keeping, or a "record book" obtainable at stationers.

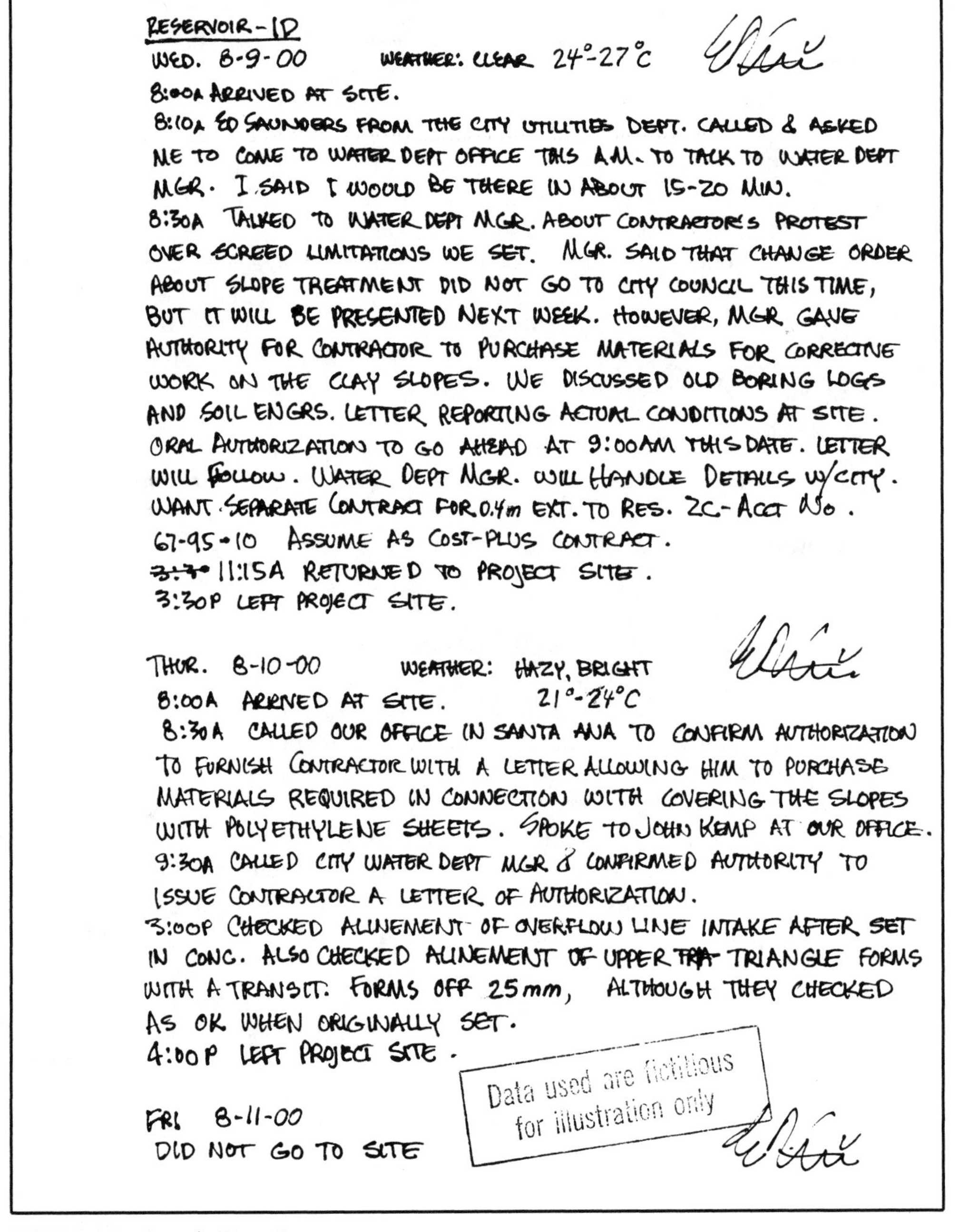

RESERVOIR-1D

WED. 8-9-00 WEATHER: CLEAR 24°-27°C

8:00A ARRIVED AT SITE.

8:10A ED SAUNDERS FROM THE CITY UTILITIES DEPT. CALLED & ASKED ME TO COME TO WATER DEPT OFFICE THIS A.M. TO TALK TO WATER DEPT MGR. I SAID I WOULD BE THERE IN ABOUT 15-20 MIN.

8:30A TALKED TO WATER DEPT MGR. ABOUT CONTRACTOR'S PROTEST OVER SCREED LIMITATIONS WE SET. MGR. SAID THAT CHANGE ORDER ABOUT SLOPE TREATMENT DID NOT GO TO CITY COUNCIL THIS TIME, BUT IT WILL BE PRESENTED NEXT WEEK. HOWEVER, MGR GAVE AUTHORITY FOR CONTRACTOR TO PURCHASE MATERIALS FOR CORRECTIVE WORK ON THE CLAY SLOPES. WE DISCUSSED OLD BORING LOGS AND SOIL ENGRS. LETTER REPORTING ACTUAL CONDITIONS AT SITE. ORAL AUTHORIZATION TO GO AHEAD AT 9:00AM THIS DATE. LETTER WILL FOLLOW. WATER DEPT MGR. WILL HANDLE DETAILS W/CITY. WANT SEPARATE CONTRACT FOR 0.4m EXT. TO RES. 2C-ACCT NO. 67-95-10 ASSUME AS COST-PLUS CONTRACT.

~~3:30~~ 11:15A RETURNED TO PROJECT SITE.

3:30P LEFT PROJECT SITE.

THUR. 8-10-00 WEATHER: HAZY, BRIGHT 21°-24°C

8:00A ARRIVED AT SITE.

8:30A CALLED OUR OFFICE IN SANTA ANA TO CONFIRM AUTHORIZATION TO FURNISH CONTRACTOR WITH A LETTER ALLOWING HIM TO PURCHASE MATERIALS REQUIRED IN CONNECTION WITH COVERING THE SLOPES WITH POLYETHYLENE SHEETS. SPOKE TO JOHN KEMP AT OUR OFFICE.

9:30A CALLED CITY WATER DEPT MGR & CONFIRMED AUTHORITY TO ISSUE CONTRACTOR A LETTER OF AUTHORIZATION.

3:00P CHECKED ALINEMENT OF OVERFLOW LINE INTAKE AFTER SET IN CONC. ALSO CHECKED ALINEMENT OF UPPER ~~TRA~~ TRIANGLE FORMS WITH A TRANSIT. FORMS OFF 25mm, ALTHOUGH THEY CHECKED AS OK WHEN ORIGINALLY SET.

4:00P LEFT PROJECT SITE.

FRI 8-11-00

DID NOT GO TO SITE

Data used are fictitious for illustration only

FIGURE 4.7 Sample Diary Page.

2. Pages should be consecutively numbered in ink, and *no* numbers should be skipped.
3. No erasures should be made. In case of error, simply cross out the incorrect information and enter the correct data next to it.
4. No pages should be torn out of the book at any time. If a page is to be voided, place a large "X" through the page and mark "void."
5. Every day should be reported, and every calendar date should be accounted for. If there is no work performed on a given date, the date should be entered on the page followed by the words "no work" or similar wording. It is still desirable to record the weather on "no work" days, as it may have later bearing on *why* no work was performed in a case involving a claim for liquidated damages.
6. All entries *must be made on the same date that they occur.* If notes are kept on separate scratch paper and later transcribed into the diary, and this fact is disclosed during a trial, the credibility of the entire diary comes into question.

Content of the Construction Diary

1. Record telephone calls made or received, and a substantial outline of the nature of such calls, including any statements or commitments made during the call. Identify the parties calling.
2. Record any work or material in place that does not correspond with the drawings or specifications, as well as the action taken. List any other problems or abnormal occurrences that arose during each day, including notations of any particular lack of activity on the part of the contractor. Note corrective actions taken.
3. Record time and the name of the contractor's representative to whom field orders are delivered, and the nature of the contents of the field order.
4. Note unforeseen conditions observed by the inspector that may cause a slowdown by the contractor.
5. Where a contractor is performing extra work because of an unforeseen underground obstruction, make a careful field count of all personnel and equipment at the site and how they are occupied. Log the number and craft of each person idled by such work, as well as any idle equipment *that would otherwise be capable of working.*
6. Record the content of all substantive conversations held with the contractor at the site, as well as any trade-offs, deals, or commitments made by either party.
7. Record all field errors made by *any* party at the site. Identify in detail and indicate probable effect.
8. Show name of the job at the head of every page.
9. Sign every diary entry and indicate your job title immediately under the last line of entry on each day's report. This will preclude claims that additional wording was added later.

Who Should Maintain Diaries and Daily Reports?

A construction diary should be kept by every individual involved in the project. The Daily Construction Report, by comparison, is prepared and submitted only by the Resident Project Representative, utilizing the Inspector's Daily Record of Work Progress reports submitted by each of the inspectors as an information source. These interim inspectors' reports serve also as a check of contractor work progress when it becomes necessary to review the contractor's partial pay request.

Documentation of Intermittent Inspection

Sometimes a project representative is responsible for monitoring the construction of several smaller projects instead of serving as a Resident Project Representative on a single, large project. In such cases, the project representative usually makes intermittent field visits to each project site, as needed, to meet the individual project requirements.

The documentation requirements for construction progress reporting of multiple projects by the same project representative calls for an approach different from the traditional daily report. A more practical approach under the circumstances is to prepare a weekly project report for each project for which the project representative is responsible (Figure 4.8). In this manner, the activities of the entire week of each project can be summarized in a single project report for each project being monitored.

Special Feedback Reports

Although not generally a contract requirement or even a common practice among most architect/engineer offices, a highly desirable practice would be the establishment of communications feedback systems between the various field forces and between the field and office. Such communications can minimize the number of repeated errors or field problems that often occur as a result of the traditional failure of communication between the construction forces and the designers of a project.

Report of Field Correction

Whenever the need for any corrective change is recognized in the field that would require a departure from the plans and specifications as originally issued, a complete detailed report should be prepared by the Resident Project Representative and submitted to the project manager for evaluation by the design and specifications departments to see if the problem can be prevented in the future by design or specifications changes.

A convenient document for this purpose is illustrated in Figure 4.9. It provides for the necessary information needed by the office forces in their evaluations and possible corrections, if merited. The report should be in sufficient detail to allow the engineer or architect to understand the problem, make a determination, and issue instructions to the field personnel. This document is intended as a source of information only, and must not be used as an authorization to perform work nor as a change order. Wherever changes are required, the normal procedures for accomplishing such

City of Thousand Oaks

INSPECTOR'S WEEKLY PROGRESS REPORT

Feature Inspected: CONCRETE RESERVOIR LINING & SITEWORK
Item No.:
Project: WILLARD RESERVOIR   Date: 10-13-99
Project No.: 99-12   Contract No.:
Work Order No.:   Purchase Order No.:
Contractor: XYZ CONSTRUCTORS, INC.   Location: THOUSAND OAKS, CA
Subcontractor: CONCRETE CONSTRUTION, LTD   Location:

CONTRACT REQ'D DELIVERY DATE: 2-18-00   ESTIMATED DELIVERY DATE: 2-18-00
(EXPLAIN DIFFERENCE):

ENGINEERING STATUS:
DESIGN COMPLETE %: 100   CONTRACTOR COMPL %:
INCOMPLETE: INCOMPLETE:

Data used are fictitious for illustration only

PRODUCTION STATUS:
CONTRACT NOT YET SCHEDULED: PARTIALLY SCHEDULED: SCHEDULED: ✓
CONTRACT IN PRODUCTION:
CONTRACT PROGRESSING TO PRESENT SCHEDULE: INSPECTOR'S ESTIMATE OF % COMPL.: 24.3

SHIPPING INFORMATION:
TOTAL SHIPPED THIS WEEK: NONE
TOTAL PREVIOUSLY SHIPPED: 4926 kg REBAR; 457m PVC WATERSTOP
TOTAL SHIPPED TO DATE: 7675 kg REBAR; 610m PVC WATERSTOP

REMARKS
BACKFILLING CONTINUED TO THE 400mm RCCP DOWN LINE. GUNITE REPAIRS TO EAST HALF OF RES. ROOF BEGUN ON FRI 10-8-99. SLOPE PANELS WERE PREPARED FOR CONC. PLACING. THREE PANELS COMPLETED ON 10-12-99 AND ANOTHER 2 COMPLETED ON 10-15-99. THIS COMPLETES ALL SLOPE PAVING.

EDISON CO. SET 4 POLES TO SERVE THE SITE, BUT TRANSFORMERS & RACKS HAVE NOT YET BEEN ERECTED. ELEC EQUIP PAD WAS CAST ON 10-11-99 USING CONC. MIX CRP-1020 AT 75mm SLUMP
WATERSTOP JOINTS IN 2ND STAGE PANELS WERE BONDED ON 10-13-99. JOINT FILLER MATL IS BEING REMOVE FROM FINISHED PANELS AND THE SPACE IS BEING PREPARED FOR THIOKOL JT. SEALANT.
IN THE VALVE PIT, THE ALTITUDE VALVES WERE LIFTED INTO PLACE AND DRAWN INTO FINAL POSITION WITH CHAIN PULLS AND EMBEDDED PULLING EYES. DRESSER COUPLINGS WERE INSTALLED OUTSIDE VALVE PIT WALLS
SITE GRADING BEGAN 10-14-99 USING A MOTOR GRADER AND A DOZER AT CURB & GUTTER AREAS ONLY

By: R.E. Barnes   Title: INSPECTOR
Inspection Agency:
Fisk Form 8-6

FIGURE 4.8 Inspector's Weekly Progress Report. (*Fisk, Edward R.*, Construction Engineer's Complete Handbook of Forms, *1st Edition, © 1993. Reprinted by permission of Pearson Education, Inc., Upper Saddle River, NJ.*)

City of Thousand Oaks

REPORT OF FIELD CORRECTION

SPECIAL REPORT NO. 2

PROJECT CHAMPLAIN RESERVOIR — PAGE 1 OF 1 PAGES

LOCATION THOUSAND OAKS — PROJECT NO. 99/90-00 — DATE 4-23-99

CONTRACTOR XYZ CONSTRUCTORS, INC. — CROSS REFERENCE TO DAILY REPORT NO. 153

INSTRUCTIONS

Whenever any corrective change is made in field construction which is at variance with the specifications and drawings as originally issued, a complete detailed report shall be filed, listing the following items, so that specifications or drawings storage data can be corrected.

1. Identify the problem: Indicate why originally specified construction was not used.
2. The Solution: Describe, in detail, the recommended change or changes that were made, as applicable.
3. Indicate whether this is an isolated case or a general condition which could be improved by changing future specifications or drawings.
4. Submit sketches as necessary.

REFERENCE DATA

SPECIFICATION SECTION No. 3B PAGE No. 2 PARAGRAPH No. 2.01

DRAWING No. 5401 ENTITLED RESERVOIR BASIN STRUCTURE

SKETCH No. ____ DATED ____ ENTITLED ____

DESCRIPTION

1. DETAILED IDENTIFICATION OF THE PROBLEM BLEEDING OF CEMENT PASTE ALONG ALL VERTICAL JOINTS OF STEEL PANEL WALL FORMS. ALL VERTICAL JOINTS IN STEEL WALL SURFACES SHOW EXTENSIVE EVIDENCE OF EXPOSED AGGREGATE AT ALL VERTICAL WALL JOINTS. EXAMINATION OF STEEL PANEL FORMS SHOWS TYPICAL 1mm GAPS AT ALL VERTICAL JOINTS, BUT TIGHT JOINTS AT ALL HORIZONTAL JOINTS. FASTENING DOGS SEEM UNABLE TO DRAW FORM JOINTS TIGHTY TOGETHER TO ELIMINATE LOSS OF CEMENT PASTE

2. DETAILED SOLUTION PROPOSED OR ACCOMPLISHED PREASSEMBLE ALL WALL PANEL UNITS WHILE CLEAN. REPLACE FASTENING DOGS WITH BOLTED JOINTS AND DRAW TIGHT BEFORE FIRST FORM USE. CALK ANY DEFECTS IN JOINTS. NO JOINTS SHOULD BE TIGHTENED AFTER USE WITHOUT DISASSEMBLING FORM PANELS AND THOROUGHLY CLEANING ALL FORM PANEL EDGES. TAPE ALL JOINTS WHICH CANNOT BE DRAWN TIGHT. ALTERNATIVE: USE PLYSCORD FORMS INSTEAD, OR REQUIRE USE OF FULL-FACE FORMS, IF STEEL FORMS ARE TO BE USED.

Data used are fictitious for illustration only

3. IS THE PROBLEM AN ISOLATED CASE OR GENERAL? GENERAL SPECIFICATIONS SHOULD BE REVISED TO ALLOW ENGR. CONTROL OVER FORMS

4. SUBMIT SKETCHES AS NECESSARY

(Attach extra sheets as necessary)

DISTRIBUTION 1. Design/Spec Depts. 2. Proj Mgr. 3. Field Office 4. File

BY [signature] TITLE RES. ENGR.

Fisk Form 8-4

FIGURE 4.9 Example Showing Proper Use of the Report of Field Correction Form. (*Fisk, Edward R.,* Construction Engineer's Complete Handbook of Forms, *1st Edition, © 1993. Reprinted by permission of Pearson Education, Inc., Upper Saddle River, NJ.)*

changes within the terms of the contract must be observed. The document should do the following:

1. Identify the problem. Indicate why originally specified construction is not recommended.
2. Offer a solution. Describe in detail the recommended change or changes that are suggested.
3. Indicate whether the case appears to be an isolated one or whether it appears to be a general condition that could be improved by changing specifications or drawings.

Whether it will facilitate an understanding of the problem or its solution, the Resident Project Representative is encouraged to submit sketches along with the Report of Field Correction.

At this point, the Resident Project Representative should be cautioned. *The foregoing instructions are not meant to imply that the inspector is to take any corrective action that will result in a variation from the plans and specifications without the approval of the architect or engineer of record.* By definition the architect or engineer of record is that individual whose signature appears on the plans as evidence that he or she, either personally or as a representative of a design firm, public agency, or owner, bears legal responsibility for such plans. The authority of anyone to take field action without consulting the architect or engineer of record is limited to cases when such action would *not* result in a variance from the approved plans and specifications. Otherwise, all such actions must be preapproved by the architect or engineer of record. A possible exception might be during emergency conditions, wherein a field decision must be made immediately. Even then, it is wise to telephone ahead to describe the condition and the solution recommended, followed by a written fax report to the architect/engineer and execution of a formal change order. In any case, if forced into a decision-making role, the inspector should inform the contractor that an inspector is not authorized to make such a determination, but that the inspector will not prevent the contractor from taking unilateral emergency action based upon the contractor's own judgment, provided that the contractor fully understands that such actions are subject to confirmation and approval by the architect/engineer; and furthermore, that in case of disapproval, the contractor may be required to take corrective action to remove portions of the work affected by the emergency at the contractor's own expense.

Concrete Batch Plant Daily Reports

Whenever critical concrete batching control is necessary, such as is often required for the construction of hydraulic structures, a separate inspector may be assigned to the concrete batch plant during all concreting operations. The data recorded by the batch plant inspector are forwarded to the Resident Project Representative and the Project Manager daily, and should be compared at the job site with the corresponding delivery ticket data. All such information may be conveniently recorded on a prepared form such as illustrated in Figure 4.10. This information should be carefully recorded and stored with the other permanent project records, as it may be utilized later to determine probable causes of field problems with concrete.

LAWRANCE, FISK & McFARLAND, INC.
CONSULTING CIVIL & ENVIRONMENTAL ENGINEERS
Santa Barbara, California

CONCRETE BATCHING PLANT DAILY REPORT

PROJECT BEAR CANYON HE FEATURE FISH BARRIER DAM SHIFT DAY DATE 4-24-95

CONTRACTOR ABC CONSTRUCTORS INC. PROJ. NO. 62-02 INSPECTOR PRS

AGGREGATE SOURCE: COARSE SAN GABRIEL RIVER FINE SAN GABRIEL RIVER

CEMENT BRAND RIVERSIDE TYPE II LOT NO. BIN-7 FALSE SET NO

NAME OF: AEA PROTEX WRA KRL-40 CACL₂ — POZZOLAN AIROX

TEMP: AIR 24°-39° WATER (ICE) CEMENT 49° SAND 28° GRAVEL 29° CONCRETE 14°

MOISTURE ADJUSTMENTS FOR CORRECT BATCH WEIGHTS

	MIX NO. 1020				MIX NO.				MIX NO.			
	Basic Batch M. S.S.D. kg	Probable Free Water %	Free Water kg	Actual Batch M. kg	Basic Batch M. S.S.D. kg	Probable Free Water %	Free Water kg	Actual Batch M. kg	Basic Batch M. S.S.D. kg	Probable Free Water %	Free Water kg	Actual Batch M. kg
AEA	0.06	—	—	0.06								
WRA	1.5	—	—	1.5								
POZZOLAN	64	—	—	64								
CEMENT	213	—	—	213								
SAND	825	+4.0	+33	858								
COARSE AGG. 19–4.75 mm	690	+0.5	+3.6	694								
38.1–19.0 mm	769	0	0	747								
4.75–38.1 mm	1193	-0.1	-1.4	1192								
WATER ICE	172	—	35	137								
WATER (LITERS)												

Data used are fictitious for illustration only

TEST BATCH DATA

Mix No.	Time	Test Cyl Nos.	Slump mm	Conc. Temp °C	Unit M. kg/m³	% Air (Meter)	Actual Mass SSD, kg: Sand	C. Agg. 19.0	C. Agg. 38.1	C. Agg. 100		Total Water kg	Net W/(C+P) (by Wt)	Yield m³/Batch	Cement/m³ kg
1020	9:20	FX-5	38	14	2629	2.4	416	346	388	600		86	0.28	1.52	2.0

CONCRETE PRODUCTION DATA

Spec. Item No.	Mix No.	Total No. Batches	Per Batch: Cement kg	Per Batch: Pozz. kg	Per Batch: Vol. Batch m³	Batched: Total m³ Conc.	Batched: Cement kg	Batched: Pozz. t	Waste: Conc. m³	Waste: Cement kg	Waste: Pozz. t	Used: Conc. m³	Used: Cement kg	Used: Pozz. t
9	1020	64	213	64	1.5				1.5					
Total														

WATER BATCHER CHECKED

AEA DISPENSER CHECKED 07:45 HRS; 10:35 HRS 13:00 HRS

WRA DISPENSER CHECKED 07:45 HRS

Over for Screen Analyses and General Remarks

FIGURE 4.10 Example of a Daily Report from the Concrete Batch Plant Inspector. *(Fisk, Edward R Construction Engineer's Complete Handbook of Forms, 1st Edition, © 1993. Reprinted by permission of Pearson Education, Inc., Upper Saddle River, NJ.)* (continued)

SCREEN ANALYSIS OF AGGREGATES - PERCENT RETAINED EACH SIZE

SCREEN SIZE	150 mm -75 mm			75 mm - 38.1 mm			38.1 mm - 19.0 mm			19.0 mm - 4.75 mm		
	M + Tare	Mass	% Each	Wt + Tare	Mass	% Each	M + Tare	Mass	% Each	M + Tare	Mass	% Each
177 mm												
152 mm												
125 mm												
90 mm				0	0	0						
75 mm				9.3	6.3	4						
63 mm				66.3	63.3	41						
45 mm				57.0	54.0	35	3.4	0.4	1			
38.1 mm				32.3	29.3	19	6.1	3.1	6			
31.5 mm				4.5	1.5	1	21.1	18.1	36			
22.4 mm							23.7	20.7	40	0	0	0
19.0 mm							11.3	8.3	16	5.2	2.2	9
16.0 mm							4.0	1.0	2	11.7	8.7	36
9.5 mm										11.5	8.5	35
4.25 mm										7.8	4.8	20
4.00 mm										3.1	0.1	0
pan				0	0	0	3.2	0.2	0	0	0	0
TOTAL			100	—	154.4	100	—	51.8	100	—	24.3	100

SCREEN SIZE	SAND											
	M + Tare	Mass	% Each	Cum. % Ret.	M + Tare	Mass	% Each	Cum. % Ret.	M + Tare	Mass	% Each	Cum. % Ret.
4.75 mm	191	6	3	3								
2.36 mm	215	24	11	14								
1.18 mm	241	26	12	26								
600 μm	287	46	21	47								
300 μm	356	69	32	79								
150 μm	395	39	18	97								
75 μm	400	5	2									
pan	402	2	1									
TOTAL	215	217	100				100				100	
F.M.												

Data used are fictitious for illustration only

Items needing attention ____________

Items of special interest ____________

General Comment ____________

Inspector ____________

Fisk Form 16-1

FIGURE 4.10 *(continued)*

Plant Inspector's Report to Field Inspector

The plant inspector should maintain regular communication with the on-site inspector, who should be informed of any special circumstances or messages in connection with the current placing operations. Ideally, this should be in a report format (Figure 4.11) and is an intermittently prepared memorandum designed for easy exchange of information between the plant inspector and the field inspector or Resident Project Representative. If a form similar to the one illustrated is used, an easy means of recording two-way communications is provided. The information should also be retained in the permanent project files.

Delivery of the memorandum is normally accomplished by sending it to the project site with one of the drivers leaving the batching plant. In this way it is even possible to send commentary that concerns the concrete load that the message was actually delivered with so that no time is lost in communications.

Field Investigation Report

On numerous occasions in construction, situations arise that cannot be resolved until more facts are known. Under such conditions, an investigation should be made by the Project Manager or Resident Project Representative, and the findings recorded. This information is then available for use in efforts to determine future courses of action where they are justified.

Documentation of Dangerous Safety Hazard Warnings

In Chapter 9, the procedure for the handling of serious contractor safety violations is described. In addition to the action described there, certain additional precautions should be taken to document the action taken by the owner's field personnel and the contractor. This is vitally important for the Resident Project Representative because failure to do so could result in serious charges being unfairly lodged against the design organization or owner for failure to take affirmative action in case of a death or serious injury resulting from the hazard.

It is recommended that in each case involving an "imminent hazard," the resident inspector take the following steps *after seeing that persons in the immediate area of the hazard are removed from danger:*

1. Notify the contractor's superintendent or foreman.
2. Issue written notice to the contractor to take immediate action to correct the hazard, and record this action in the inspector's diary, including the exact time of day that the notice was given. Also, inform the contractor that unless immediate action is taken to correct or remove the hazard, the matter will be immediately referred to the OSHA compliance officer serving that area.

PLANT INSPECTOR'S REPORT TO FIELD INSPECTOR

PLANT NAME: UNITED ROCK & SAND

PLANT LOCATION: ORANGE, CA

INSTRUCTIONS TO PLANT INSPECTOR: Fill out the upper half of this form, attach to load ticket, and send to Field Inspector with first load.

FOR CONCRETE MIXTURES:

CLASS OF CONCRETE		SOURCE OF COARSE AGGREGATE
A-3500		SAN GABRIEL
BRAND OF CEMENT	**TYPE**	**SOURCE OF FINE AGGREGATE**
RIVERSIDE	I/II	SAN GABRIEL
BRAND OF ADMIXTURE		
NONE		

FOR BITUMINOUS MIXTURES

TYPE OF MIX	GRADING
AR GRADE OF ASPHALT	**SOURCE OF BITUMEN**

Data used are fictitious for illustration only

PLANT INSPECTOR: M. COLLINS — DATE: 6-07-94

REMARKS: HELD 60L WATER FROM EA. 5m³ BATCH — WILL YOU BE RUNNING TOMORROW?

PROJECT TITLE: RESERVOIR 1-D

LOCATION: FULLERTON, CA

INSTRUCTIONS TO FIELD INSPECTOR: Fill out this lower portion and return to the plant inspector immediately.

PROJECT NO.: 95-90-00 — SPECIFICATIONS OR CONTRACT NO.: F 3057-556

(CONCRETE)(~~ASPHALT~~) USED FOR: VALVE VAULT STRUCTURE

GALLONS OF WATER ADDED ON THE JOB: 45L/5m³ LOAD

FIELD INSPECTOR: R.E. BARNES — DATE: 6-07-94

NOTES: HAD ALL DRIVERS RUN AT MIXING SPEED FOR 3 MIN. AFTER ARRIVAL AT SITE TO ASSURE FULL MIXING TIME AFTER RECEIVING TIP FROM ONE OF THE DRIVERS THAT MOST LOADS ARE FORCED TO LEAVE PLANT WITH INADEQUATE MIXING TIME.
BE SURE TO CHECK DRUMS BEFORE BATCHING TO BE SURE THAT ALL WASHDOWN WATER HAS BEEN REMOVED.
WE EXPECT TO PLACE 150m³ TOMORROW

Fisk Form 16-2

FIGURE 4.11 Plant Inspector's Report to Field Inspector. (*Fisk, Edward R.,* Construction Engineer's Complete Handbook of Forms, *1st Edition, © 1993. Reprinted by permission of Pearson Education, Inc., Upper Saddle River, NJ.*)

3. Upon failure or refusal of the contractor to take immediate steps to correct or remove an "imminent hazard," note the exact time of day, and telephone the OSHA compliance officer and make a full oral report. After completing the call, enter into the diary or log that the contractor either failed or refused to effect immediate correction of the hazard; describe all steps taken to alleviate the hazard, including orders given to remove personnel from the danger area; record all field orders (written and oral) given to the contractor; record the exact time of the day that (1) persons were ordered out of the danger area, (2) a correction order was issued to the contractor, and (3) the OSHA compliance officer was notified.
4. Upon completion of the foregoing, write a full report to the design firm or owner, including a summary of all pertinent data recorded in the construction diary.
5. Upon mailing the field report to the design firm or owner, telephone the firm or owner to advise of the forthcoming report, describe the incident briefly, indicate the action taken, and record the call in the construction diary.

MISCELLANEOUS RECORDS

There are numerous types of individual records that are important to log and retain for future reference. Many of the records that must be maintained are primarily of a technical, not an administrative nature. Thus, no detailed coverage will be attempted here. However, as a reminder, the following will serve as a partial list of some of the many technical records that must be maintained on a job, as applicable (once compiled, the handling of these records becomes an administrative matter):

1. Manufacturers' certificates for products
2. Laboratory test certificates (Figure 4.3)
3. Concrete transit-mix delivery tickets (Figure 4.12)
4. Records of pile driving (Figure 4.13)
5. Record of inspection of structural welding
6. Sewer infiltration test reports
7. Fabricating plant inspection reports
8. Special inspector reports
9. Weld radiographs
10. Acceptance certificates by public agency inspectors
11. Discrepancy reports
12. Deviation requests and action taken
13. Concrete mix designs

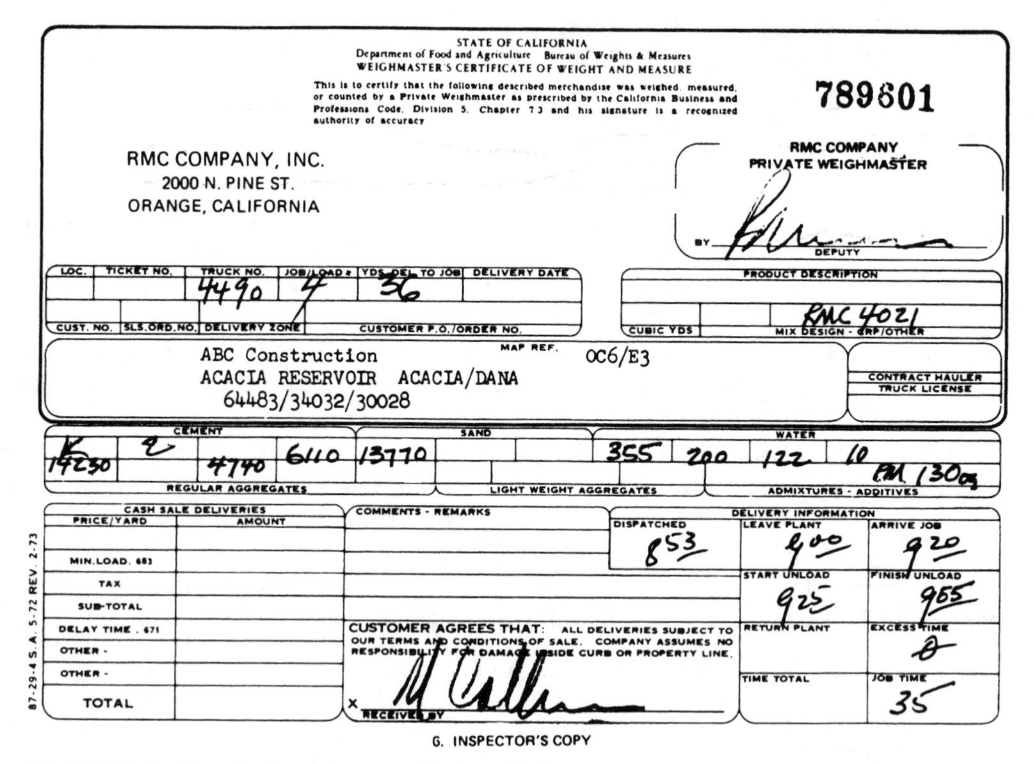

STATE OF CALIFORNIA
Department of Food and Agriculture Bureau of Weights & Measures
WEIGHMASTER'S CERTIFICATE OF WEIGHT AND MEASURE
This is to certify that the following described merchandise was weighed, measured, or counted by a Private Weighmaster as prescribed by the California Business and Professions Code, Division 5, Chapter 7.3 and his signature is a recognized authority of accuracy

789601

RMC COMPANY, INC.
2000 N. PINE ST.
ORANGE, CALIFORNIA

RMC COMPANY
PRIVATE WEIGHMASTER
BY ______ DEPUTY

LOC.	TICKET NO.	TRUCK NO.	JOB/LOAD #	YDS. DEL. TO JOB	DELIVERY DATE
		4490	4	36	

CUST. NO. | SLS.ORD.NO. | DELIVERY ZONE | CUSTOMER P.O./ORDER NO.

PRODUCT DESCRIPTION
CUBIC YDS | MIX DESIGN - GRP/OTHER: RMC 4021

ABC Construction
ACACIA RESERVOIR ACACIA/DANA
64483/34032/30028
MAP REF. OC6/E3

CONTRACT HAULER
TRUCK LICENSE

CEMENT | SAND | WATER
K 14250 | 2 | 4740 | 6110 | 13770 | | | 355 | 200 | 122 | 10 | EM 130oz
REGULAR AGGREGATES | LIGHT WEIGHT AGGREGATES | ADMIXTURES - ADDITIVES

CASH SALE DELIVERIES	
PRICE/YARD	AMOUNT
MIN.LOAD. 683	
TAX	
SUB-TOTAL	
DELAY TIME . 671	
OTHER -	
OTHER -	
TOTAL	

COMMENTS - REMARKS

CUSTOMER AGREES THAT: ALL DELIVERIES SUBJECT TO OUR TERMS AND CONDITIONS OF SALE. COMPANY ASSUMES NO RESPONSIBILITY FOR DAMAGE INSIDE CURB OR PROPERTY LINE.
X ______ RECEIVED BY

DELIVERY INFORMATION		
DISPATCHED 853	LEAVE PLANT 900	ARRIVE JOB 920
	START UNLOAD 925	FINISH UNLOAD 955
	RETURN PLANT	EXCESS TIME 0
	TIME TOTAL	JOB TIME 35

87-29-4 S.A. 5-72 REV. 2-73

6. INSPECTOR'S COPY

FIGURE 4.12 Transit-Mixed Concrete Delivery Ticket.

LABOR STANDARDS REVIEW RECORDS

On contracts where any federal funds are to be used, the Davis-Bacon Act applies. There are two types of situations where this will occur: on contracts with local cities, counties, or special districts where the projects are being constructed with federal grant money, such as the EPA Clean Water Grant Program, EDA, HUD; or on contracts directly with a federal agency, such as military construction, the Postal Service, or GSA projects.

The project administrator has additional responsibilities on all such work, and checks with the contractor must be made for the following items:

1. On military projects, employee interviews must be conducted to determine classifications and rates of pay, including fringe benefits. All such data must be recorded on "Labor Standards Interview," Form No. DD-1567.
2. On federal grant programs, certified payrolls must be obtained from the contractor on Labor Department Form WH-347 (or similar forms), and the Resident Project Representative must check them for compliance with wage rates in the specifications and report any errors to the contractor.
3. On-site checks must be made of the type and classification of all work performed at the site and the number of workers in each category. Payroll reviews of prime and subcontractors' payroll submittals should be made.

City of Thousand Oaks

PILE DRIVING RECORD

1 Project Title: CHLORINATION FACILITY	2 Structure: ADMIN BLDG/LABS			Date: 4-17-90
3 Pile Material and Type: STEEL "H"	6 Hammer (Make and Size): McKIERNAN-TERRY DIESEL DE-20		7 Contract No.: 89/90-00	Inspector Signature: O.R. Bailey
4. Minimum Penetration: 50 FT	6a. Blows per minute: 48-52	6b. Stroke: 6-FT	8. Reference Dwg.: C-314122	Shift: DAY
5 Resistance Required: 40 BLOWS/FT THRU LAST 4 FT	6c. Weight of ram: 2000 #	6d. Rated Striking Energy: 16,000 FT LB	9. Datum Elevation: +10' ABOVE MLW	SHEET NO. 1 OF 3

PILE DATA				ELEVATIONS			PENETRATION (BLOWS/F.T.) Last 4 Feet					
COL. OR BENT LINE	ROW	SIZE (BUTT TO TIP)	ORIGINAL LENGTH	GROUND	TIP	CUT-OFF	2" to 3"	3" to 4"	4" to 1'	1' to Tip	DATE DRIVEN	REMARKS
A	B	C	D	E	F	G	H	I	J	K	L	M
C3	3	14x14	40'	283	248	4'					4-17-90	
C4	4	14x14	40'	283	247	3'					4-17-90	
C5	5	14x14	40'	283	253	9'					4-17-90	TO REFUSAL

Wiley-Fisk Form 12-4

FIGURE 4.13 Pile Driving Record.

4. On military construction, the previously described information with all available data should be recorded on the daily Report to Inspector and the CQC report to assure consistency.
5. On grant-funded projects, the agency administering the grant program will make periodic inspections to check for compliance with labor laws. This inspection includes checking payrolls and interviewing employees to ascertain working hours and actual payment received.
6. Checks must be made that all equal employment opportunity laws are being followed.

Normally, tasks 1, 2, and 3 are the responsibility of the on-site project representative (CQC representative or Resident Project Representative), whereas tasks 4, 5, and 6 are normally performed by the agency administering the grant program, or in the case of direct federal contracts, by the federal agency construction office staff.

Job Conferences

Often overlooked, but no less important, is the task of recording the proceedings of job conferences. This includes management meetings, safety meetings, coordination and scheduling meetings, and similar functions. The Resident Project Representative should attend all such functions, fully prepared to document for personal information the business transacted at each such meeting. The notes should include the time, date, and location of the meeting, the name and employer of each person in attendance, and the time the meeting ended. During the meeting, careful notes should be taken to have a complete record of the substance of all important statements made at the meeting. Where statements are made by more than one person, the identities of all speakers should be listed.

Contractor Submittals

The normal recommended procedure for the handling of submittals from the contractor is to require that *all* such submittals (e.g., shop drawings, samples, certificates, or other similar items) be submitted directly to the Resident Project Representative by the general contractor (see Figure 3.6). Thus, if a submittal is mailed directly to the design firm or owner, it should be returned unopened, with the request that it be transmitted through the Resident Project Representative's field office. Similarly, if a submittal is made to the Resident Project Representative directly from a subcontractor, it should be returned to the subcontractor unopened, with the request that it be transmitted through the general contractor. Such practices are not created out of a love of red tape, but rather to make use of a proven system that can prevent disputes or even field errors resulting from lack of communication between the parties to the contract (see "Handling Job-Related Information" in Chapter 3).

Upon receipt by the Resident Project Representative, each such submittal should be logged in before forwarding to the office of the design firm or owner. All such material should be forwarded to the design firm or owner without action by the Resident Project Representative, unless specifically directed to do so in any particular case. As far as contractor submittals are concerned, the Resident Project Representative simply serves as a central receiving point and recorder for such information. As long as all submittals follow this orderly procedure, there should be no excuse for loss or delay in a transmittal. If all such material is properly logged in and out, claims by the contractor of excessive holding time by the design firm or owner, or claims of submittals that were in fact never made, can be quickly and effectively confirmed at the site.

There is often some confusion during the setting up of a contractor submittal log sheet to decide whether to assign transmittal numbers on the basis of the order *received* or on the order *returned* to the contractor. Most offices seem to find that less confusion results from assigning a transmittal number immediately upon *receipt* of a submittal. In the illustrated submittal log (Figure 4.4), spaces are provided for indicating the action taken on each submittal as well. Although this may seem to be an unnecessary bit of added work, as such information can also be found by searching for a copy of the transmittal itself, the time saved from making searches of the files and wading through the many drawing sheets that may form a part of each such transmittal may soon make it all seem worthwhile. In the submittal log, it is important that the "action taken" wording conform *exactly* to the term used on the submittal being returned to the contractor.

It is generally undesirable to attempt to utilize the contractor's submittal numbers as a "convenience"; often the contractor may submit several dissimilar items all grouped under a single transmittal, with a single reference number. For efficient processing by the engineer/architect, it is desirable to separate each such submittal item and assign each item a separate incoming transmittal number. In this manner, if only a portion of the contractor's submittal is subject to delay, it will not negatively affect the other items that were grouped into the single transmittal. This will avoid the confusion that might result if an attempt was made to describe the various actions taken by the engineer/architect relative to each item submitted when some of the items are ready to return and others must be held for further evaluation.

Until the advent of electronic project management tools, contractor submittals was largely a manual bookkeeping job, with documentation such as that illustrated in Figure 4.4. However, with the advent of computerized project management systems, instead of a series of static records, it has become interactive as well. The computer can be used to track the submission and routing of contractor submittals and samples.

The project team just sets up a routing track for each type of submittal that is used as the work flow for the approval process for submittals. Involved parties can be notified of the submittals currently in progress. Users can begin with a master listing of expected submittals and samples, and use it to develop the specific list for the project. Scanned or photographic images can be used as attachments to process the submittals electronically.

Construction Photographs

More and more reliance is being placed upon the use of photography to document construction progress, damage, technical detail, types of materials, methods of installation, evidence of site conditions before starting work, and similar tasks. Basically, it is the Resident Project Representative who will probably be called upon to do the photography. Thus, an understanding of the types of photographs normally used in construction as well as the purpose of each is beneficial.

There are four basic types of construction photography that the Resident Project Representative may be called upon to provide: public relations photography, progress photography, documentation of safety hazards, and claims support.

Public relations photography and progress photography, including time-lapse applications, are described here in more detail.

1. *Public relations photography.* Subject matter as well as the composition of the picture and the lighting are selected on the basis of artistic composition. The photograph is intended to appeal to the layperson and show what

FIGURE 4.14 Typical Public Relations Photograph of Construction. Greystone Reservoir, City of Beverly Hills, California. Engineers: Montgomery-Watson Consulting Engineers, Pasadena, California. *(Photo courtesy of MWH Global, Broomfield, CO; photo by J. Allen Hawkins, Pasadena, CA.)*

an impressive structure is being built. An example is a view inside a reservoir under construction: The photographer will use an extra-wide-angle lens, accent the perspective, and may even use filters and special lighting to create striking lighting effects. No technical details can normally be noted, but it does produce a beautiful photograph (Figure 4.14). In Figure 4.15, the positioning of the aircraft in front of the hangar was for visual effect. The actual project for which the author was the Construction Manager appears partially behind the aircraft. The total project consisted of the hangar and repair shops, tow-way, fire pump station, underground piping, the mating device, fuel and oxidizer facilities, and an office building. However, this approach is typical of public relations photographs, as all that is necessary is to suggest the project concept, not to show all of its components or details.

2. *Progress photography.* Selection of the subject matter is based upon the need to show as much detail of the construction as possible. Photo composition is secondary, as the primary intent is to disclose the quantity and kind of work that was completed since the last progress photographs were taken. In addition, some photos are intended simply to document kinds of materials used and the method of installation. Thus, the photographer may sacrifice beauty and composition in the photograph to assure legibility of a material label or an identifying mark in a finished photograph. It is considered undesirable to strive for striking lighting effects in progress photographs as they all too often obscure details in shadow areas of the picture (Figure 4.16).

FIGURE 4.15 Public Relations Photograph of the NASA ALT Facilities at the Hugh L. Dryden Flight Research Center, Edwards AFB, California. (*Photo courtesy of NASA.*)

3. *Time-lapse photography.* The subject of job progress photography would not be complete without some mention of "time-lapse" or interval photography. Simply stated, it is just a means of using automatic equipment to take photographs at regular intervals from the same point each time. This is normally accomplished with professional equipment, using an electric motor-driven camera connected to an electric interval timer or "intervalometer." Such equipment can be used with either still cameras or movie cameras, and by setting the delay interval between subsequent pictures to anywhere from seconds to days, a unique sequence of construction events can be recorded to show the exact nature and amount of construction work completed as of any given day or other interval. When adapted to a movie camera, this technique can also be used to simulate a speeded-up construction operation. Time-lapse photography can also be accomplished manually, using inexpensive equipment, provided that a camera can be fixed on a tripod and allowed to remain in this position during the entire sequence time. The camera can be operated manually, but must be operated faithfully on a predetermined schedule each day (Figure 4.17).

Identification of Photographs

Although the data required for a public relations photograph may be minimal, the progress photograph must be considered as a reference source or even as potential

FIGURE 4.16 Typical Construction Progress Photograph.

evidence in case of later claims or disputes. To be of greatest value, certain information should be recorded on the back of each progress photograph taken:

Identification of Photographic Prints

1. Date photo was taken
2. Identity of the subject
3. Photo or negative number (for reorder)

If it is known that the specific photographs being taken are going to be used in evidence as claims-defense photographs, the list should be continued to indicate the following additional data that should be recorded on each print:

Claims-Defense Photographs: Additional Data

4. Time of day, name of photographer
5. Direction in which camera was pointed
6. Where photographer was standing

FIGURE 4.17 Long-Interval Time-Lapse Progress Photograph of the Harry Cochrane Hydroelectric Project for the Montana Power Company.

An essential complement to accurate identification of individual photographs is a construction log of all photographs taken. The identity of all photographs should be recorded in sequence in a photo log as soon as they are exposed. The effectiveness of a photographic record can be diminished as a defense against claims if contemporaneous records are not kept to graphically validate the appearance of any portion of a project at any given day. Data backs on cameras can go a long way toward verifying the time and date that a particular photograph was exposed, but data back information may not be conclusive if allegations of tampering with the data back are made. A contemporaneous photographic record or "construction photograph log" (Figure 4.18) can be valuable as a supplement to the diary if kept as a hard-copy record.

Photographs as a Defense against Claims

A camera is an important and often vital tool of all field personnel, whether in the employ of the design firm, the owner, or the contractor. It is often the only means of establishing a condition of fact at any given occasion prior to its being permanently covered. The camera could well be the tool that might save the Resident Project Representative, the design firm, the owner, or the contractor from charges that might be based upon one person's word against another's had it not been for a photographic record.

CONSTRUCTION PHOTOGRAPH LOG

Project PYRAMID POWER PLANT Job No. 2400.0000

Roll No. 12	By FISK	Date Printed 5/14/99	Roll No. 12	By FISK	Date Printed 5/14/99
No.	Date	Subject and Location	No.	Date	Subject and Location
1	5/3 1999	PYRAMID POWER PLANT OVERVIEW FROM SOUTH EMBANKMENT	13	5/5	PIPE DELIVERY & STORAGE AREA ADJ. TO OLD HWY 99
2	5/4	INSTALLATION OF TURBINE PIT LINERS LOOKING WEST AT UNIT #2	14	5/5	PIPEMOBILE DISCHARGING LOAD 1/4 MILE SOUTH OF FIELD OFFICE #2
3	5/4	EARTH MOVING IN OUTLET CHANNEL FROM P.H. LOOKING SOUTH	15	5/6	PIPE ZONE BACKFILL CONSOLIDATION TEST TEST PIT ADJACENT TO TUNNEL
4	5/4	PENSTOCK WELDING - UNIT No. 2	16	5/6	CHANNEL PAVING W/BIDWELL PAVER NORTH OF TEMPORARY DIVERSION
5	5/4	SHIELD TUNNELLING UNDER I-5 LOOKING WEST			
6	5/4	QUAIL LAKE OUTLET WORKS LOOKING SOUTH AT GATES			
7	5/4	LOOKING NORTH FROM CHANNEL QUAIL LAKE OUTLET WORKS			
8	5/4	QUAIL LAKE OUTLET GATE HOISTS			
9	5/4	LPG TANK - PAINT DAMAGE QUAIL LAKE OUTLET WORKS			Data used are fictitious for illustration only
10	5/4	STAND-BY ENGINE GENERATOR QUAIL LAKE OUTLET WORKS			
11	5/5	PLACING CONCRETE AT POWERHOUSE EAST WALL			
12	5/5	CONCRETE BATCH PLANT NORTH OF POWER PLANT			

Fisk Form 8-12

FIGURE 4.18 Example of a Construction Photograph Log Showing Roll Number, Date, and Description of Each Photograph and the Identity of the Photographer.

A good example of the use of photography to prevent a potential claim is illustrated in Figure 4.19. A 150-ton turbine-generator shaft, manufactured by Hitachi in Japan, was destined for installation at the Brownlee hydroelectric plant in Hells Canyon, Idaho. The shaft was shipped by sea to Oakland, California, where it was placed on the railcar shown in Figure 4.19 and transported to the end of the rail line in Cambridge, Idaho. From there it was to be transported by the contractor to the project site. However, when the shipment arrived at the end of the rail line, it was evident by the condition of the load that something took place between Oakland and Cambridge that caused the 150-ton load to shift. This was evidenced by observing the slope angle of the tie-down cables and the dip observed in the lagging from one of the tie-down cables. Also, it could be noted that the steel supporting cradles had broken loose.

At the time of arrival of the shipment, only an insurance agent and the author were present. Both of us took photographs of the load before anyone was permitted to move or unload it. These proved to be invaluable, as at first the railroad denied that anything had happened to the load when the incident was reported to them. Upon seeing the photographs, however, it was evident that some great force moved the load on the railcar while in transit. No amount of denial could dispute the photographic evidence.

FIGURE 4.19 Claims-Defense Photo of 150-Ton Turbine-Generator Shaft as It Arrived at the End of the Rail Line Prior to Release to Contractor for Transport to Project Site.

After documenting the condition of the load, the contractor was notified to pick up the load, and it was transported and set up on the powerhouse erection deck where a shelter was constructed around the shaft to protect it while instruments were used to determine if there had been any damage to the shaft as a result of the load shift on the railcar. Fortunately, there was no damage and the issue was closed. The photographs, however, served their purpose by providing a potential defense for the contractor.

In another example, some pipeline contractors who regularly construct large underground pipelines beneath city streets in residential and business areas send a photographic crew to the site before the start of any work to photograph all of the curbs, sidewalks, and frontage of every residence and business place along the pipeline route. Every foot of frontage is photographed as a permanent record of the condition that existed prior to the beginning of construction as evidence against frivolous damage claims filed after work is completed. It is remarkable that so many honest people can have cracks in their sidewalks or curbs and never notice them until a contractor starts tearing up the street in front of the house. For some reason, the homeowner is invariably convinced that all such cracks or other damage were caused by the new construction. The Resident Project Representative or the CQC representative of the contractor should be able to find a lesson in this example that will indicate a means of protection for the design firm, the owner, or the contractor from the hazards of frivolous claims.

Challenges of Digital Images

In rare cases, the photographic image produced by a digital camera may be subject to challenge, as the attributes that make a digital camera and its resulting images so versatile may be the same features that make such images subject to challenge in case of dispute. Electronic media storage lends itself to easy editorial changes or data corruption, either intentional or unintentional. Film cameras still have the edge in such cases.

Photographic Equipment and Materials

Types of Equipment Used

Where field personnel are expected to take the progress photographs, they are generally asked to use a digital camera or a 35-mm camera and possibly a Polaroid camera. Usually, use of a Polaroid film or Polaroid digital printing camera is recommended for photographing field problem areas so that they may immediately be available to discuss with the architect/engineer. Some firms will allow the use of a Polaroid camera for progress photography, but its disadvantages may outweigh its advantages. There is no negative possible with the older color-pack Polaroid. Although the older Polaroid offers a P/N (positive/negative) film for some models, it is not commonly available for most popular Polaroid cameras; thus, copying of an original Polaroid film print photograph must be done by rephotographing the print itself—seldom an ideal process. The digital Polaroid avoids this problem. If the architect/engineer wishes to retain copies of the prints and forward other print copies to the owner, both the versatility of a negative film plus the higher quality of the photographs taken with the 35-mm film camera or a high-resolution digital camera (3- to 4-megapixels) will pay off.

In recent years, new developments have been made in digital imaging that include digital camera models in the megapixel range, thus making them competitive with 35-mm cameras for print resolution. Although digital cameras in the 3- to 4-megapixel range are still comparatively pricey, they do offer the capability of instant playback of still photographs that can be viewed or reproduced on any PC or MAC computer (see "Digital Imaging" later in this chapter). Although the concept has been around for some years, recently the prices of consumer models of digital cameras have dropped to within reach; however, the resolution of the digital cameras in the lower price range still leaves something to be desired. Higher-priced models are capable of delivering satisfactory image sharpness, but still are a bit pricey for use on a construction job. By now all major camera manufacturers offer digital cameras, and as soon as the digital technology achieves an image resolution comparable to that achievable with film cameras, within a reasonable price and convenience, I see this as the future mainstay of photography.

Camera Handling

Although the subject of photographic technique is somewhat beyond the scope of this book, some helpful hints are offered that may assist the construction inspector in resolving some of the problems of recording his or her project progress on film.

Exposure times are important, as an error in estimating the light conditions can lead to the loss of a picture, and in construction you will seldom get a second chance to repeat the scene. Because of this, it is recommended that field personnel have either an automatic exposure camera or at least one with a built-in light meter to assure proper shutter speed and lens aperture. This is especially important in the case of color films, as an error of as little as one shutter speed or one f-stop can sometimes result in an unsuccessful photograph.

Selection of Still Camera Equipment

The most commonly used cameras for construction are still cameras. Video cameras, while quite useful in construction, are not that widely used, nor are they a valid substitute for the permanent record provided by still-camera photographic prints.

In the choice of still (nonvideo) cameras, do not be misled by cameras with extremely "fast" lenses, because the measure of the quality of a camera, or of a lens for that matter, is not in the speed of the lens but in its quality. Although the added lens speed may allow the taking of pictures under extremely low light conditions, it is not worthwhile if the results are not sharp and clear, or if the sharpness falls off at the edges.

A camera with an f3.5 lens is quite adequate for most field uses and can generally be obtained at a reasonable cost. Also, as the ability of the average person to estimate distances accurately is somewhat less than that person is usually willing to admit, it is essential that the camera selected either be an autofocus camera or have a range finder that allows the Resident Project Representative or inspector to focus the camera accurately.

If the Resident Project Representative or inspector plans to select a digital or 35-mm film camera for work progress photography, it may be desirable to obtain one that allows the use of interchangeable or zoom lenses, if the somewhat added cost of this feature is not objectionable. Many of the subjects that an inspector must photograph will involve the inclusion of wide viewing areas, which if a normal 50-mm lens is used (standard on 35-mm cameras) would require backing up too far, or there may be insufficient space to back up any farther. A wide-angle lens of 28-mm or 35-mm is better for such conditions, as it will allow the inspector to cover adequately the entire project area when necessary.

Occasionally, a close-up detail may be needed of a portion of the work that is inaccessible to the photographer without turning into a human fly and walking up the side of a tall building. A long-focal-length (telephoto) lens can allow the same effect as actually being up close. A 135-mm lens is an ideal telephoto for this purpose. If the Resident Project Representative or inspector plans to have only one lens, however, a 35-mm focal-length lens is ideal. If two lenses are to be used, both the 50-mm and the 35-mm lens are a good combination, although a 28-mm would be a good substitute for the 35-mm. An example of the same subject as photographed from the same location, but using three different focal-length lenses, is illustrated in the three photographs of the NASA mating device structure at the Hugh M. Dryden Flight Research Center in California. In this example, the structure was photographed first with a 35-mm wide-angle lens, then with a 50-mm normal lens, and finally with a 135-mm telephoto lens (Figure 4.20).

When selecting a camera for use in construction, do not underestimate the capability of the inexpensive 35-mm autoexposure, autofocus cameras currently on the market. It will be far less costly than any digital camera of comparable resolution. Many such cameras have been used to produce excellent progress photos. For a few dollars more, you can get a fully automatic camera with a lens that adjusts to two preselected focal lengths. The usual combination is 35-mm to 70-mm. One manufacturer, however, offers a 28-mm to 48-mm lens instead. You will find that a wide-angle lens offers more opportunities for good photographs than a telephoto lens, so the 28-mm to 48-mm might be a good buy. The 48-mm covers the angle of view that reasonably approximates that of the human eye.

A helpful hint: when you are working with any automatic camera, remember that they are battery powered. Therefore, when the batteries lose their power, the photography session is over. Find out what size and type of batteries are needed for your camera and keep several spares in the field office, as they may not be readily available everywhere.

If you are a camera buff and choose to bring your own high-priced, 35-mm reflex film camera or high-resolution digital camera into the field, it is suggested that you budget from $300 to $400 for cleaning by an authorized professional camera repair shop after the project is over. Such cameras are sensitive to dust and moisture and will probably have to be completely disassembled, cleaned, readjusted, and reassembled in order to operate properly in the future.

With 35-mm lens

With 50-mm lens

FIGURE 4.20 Same Subject Photographed with Three Different Lenses.

With 135-mm lens

FIGURE 4.20 *(continued)*

Transporting Films through Airport Security

Often, construction people are required to travel by air and frequently bring their cameras and film with them. Thus, they will experience firsthand the intricacies of protecting their unexposed film during a security check of their carry-on baggage. There appears to be a considerable amount of misinformation about the effect of airport X-ray equipment on photographic films. They may tell you that the amount is so slight that it will have no effect on the film. This is patently false. The person who tells you this is usually the security person at the screening location, who is merely repeating what he or she has been told by superiors. They are not scientists or photographic engineers, and are generally unqualified to make such statements.

In 1998, an industry group alerted travelers that photographic film can be damaged by new bomb-detection machines being used at major airports worldwide. The bottom line is this: "If you are travelling with film and a camera, keep them with you, and don't check them in," says Tom Dufficy of the Photographic Imaging Manufacturing Association. You should hand-carry your film and do not carry it in checked baggage. There were tests conducted in the summer of 1997 at the Federal Aviation Administration on the In Vision Technology's CTX 5000. This machine is widely used around the world and is being installed in more and more airports in the United States. The tests proved that the machine can leave a line or a streak on the film. It can also cause film to lose color or become foggy or grainy.

Guidelines released by Eastman Kodak Company some years ago state that a cumulative X-ray exposure not exceeding 5 milliroentgens will not affect most films provided that the exposure is made in increments not exceeding 1 milliroentgen, and that the orientation of the film or luggage in which the film is placed is changed between exposures. However, the effect of small amounts of radiation will be greatly influenced by the film speed as well. High-speed films such as any film with a speed of ISO 400/din 27° or especially ISO 1000/din 31° should always be hand-checked and never allowed to pass through X-ray at any radiation level.

Some airport X-ray machines only expose film to a dosage of approximately 0.05 milliroentgen for each inspection. The maximum possible dosage that might occur for each inspection of carry-on baggage with such equipment could possibly be as low as 0.01 milliroentgen. However, do not be misled into thinking that you can protect your film by storing it in the checked baggage. On many flights, all checked baggage is X-rayed routinely, and the dosage used for checked baggage is frequently far higher than that of inspection stations at the airport gates.

An important fact to remember, however, is that the effect of X-ray exposure on film is cumulative, so that the total number of X-rays received by the film during each inspection is added to the X-ray count received by the same film on other inspections. Keep in mind that some X-ray equipment exposes the film to a much larger dosage at each inspection. This is particularly true of international inspection points. Even atmospheric radiation has a cumulative X-ray effect on film, though it is minimal. Of special concern to the discriminating photographer is the fact that while some exposure to X-rays may not cause a visible blemish on the film, it definitely can affect the color balance and thus the final effect on the prints or slides. Only undeveloped film, exposed or not, can be affected by X-ray radiation.

Another recommended precaution is that if you go on more than 10 flights with the same roll of film, you should ask the baggage inspector to hand-inspect your camera case and film. In the United States, on all flights regulated by the FAA, you have a legal right to demand that your cameras and film be hand-checked. Storage of film in lead containers has not produced consistently satisfactory results, so reliance on such precautions should be seriously reconsidered.

Unfortunately, in the aftermath of the 9/11 disaster, any attempts to clear an airport security area without running cameras and film through the X-ray machine may well be doomed to failure. If you must use film cameras, the only solution is to obtain film at time of arrival and either process all film before returning to your point of origin or ship all nonprocessed film by FedEx before you return. The preferred method for travellers is to utilize digital cameras when travelling, as no harm will come to either the equipment or your photographic images by running it all through the X-ray equipment. Also, you will spare yourself a lot of confrontations with airport security personnel . . . confrontations you would probably lose.

Video Cameras in Construction

A video camera can be an extremely useful tool in construction management. Used properly, it can be used to document field problem areas for showing and discussion at regular job management meetings in the field, it can be used as a training tool for inspectors, it can be used as a marketing tool during proposal presentations, and it can serve to document work progress and delays in anticipation of possible future litigation. This is especially true in cases involving delay claims.

On the downside, however, a video camera is somewhat bulky to carry around, and it is also quite demanding on the available time of the Resident Project Representative, and thus there is often a tendency to pass by video opportunities that might otherwise be a valuable record.

Although the selection of video camera equipment is beyond the scope of this book, it should be kept in mind that most facilities for playback of videotapes are designed to play VHS videocassettes. Smaller cameras are convenient, but smaller cassettes will require special adapters to play on VHS equipment.

Video cameras are generally capable of fully automatic operation, including exposure control and focusing, and most also have power-operated zoom lenses as standard equipment. Beyond the automatic features, modern video cameras seem to be capable of operations that are well beyond human comprehension except, perhaps, to a computer programmer.

Selection of Film

The choice of color films is often based on the answer to the question, "Do you want slides or color prints?" Generally, slides cost less initially, because there are no prints to make—all that you get back is the original film that was in the camera. In case of color print film, however, you receive both a color negative (the film that was in the camera) and color prints that were made from those negatives—thus the added cost. It is wise to choose color negative film over color positive (slide) film for a very practical reason. Exposure errors in color negative film can be corrected in the final prints; however, in the case of slide film, there is no chance for compensating for exposure errors. For construction, color prints are recommended. From the color negatives, slides can still be made if needed; color prints and enlargements can be made, and even black-and-white prints can be made if desired. Thus, it is a truly versatile film. However, if prints are wanted from a slide, the processor must first rephotograph the slide to make a negative, then make prints in the ordinary way. This is a considerably more costly process than if color print film had been used in the first place.

Film and Camera Storage

Security is always a problem on a construction site, and, in particular, items such as cameras, pocket calculators, and similar pocket-sized items of considerable value are always in jeopardy. Cameras should be locked up; however, this usually presents a problem. They are sensitive to temperature, particularly when they have film inside. Usually, the only available secure areas at a construction site are locked file cabinets in the field office (a rarity) or the locked trunk of the inspector's automobile. If the field office is cooled in the summer and heated in the winter, the file cabinet is best. The auto trunk is a high-risk area for film and cameras unless special precautions are taken to prevent damage from the high heat concentrations usually present there. One method of protecting the camera and film in an automobile trunk in hot weather is to keep the camera and film in one of the popular Styrofoam beverage containers designed to keep a six-pack cool on a picnic. If film or cameras are subject to such conditions for several hours, serious consideration should be given to the use of one of the popular Styrofoam containers that are provided with special freeze-pack covers. When stored in the refrigerator freezer overnight, these will maintain safe film and camera temperatures all day long, even under the most severe conditions.

In any case, neither film nor cameras should be left in the trunk without taking special precautions. In winter months, there is little substitute for a heated field office, as nothing you can do to your auto trunk is likely to keep the camera and film from freezing when the temperature drops below 0° (32°F).

If a camera is taken out of a heated enclosure into a very cold atmosphere, be sure to let the camera adjust to the cold air for awhile before attempting to take pictures; otherwise, the condensation that may form will cause problems.

Color films require storage in cool, stable temperatures if the color balance is to be maintained. Try to keep them about 18° to 21°C (65° to 70°F) at all times, if possible. Polaroid film can be an even greater problem in the field. It has been the author's experience that Polaroid color film requires special care and handling during extremely hot weather and during extremely cold weather. As with hot-temperature photography, the ideal arrangement is to keep the camera in a controlled temperature inside the Resident Project Representative's or inspector's automobile or field office until the actual time of exposure.

Digital Cameras for Construction

Digital Imaging

The current technology in the photographic industry is digital cameras. Instead of a sensitized film upon which to record an image, it utilizes computer technology to record an image digitally on electronic media.

Digital cameras have only been available to consumers since the early 1990s. Externally, both digital and film cameras are similar in appearance, but that is where the similarity ends.

Most new cameras also have both an optical and an electronic LCD viewfinder. Get one with an optical viewfinder and save yourself a bundle of money for batteries.

Resolution. Image quality also depends on a camera's resolution, or the amount of pixels (or picture elements) the CCD uses to capture an image. The greater the resolution a sensor has, the more pixels it has to capture an image with. The more pixels it uses, the better the image quality. Early consumer digital cameras typically had resolutions that topped out at less than 1 megapixel. Recently, high-end consumer cameras have been made available with resolutions of 6 and 11 million pixels.

In my opinion, to justify the economic risk of placing a digital camera in the field where it is subject to exposure, to extremes of heat, cold, dust, and moisture, as well as to potential damage, it would seem desirable to keep the price under $300, but for the present that would result in a compromise on print image quality.

As a general guideline in selection of a digital camera, if you will not require final prints greater than 4"-by-6" size a camera rated at 2 megapixels should do it for you. However, if you want to go to 8"-by-10" prints you will need 4 or more megapixels to achieve acceptable sharpness. If you are just going to post the pictures to your Web site, a 2 or 3 megapixel camera should do.

Digital Camera Storage Media

When considering the purchase of a digital camera for construction, some consideration should be given the data storage media being used by the camera of your choice. The two storage cards used by most cameras are CompactFlash and Smart Media.

Once an image has been stored, the user can immediately view it using the LCD. Unlike film, there is no need for developing and processing. The results are nearly instantaneous but not permanent. Unlike film, digital images can be immediately transferred to a computer where the user can use software to enhance and manipulate them.

PROBLEMS

1. True or false? An inspector's diary may be in loose-leaf form as long as the pages are numbered.
2. List at least four types of information that should be documented in the Daily Construction Report.
3. Who should keep a construction diary?
4. What is the purpose of preparing a "feedback" report?
5. What types of transactions between the contractor and the owner or engineer/architect should be documented?
6. How long should project records be saved by the owner?
7. Name the four principal applications of photography to construction documentation.
8. What type of information is recommended to be documented in the Daily Construction Report and in the diary?
9. What are the six principal concerns in the format of a construction diary?
10. If the Daily Construction Report is electronically stored in a computer, what must be done to protect the integrity of the document?

5

ELECTRONIC PROJECT ADMINISTRATION

In the past decade great strides have been made in computer utilization as a tool for project administration documentation. However, it was not until the early twenty-first century that it became more than simply another tool for keeping records.

The rapid movement into wireless communications that followed offered new opportunities not only for documentation, but for interactive communications as well. Most of the good software available to accomplish this was written for contractors; very little addressed the specific needs of the architect/engineer or owner as a contract administration tool.

Only a few programs were available for the owner or architect/engineer as an aid in project management. At a recent conference the author was able to see many of these programs demonstrated. As an engineer, it was disappointing to find that the construction contractors were light years ahead of them in this field, and that only a few programs came even close to meeting their contract administration needs. In fact, most software vendors backed off when asked about applications slanted toward the needs of the owner and engineer.

Only two vendors at the trade show, ProjectEDGE and FieldManager, offered a software program that seemed to be suited to the needs of the architect/engineer.

After discussion with other construction professionals, ProjectEDGE was selected as a demonstration program for this book because it offers the most thorough coverage consistent with the principles addressed in this book. It was written for owners/engineers by a professional civil engineer.

Other programs that the author investigated included Project King, Microsoft Project, Primavera Expedition, Eclectus Simply Construction, and Meridian Project Systems. Many other popular-name programs also were evaluated but rejected as being too contractor oriented (i.e., emphasizing contractor applications of estimating, subcontracting, cost accounting, payroll, expediting, and similar contractor-related subjects). Several of the contractor-oriented programs were excellent for that application.

Using Computers for Project Administration[1]

Of the few programs on the market specifically dedicated to the concerns of the owner and its architect/engineer or Resident Project Representative, the one selected by the author is an example of a program that is designed to meet the needs of the owner's contract administrator.

As a means of further illustrating the versatility of a computerized project management program, each of the tasks described in this chapter is available on the program.

Design and construction projects typically involve many individuals in several different firms all working toward the task of completing a project. Each person may have his or her own method of tracking items under his or her responsibility, but seldom will everyone use a coordinated system to track all related items for the entire project. Team members using an e-mail application such as Microsoft Exchange or Lotus Notes can maintain their own calendars and task lists.

A project extranet can provide an additional tool to integrate the action items for the entire team. Individuals receiving notice of a required current action can perform the necessary task upon receipt. Items with future due dates can be copied into calendar tasks along with a direct Universal Resource Locator (URL) link for the project task. At the appropriate time, the responsible person would notice the task on his or her Exchange or Notes calendar and click on the link while connected to the Internet. The URL would take the person directly to the task where he or she could complete the transaction.

Action Items

The opening view of ProjectEDGE is a listing of all of the open action items for the project together with due dates and the names of the persons responsible for completion of each of those action items, as illustrated in Figure 5.1. The design of the program focuses the persons and their tasks while providing an overview to the person who assigned the task. Under the program, anyone who creates an action item, who is responsible for its completion, or who is a copied interested party is given e-mail notice to the address that they have entered in the project phonebook. Other persons involved in a multistep workflow activity, such as processing RFIs, are given notice when it is their turn to act. As the responsible party changes, the action item view also changes to reflect the current situation. When the action item is completed, all parties are notified, and the item is taken off the list.

Action items may be created directly or may be a result of a follow-up from meetings, contract or change order executions, expiring insurance certificates, and any other time-related item. The views also use a system of colored status icons to help users readily identify those that are past due from those coming due in the next 5 days as well as those with due dates beyond 5 days.

[1]Contributed by W. Gary Craig, PE, President, ProjectEDGE, Liverpool, New York 13088.

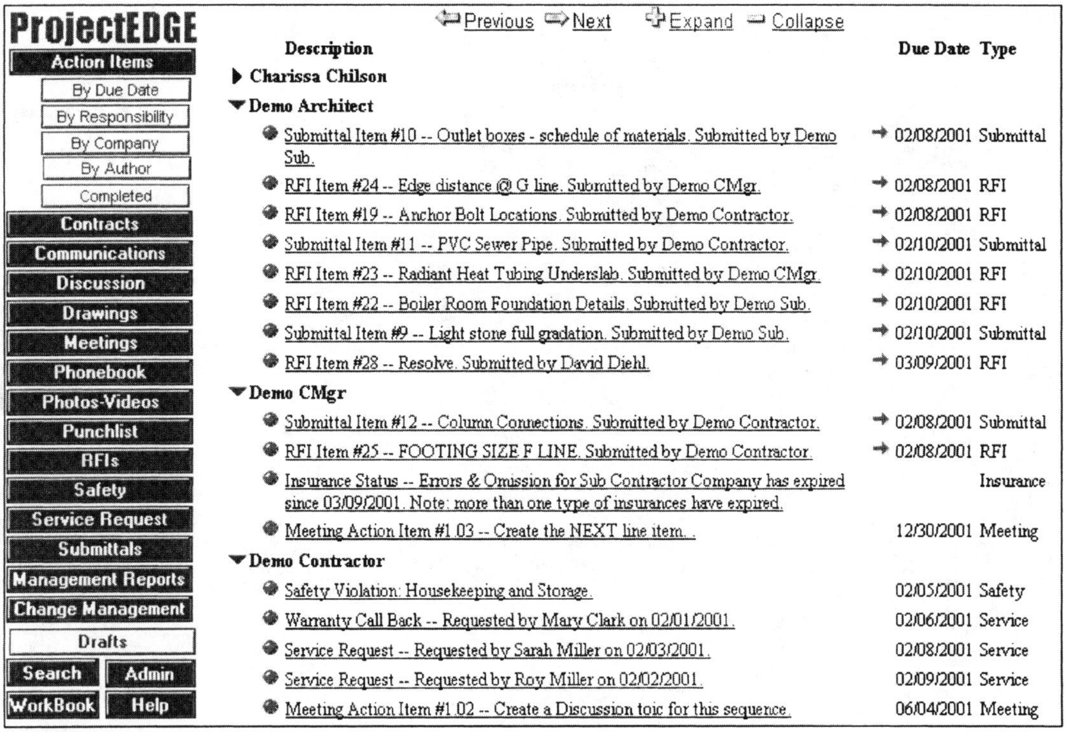

ProjectEDGE

Action Items | By Due Date | By Responsibility | By Company | By Author | Completed | Contracts | Communications | Discussion | Drawings | Meetings | Phonebook | Photos-Videos | Punchlist | RFIs | Safety | Service Request | Submittals | Management Reports | Change Management | Drafts | Search | Admin | WorkBook | Help

Previous Next Expand Collapse

Description	Due Date	Type
▶ Charissa Chilson		
▼ Demo Architect		
Submittal Item #10 -- Outlet boxes - schedule of materials. Submitted by Demo Sub.	02/08/2001	Submittal
RFI Item #24 -- Edge distance @ G line. Submitted by Demo CMgr.	02/08/2001	RFI
RFI Item #19 -- Anchor Bolt Locations. Submitted by Demo Contractor.	02/08/2001	RFI
Submittal Item #11 -- PVC Sewer Pipe. Submitted by Demo Contractor.	02/10/2001	Submittal
RFI Item #23 -- Radiant Heat Tubing Underslab. Submitted by Demo CMgr.	02/10/2001	RFI
RFI Item #22 -- Boiler Room Foundation Details. Submitted by Demo Sub.	02/10/2001	RFI
Submittal Item #9 -- Light stone full gradation. Submitted by Demo Sub.	02/10/2001	Submittal
RFI Item #28 -- Resolve. Submitted by David Diehl.	03/09/2001	RFI
▼ Demo CMgr		
Submittal Item #12 -- Column Connections. Submitted by Demo Contractor.	02/08/2001	Submittal
RFI Item #25 -- FOOTING SIZE F LINE. Submitted by Demo Contractor.	02/08/2001	RFI
Insurance Status -- Errors & Omission for Sub Contractor Company has expired since 03/09/2001. Note: more than one type of insurances have expired.		Insurance
Meeting Action Item #1.03 -- Create the NEXT line item..	12/30/2001	Meeting
▼ Demo Contractor		
Safety Violation: Housekeeping and Storage.	02/05/2001	Safety
Warranty Call Back -- Requested by Mary Clark on 02/01/2001.	02/06/2001	Service
Service Request -- Requested by Sarah Miller on 02/03/2001.	02/08/2001	Service
Service Request -- Requested by Roy Miller on 02/02/2001.	02/09/2001	Service
Meeting Action Item #1.02 -- Create a Discussion toic for this sequence.	06/04/2001	Meeting

FIGURE 5.1 Action Items Sorted by Responsibility.

Contracts

The bid and award phase of a project can be very document intensive especially when the project delivery method involves multiple-prime contracts. With such a delivery method, there could be 20 or more bid packages and resulting prime contracts, many or all of which could also have subsequent change orders.

Each contract could have the following elements to manage: bid package, contract form, insurance certificates, payment and performance bonds, progress invoices, close-out documents, and similar items, as illustrated in Figure 5.2. Each contract, including any professional service or consultant contract agreements, is assigned a control number. Depending upon the contract form being used, the actual contract could be prepared from the contract profile that is an abstract of the contract variable terms. Using a process similar to a mail merge, the profile and the contract boilerplate can be used to generate the contract for negotiation and execution. Once generated, a workflow approval process can be used to move the contract through the approval and execution steps required for the owner organization or by its lenders.

A schedule of values or list of unit prices can be used to prepare a standard invoice precoded using the CSI (Construction Specifications Institute) or other system of identification used for the periodic invoices. A workflow process (Figure 5.3) may be set up

Contract | PO/Work Order | Action Item

ProjectEDGE

Action Items
Contracts
Insurance
PO's by Company
PO's by Number
With No Amt
Communication
Discussion
Drawings
Meetings
Phonebook
Photos-Videos
Punchlist
RFIs
Safety
Service Request
Submittals
Management Reports
Change Management
Drafts
Search | Admin
WorkBook | Help

Previous | Next | Expand | Collapse

	Committed	Pending Approval
Demo2000-0 Architect Company		
Contract Profile #Demo2000-0: Created on 07/19/2000. Scope:	$5,000	
Demo2000-02 Construction Manager Company (Construction Manager)		
Insurance Status: Insurance status is current. The next Insurance expiration date is 11/11/2001.		
Contract Profile #Demo2000-02: Created on 06/18/99. Scope: Construction Mgmnt	$70,000	
Contract Tracking - Construction Mgmnt. Status: No information has been entered.		
Demo2000-04 Contractor Company (Contractor)		
Insurance Status: Insurance status is current. The next Insurance expiration date is 10/20/2001.		
Contract Profile #Demo2000-04: Created on 06/18/99. Scope: General Requirements	$50,000	
Demo2000-05 Sub Contractor Company		
Insurance Status: Errors & Omission expired on 03/09/2001. Note: the following Insurance policies have expired: Errors & Omission, Bonding Performance, Bonding Payment.		
Contract Profile #Demo2000-05: Created on 08/01/2000. Scope: Doors & Windows	$50,000	
Partial Release Retention for Contract Profile #Demo2000 - 05: Created on 02/06/2001.		
Change Order #02 - Doors & Windows. Status: Change Order for Sub Contractor Company is due back on 02/07/2001		($1,500)
Change Order #01 - Doors & Windows. Status: Fully Executed	$5,000	
Contract Tracking - Doors & Windows. Status: Fully Executed		
Payment Request #1		
PO Demo2000-CDC-M00523 {w/General Conditions} to Vendor Company		
Purchase Order #-CDC-M00523 to Vendor Company created on June 21, 1999	$24,365	
Purchase/Work Orders/Backcharges		
Work Order #-CDC-W0023 to Sub Contractor Company created on June 21, 1999	$0	

FIGURE 5.2 Listing of Contractors, Contacts, Change Orders, Tracking Documents, and Invoices, as Well as Amounts as Seen by a User with Overall Access Rights.

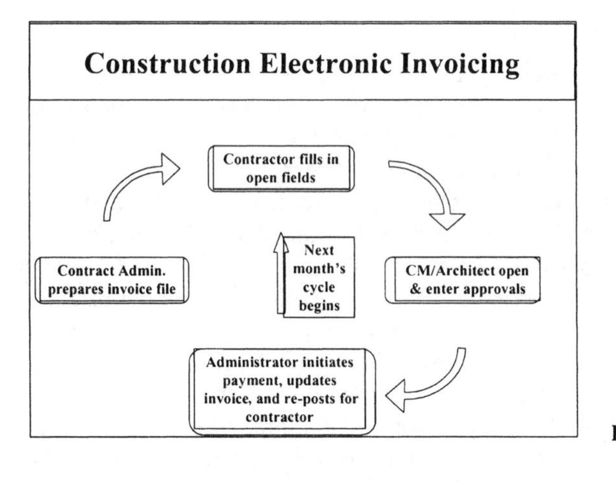

FIGURE 5.3 Contractor Progress Invoice Workflow.

to route the invoices through the review and approval steps required by the contract and/or by the lender. Once the prime contractors or general contractor submits an invoice, each reviewing person in order would be notified by e-mail when the invoice was ready for their review. At the end of the review cycle, approved invoices would be transferred to the accounting function for payment and the invoice form for the next cycle prepared including being updated for any approved change orders. The underlying form allows for either a schedule of values or unit-priced progress billing format. Stored materials and retainage, as well as releases, are all supported under the program.

Each type and version of insurance certificates for each contract can be scanned into the database, and its policy information including expiration date would be recorded. Any subsequent renewal certificates should also be attached to provide a record of coverage throughout the project in the event of a claim being filed for a period covered by a previous certificate. Similar information can be entered for each type of bond required under the contract. The expiration dates of insurance policies and bonds are automatically tracked and advance notice given before any expiration dates to assure continuous coverage.

Purchase orders, work orders, and back-charges can also be prepared and administered within the program using templates containing information such as vendor, cost codes, descriptions, and prices. An example of a contractor progress requisition form showing base contract and a change order is shown in Figure 5.4. A workflow approach can be used if any of these documents must be approved before it is issued. Once issued, the workflow approach can be used to notify the vendor of the order as well as notifying the accounting function of the commitment made under the underlying document.

Change or issue management is a critical function on most projects due to the need to address changes in a timely manner. Issue management assigns a tracking number to each issue as soon as it is identified. The workflow concept shown in Figure 5.5 begins with the entry of the issue and ends when all the solutions to the

Requested by: Sam Jones
Contractor: Demo Sub Contractor
Contract Number: 06
Project: Demo Project
Architect's Number: 3000
Requisition Number: 02
Period Ending Date: 09/01/00

PROGRESS REQUEST

	Requested	Approved
Amount:	$0.00	$0.00
Less Retention:	$0.00	$0.00
+Retention Rel:	$0.00	$0.00
	$0.00	$0.00

PROJECT: Demo Project

Line #	Phase	Cost Code	Cost Type	Description of the Work	Original Scheduled Value	Previously Approved Work In-Place and Stored Material Amounts (Computed)	Work in Place this Period (Requested)	Previously Stored Material Balance (Computed)	This Period New Material Stored (Requested)	Less This Period Material Removed from Storage (Requested)	This Period Completed Work plus Change in Stored Material (Computed)
				*** BASE CONTRACT ***							
1	A	33321	5	FOUNDATIONS	$1,000.00	$1,000.00		$0.00			$0.00
2	A	33245	5	FLOOR SLABS	$2,000.00	$0.00		$0.00			$0.00
3	A	45000	5	COMMON BLOCK	$3,000.00	$0.00		$0.00			$0.00
4	A	45800	5	FRACTURED FACE BLOCK	$4,000.00	$0.00		$0.00			$0.00
5	A	56780	5	METAL JOISTS	$5,000.00	$0.00		$0.00			$0.00
6	A	72300	5	ROOFING	$6,000.00	$0.00		$0.00			$0.00
7	A	81000	5	DOORS	$7,000.00	$0.00		$0.00			$0.00
8	A	91000	5	FINISHES	$8,000.00	$0.00		$0.00			$0.00
9	B	10100	5	SPECIALTIES	$9,000.00	$0.00		$0.00			$0.00
10	B	15000	5	MECHANICAL	$10,000.00	$0.00		$0.00			$0.00
11	B	16000	5	ELECTRICAL	$11,000.00	$0.00		$0.00			$0.00
				TOTAL BASE CONTRACT	$66,000.00	$1,000.00	$0.00	$0.00	$0.00	$0.00	$0.00
				*** CHANGE ORDER #1 ***							
1	A	33321	5	FOUNDATIONS	$1,000.00	$0.00		$0.00			$0.00
2	A	33245	5	FLOOR SLABS	$1,000.00	$0.00		$0.00			$0.00
3	A	45000	5	COMMON BLOCK	$1,000.00	$0.00		$0.00			$0.00
4					$0.00	$0.00		$0.00			$0.00
5					$0.00	$0.00		$0.00			$0.00
6					$0.00	$0.00		$0.00			$0.00
7					$0.00	$0.00		$0.00			$0.00
8					$0.00	$0.00		$0.00			$0.00
9					$0.00	$0.00		$0.00			$0.00
10					$0.00	$0.00		$0.00			$0.00
				TOTAL CHANGE ORDER #1	$3,000.00	$0.00	$0.00	$0.00	$0.00	$0.00	$0.00
				TOTAL BASE CONTRACT AND ABOVE CHANGE ORDERS	$69,000.00	$1,000.00	$0.00	$0.00	$0.00	$0.00	$0.00

FIGURE 5.4 Sample Contractor Requisition Form Showing Base Contract and a Change Order.

ISSUE MANAGEMENT
Issue Management over an Extranet

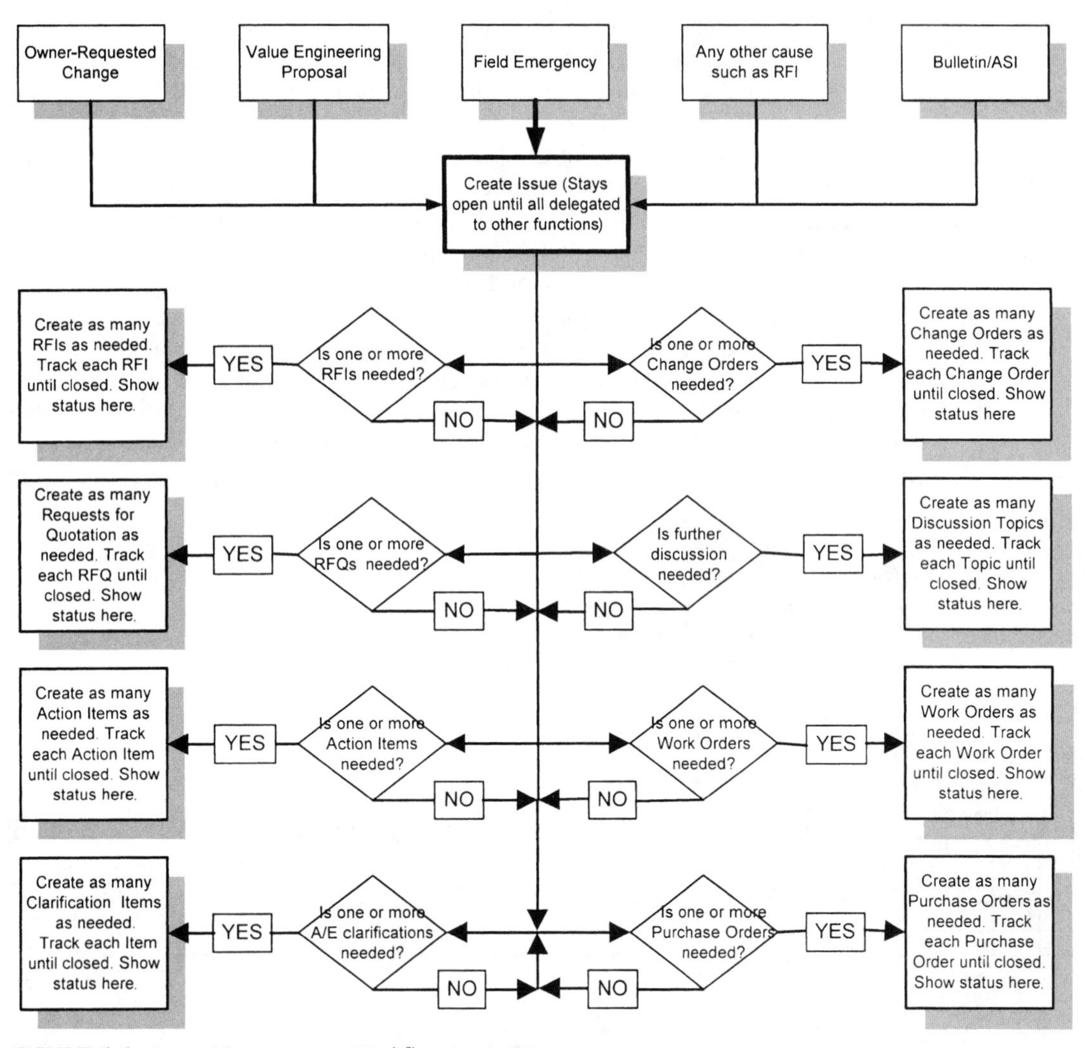
Owner-Requested Change
Value Engineering Proposal
Field Emergency
Any other cause such as RFI
Bulletin/ASI
Create Issue (Stays open until all delegated to other functions)
Is one or more RFIs needed?
Create as many RFIs as needed. Track each RFI until closed. Show status here.
Is one or more Change Orders needed?
Create as many Change Orders as needed. Track each Change Order until closed. Show status here
Is one or more RFQs needed?
Create as many Requests for Quotation as needed. Track each RFQ until closed. Show status here.
Is further discussion needed?
Create as many Discussion Topics as needed. Track each Topic until closed. Show status here.
Is one or more Action Items needed?
Create as many Action Items as needed. Track each Action Item until closed. Show status here.
Is one or more Work Orders needed?
Create as many Work Orders as needed. Track each Work Order until closed. Show status here.
Is one or more A/E clarifications needed?
Create as many Clarification Items as needed. Track each Item until closed. Show status here.
Is one or more Purchase Orders needed?
Create as many Purchase Orders as needed. Track each Purchase Order until closed. Show status here.
YES
NO

FIGURE 5.5 Issue Management Workflow Logic Diagram.

issue have been identified and are assigned to the appropriate parties for completion. All components of the issue can be seen in one view. Each respective solution component is also shown in its respective module and in the action item view until it is closed out.

Communications

The Communications Module was designed to provide a means of creating and capturing all written communication so that it could be shared with those members of the project team who were either the originator, the addressee(s), or copy addressee(s). The types of documents that can be created include letters, faxes, memos, transmittals, form letters, and records of telephone conversations. All of these document types can be written and saved as drafts. If other team members' input is needed, those individuals can be designated as collaborators until the document is finalized and locked to prevent change by anyone. Incoming e-mail or paper communications can also be captured and given keyword descriptions.

Action times can be launched from these communications to provide a follow-up reminder. The database phonebook is used to look up all the contact information for all documents. The e-mail addresses in the phonebook are used to provide notification to all the referenced parties whenever the originator selects that option, which is recommended so all affected parties will be made aware that the document is ready for them.

The design of the ProjectEDGE application creates an audit trail (Figure 5.6) of when and by whom the document was created, edited, and read in addition to displaying the entire record of any e-notices. However, the concept of delivering an e-notice to someone instead of the actual document has not been tested in the courts yet. Until it is, it is recommended that the parties to all contracts in a project using an extranet settle the issue using contract modifications in which all parties acknowledge in writing that an e-notice sent to someone and a subsequent recording of their having read the document are equivalent to their receipt of an original document.

The use of an extranet for communications means that parties in remote locations can initiate or collaborate with communications without the need for support staff. Actual printing, filing, and mailing can be reduced to nothing if the parties are comfortable with that approach. Those who are not may continue to print, file, and mail as appropriate for their needs.

All communications can be viewed in sorts listed by date, originator, or recipient and any copy addressee(s), as shown in Figure 5.7. Drafts can be viewed only by the originator and any designated collaborators during the draft stage.

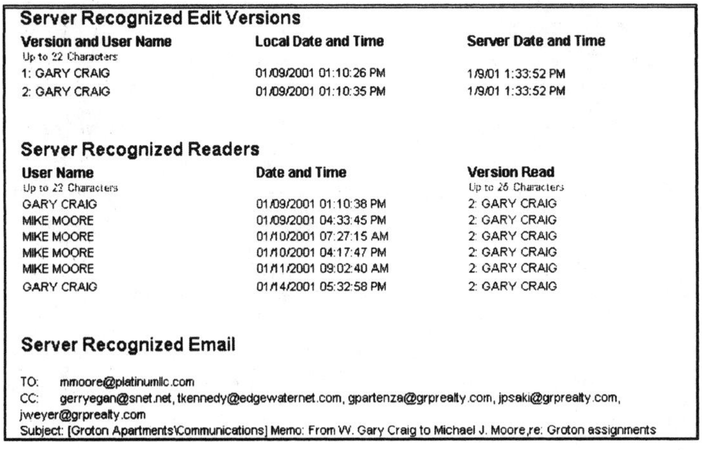

Server Recognized Edit Versions

Version and User Name (Up to 22 Characters)	Local Date and Time	Server Date and Time
1: GARY CRAIG	01/09/2001 01:10:26 PM	1/9/01 1:33:52 PM
2: GARY CRAIG	01/09/2001 01:10:35 PM	1/9/01 1:33:52 PM

Server Recognized Readers

User Name (Up to 22 Characters)	Date and Time	Version Read (Up to 26 Characters)
GARY CRAIG	01/09/2001 01:10:38 PM	2: GARY CRAIG
MIKE MOORE	01/09/2001 04:33:45 PM	2: GARY CRAIG
MIKE MOORE	01/10/2001 07:27:15 AM	2: GARY CRAIG
MIKE MOORE	01/10/2001 04:17:47 PM	2: GARY CRAIG
MIKE MOORE	01/11/2001 09:02:40 AM	2: GARY CRAIG
GARY CRAIG	01/14/2001 05:32:58 PM	2: GARY CRAIG

Server Recognized Email

TO: mmoore@platinumllc.com
CC: gerryegan@snet.net, tkennedy@edgewaternet.com, gpartenza@grprealty.com, jpsaki@grprealty.com, jweyer@grprealty.com
Subject: [Groton Apartments\Communications] Memo: From W. Gary Craig to Michael J. Moore,re: Groton assignments

FIGURE 5.6 Audit Trail of a Particular Document Showing Its Creation as Well as Its Readers and E-mail Notifications.

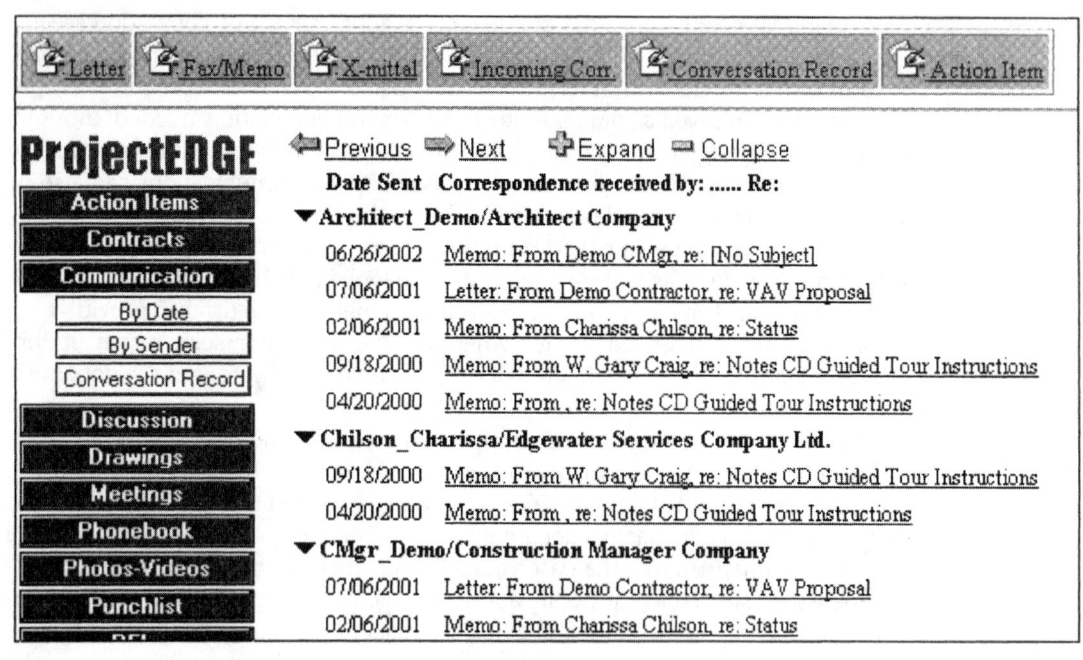

FIGURE 5.7 Correspondence Sorted by Name of Person Receiving the Document.

Discussions

Threaded discussions begin with a topic proposed by an originator who might have incorporated input from collaborators before publishing the discussion for comments by other team members. The entire team or selected invited team members can provide their comments in as many response documents as needed. The originator or others can further respond with responses to the responses. All the documents are "threaded" together, as illustrated in Figure 5.8, as long as the response and responses to responses are continued. Discussions can be used for engineering or architectural programming, value engineering, creating a decision log, process improvement, and building a knowledge base.

Drawings

The Drawing Module provides a catalog of drawings as well as listings of both current and historical drawings. Drawings, sketches, consultant reports, supplemental instructions, specification sections, and other similar documents can be attached to the database or to a companion drawing warehouse. At the end of the project, the record drawings can also be attached.

All authorized database users can use a Web browser to view the drawings and mark them up nondestructibly as shown in Figure 5.9. The markup can then be saved and distributed to other team members for their information. These markups can be used in any phase of the project. Typical uses are for conveying design or programming intent, value engineering, and a support for Requests for Information.

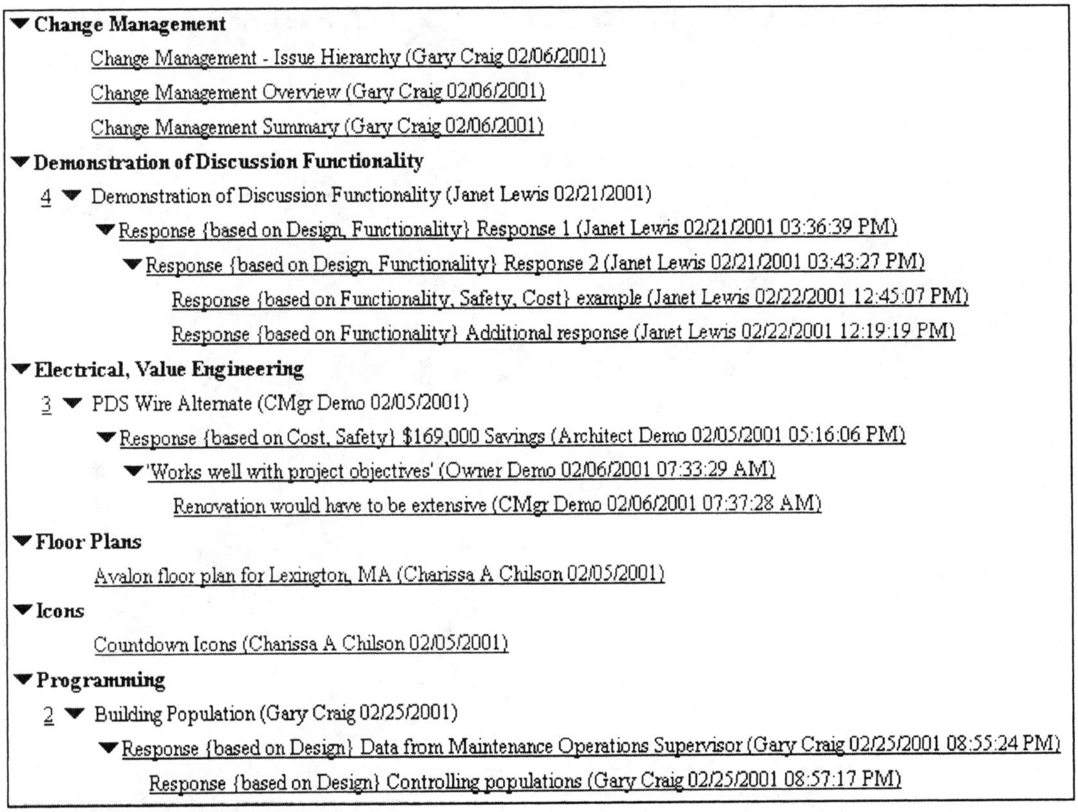

▼ **Change Management**

Change Management - Issue Hierarchy (Gary Craig 02/06/2001)

Change Management Overview (Gary Craig 02/06/2001)

Change Management Summary (Gary Craig 02/06/2001)

▼ **Demonstration of Discussion Functionality**

4 ▼ Demonstration of Discussion Functionality (Janet Lewis 02/21/2001)

▼ Response {based on Design, Functionality} Response 1 (Janet Lewis 02/21/2001 03:36:39 PM)

▼ Response {based on Design, Functionality} Response 2 (Janet Lewis 02/21/2001 03:43:27 PM)

Response {based on Functionality, Safety, Cost} example (Janet Lewis 02/22/2001 12:45:07 PM)

Response {based on Functionality} Additional response (Janet Lewis 02/22/2001 12:19:19 PM)

▼ **Electrical, Value Engineering**

3 ▼ PDS Wire Alternate (CMgr Demo 02/05/2001)

▼ Response {based on Cost, Safety} $169,000 Savings (Architect Demo 02/05/2001 05:16:06 PM)

▼ 'Works well with project objectives' (Owner Demo 02/06/2001 07:33:29 AM)

Renovation would have to be extensive (CMgr Demo 02/06/2001 07:37:28 AM)

▼ **Floor Plans**

Avalon floor plan for Lexington, MA (Charissa A Chilson 02/05/2001)

▼ **Icons**

Countdown Icons (Charissa A Chilson 02/05/2001)

▼ **Programming**

2 ▼ Building Population (Gary Craig 02/25/2001)

▼ Response {based on Design} Data from Maintenance Operations Supervisor (Gary Craig 02/25/2001 08:55:24 PM)

Response {based on Design} Controlling populations (Gary Craig 02/25/2001 08:57:17 PM)

FIGURE 5.8 Threaded Discussions by Category.

Users who subscribe to individual databases, such as bid a package database, are automatically notified when a drawing is updated. Project managers can verify who received notice and whether they read the notice. The subscription feature can also be used with a third-party reprographics firm to place orders for prints of the drawings.

Meetings

The Meeting section provides an agenda planning feature that allows users to arrange their meeting topics in the desired sequence and assign time allocations. The users may specify different meeting tracks or categories to separate meetings by their similar topics such as Kick-off, Core Team, Periodic Progress, and Commissioning meetings. In each meeting track, users can define which items are closed and which items are carried over to the next meeting. Once closed, a meeting topic can be reopened if the topic needs further discussion. Closed items are retained in the history for searching or reference, whenever needed.

Some users prefer to carry the past meeting content forward into the subsequent meeting so all the history is in one place. Others prefer to document each meeting with its own specific content. Either approach is valid and can be accommodated under the program.

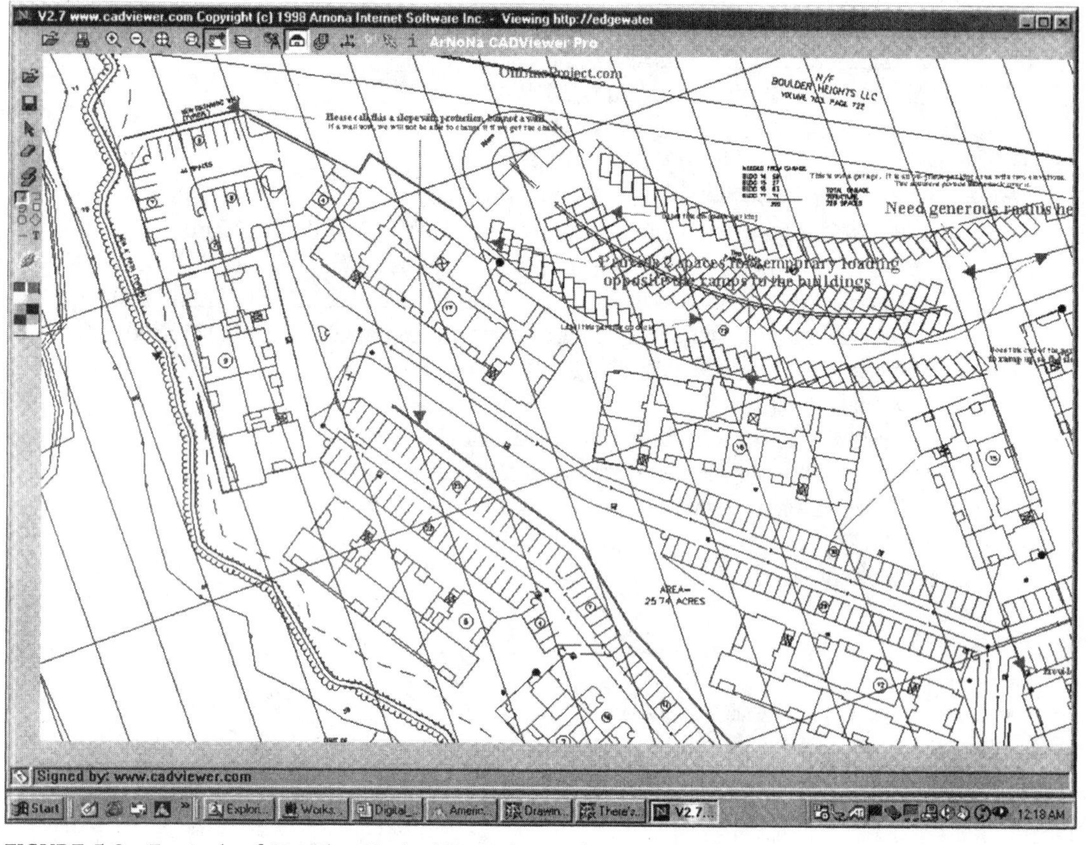

FIGURE 5.9 Example of Site Plan Marked Up with a Web Browser.

The meeting topics are numbered in a user-defined scheme. It is recommend that each subsequent meeting be numbered with a unique number with all new topics in that meeting having a decimal that is random or follows some user-defined standard. Typical meetings could have topics with the following meaning: 1.02 (an open item from the first meeting); 2.05 and 2.15 (open items from the second meeting); and 3.05, 3.10, and 3.15 (topics in the current meeting).

The meeting minute topics are each threaded discussions that can have numerous related action items and user comments. Both action items and comments are independent and can be made at any time to an open meeting topic. Action items and comments can also be individually e-noticed to interested or responsible parties. Virtual meetings can be held by a a facilitator setting up an agenda with content outlines that are then e-noticed to each invited attendee or interested party along with a deadline for their reading the proposed topic and making comments and/or creating action items. After the time for input has passed, the facilitator can read the input and close any items that have been completed. If further action is required, the item can be left open for comment. A totally virtual meeting requires discipline on the part of all the participants, but the benefits of the time saved in travel or sitting in a meeting waiting to provide their input can be significant if everyone does their part.

Some users have sufficient typing skills to enter their minutes into a laptop computer while in the meeting. Others will enter everything later using a preprinted note-taker form. Often the best solution is a combined approach in which the action items are entered during the meeting and e-notice is sent to the responsible parties. When Web conferencing applications such as NetMeeting or Lotus Sametime are used, participants can follow along with both the discussions and the minutes as they are entered (Figure 5.10). Participants can also be adding their own action items or comments at the same time.

The setup of the meeting minutes allows the minutes to be read on-line or to be printed off in a conventional meeting format.

Phonebook

The Phonebook section contains all of the contact information about companies and persons with which to provide a lookup for all of the other sections of the Project-EDGE program. Whenever any information about a person or a company is required, it may be found by a simple recall action rather than by reentry. All of the items in the phonebook may be categorized by any number of user-defined categories that promote the searchability of the entries. For example, companies can be coded by their characteristics such as scope of work, design–build, or disadvantaged status.

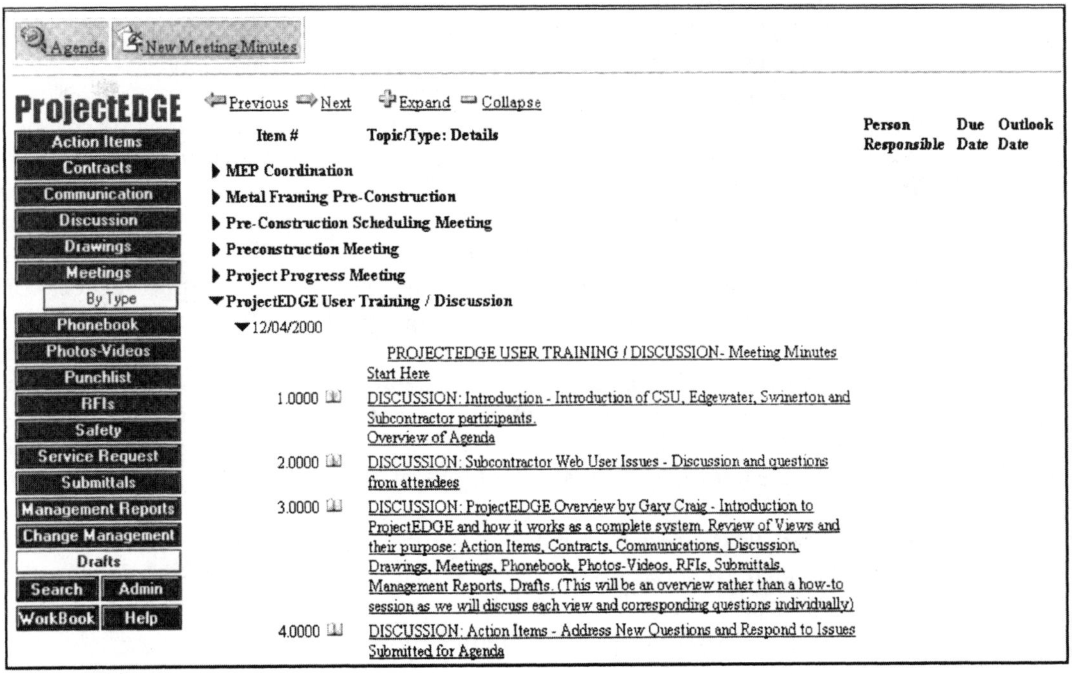

FIGURE 5.10 Listing of Meeting Minute Tracks and Expanded Detail of One Meeting.

A Core Team roster is also provided for identifying all the persons responsible for specific tasks. For example, the engineer-of-record, the construction administrator, and the gateway person for RFI and submittals are in the Core Team roster.

Phonebook entries can be readily transferred from master phonebooks or from project to project.

Photos and Web Cam Videos

The Photos and Video section provides the means of cataloging project photographs obtained from conventional photography, scanned images, digital photographs, or video. Conventional photographs can be scanned or sent to a processing laboratory that can provide the digital images as well as the finished paper prints.

Web Cam can be used to document site conditions, conduct productivity analyses, provide a collaboration tool, and reduce travel time by allowing remote viewing of progress or problems. Periodic safety or workmanship inspections are also possible. On a well-run project, the Web Cam can also be an effective marketing tool.

Punch Lists

Developing punch lists of items to be completed or corrected can be a time-consuming task without an efficient process. The best process is continuous inspection and monitoring throughout the project so there is nothing to correct or complete at the end. Often projects come close to achieving that goal through the use of model mock-up areas or rooms constructed early in the project as the agreed standard for the remainder of the work.

When punch lists are necessary, begin by developing a standard list of items that most experienced persons can expect to encounter on similar projects. Standard lists promote a uniform description of problems and provide a checklist to ensure that a tedious process is consistent from beginning to end. The standard list should assign responsibility and time frames for the corrections or completions. Armed with the preprinted list, the reviewer only has to enter the room number or location of the item along with any necessary clarifying remarks and a check mark next to each item. If carbon paper is used, a copy of the list can be posted in the location for the tradesmen to make the correction. The list is used along with a spreadsheet containing all the responsibility, duration, and other related information about the items to enter the items in the database.

Once entered (Figure 5.11), the items are available for the responsible contractors to view and act upon. Review and acceptance can be done with a PDA or laptop computer.

Any punch-list item that is not completed within a user-defined time after the due date becomes an action item, and e-notice is generated to the responsible party.

If needed, the punch-list information can be exported to a spreadsheet for analysis of recurring problems or any other pattern that might exist. Sometimes cumulative graphs of numbers of punch-list items verses identification and completion dates can be used to isolate problems in performance.

Previous Next Expand Collapse

Responsibility	Due Date	Item	Description	Where / Comments
▼Room: 02				
Contractor Company - Demo Contractor	02/07/2002	12	Touch up paint near door	-
▼Room: 07				
Contractor Company - Demo Contractor	02/07/2002	11	Repair holes, repaint walls at relocated overhead cabinet.	-
Sub Contractor Company - Demo Sub	02/17/2002	10	Install electrical power strips - see attached sketch	-
▼Room: 501				
Contractor Company - Demo Contractor	02/07/2002	13	Washing Machine Leakage	- Repair ASAP

FIGURE 5.11 Punch-List Items Listed by Location.

Requests for Information (RFIs)

Requests for Information can be tracked with user-defined workflow routes that specify all the review steps and the time allotted for each. RFIs are coded according to the design discipline of the scope of the request. Each route category may have different reviewing parties, alternative parties, and allowed durations. For example, a request about landscaping should be reviewed by different parties than one concerning telecommunications. Each route might also have different time frames for the reviews, but often the total time is dictated by contract terms. An RFI log can be viewed on the Web screen as shown in Figure 5.12, or on hard copy as shown in Figure 5.13.

The RFI routing allows interested parties to be copied when an RFI of a particular type is in process. For example, the fire marshall might want to be notified when a life safety times issue is being reviewed. If so, starting a request along the review

Previous Next Expand Collapse

	No.	2nd No.	Subject	Assigned To	Step Due	Final Due	Resolved	Submitted	Company Submitted By	Route
▼Open										
●	28		Resolve	Demo Architect	03/09	03/13		03/07	Edgewater Services Company, Ltd.	Architectural
●	25	MOR003	FOOTING SIZE F LINE	Demo CMgr	02/08	02/12		02/06	Contractor Company	Architectural
●	24	ART006	Edge distance @ G line	Demo Architect	02/08	02/12		02/06	Construction Manager Company	Architectural
●	23		Radiant Heat Tubing Underslab	Demo Architect	02/10	02/14		02/06	Construction Manager Company	Mechanical
●	22		Boiler Room Foundation Details	Demo Architect	02/10	02/14		02/06	Sub Contractor Company	Mechanical
●	19		Anchor Bolt Locations	Demo Architect	02/08	02/12		02/06	Contractor Company	Architectural
▼Closed										
■	27	Mor004	Rock location			02/12	02/06	02/06	Contractor Company	Architectural
■	26		Color or roof			02/11	02/02	02/05	Contractor Company	Architectural
■	21	DCC-001	Scupper near 45 Deg. Glass			02/12	03/17	02/06	Contractor Company	Architectural
■	20	MOR001	concrete at 37			02/12	03/17	02/06	Contractor Company	Architectural
■	18	ART001	Material grade specs			02/12	03/17	02/06	Sub Contractor Company	Architectural

FIGURE 5.12 Web Screen Capture of RFI Log.

workflow will trigger an e-notice to the fire marshall along with a link to use to access the item in the database. When the review of the item is completed, the fire marshall would be given notice and a link to the result.

Alternative review persons are recommended for each route in case the primary person does not respond in a timely manner. When the item becomes overdue, the alternative party is sent an e-notice. If a primary person is sick or away, another person can be assigned the route for the duration of the absence. Any new reviews will be automatically rerouted to the substitute person until changed. When required, the RFI routing can be changed by selecting another workflow route or by developing a specialized route for the remaining steps of the RFI.

RFIs can be used during design as a means of clarifying program requirements. The routes for planning RFIs could include the owner, user, or tenants.

All RFI routes should have a gateway person who initially accepts the RFI for processing or rejects it as being incomplete or unnecessary. The gateway person would be indicated in the Core Team roster for both planning and construction RFIs. In addition to the gateway person, a master editor is recommended for each route in the event that someone is unable to update his or her own transaction. For example, a review and approval is done during a meeting where the reviewer does not have a computer available. The master editor could enter the information even if using Web conferencing. As soon as the entry has been made, all the participants will see the updated information, and the person on the next step will receive notice of their pending action.

The most common use of RFIs occurs in construction where contractor, subcontractors, and vendors normally initiate them. However, inspectors and other parties can also use the process to receive answers to their questions. The exact routes used will depend on the method of project delivery being used. Multiple prime contracts will probably use different routes than general contracts. There is also a difference between the RFI routes used by an owner with its general contractor from the routes used by that contractor with its subcontractors and vendors. Some RFIs initiated by the subcontractors or vendors might only be the business of the general contractor. A contract scope issue between the general contractor and one of its subcontractors should not be reviewed by the owner unless it also involves a design question. Assuming that it does not, that RFI should follow a route that ends with the general contractor.

RFI Log for DemoV3B

Printed on	03/17/2001 at 03:29 PM	Project:	DemoV3B
Printed By:	Contractor Demo	Project Number:	Demo2000

RFI#	2nd RFI#	Status	Date Received	Step Due Date	Final Due Date	Date Resolved	RFI Title	Assigned To	Potential CO	Route
0	DCC-002	Open	03/17/01	03/19/01	03/23/01		Curtain wall member sizes	Demo Architect		Architectural
19		Open	02/06/01	02/08/01	02/12/01		Anchor Bolt Locations	Demo Architect		Architectural
20	MOR001	Closed	02/06/01	02/12/01	02/12/01	03/17/01	concrete at 37		No	Architectural
21	DCC-001	Closed	02/06/01	02/12/01	02/12/01	03/17/01	Scupper near 45 Deg. Glass		No	Architectural
25	MOR003	Open	02/06/01	02/08/01	02/12/01		FOOTING SIZE F LINE	Demo CMgr		Architectural
26		Closed	02/05/01	02/11/01	02/11/01	02/02/01	Color or roof			Architectural
27	Mor004	Closed	02/06/01	02/12/01	02/12/01	02/06/01	Rock location			Architectural

FIGURE 5.13 Print Version of RFI Log.

If the general contractor "elevates" an RFI into the owner's process, it should follow a route based on its category. If the RFI was originated by a subcontractor, the general contractor might have to reword the request to a description suitable for the intended route. Many times the subcontractor does not have sufficient information to pose the question for review by the design team.

RFI information should be reported on-line and in print form, as shown in Figure 5.13, for use by persons who prefer working on paper or may need to receive the notice by fax. Many users also prefer working with logs of data organized in columns like a spreadsheet.

Safety

The Safety section provides a format for collecting information on safety plans, documenting OSHA site visits, collecting information on MSDS, and tracking accidents and safety violations. Such events also trigger action items with e-notices that are used to follow up until the situation is remedied. The daily field report automatically triggers a safety notice prompt if an incident is reported when filling out the report.

Some owners may choose not to use all the features of this section based on the perception of increased risk of exposure to litigation due to hosting or having access to a source of knowledge about an incident or a potential incident and doing nothing about it. The risk can be minimized with contract language and an understanding with all users that the section is provided only for their convenience and not to place any added responsibility on any party.

We have seen instances where the construction manager's having generated violation notices to subcontractors was sufficient to avoid an OSHA penalty.

Shop Drawings and Submittals

The processing of shop drawings and submittals is monitored by user-defined workflow routes like those used for RFIs. There are many similarities, but there are also two important differences. The first is that all shop drawings and submittals do not have to be submitted for review and approval. Some are for record only, but still need to be tracked at least to the extent of recording their submission. Early in the design phase, the consultants and the owner's team should determine what is required for record only. The second difference is the need for a resubmission process for any shop drawings or submittals that are rejected and resubmitted. Any resubmitted documents should be numbered as a subnumber of the original number. For example, a resubmission of shop drawing #5 should be numbered 5.01 to tie them together.

When submissions are made with electronic files, any mark-up and approval can be noted right in the document file. Multiple hand-marked copies can be eliminated. Each party can print a record copy, and an electronic file can be left posted in the database. When submissions are not made with electronic files, it might be possible to scan the submissions for electronic processing. Color and mock-up samples that do not have an electronic equivalent can still use the workflow process by generating transmittals to accompany the physical samples as they move along the route.

Previous Next Expand Collapse

No.	2nd No.	Subject	Assigned To	Step Due	Final Due	Resolved	Resolution	Submitted
Open								
12		Column Connections	Demo CMgr	02/08	02/10		Open	02/06
11		PVC Sewer Pipe	Demo Architect	02/10	02/14		Open	02/06
10		Outlet boxes - schedule of materials	Demo Architect	02/08	02/12		Open	02/06
9		Light stone full gradation	Demo Architect	02/10	02/14		Open	02/06
0		Manhole Covers	Demo Architect	03/22	03/26		Open	03/18

FIGURE 5.14 Sample Listing of Open Submittal or Shop Drawing Items.

Like RFIs and punch lists, submittals can be analyzed to determine the metrics of their processing. The reporting of submittals as shown in Figure 5.14 is similar to that of RFIs, shown in Figures 5.12 and 5.13.

Management Reporting

Management reporting is separate from the "reporting" that occurs in each section. There the reporting is both exception reporting and detailed reporting of each item's status. Management reporting concerns the entire project—its schedule, budget, and status.

Customary reporting is done monthly, and sometimes it can be as much as two weeks past the end of the month before the report is published. Electronic reports can provide continuous status updates rather than waiting for news that could be as much as six weeks old. Such reports can also eliminate the need to print and distribute reports that are seldom read and are merely filed for reference.

A 12-section status report is set up for management updating. Users can select anywhere from one to twelve sections as areas to report about the project status. The number of sections used and their content can vary throughout the project as needs change. For example, early in the project, the emphasis could be on programming, value engineering, and design concepts. During construction, the emphasis might shift to issues to be resolved, potential risks to schedule or budget, cash-flow forecasting, and progress reporting. Users can change the suggested template topics to ones that meet their needs. Separate status reports could also be prepared for different audiences such as tenants and lenders. In addition to changing the sections, the users can also use links to more detailed descriptions if the user wants more information. A management overview could direct the user to "click in the link for more details." The linked document could be any text, graphic, or electronic file.

Daily field reporting is performed by site personnel as well as by consultants, the Resident Project Representative, or inspectors. The requirements for each group are similar enough that one common form can be used. The specific content will vary according to their individual needs. The Resident Project Representative is likely to use the report as a means of attaching to his or her handwritten Daily Construction Reports and to summarize the results of testing. Architects and engineers, as parties usually directed more toward engineer-owner relations, are likely to use the report less frequently unless they are charged with daily on-site construction administration or project oversight. Except for special site visits, the consultants probably will use the status report more than the daily report form.

A budget reporting form is provided that can use a spreadsheet for summarizing budgets from many sources such as owners, users, contractors, and construction managers if there is no common cost-reporting application used by all these parties. Each party might have a system for the reporting of the cost of their commitments and expenditures, but a unified summary by spreadsheet is often the only practical means of analyzing the total project. Besides the lack of a common reporting method in the individual applications, there is also the possibility that each system might be based on different accounting rules.

WEB-ENABLED PROJECT MANAGEMENT APPLICATIONS[2]

Electronic Posting of Bidding Opportunity Advertisements

Many states have adopted systems that utilize Web applications. As an example, the California State Contracts Register (CSCR) facilitates state contracting by centralizing and electronically disseminating advertisements for state construction and service contracting opportunities. Local governments and special districts located within the state may also choose to utilize the system.

The Procurement Division of the Department of General Services publishes the CSCR (Figure 5.15). State law requires state agencies to advertise their construction and service opportunities of $5,000 or more in the CSCR. Each advertisement contains key contract information including the contact agency/person to contact to obtain a full package of bidding documents (Figure 5.16).

The state of California no longer publishes the CSCR in hard-copy form. Since advertising on the electronic system from the state's Web site constitutes "legal notice" for all state agencies in California, the system has been chosen for presentation as illustrative of electronic bid advertising. Third-party systems or government sites that supplement legal advertising in local newspapers may, but do not have to, adhere to the same standards for providing information. Construction projects are advertised according to project categories and subcategories as illustrated in Figures 5.17 and 5.18.

An electrical contractor may choose to browse through three categories: the categories for general engineering contractor (California Contractor License Class A) and general contractor (California Contractor License Class B), for bidding opportunities as a subcontractor, and the category for electrical work (California Contractor License Class C-10), for trade work in which the bid is submitted to the public agency as a prime contractor. Contractors who visit the Web site daily do not have to scroll through the listing of bidding opportunities they have already seen. They may choose an option to view only "new listings."

[2]Contributed by Harold Good, CPPO, Director of Procurement and Contracting, City of Palm Springs, California.

Procurement DGS

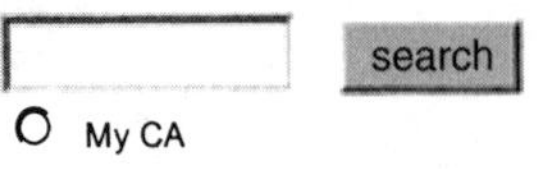

O My CA

California State Contracts Register

General Explanation

The California State Contracts Register (CSCR) is your one-stop information source for state contracting opportunities. The CSCR connects business to government, providing an easy-to-use, on-line publication which lists services and construction contracts over $5,000, commodity contracts over $25,000, and IT goods contracts over $100,000.

Contracts Register (CSCR) - Find Contracting Opportunities

- View By Category
- View By Contract Number Or Agency

  No Password Required

CSCR Tools

- View Contracts Exempt From Advertising
- View Contracts Awarded/Intent to Award
- View Construction Progress Payments
- Special Announcements
- View World Trade Agreement Awards
- CSCR Download Site (Alternate Access Providers)

What? Why? How? - The Ins and Outs of the CSCR

- What is The CSCR?
- How Do I View The Bids?
- What Additional Services Does The CSCR Offer?
- FAQS

Best Viewed with Internet Explorer. Download a free copy from Microsoft.

default.asp

User Login

CSCR Advertisement Submittal

Government Users Only

Subscription Outreach Services

Contractor Advertisements

Bid Package Download

Returning Visitors

New Users

CSCR Home

Disclaimer

FIGURE 5.15 Entry Page for California State Contracts Register (CSCR).

Procurement DGS

search

My CA

Contracts Register - Contract Advertisement Detail

ATTACH CONTRACTOR AD

CATEGORY LIST HELP

Contract Advertisement Category(s): 10 - Architectural and Engineering; 18 - Construction

Advertisement Appears in the following Sub Categories:
10 - Architectural Consultants/Civil Engr/Design&Planning/Drafting/Environmental Svcs; 18 - B General Contractor; 10 - Construction Management

View Contractor Advertisement

Contract Advertisement	**DSH-04-09 CSULB - Peterson Hall 3 Replacement - Construction Manager at Risk**
Advertisement Publication Date	7/22/2004 12:01 AM
Branch of Government	State Of California
Agency	CSU, Long Beach
Estimated Value	$10,000,000 and Over
Duration	976 Calendar Days
License(s) Required	B
Location(s):	Los Angeles

In general, the scope of work consists of providing preconstruction services and construction management/administration services at risk with a guaranteed maximum price for the construction of the Peterson Hall 3 Replacement building.

The project includes the demolition of an existing Science Building and the construction of a new 154,000 GSF Science Building.

Each Proposer will submit a technical proposal and a fee percentage proposal for design services, preconstruction services, and construction phase services, contingency and fee for the total project.

A mandatory Pre-bid meeting is scheduled for 10 a.m. on August 19, 2004. Proposals are due by close of business (5 p.m.) on September 14, 2004.

The Construction Budget for this project is $48,108,000.00.

Bid Submittal Deadline **9/14/2004 5:00 PM**

Pre Bid Conference Information

Conference Attendance	Mandatory
Conference Date/Time	8/19/2004 10:00 AM
Conference Location	Psychology Building - Room 153
Location Address	CSULB 1250 Bellflower Blvd. Long Beach, CA 90840

Contract Information Contact

Name	Genieve Maestas
Phone Number	562.985.8445
Fax Number	562.985.7647

FIGURE 5.16 Example of Complete Advertisement for Construction.

Contracts Register - Contracts Awarded/Intent to Award

EXIT HELP

CLICK on Category to select

Categories Sorted by Number

1 Janitorial
2 Printing and Publishing
3 Extermination
4 Transporting and Warehousing
5 Consulting
6 Security
7 Information Technology
8 Mailing
9 Refuse and Sewage Disposal
10 Architectural and Engineering
11 Photography and Reproduction
12 Equipment Service, Repair or Installation
13 Equipment Rental and Leasing
14 Automotive/Aircraft Repair, Rental and Leasing
15 Gardening and Agricultural
16 Medical/HealthCare
17 Office Leasing
18 Construction
19 Miscellaneous
20 Commodities

SORT LIST BY CATEGORY NAME

FIGURE 5.17 Initially, Contract Advertisements Are Grouped into Separate Main Categories.

Another convenient option provided to contractors is to provide profile information to the system so that they are notified by e-mail of bidding opportunities that fit their profile, eliminating the need to search the system for bidding opportunities. For example, a general contractor may elect to be notified only of bidding opportunities for projects with an estimated value of a million dollars, located in San Diego County, and requiring a B license. Because the project's estimated value, its location, and the contractor's license requirements must be provided by the listing agency in order for the advertisement to be accepted, it is easy for the system to match projects to contractor profiles and to generate an e-mail notification to contractors whose submitted profiles match the project.

Contractor Advertisements

Private sector businesses may attach their own advertisements to existing government agency advertisements in the CSCR. Firms wishing to bid the project can

Procurement DGS

search

My CA

Contracts Register - Contracts Awarded/Intent to Award

FULL AWARD INDEX NEW AWARD INDEX CATEGORY LIST HELP

Category 18 - Construction
Select box next to Sub Category desired.

Sub Categories

- ☐ * **Select All Sub Categories Listed** - *Just check this to select them all*
- ☐ - A General Engineering
- ☐ - ASB Asbestos
- ☐ - Asbestos, Lead Abatement/Tank Removal, Installation/Fireproofing, Fire Safety
- ☑ - B General Contractor
- ☐ - Building Alterations/Painting/Flooring/Decorating/Doors/Roofing/Plumbing
- ☐ - Building Constr/Struct Steel Erection, Reinforcing/Seismic Retrofit/Demolition
- ☐ - C-02 Insulation and Acoustical
- ☐ - C-04 Boiler, Hot Water, Steam Fitting
- ☐ - C-05 Carpentry
- ☐ - C-06 Cabinet and Mill Work
- ☐ - C-07 Low Voltage Electrical
- ☐ - C-08 Concrete
- ☐ - C-09 Drywall

FIGURE 5.18 Example of Grouping into Subcategories within Each Main Category (Partial List).

advertise for subcontractors and suppliers, or, conversely, subcontractors and suppliers can announce their availability to contribute to the project.

Electronic Bid Packages

The public agency posting the bid advertisement may opt to make the entire bid package available for electronic download. Caltrans, the state transportation department, is an agency that increasingly utilizes this option. Before a bid package can be downloaded, the requestor must provide detailed information concerning its firm and full contact information (Figure 5.19). Public agencies that provide electronic bid package downloads also post all project addenda bid packages on-line.

The system maintains a listing of all individuals or firms that have either downloaded the electronic bid package or requested it by mail and provides access to a fully updated list for the public agency (Figure 5.17).

Standard Electronic Ad Text

The CSCR allows public agencies to submit, maintain, and update standard advertising text. This text appears in all agency ads unless modified or deleted.

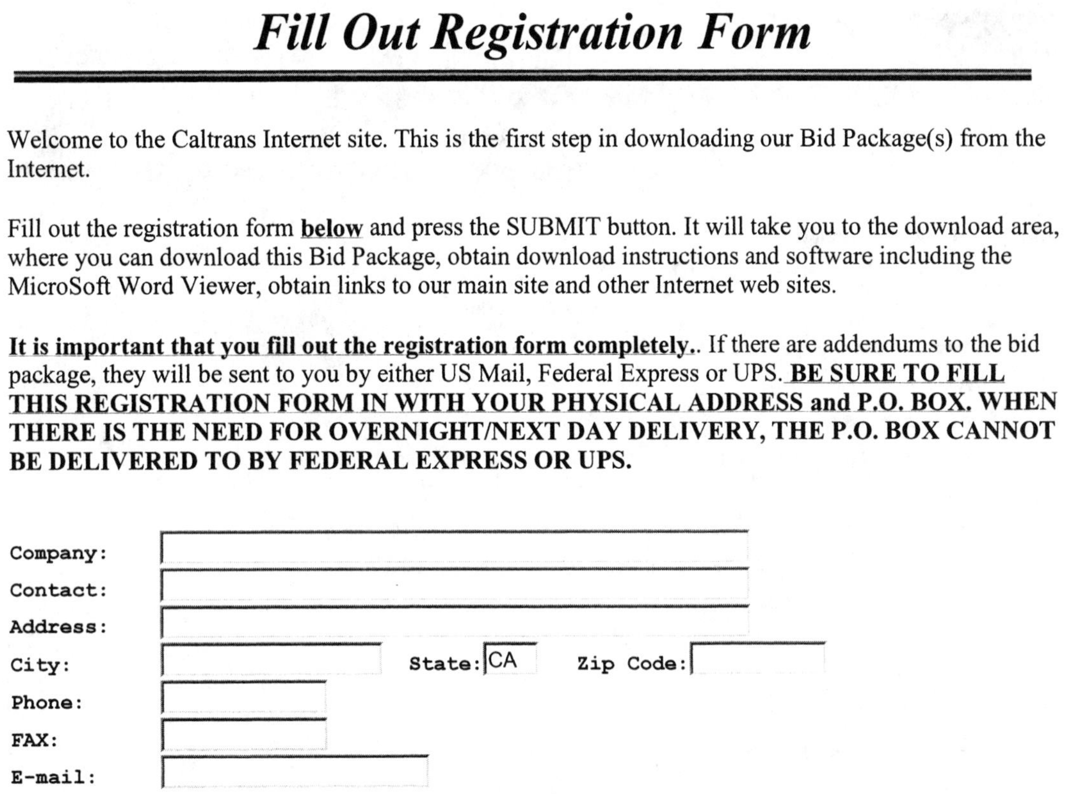

Fill Out Registration Form

Welcome to the Caltrans Internet site. This is the first step in downloading our Bid Package(s) from the Internet.

Fill out the registration form **below** and press the SUBMIT button. It will take you to the download area, where you can download this Bid Package, obtain download instructions and software including the MicroSoft Word Viewer, obtain links to our main site and other Internet web sites.

It is important that you fill out the registration form completely.. If there are addendums to the bid package, they will be sent to you by either US Mail, Federal Express or UPS. **BE SURE TO FILL THIS REGISTRATION FORM IN WITH YOUR PHYSICAL ADDRESS and P.O. BOX. WHEN THERE IS THE NEED FOR OVERNIGHT/NEXT DAY DELIVERY, THE P.O. BOX CANNOT BE DELIVERED TO BY FEDERAL EXPRESS OR UPS.**

Company:

Contact:

Address:

City: State: CA Zip Code:

Phone:

FAX:

E-mail:

FIGURE 5.19 Example of Caltrans (DOT) Registration Form for Electronic Download of Bid Packages.

Notification of Prebid or Preproposal Meetings

Each time an advertisement is placed, the option appears for advertising either a mandatory or nonmandatory prebid or preproposal meeting. If such a meeting is scheduled, it is placed prominently within the advertisement. Additionally, the system will not accept a bid opening date that does not conform to the state requirement to provide adequate time between the prebid meeting and the bid close date.

PROBLEMS

1. What type of information identifies an action item?
2. What do accountability and responsibility mean in a project setting?
3. Give an example of a process to demonstrate how workflow could be used for contract administration.
4. Why are workflow processes beneficial?
5. What is the value of having a project extranet track outstanding issues?

6. What purpose does an audit trail of documents or transactions provide?
7. What value is there in having project communications in a database as opposed to individual e-mail accounts?
8. What is the value of having project drawings associated with the extranet?
9. Why use an agenda planning feature for a meeting?
10. Why bother creating a punch list on an extranet?
11. What is the role of a gateway person in contract administration?

6

SPECIFICATIONS AND DRAWINGS

WHAT IS A SPECIFICATION?

The specifications are the part of the contract documents that define the qualitative requirements of the project that is to be built. The dictionary defines *specification* as "a detailed description of requirements, dimensions, materials, etc., as of a proposed building, machine, bridge, etc.," and further as "the act of making specific."

The role of the drawings is to define the geometry of a project, including dimensions, form, and details. The specifications are intended to complement this by defining the nature of the materials that are to be used and the description of the workmanship and procedures to be followed in constructing the project.

All too often an inspector, just like many people in the building trades, expects the drawings to provide all the information required, incorrectly assuming that the specifications are needed by the lawyers only in case of dispute. To be sure, the specifications may be needed in cases of dispute, but if used properly and referred to throughout the construction work, they can also serve to minimize disputes. Even more important to the Resident Project Representative and the contractor is the fact that the specifications are the only documents that will spell out the obligations for administration of the project during its construction. By far the majority of the administrative tasks that the Resident Project Representative will be required to perform are covered by the specific terms of the *General Conditions* of the contract and by nothing else. Even years of past experience cannot serve as a substitute, as the rules of the game change from project to project. What may have been proper on a previous job may be wrong on the next, and only the specifications will tell it as it should be.

Conflicts Due to Drawings and Specifications

It should be brought out here that neglecting the specifications can lead to serious problems. In case something is shown or noted one way on the drawings and described differently in the specifications, which will govern? The answer to that question is easy. The specifications will normally take precedence unless it says *in the specifications* that the plans will govern. Thus, it is still the specifications that set the controlling criteria. Normally, it is easy to determine the relative importance of one document over another, as most specifications specify the relative order of importance of the different parts of the contract documents in the General Conditions of the construction contract. However, it should be of interest that in the absence of such a specific provision, the courts have repeatedly held that the provisions of the specifications will take precedence over the drawings in case of a conflict between the two [*Appeal of Florida Builders, Inc.,* ASBCA No. 9013, 69-2 BCA 8014 (1969)].

Therefore, if the specifications are the most important single document, the inspector can hardly perform in a competent manner without being thoroughly familiar with both the specifications and the construction drawings.

Conflicts between Drawings and Specifications

In some cases, the same data are covered in both the drawings and the specifications—not a great arrangement, but it happens often enough. The problem here is that frequently one document is changed during design and the other is overlooked. This generally creates the problem just referred to. The unfortunate situation is that usually where such a problem exists, it is the drawings that were updated to receive the latest changes or corrections, and the specifications may in fact be outdated and incorrect. The basic philosophy still controls, however, and the inspector has no authority to force the contractor to provide that which is shown on the drawings when the bid may have been based upon the article contained in the specifications. In case of any such conflict, the contractor is obligated to notify the owner's representative before continuing. However, it is well for the inspector to monitor carefully any such possibilities personally, as the contractor may honestly miss recognizing the presence of a conflict. It would also be possible for a dishonest contractor merely to *claim* to miss the conflict so as to furnish the cheaper of the two items, knowing full well that if the design firm wants it changed after the contractor has already built it in accordance with the specifications, in all likelihood, he or she will be able to claim successfully extra compensation for such additional work.

It should be remembered that some items will appear only in the specifications and not on the drawings; others will appear only on the drawings and not be mentioned in the specifications. This is not necessarily an oversight, nor is it to be considered as a flaw in the specifications or drawings. Many architectural and engineering firms as a matter of policy prefer not to repeat data on both documents. This is done intentionally as a means of preventing conflicts due to late changes that may be made to one document alone and not to the other. The Resident

Project Representative should make certain that the trade supervisor doing the work uses the specifications also, as it will minimize construction problems and conflicts.

In one case involving a conflict between the general notes on the drawings and the provisions of the specifications, the question was whether the general notes on the drawings were an extension of the specifications. In this case, a federal agency awarded a contract for construction of a preengineered metal building. The specifications required the roof panels to have a flat profile with no corrugations between the interlocking seams. A note on the roof drawing referred to the acceptability of certain manufacturer's standard designs.

The contractor attempted to furnish a standard roof of a certain roof manufacturer. The government refused to approve it because it did not have a flat profile. The contractor argued that the drawing note was actually a specification that authorized the use of this standard design. The contractor claimed that this particular note should prevail over the more general language of the specifications.

The Board of Contract Appeals ruled that the note was actually a part of the drawings, not the specifications. The contract stated that in the event of conflict, the specifications should govern over the drawings. The government was therefore entitled to enforce the specifications calling for a flat roof profile [*Appeal of Abco Builders, Inc.*, ASBCA No. 47413 (April 20, 1995)].

Scope-of-Work Disputes

The contract documents for a typical construction project consist of the Agreement itself, the General Conditions, Technical Specifications, and numerous Drawings. Additionally, some provisions are covered by reference to another document. Unfortunately, contract documents often fail to adequately describe, define, or delineate the work to be performed. This generates some of the so-called "scope-of-work" disputes. Such disputes center on the nature and extent of the performance obligation.

The most frequent cause of such disputes is the lack of detail in drawings or lack of specifics in specifications. The problem with drawings is most common on small projects where design costs, and therefore the number of drawings, are held to a minimum. If the drawings lack sufficient detail, the contractor will have to rely on what it understands is expected. This may or may not coincide with the intentions or expectations of the owner or designer.

Specifications problems usually result from lack of attention during the specification process. The widespread use of "canned" or "off-the-shelf" specifications and manufacturer's specifications is another large contributor to the number of disputes occurring. Errors of omission or ambiguity are frequent, and unfortunately for the owner, are generally interpreted in favor of the contractor.

Problems also occur when specifiers resort to broad, subjective generalities (e.g., "in accordance with highest industry standards") rather than describing the work objectively and in detail.

Avoiding Scope-of-Work Problems

The best but most difficult way to avoid scope-of-work problems is outlined in the following guidelines:

1. Provide adequate budget for drafting specifications.
2. Work with a set of carefully prepared, carefully coordinated front-end documents.
3. Place control of each set of project specifications in the hands of a single, qualified specifications engineer, or require that all specifications be subject to his or her review and editorial control.
4. Do *not* mark up previous job specifications to create another set of job specifications; always work from the same set of master specification documents for every job.
5. Update the master specification at least annually to keep current with industry standards, codes, and laws affecting construction.
6. Use the same set of front-end documents on all projects to minimize contract administration problems and to ensure that there has been no contractual variation in the owner's risk posture.
7. Front-end documents should not be edited by project engineers. All changes to front-end documents should be subject to review of both legal and construction management personnel with specific experience in construction contracts management.

While the foregoing will go a long way toward improving the specifications product, it is important that every office either establish a specifications department or at the very least designate a Specifications Engineer to control specification standards and policies.

The Use of Generalities in Specifications

It is in the owner's best interest to keep designers from indulging in generalities in specifications. It may be tempting for designers to use vague, catchall language or to graft general performance standards onto what are otherwise proprietary specifications.

It is equally important to avoid the use of vague and unenforceable subjective terms in specifications. When preparing specifications for some federal projects, the author was furnished with a list of phrases typically found in specifications, which we reproduce in the list in the next section. The agencies' instructions were that such terms would not be permitted on their specifications. Needless to say, there are more specifications that contain such phrases than there are examples that avoid them.

The specifications are one of the most important tools of the inspector or Resident Project Representative. Often, professional engineers have taken the position that the sole mission of their Resident Project Representative and inspectors is to provide a project constructed in strict accordance with the plans and specifications. Upon issuing this profound proclamation, they lean back and assume that all should go well if their inspector or Resident Project Representative simply follows the plans and specifications.

Perhaps in many cases, he or she can do just that. However, as is often the case, the problem began in the office. All too often, specifications are not properly prepared, nor are they often prepared by engineers or architects who specialize in such work, so problems are just built-in from the outset.

Unenforceable Phrases

There are many pitfalls that can be encountered in the preparation of specifications, and following is a simple list of words and phrases often encountered that have no enforceable meaning and should not be allowed to be used in specifications. A few are actually unbiddable. There are two or three others that can be used in proper context but must be used only with great care. The list was originally compiled by a federal agency and may include some phrases used only by them. If a Resident Project Representative or inspector encounters this type of specification language, it is suggested that diplomacy and compromise are in order, as the terms are virtually impossible to administer in an objective, literal sense:

1. To the satisfaction of the engineer
2. As determined by the engineer
3. In accordance with the instructions of the engineer
4. As directed by the engineer
5. In the judgment of the engineer
6. In the opinion of the engineer
7. Unless otherwise directed by the engineer (unbiddable)
8. To be furnished if requested by the engineer (unbiddable)
9. In strict accordance with
10. In accordance with the best commercial practice
11. In accordance with the best modern standard practice
12. In accordance with the best engineering practice
13. Workmanship shall be of the highest quality
14. Workmanship shall be of the highest grade
15. Accurate workmanship
16. Securely mounted
17. Installed in a neat and workmanlike manner
18. Skillfully fitted
19. Properly connected
20. Properly assembled
21. Good working order
22. Good materials

23. In accordance with applicable published specifications
24. Products of a recognized reputable manufacturer
25. Test will be made unless waived (unbiddable)
26. Materials shall be of the highest grade, free from defects or imperfections, and shall be of grades approved by the engineer
27. Links and bends may be cause for rejection
28. Carefully performed
29. Neatly finished
30. Metal parts shall be cleaned before painting
31. Suitably housed
32. Smooth surfaces
33. Pleasing lines
34. Of an approved type
35. Of a standard type
36. When required by the engineer (unbiddable)
37. As the engineer may require (unbiddable)
38. In accordance with the standards of the industry

If phrases such as those in the foregoing list are encountered in a specification draft, a careful study should be made of methods of rewording the document to avoid the probability of disputes with the contractor over the interpretation of what the terms mean. Remember, too, that the basic rule of contracts is that in case of ambiguity, the intent of the contract will generally be interpreted in favor of the party who did not draft the contract—which, in short, means that the contractor's interpretation will carry greater weight than that of the owner or the architect/engineer. It is wise to avoid the habit of using such terms. Although some of the foregoing phrases can occasionally be justified, their use is usually a sign of a specifications engineer who either does not understand what he or she is calling for, is working under severe budget or time constraints, or is too lazy to research the issues properly and establish more specific, unambiguous requirements.

Content and Component Parts of a Specification

Content of the Specifications

In addition to the well-known *technical* provisions contained in the specifications, it should be clear that the term *specifications* is not necessarily limited to the technical portions alone. In many organizations, everything that is bound into the specifications document is referred to as the "specifications." This may include the notice of invitation to bid on the project; the bidding documents and forms, including the bid bond where required; contract (agreement) forms, including performance and payment bonds where required and noncollusion affidavits where required; the conditions of the contract, often referred to simply as the "boilerplate" because it provides a protective shield around the contract by anticipating most of the areas of discussion or dispute that might arise and provides for an orderly way of resolving each such case; and finally, the technical provisions. If the term *contract documents* is used, it legitimately includes everything, including the drawings, and sometimes includes a book of "standard specifications" by reference, as well. Usually, some of the boilerplate documents will specify a list of all items that are to be officially classed as a part of the contract documents.

Another often misunderstood characteristic of a set of specifications is that the size of a specification is in no way related directly to either the size or cost of a project, but is actually more influenced by how many different trades or materials are involved in the work. Thus a public restroom building in a park with only one room and a single set of plumbing fixtures may require as thick a set of specifications as a two-story building. On the other hand, a highway construction job costing 10 times the price of either of the two buildings described may involve a specification of only three or four sections and possibly as few as 8 or 10 pages of technical provisions.

Most contractors, inspectors, or other construction administrators have, at some time or another, questioned the wisdom of the specifications writer. If it will give the inspector any peace of mind, the author readily believes that all specifications writers are not necessarily knowledgeable in some of the subjects about which they write (sometimes an understatement). In a recent ASCE (American Society of Civil Engineers) questionnaire circulated on a national basis by its National Task Committee on Specifications to engineers, contractors, public agencies, owner-developers, suppliers, and attorneys, the general response from the contractors was that a specifications writer should have field construction experience before becoming a specifications writer. Every construction worker has undoubtedly run into specifications at some time where it seemed obvious that the specifications writer did not possess this background. Part of the problem lies in the procedures often used by architects and engineers in selecting personnel for, and budgeting time and costs for, the production of specifications. All the standardized specification formats in the world cannot cure the problem of a skimpy budget. If each of the items specified is properly covered, the Resident Project Representative or CQC representative will have the tools for and sufficient authority to assure the owner of quality in construction. The inspector is deprived of a primary tool if these important considerations are neglected.

Component Parts of a Specification

Generally, most specifications can be divided into three main elements, or parts. Although these parts are not necessarily arranged on each job in the same order in which the resident engineer or inspector will encounter them, the various design firms or public agencies responsible will generally keep the content of the specifications within the classifications shown in Figure 6.1.

In addition to the classifications indicated in Figure 6.1, all publicly funded projects require a listing of minimum wage rates. These are normally listed in the specifications.

If a public project has federal funding as well as being subject to state labor code requirements, then, in addition to the requirements for state wage rates, a complete copy of the applicable federal wage rates must also be bound into the specifications,

PART I - BIDDING AND CONTRACTUAL DOCUMENTS AND FORMS

Notice Inviting Bids	(Private may be informal)
Instructions to Bidders	
Proposal (Bid) Forms	
Proposal (Bid)	
Bid Schedule	
Contractor Certificates	(Applies to public works contracts)
List of Subcontractors	(Applies to public works contracts)
Non-Collusion Affidavit	(Applies to public works contracts)
Non-Discrimination Clause	(Applies to public works contracts)
Bidder's General Information	
Bid Bond (Bid Security Form)	(Applies to public works contracts)
Agreement and Bonds	
Agreement (Contract)	
Workers Compensation Certificate	(Public or private)
Performance Bond	(Public or private)
Payment Bond	(Public or private)
Certificate of Insurance	

PART II - CONDITIONS OF THE CONTRACT

General Conditions of the Construction Contract	
Supplementary General Conditions (Special for the project)	
Federal Supplement	(Federally funded projects only)

PART III - TECHNICAL SPECIFICATIONS

(From this point on, the architect-engineer provides technical sections covering the various parts of the project. Generally, this will be in CSI 16-Division format)

FIGURE 6.1 Three-Part Specifications Format.

and the contractor is obligated to pay the higher of the two rates if there is a difference. The federal wage rates are normally reproduced directly from the *Federal Register,* which lists wage rates all over the nation. Every Friday, new listings appear in the *Federal Register* of all wage rate schedules or changes. Any specifications containing federal wage rates should also have the sheets containing the current modifications to the general wage rate determination.

Instructions to Bidders

The Instructions to Bidders is usually a preprinted document and on public works projects is normally considered as one of the contract documents. Thus, the provisions of the Instructions to Bidders are as binding upon the bidder and the contractor as are the provisions of the technical specifications. Failure to comply with its terms can render a contractor's bid as "informal," or "nonresponsive," which may be used as justification for rejecting it. The general subject area usually covered by an Instructions to Bidders document includes the following:

Form of bid and signature

Interpretation of drawings and specifications

Preparation of the proposal
List of documents to be submitted with the bid
Bonding requirements
What is expected of the successful bidder
Insurance policies required
Basis for selection of the successful bidder

General Conditions

The General Conditions, or "boilerplate" as it is often called, is the most overlooked, yet one of the most important documents in the specifications to the resident inspector. It is this document that establishes the ground rules for administration of the construction phase of the project. The subject matter generally covered in General Conditions is fairly consistent from job to job wherever "standard" preprinted documents of a governmental agency or AIA, EJCDC, FIDIC, or similar organizations are used. Whenever an architect or engineer or a public agency elects to prepare its own General Conditions, the resulting document frequently lacks many of the essentials and, worse yet, has never been "proved" in court. The general subjects covered in most so-called "standard" General Conditions documents include:

Legal definitions of terms used in the contract
Correlation and intent of the documents
Time and order of the work
Assignment of contracts
Subcontracts
Where to serve legal notices
Authority of the architect/engineer
Change orders and extra work
Extensions of time for delays
Right of the owner to terminate the contract
Right of the contractor to terminate the contract
Right of the owner to take over work
Obligations of the contractor
Supervision by the contractor
Handling of claims and protests
Lines, grades, and surveys; who performs and who pays
Defective work or materials
Materials and workmanship
Provisions to allow access to all parts of the work
Inspection and tests; how administered and who pays
Coordination with other contractors at the site or nearby
Suspension of all or part of the work

Liquidated damages for delay
Stop notice procedures
Right of owner to withhold payment
Provisions for public safety
Changed conditions (sometimes called "unforeseen conditions")
Estimates and progress payments
Final payment and termination of liability
Protection and insurance
Disputes; settlement by arbitration

Technical Provisions

Part III of the specifications, as outlined in Figure 6.1, refers to that portion of the specifications that a layperson usually thinks of when one speaks of specifications. In this portion of the document are the detailed technical provisions that relate to the installation or construction of the various parts of the work and to the materials used in the work. There are several ways of logically dividing these sections into subject areas so as to lend some sort of order to the final document. Most of the systems, however, generally group specifications sections into trade-related functions as a means of easy grouping. This sometimes prompts the complaint from contractors that the sections do not accurately represent the responsibility areas of the various trades. Actually, there is usually no attempt made to conform exactly to trade jurisdictions, as they vary significantly from one part of the nation to the other; in fact, in some cases, jurisdictional differences may be evident from one adjacent county to the other within the same state.

It should be recognized that on any project there are usually a few technical requirements that would apply to *all* sections equally. In such cases, it has been found desirable to provide a section of the technical specifications, usually at the front of the technical portion, that may be entitled "General Requirements" and that spells out the various requirements of a *technical* nature that apply generally to the entire project. This section should not be confused with any of the text of Part II—Conditions of the Contract; there the General Conditions and Supplementary General Conditions are matters of a legal/contractual nature—*not technical provisions.* A common error made by general contractors in their dealings with subcontractors is to hand them a single technical specifications section relating to their trade, without copies of either the General Requirements or the General and Supplementary General Conditions of the Contract. This has often created serious problems in the conduct of the work, as all of the boilerplate sections apply to all of the work of each subcontractor as well.

What Do the Specifications Mean to the Inspector?

The specifications, in short, are one of the inspector's vital tools. Without them, an inspector cannot possibly perform in a competent manner. To be able to use these tools effectively, however, the inspector should have an idea of the relative importance of each of the various component parts of the contract documents. The following is a condensed list of some of the contract document components and their relative importance. For a more complete list, refer to Chapter 18.

Agreement governs over specifications.

Specifications govern over drawings.

Detail specifications govern over general specifications.

Each month, on larger projects, the contractor normally applies for and receives monthly partial payments (progress payments) for work completed thus far. The amount of each payment must be in direct proportion to the amount of work completed during the preceding month. It is the responsibility of the Resident Project Representative to check the quantities of such work completed, estimate its value, and review the monthly payment request of the contractor prior to submitting it to the design firm's project manager or sometimes the owner, along with a recommendation for payment, if justified. In the handling of such matters, the terms of the General Conditions must be strictly followed, as the procedures for handling such payment claims must follow an orderly, prearranged plan. There are no provisions for allowing terms or creating restrictions that were not written into the original contract.

In short, the entire policy for the administration and conduct of the work at the job site is established under the terms and conditions of the General and Supplementary General Conditions of the construction contract. The remaining portions of the specifications more properly relate to quality control or quality assurance functions, which are a subsidiary function of the Resident Project Representative.

CSI Specifications Format—Its Meaning and Importance

Briefly mentioned in earlier paragraphs was the fact that the technical portions of the specifications were generally structured in whatever manner suited the architect or engineer who prepared them. In the past years, this problem was even worse, and a contractor would indeed have to be versatile to be required to work from one type of contract documents on one job and at the same time be constructing another similar project nearby, from another set of documents that bore no resemblance to the first.

In recent years, an organization called the Construction Specifications Institute (CSI) tackled the task of attempting to inject some degree of uniformity and standardization into the general arrangement and method of writing construction specifications. To this end it has been enormously successful, although it can be seen that the format or arrangement and classification system it has devised was created by architects and engineers whose experience was limited to the construction of *buildings,* and the resulting format shows very little influence of engineers engaged in heavy engineering and similar types of construction. The CSI Format was indeed intended for buildings, but once entrenched, no one seemed to be able to change or even alter the system when it became desirable to extend it to other types of engineered construction as well. Thus, when the system is used for certain types of heavy engineering projects, some serious formatting conflicts present themselves. In any case, the system *did* create order where none existed before by setting forth a list of 16 standardized "divisions," which were supposed to work for everything, and with a little imagination it could indeed be adapted to many, though not all, nonbuilding construction projects, even though it may be somewhat cumbersome and awkward in certain types of work.

Original CSI 16-Division Format

The Original CSI 16-Division Format was adopted by the AGC (Associated General Contractors), the AIA (American Institute of Architects), the NSPE (National Society of Professional Engineers), and others in the United States and Canada in the form of a document entitled *Uniform System for Building Specifications.* Note that word *building* again—no mention of heavy engineering projects. Nevertheless, it is widely used both for building and some engineering work, and is popularly known as the CSI Format. This system has been officially adopted for *all* construction work by the U.S. Army, Corps of Engineers; the U.S. Navy (NAVFAC); National Aeronautics and Space Administration (NASA); the state of New York for public works projects; and by numerous other public and private agencies; however, it was not adopted nor endorsed by the ASCE (American Society of Civil Engineers) for heavy construction projects. In fact, with the release of MasterFormat™ 2004, the new 50-Division Format will be better suited to heavy construction.

Eventually, all the manufacturers followed suit, and now most, if not all, building materials are identified with the CSI classification number for filing purposes, which corresponds to the CSI division number under which each such product is intended to be grouped. Thus, most of the time, if you pick up a specification, it will be under CSI Format, and even without a table of contents, you should be able to find the section you are searching for. As an example, you should automatically turn to Division 3 if you are looking for concrete, or Division 16 if you are looking for electrical work (Figure 6.2). Note that under the CSI 16-Division Format, whenever a job does not use a certain division, it is simply skipped, but the numbers of the remaining divisions never change.

Unofficially, many civil/sanitary engineers who design water and wastewater treatment plants added a Division 17 to their specifications for Instrumentation and Control. Under the CSI manual, this subject is split between Divisions 15 and 16; unfortunately, this was often unworkable for the engineer that must sign for that portion of the specifications. Technically, the subject of instrumentation and control is neither mechanical nor electrical, but actually involves electronic systems.

The CSI moved slowly at first when out of its original element (buildings), but with the adoption of the new 50-Division MasterFormat 2004, I believe that we can look forward to a universal specification formatting system that may well serve the needs of those professionals who design buildings as well as heavy engineering works or systems. As the new MasterFormat becomes more widely adopted by the industry, all inspectors should learn to use it. Memorize all 50 Division titles, as they should never change even from one job to another.

1. General Requirements
2. Site Work and Utilities (includes all civil work)
3. Concrete
4. Masonry
5. Metals
6. Wood and Plastics
7. Thermal and Moisture Protection
8. Doors and Windows
9. Finishes
10. Specialities
11. Equipment
12. Furnishings
13. Special Construction
14. Conveying Systems
15. Mechanical
16. Electrical

Note that whenever a job does not use a certain division, it is simply skipped . . . but the numbers of the remaining divisions still never change.

FIGURE 6.2 List of Original CSI 16 Divisions.

CSI Division/Section Concept

Whenever the term *division/section concept* is heard with regard to the CSI Format, it is an expression of the relationship of the fixed-title divisions to the subclassifications under each division called "sections." Although division titles *never* change from job to job (although some divisions may be omitted from a project if they are not applicable), the titles of the sections that are grouped under them are adapted to the specific needs of each individual project. In Figure 6.3 all 16 fixed-division titles may be used under each appropriate division. In addition to the short list illustrated, the CSI also publishes a document that provides a ready index of section titles to fit the CSI 16-Division Format. The latest version of the CSI 16-Division format was the 1995 edition. Subsequent to that the new 50-Division MasterFormat™ 2004 was released. It is currently undergoing a transition to replace the 1995 edition previously used. This transition will take time, but it may yet resolve many of the previous complaints by engineers designing heavy construction projects.

CSI Three-Part Section Format

One of the most valuable contributions of the CSI to the work of the contractor and the inspector is the adoption of the three-part *section format* (Figure 6.4). It is a time-honored concept first observed by the author on the published specifications standards for a Federal Aid Road Act project dated in 1917. Under this arrangement each *section* is divided into three parts, each containing one type of information only. With this system, fewer items are overlooked simply because the specifications for a particular product were sandwiched between some unlikely paragraphs dealing with the

MasterFormat™

LEVEL TWO NUMBERS AND TITLES

INTRODUCTORY INFORMATION

00001 PROJECT TITLE PAGE
00005 CERTIFICATIONS PAGE
00007 SEALS PAGE
00010 TABLE OF CONTENTS
00015 LIST OF DRAWINGS
00020 LIST OF SCHEDULES

BIDDING REQUIREMENTS

00100 BID SOLICITATION
00200 INSTRUCTIONS TO BIDDERS
00300 INFORMATION AVAILABLE TO BIDDERS
00400 BID FORMS AND SUPPLEMENTS
00490 BIDDING ADDENDA

CONTRACTING REQUIREMENTS

00500 AGREEMENT
00600 BONDS AND CERTIFICATES
00700 GENERAL CONDITIONS
00800 SUPPLEMENTARY CONDITIONS
00900 ADDENDA AND MODIFICATIONS

FACILITIES AND SPACES

FACILITIES AND SPACES

SYSTEMS AND ASSEMBLIES

SYSTEMS AND ASSEMBLIES

CONSTRUCTION PRODUCTS AND ACTIVITIES

DIVISION 1—GENERAL REQUIREMENTS

01100 SUMMARY
01200 PRICE AND PAYMENT PROCEDURES
01300 ADMINISTRATIVE REQUIREMENTS
01400 QUALITY REQUIREMENTS
01500 TEMPORARY FACILITIES AND CONTROLS
01600 PRODUCT REQUIREMENTS
01700 EXECUTION REQUIREMENTS
01800 FACILITY OPERATION
01900 FACILITY DECOMMISSIONING

DIVISION 2—SITE CONSTRUCTION

02050 BASIC SITE MATERIALS AND METHODS
02100 SITE REMEDIATION
02200 SITE PREPARATION
02300 EARTHWORK
02400 TUNNELING, BORING, AND JACKING
02450 FOUNDATION AND LOAD-BEARING ELEMENTS
02500 UTILITY SERVICES
02600 DRAINAGE AND CONTAINMENT
02700 BASES, BALLASTS, PAVEMENTS, AND APPURTENANCES
02800 SITE IMPROVEMENTS AND AMENITIES
02900 PLANTING
02950 SITE RESTORATION AND REHABILITATION

DIVISION 3—CONCRETE

03050 BASIC CONCRETE MATERIALS AND METHODS
03100 CONCRETE FORMS AND ACCESSORIES
03200 CONCRETE REINFORCEMENT
03300 CAST-IN-PLACE CONCRETE
03400 PRECAST CONCRETE
03500 CEMENTITIOUS DECKS AND UNDERLAYMENT
03600 GROUTS
03700 MASS CONCRETE
03900 CONCRETE RESTORATION AND CLEANING

DIVISION 4—MASONRY

04050 BASIC MASONRY MATERIALS AND METHODS
04200 MASONRY UNITS
04400 STONE
04500 REFRACTORIES
04600 CORROSION-RESISTANT MASONRY
04700 SIMULATED MASONRY
04800 MASONRY ASSEMBLIES
04900 MASONRY RESTORATION AND CLEANING

DIVISION 5—METALS

05050 BASIC METAL MATERIALS AND METHODS
05100 STRUCTURAL METAL FRAMING
05200 METAL JOISTS
05300 METAL DECK
05400 COLD-FORMED METAL FRAMING
05500 METAL FABRICATIONS
05600 HYDRAULIC FABRICATIONS
05650 RAILROAD TRACK AND ACCESSORIES
05700 ORNAMENTAL METAL
05800 EXPANSION CONTROL
05900 METAL RESTORATION AND CLEANING

DIVISION 6—WOOD AND PLASTICS

06050 BASIC WOOD AND PLASTIC MATERIALS AND METHODS
06100 ROUGH CARPENTRY

FIGURE 6.3 CSI *MasterFormat*™ (1995 Edition) Master List of Section Titles and Numbers based upon the original CSI 16-Division Format as used prior to 2004.

06200 FINISH CARPENTRY
06400 ARCHITECTURAL WOODWORK
06500 STRUCTURAL PLASTICS
06600 PLASTIC FABRICATIONS
06900 WOOD AND PLASTIC RESTORATION AND CLEANING

DIVISION 7—THERMAL AND MOISTURE PROTECTION

07050 BASIC THERMAL AND MOISTURE PROTECTION MATERIALS AND METHODS
07100 DAMPPROOFING AND WATERPROOFING
07200 THERMAL PROTECTION
07300 SHINGLES, ROOF TILES, AND ROOF COVERINGS
07400 ROOFING AND SIDING PANELS
07500 MEMBRANE ROOFING
07600 FLASHING AND SHEET METAL
07700 ROOF SPECIALTIES AND ACCESSORIES
07800 FIRE AND SMOKE PROTECTION
07900 JOINT SEALERS

DIVISION 8—DOORS AND WINDOWS

08050 BASIC DOOR AND WINDOW MATERIALS AND METHODS
08100 METAL DOORS AND FRAMES
08200 WOOD AND PLASTIC DOORS
08300 SPECIALTY DOORS
08400 ENTRANCES AND STOREFRONTS
08500 WINDOWS
08600 SKYLIGHTS
08700 HARDWARE
08800 GLAZING
08900 GLAZED CURTAIN WALL

DIVISION 9—FINISHES

09050 BASIC FINISH MATERIALS AND METHODS
09100 METAL SUPPORT ASSEMBLIES
09200 PLASTER AND GYPSUM BOARD
09300 TILE
09400 TERRAZZO
09500 CEILINGS
09600 FLOORING
09700 WALL FINISHES
09800 ACOUSTICAL TREATMENT
09900 PAINTS AND COATINGS

DIVISION 10—SPECIALTIES

10100 VISUAL DISPLAY BOARDS
10150 COMPARTMENTS AND CUBICLES
10200 LOUVERS AND VENTS
10240 GRILLES AND SCREENS
10250 SERVICE WALLS
10260 WALL AND CORNER GUARDS
10270 ACCESS FLOORING
10290 PEST CONTROL
10300 FIREPLACES AND STOVES
10340 MANUFACTURED EXTERIOR SPECIALTIES
10350 FLAGPOLES
10400 IDENTIFICATION DEVICES
10450 PEDESTRIAN CONTROL DEVICES
10500 LOCKERS
10520 FIRE PROTECTION SPECIALTIES
10530 PROTECTIVE COVERS
10550 POSTAL SPECIALTIES
10600 PARTITIONS
10670 STORAGE SHELVING
10700 EXTERIOR PROTECTION
10750 TELEPHONE SPECIALTIES
10800 TOILET, BATH, AND LAUNDRY ACCESSORIES
10880 SCALES
10900 WARDROBE AND CLOSET SPECIALTIES

DIVISION 11—EQUIPMENT

11010 MAINTENANCE EQUIPMENT
11020 SECURITY AND VAULT EQUIPMENT
11030 TELLER AND SERVICE EQUIPMENT
11040 ECCLESIASTICAL EQUIPMENT
11050 LIBRARY EQUIPMENT
11060 THEATER AND STAGE EQUIPMENT
11070 INSTRUMENTAL EQUIPMENT
11080 REGISTRATION EQUIPMENT
11090 CHECKROOM EQUIPMENT
11100 MERCANTILE EQUIPMENT
11110 COMMERCIAL LAUNDRY AND DRY CLEANING EQUIPMENT
11120 VENDING EQUIPMENT
11130 AUDIO-VISUAL EQUIPMENT
11140 VEHICLE SERVICE EQUIPMENT
11150 PARKING CONTROL EQUIPMENT
11160 LOADING DOCK EQUIPMENT
11170 SOLID WASTE HANDLING EQUIPMENT
11190 DETENTION EQUIPMENT
11200 WATER SUPPLY AND TREATMENT EQUIPMENT
11280 HYDRAULIC GATES AND VALVES
11300 FLUID WASTE TREATMENT AND DISPOSAL EQUIPMENT
11400 FOOD SERVICE EQUIPMENT
11450 RESIDENTIAL EQUIPMENT
11460 UNIT KITCHENS
11470 DARKROOM EQUIPMENT
11480 ATHLETIC, RECREATIONAL, AND THERAPEUTIC EQUIPMENT

FIGURE 6.3 (Continued)

11500 INDUSTRIAL AND PROCESS EQUIPMENT
11600 LABORATORY EQUIPMENT
11650 PLANETARIUM EQUIPMENT
11660 OBSERVATORY EQUIPMENT
11680 OFFICE EQUIPMENT
11700 MEDICAL EQUIPMENT
11780 MORTUARY EQUIPMENT
11850 NAVIGATION EQUIPMENT
11870 AGRICULTURAL EQUIPMENT
11900 EXHIBIT EQUIPMENT

DIVISION 12—FURNISHINGS

12050 FABRICS
12100 ART
12300 MANUFACTURED CASEWORK
12400 FURNISHINGS AND ACCESSORIES
12500 FURNITURE
12600 MULTIPLE SEATING
12700 SYSTEMS FURNITURE
12800 INTERIOR PLANTS AND PLANTERS
12900 FURNISHINGS RESTORATION AND REPAIR

DIVISION 13—SPECIAL CONSTRUCTION

13010 AIR-SUPPORTED STRUCTURES
13020 BUILDING MODULES
13030 SPECIAL PURPOSE ROOMS
13080 SOUND, VIBRATION, AND SEISMIC CONTROL
13090 RADIATION PROTECTION
13100 LIGHTNING PROTECTION
13110 CATHODIC PROTECTION
13120 PRE-ENGINEERED STRUCTURES
13150 SWIMMING POOLS
13160 AQUARIUMS
13165 AQUATIC PARK FACILITIES
13170 TUBS AND POOLS
13175 ICE RINKS
13185 KENNELS AND ANIMAL SHELTERS
13190 SITE-CONSTRUCTED INCINERATORS
13200 STORAGE TANKS
13220 FILTER UNDERDRAINS AND MEDIA
13230 DIGESTER COVERS AND APPURTENANCES
13240 OXYGENATION SYSTEMS
13260 SLUDGE CONDITIONING SYSTEMS
13280 HAZARDOUS MATERIAL REMEDIATION
13400 MEASUREMENT AND CONTROL INSTRUMENTATION
13500 RECORDING INSTRUMENTATION
13550 TRANSPORTATION CONTROL INSTRUMENTATION
13600 SOLAR AND WIND ENERGY EQUIPMENT
13700 SECURITY ACCESS AND SURVEILLANCE
13800 BUILDING AUTOMATION AND CONTROL
13850 DETECTION AND ALARM
13900 FIRE SUPPRESSION

DIVISION 14—CONVEYING SYSTEMS

14100 DUMBWAITERS
14200 ELEVATORS
14300 ESCALATORS AND MOVING WALKS
14400 LIFTS
14500 MATERIAL HANDLING
14600 HOISTS AND CRANES
14700 TURNTABLES
14800 SCAFFOLDING
14900 TRANSPORTATION

DIVISION 15—MECHANICAL

15050 BASIC MECHANICAL MATERIALS AND METHODS
15100 BUILDING SERVICES PIPING
15200 PROCESS PIPING
15300 FIRE PROTECTION PIPING
15400 PLUMBING FIXTURES AND EQUIPMENT
15500 HEAT-GENERATION EQUIPMENT
15600 REFRIGERATION EQUIPMENT
15700 HEATING, VENTILATING, AND AIR CONDITIONING EQUIPMENT
15800 AIR DISTRIBUTION
15900 HVAC INSTRUMENTATION AND CONTROLS
15950 TESTING, ADJUSTING, AND BALANCING

DIVISION 16—ELECTRICAL

16050 BASIC ELECTRICAL MATERIALS AND METHODS
16100 WIRING METHODS
16200 ELECTRICAL POWER
16300 TRANSMISSION AND DISTRIBUTION
16400 LOW-VOLTAGE DISTRIBUTION
16500 LIGHTING
16700 COMMUNICATIONS
16800 SOUND AND VIDEO

FIGURE 6.3 (Continued)

PART I - GENERAL

Description of work; related work; submittals; inspection requirements; testing; certificates; etc.

PART II - PRODUCTS

Technical specifications for all materials, equipment, fabricated items; etc. In no case is it appropriate to describe any installation requirements in this part, nor to specify quality of workmanship in this part.

PART III - EXECUTION

Qualitative requirements relating to workmanship, approved methods, etc. This Part III covers installation, erection, construction, etc. It is not appropriate to cover any product, material, equipment, or items fabricated off-site and delivered to the contractor as a finished component.

FIGURE 6.4 Three-Part Section Format.

installation of some totally unrelated item—which just happened to be located there because some architect or engineer happened to think of it while writing that portion of the section.

In the three-part section format, all technical sections of the specification are divided into three distinct parts, always in the same order: (1) general, (2) products, and (3) execution. If followed faithfully, as most users of the system will do, it makes the reading of the specifications a simple, orderly process and eliminates many an error due to oversight.

Heavy Construction Engineering Specification Format

For many years there was some negative reaction from engineers in the heavy engineering construction who felt that the 16-Division CSI format did not adequately serve their needs and in 1994 the ASCE formed a committee to study the needs of the heavy construction industry and develop a format that was specifically designed for heavy construction such as airports, treatment plants, landfills, waterways, tunnels, power plants, electrical power facilities, industrial process plants, railroads, dams, and underground utilities.

Some effort was made to work with CSI to accomplish this, and with the release by CSI of the new MasterFormat™ 2004, the older 16-Division Format was discontinued and it was replaced by an entirely new 50 Division format. However, the use of Federal Highway Administration and State Highway Department formats remained the standard for public agencies doing bridge and highway construction throughout the United States.

State Highway Department Formats

Long before the coming of the CSI Format, the federal government and various state highway departments established specification formats of their own in response to the needs of the type of construction in which they were engaged. Most states have settled on a uniform format based on the AASHTO (American Association of State Highway and Transportation Officials) model.

The basic similarity among all state highway specifications is the fact that they all use a published, bound book of "standard specifications" that covers in detail all general contract conditions as well as the technical specifications for all types of construction that could reasonably be anticipated in any highway or bridge project. The subject matter covered is not as narrow as one might at first expect, and to add complications to the specification, it frequently covers several alternative methods of completing the work.

To adapt these standard specifications to a specific project requires an additional document, for the standard specifications themselves cover far too broad a subject area. Furthermore, they do not indicate whether a specific method should be used on a particular project. This adaptation is accomplished by the preparation of a small supplementary specification called the *Special Provisions* or *Supplemental Specifications,* or, in some cases, *Contract Provisions* (Figure 6.5). This document clearly defines the changes to the Standard Specifications or any additions to or deletions from the Standard Specifications that might be necessary to adapt it to the specific project being constructed. For the sake of uniformity, the Special Provisions follow a

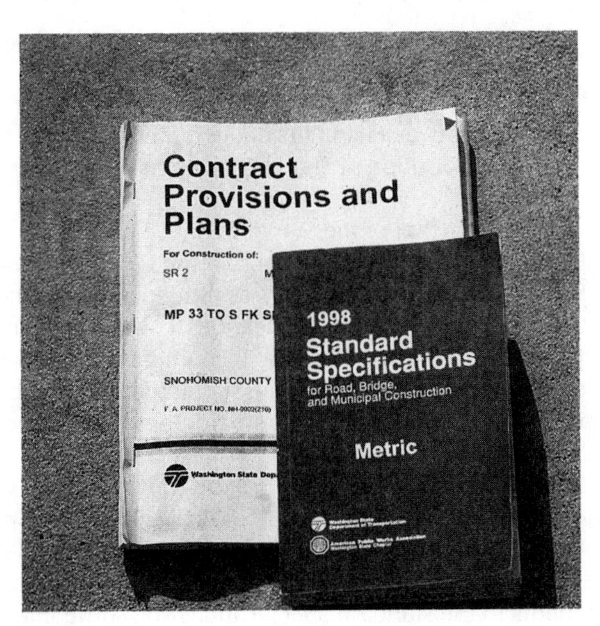

FIGURE 6.5 Example of a *Contract Provisions* Supplement to Standard Specifications for a Bridge Project in Snohomish County in the State of Washington.

"standard" format adopted by each using agency, so that all users throughout that state will produce a document that is in the same format, and all contractors will have prior knowledge of the basic requirements and conditions for highway and bridge construction in their state.

In the previous cases referred to, when "specifications" were mentioned, the term was interpreted to include all documents, General Conditions, and technical provisions. On a state highway project, this definition must be revised. Here the "specifications" are the Standard Specifications, and the document issued for the specific project contains only supplementary material. The usual title for the book containing the documents plus the special technical conditions is "Special Provisions."

Adaptation to City and County Civil Engineering Projects

Although originally developed for use by the Federal Highway Administration and the various state transportation agencies, the concept of utilizing a book of Standard Specifications, supplemented by a brief Special Provisions document prepared for each specific project, is widely used by cities, counties, and other public agencies whose work parallels the technical provisions contained in the Standard Specifications. Use by other agencies requires only minor changes to the General Provisions of the Standard Specifications to adapt it to local agency use. The concept is popular because of the considerably reduced cost of specification preparation using this approach.

The Special Provisions concept is not limited to public agencies, either. Wherever the work of a private project parallels the technical provisions contained in the Standard Specifications, it can be readily adapted to private use as well by modifying the General Provisions portion of the Standard Specifications in the Special Provisions document.

AASHTO Standard Format for Highway Construction Specifications

The majority of the state highway departments of the United States closely follow the standards of the American Association of State Highway and Transportation Officials (AASHTO).

The example used here is that of the Standard Specifications for Construction of Roads and Bridges on Federal Highway Projects of the U.S. Department of Transportation, Federal Highway Administration. Just as used in all of the states, the Federal Highway Administration Standard Specifications cover all potential types of highway and bridge construction that may be encountered, along with all of the possible acceptable alternatives. These must be supplemented by a specially prepared set of Special Provisions to adapt the Standard Specifications to a particular project.

As a means of keeping the Standard Specifications up to date, the various states also issue Supplemental Specifications, which are modifications sheets needed to update the Standard Specifications. These are normally available on the Web, or in hard copy from the state involved. These are sometimes termed as "standard special provisions" or "amendments to the standard specifications" or, sometimes, a combination of both names.

As with the Federal Highway Standard Specifications and the various state highway specifications, the selection and arrangement of the subject matter in the General Provisions, often called Division one or Section one, is virtually identical from state to state. Technical Provisions of Standard Specifications, however, are similar but not identical from state to state.

The Federal Highway Administration Standard Specifications are initially divided into 10 main divisions.

Div. 100 General Requirements (Sections 101 to 109; contractual relationships)
Div. 150 Project Requirements
Div. 200 Earthwork
Div. 250 Structural Embankments
Div. 300 Aggregate Courses
Div. 400 Asphalt Pavements and Surface Treatments
Div. 500 Portland Cement Concrete Pavement
Div. 550 Bridge Construction
Div. 600 Incidental Construction
Div. 700 Materials

Of these, the Division 100 subjects are common to virtually all DOT Standard Specifications. Typically, the subjects covered in the first nine sections of the Standard Specifications form the main portion of the Special Provisions, when written to be job-specific. Then by adding one more division or section, usually a section 10 commonly entitled Construction Details, all technical additions, deletions, or other modifications to the Standard Specifications can be covered in one section.

Typical Division 100 General Provisions format and content of the FHWA Standard Specifications, which is common to all state DOT Standard Specifications, follows. These titles then are used as a format for preparing the Special Provisions:

FHWA Division 100—General Requirements

1. 101. Definitions and Terms
2. 102. Proposal Requirements and Conditions
3. 103. Award and Execution of Contract
4. 104. Scope of Work
5. 105. Control of Work
6. 106. Control of Materials
7. 107. Legal Relations and Responsibility
8. 108. Prosecution and Progress
9. 109. Measurement and Payment
10. 150. Project Requirements

In the foregoing illustration, Division 150 of the FHWA Standard Specifications is seldom used in state DOT specifications; however, all other listed titles are applicable. In addition to the General Provisions, which are common to the Federal Highway

Administration (FHWA), the California Department of Transportation (Caltrans), the State of Washington (WSDOT), and other state departments of transportation, the Standard Specifications are further divided into sections or divisions relating to the principal construction materials and methods expected to be encountered in normal road and bridge construction. Anything needed for a project that is not contained in the standard specifications must be specified in the Special Provisions. With few exceptions, the FHWA and most state DOT Standard Specifications are published in metric units only, with no English units cited.

What most state highway specifications have in common with the CSI Format (long before the formation of the CSI) is the strict separation of materials and execution in their specifications. Thus, in all execution portions of a state highway specification, the materials are covered by reference to the detailed specifications in the division provided for that purpose.

State Department of Transportation Formats

The State of Washington Standard Specifications for Road, Bridge, and Municipal Construction is somewhat of a departure from the AASHTO standard, as they do not separate the technical sections into separate discrete sections or divisions. In some respects they follow the concept used by the CSI in that they utilize separate divisions to cover the various basic construction materials or types. They further depart from typical state highway department specifications by addressing municipal construction as well as state highway work.

The Washington Standard Specifications are separated into nine divisions. Division 1, General Requirements, is further divided into 10 sections, similar to that of the Federal Highway Administration (FHWA).

Div. 1. General Requirements (Sections 1–01 to 1–10 contractual relationships)
Div. 1. APWA Supplement
Div. 2. Earthwork
Div. 3. Production from Quarry and Pit Sites and Stockpiling
Div. 4. Bases
Div. 5. Surface Treatments and Pavements
Div. 6. Structures
Div. 7. Drainage Structures, Storm Sewers, Sanitary Sewers, Water Mains, and Conduits
Div. 8. Miscellaneous Construction
Div. 9. Materials

The principal difference is in the organization of the technical subject matter. Otherwise, as with all other State Highway specifications, they are formatted for the Special Provisions concept rather than Project Specifications or CSI Project Manual.

The California Department of Transportation, or "Caltrans," standard specifications also are a departure from the AASHTO standard, having predated AASHTO. They were designed to meet the specific needs of the construction industry in California for street, highway, and bridge construction projects and, as such, were adopted by many cities and counties throughout that state that administer similar projects at local levels. As with the AASHTO standard specification format, they must be supplemented by a book of Special Provisions. The format used in these Special Provisions has been standardized, and a uniform format with fixed-number sections and standard section titles has been adopted.

The Caltrans Standard Specifications are divided into nine divisions:

1. General Provisions (Sections 1 to 9; contractual relationships)
2. Miscellaneous
3. Grading
4. Subbases and Bases
5. Surfacings and Pavements
6. Structures
7. Drainage Facilities
8. Right of Way and Traffic Control Facilities
9. Materials

Of the above, Division 1 General Provisions follows the industry standard of dividing the General Requirements into nine standard section titles. The list of subjects covered, as well as the standardized title of each section, remains unchanged from one project to the next.

CALTRANS DIVISION 1—GENERAL PROVISIONS

Section 1. Definitions and Terms
Section 2. Proposal Requirements and Conditions
Section 3. Award and Execution of Contract
Section 4. Scope of Work
Section 5. Control of Work
Section 6. Control of Materials
Section 7. Legal Relations and Responsibility
Section 8. Prosecution and Progress
Section 9. Measurement and Payment

The arrangement of a typical California Department of Transportation "Special Provisions" document is shown in Figure 6.6.

Non-DOT Standard Specifications

Various non-DOT versions of standard specifications exist in various agencies throughout the public project sector. Many of these are locally prepared, but soon fall into obsolescence due to failure to keep updating the material. The accepted industry standard for updating codes and standards is approximately three-year intervals, or sooner.

NOTICE TO CONTRACTORS

COPY OF ENGINEER'S ESTIMATE

SPECIAL PROVISIONS

- Section 1 - Specifications and Plans
- Section 2 - Proposal Requirements and Conditions
- Section 3 - Award and Execution of Contract
- Section 4 - Scope of Work
- Section 5 - Control of Work
- Section 6 - Control of Materials
- Section 7 - Legal Relations and Responsibility
- Section 8 - Prosecution and Progress
- Section 9 - Measurement and Payment
- Section 10 - Construction Details (Cites only additions, deletions, or variations from the Standard Specifications)
- Section 11- (Blank)
- Section 12- (Blank)
- Section 13- (Blank)
- Section 14 - Federal Requirements for Federal Aid Construction Projects

PROPOSAL AND CONTRACT

Bid Forms
- Proposal (Bid)
- Bid Schedule
- List of Subcontractors
- Non-Collusion Affidavit
- Bidder's General Information
- Bid Bond (Bid Security Form)

Agreement and Bonds
- Agreement Form
- Workers Compensation Certificate
- Performance Bond
- Payment Bond
- Certificate of Insurance

FIGURE 6.6 California Department of Transportation Format for Special Provisions. (Other State DOT Special Provisions Are Similar.)

The American Public Works Association (APWA) in various regions has participated in the development of either local regional standard specifications, or supplements to existing state DOT standard specifications to enable the state DOT standard to be readily adaptable to local public agency use. The APWA in one such region has produced the Standard Specifications for Public Works Construction, more commonly called the "Greenbook." The publication is the work product of the

APWA Southern California Chapter and the Southern California Districts of the Associated General Contractors of America (AGC), and is updated every third year. It is used in the same manner as a DOT standard specification (although broader in coverage), by preparation of a special provisions document to supplement the standard Specifications.

APWA Standard Specifications differ primarily in that they address municipal engineering types of projects usually associated with city or county public works, whereas the state DOTs, with few exceptions, address only highway and bridge construction.

Other Nonstandard Construction Specifications Formats in Use

There are different approaches to the problem of separating a project specification into seemingly logical units of construction. The two most common concepts are:

1. Separation into trade-group and material classifications, as in the CSI Format
2. Separation by project features, wherein each significant feature of a project is described completely within a single section, including all the materials and methods involved to complete the specified structure or feature

Depending upon the nature of a specific project, each concept may have something good to be said of it, and each could serve to special advantage if used properly. If the two systems would ever be mixed in the same specification, however, the job of field administration could become chaotic! The problems resulting from such an unwise choice include the inability reasonably to control payments to the contractor for the various portions of the work, difficulty in defining interfaces in construction, and duplication of specification provisions in different parts of the work—often with varying and conflicting requirements. An engineer or inspector who is assigned to a project like this should be prepared for a lion-sized job of maintaining cost control, especially if it turns out to be a unit-price job.

A graphic example of the complicated relationships involved can be determined from Figure 6.7, which takes the various project features of a recreational park project as an example and superimposes them in matrix form with the various trade-group and material classifications that apply to the same work.

It becomes obvious from the chart that any project that included all these various project features would benefit most by using the CSI Format, which is based upon the separation of the specification technical provisions into sections corresponding to the vertical columns in the chart in Figure 6.7. Thus, there would be only one concrete section, one metals section, one electrical section, and so forth, no matter how many separate structures or units of construction were included in the project.

	PROJECT FEATURE	ORIGINAL CSI 16 DIVISIONS														
		2	3	4	5	6	7	8	9	10	11	12	13	14	15	16
A	Site Grading	X														
B	Site Utility Lines	X	X		X										X	
C	Storm Drain System	X	X		X											
D	Valve Chamber	X	X	X	X		X		X						X	X
E	Streets and Parking	X	X		X				X						X	X
F	Snack Bar Building	X	X	X	X	X	X	X	X	X	X				X	X
G	Rest Room Building	X	X	X	X	X	X	X	X	X	X				X	X
H	Administration Bldg	X	X	X	X	X	X	X	X	X	X				X	X
I	Landscaping & Irrig	X	X												X	

FIGURE 6.7 Comparison of Trade-Group versus Project-Feature Formatting.

However, if a project involved only street paving and parking lots, street lighting and signals, landscaping and irrigation, and a storm drain system, as many municipal improvement projects do, it might be perfectly practical to divide the specifications into sections describing project features corresponding to the horizontal lines on the chart in Figure 6.7. This is primarily because the materials of construction and the trades used in this case are peculiar to the unit being built, and the material specification for one of these units is not equally applicable to any of the other sections.

Project Specifications (Project Manual) versus Special Provisions Concept

This subject was touched on slightly in earlier paragraphs, but it is of great importance for the resident engineer or inspector to know the relative importance of the documents that must be used in the field and to understand the very important difference between these two concepts. Figure 6.8 illustrates the choices to be made.

Project Specifications (CSI Project Manual)

The *project specifications* concept (or *Project Manual* in CSI terminology) is based upon the issuance of a single, all-inclusive project specifications book containing all of the contract provisions that apply to the job, although *references to* outside sources are permissible. The effect of such outside references in this type of specification,

CHOOSING A SPECIFICATION FORMAT

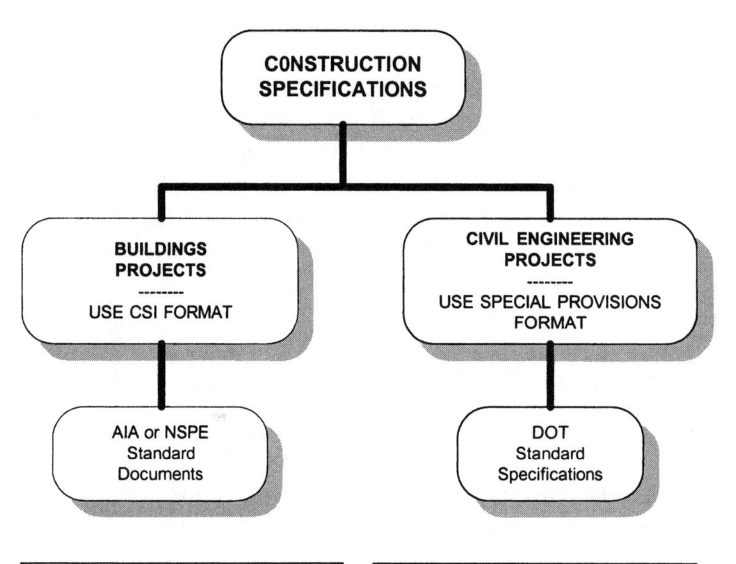

The primary difference between the AIA and the NSPE Standard Documents is that AIA documents are designed for use by architects for the construction of buildings. They are primarily slanted toward the private sector. The NSPE Standard Documents are the product of the Engineers Joint Contract Documents Committee (EJCDC) and are directed more toward engineering projects. Neither document meets the needs of public agencies for public works projects without extensive supplementary provisions.

The principle of Special Provisions is based upon the acceptance of a bound, preprinted book of Standard Specifications as a base document. The Special Provisions are a job-specific modification of various portions of the Standard Specifications. The Special Provisions should only contain additions, deletions, or modifications to the Standard Specifications. Unless specifically modified in the Special Provisions, all applicable provisions of the Standard Specifications will govern the Work. Under DOT Standard Specifications, Sections 1 through 9 contain the General Provisions.

FIGURE 6.8 Basis for Determining Choice of a Specifications Format.

however, is to bind the contractor only to the extent of the specific reference specification named. Thus, if a project specification states that "portland cement shall be Type II cement as specified in Section 90 of the Standard Specifications," it limits control of the cited reference to the *portland cement specifications only* and does not bind the contractor to any other provisions of that same section, such as the grading of aggregates for concrete.

Special Provisions or Supplemental Specifications

The *Special Provisions* or *Supplemental Specifications* concept is based upon the idea that a previously published book of Standard Specifications is the actual detailed specification for all applicable work on the project, and that the Special Provisions or Supplemental Specifications are merely a supplemental document to provide for those items on a particular project that the design engineer wanted changed from the provisions of the Standard Specifications, or where he or she made a specific selection of options provided in the Standard Specifications (Figure 6.9). Thus, for example, although it has already been established that the entire concrete section of the Standard Specifications will control the project (insofar as it is applicable, of course), the reference in the Special Provisions or Supplemental Specifications that "portland cement shall be Type II cement as specified in Section 90 of the Standard Specifications" merely controls the choice of option as to the *type of cement* required for the work. The total provisions of the rest of "Section 90" still apply to the concrete work. The exact specification phrase used in the *project specifications* can have a vastly different meaning when used in a *Special Provisions* or *Supplemental Specifications* document.

Inspector Training and Knowledge of Specifications

Inability to interpret specifications was next to the top of the contractor's list of complaints about inspectors in an ASCE study. These were generally felt to be "go-by-the-book" type inspectors. The usual reason offered for the go-by-the-book philosophy is lack of sufficient field experience to risk the slightest departure from the exact wording of the specifications—right or wrong.

Anyone who has ever had to work both in specifications and in field construction will know that there is little possibility that the information generated in the "sterile" design office environment can be followed literally during the construction phase of a contract. Field conditions vary frequently from the assumed design conditions, and many adjustments must be made. To accomplish this requires not only an experienced, well-qualified inspector, but also an enlightened project manager and specifications staff who can recognize the validity of these claims, respond to field suggestions to improve their own product, and work effectively with the inspectors as team members instead of as adversaries.

Of the contractors who claimed to have had to work with incompetent inspectors, some felt that their construction problems were the result of the inspectors' attempts to impose excessive requirements upon them as a form of "play-it-safe" engineering.

The two principal items that the contractors named as having been the cause of the greatest number of problems were (1) specifications interpretation and (2) lack of inspector experience and proper training.

The offshoot of the study was that inspection should be recognized as the principal problem and that poor inspection is prevalent. It was the recommendation of the contractors contacted that inspectors should improve themselves through education, training, certification, and frequent liaison with contractor organizations. It was also stated, in all fairness to inspectors, that the specifications were also in great need of improvement and that the engineers and architects should be held more

PROJECT SPECIFICATIONS (CSI)	SPECIAL PROVISIONS (Federal/State DOT)
PART I - PROPOSAL AND CONTRACT Notice Inviting Bids Instructions to Bidders (Preprinted) Proposal (Bid) and related forms Agreement and related forms Contract Administration forms (optional)	**STANDARD SPECIFICATIONS** DOT Standard Specifications (Includes General Provisions and technical specifications for typical highway, bridge, or street improvement project).
PART II - CONDITIONS OF THE CONTRACT General Conditions (Preprinted) Supplementary General Conditions Federal Requirements (if applicable)	**PART I - PROPOSAL AND CONTRACT** Notice Inviting Bids Proposal (Bid) and related forms Agreement and related forms (No separate Instructions to Bidders).
PART III - TECHNICAL SPECIFICATIONS Division 1 - General Requirements Division 2 - Site Work and Utilities Division 3 - Concrete Division 4 - Masonry Division 5 - Metals Division 6 - Wood and Plastics Division 7 - Thermal and Moisture Protection Division 8 - Doors and Windows Division 9 - Finishes Division 10 - Specialties Division 11 - Equipment Division 12 - Furnishings Division 13 - Special Construction Division 14 - Conveying Systems Division 15 - Mechanical Division 16 - Electrical End of standard Division titles; the specifier may add additional Divisions as necessary.	**PART II - SPECIAL PROVISIONS** 1. Definitions and Terms 2. Bidding Requirements and Conditions 3. Award and Execution of Contract 4. Scope of Work 5. Control of Work 6. Control of Material 7. Legal Relations and Responsibility to Public 8. Prosecution and Progress 9. Measurement and Payment 10. Construction Details (Technical) Section 10 is intended to cover ONLY those technical provisions that modify, delete, or add to the provisions of the Standard Specifications. Otherwise the Standard Specifications will govern all work. Sections 1 through 9 only require sufficient editing to make them job-specific.

COMMENTARY

1. Widely used nationally in public & private sectors for all types of construction; primarily for buildings.
2. Adopted by all federal agencies except Transportation
3. Not used by any state DOT agencies, nor by most local agencies on street & bridge work.
4. Uniform national standard format recognized by EJCDC, ACEC, ASCE, NSPE, AIA, AGC, ABC, USCE, USBR, NAVFAC, NASA, and many others.
5. Readily adaptable to published national front-end documents of AIA and EJCDC (NSPE)

1. Concept used by Federal Highway Administration and all state DOT's for roads & bridges.
2. Widely used by cities & counties for street improvements, utilities, and bridge construction.
3. Not suitable for building construction, water & wastewater treatment facilities, pump stations, power projects, etc.
4. Generally limits specifier to materials covered in Standard Specifications and contract admin. principles adapted to roads & bridges.
5. Format is a uniform national standard, but specific terms vary from state to state.

FIGURE 6.9 Comparison of Documents of Two Commonly Used Formats: Project Specifications versus DOT Special Provisions.

accountable for their own professional errors. Finally, the study concluded, a means should be developed to provide for more effective coordination and communication between the inspectors and the contractors, and between the inspectors and the specifications engineers.

Specifications and the Inspector

One of the principal weaknesses of some inspectors is their general lack of a good working knowledge of specifications. It is only fair at this point to mention also that some of the principal weaknesses among the specifications engineers are their general lack of specific knowledge of construction methods.

As an arbitrator over a number of years, the author has often discovered situations where the case involved a contractor versus a subcontractor, or an owner versus a contractor on a case where the root cause was flaws in the specifications. In such cases, the fault lay not with the parties to the arbitration or litigation, but with the engineer or architect who was responsible for the preparation of the specifications.

For many years the remark has been heard from specifications writers that "if only those people in the field would follow the specifications, we wouldn't have gotten into this problem." At the same time, the field people were saying "someone ought to send those specifications writers into the field for a while where they can be introduced to the real world." Therein lies the essence of a serious problem involving a failure to communicate. Each side suspects the other and challenges their competence.

Actually, there is some justification for each of their viewpoints. Field situations involving design or specifications difficulties should be reported in writing to the specifications department, and, wherever justified, changes should be made in the master specifications (guide specifications) to assure that the problem will not be continually repeated.

The fault does not lie wholly with either the inspectors or the specifications engineers. The point is that there must be an effective channel of communications between them. There must also be provisions for educating each group to the needs and problems of the other. In the author's own office, this was accomplished by occasionally transferring specifications writers into the field to work under proper supervision as an inspector. This was done during periods of slowdown in the specifications department, which usually occurs during periods of increased activity in the field.

The logic of this approach is based upon the fact that one of the major elements of each technical specification is that portion dealing with the "execution" of the work. It is in this area that many specifications writers are weakest, since the knowledge necessary to write an effective execution provision can only be obtained through actual field experience. Yet firms continue to have specifications written by people who have never worked in the field and who are simply copying material from old specifications that, to their eyes, appear to be from a "similar" project. Remember the old adage: "Methods used on other projects only work on other projects."

Specification General Conditions, or Boilerplate

The education of the inspector must be expanded to include careful study and understanding of the General Conditions of the Contract. Unfortunately, many seem to believe that the General Conditions were created by the lawyers for other lawyers to read only if, and when, a dispute goes to court.

Actually, the entire administration of the contract is based upon the provisions contained in this document. While the technical portions of the specifications, or "project manual" as it is called by the Construction Specifications Institute, describe the qualitative requirements of the final product that is to be constructed, the General Conditions provide for an orderly means of handling situations or problems that arise in the contractual relationship with the contractor. In preparing the General Conditions, an attempt is made to anticipate the various types of contractual situations that might arise during the course of the project, and by signing the agreement the owner and the contractor have agreed that should any of the described problems arise, they will be handled as set forth in the general conditions. This provides for the orderly resolution of construction difficulties without the necessity of resorting to the filing of claims that must be resolved through arbitration, litigation, or other dispute resolution alternative.

There is a need in today's complex construction environment for continuing education programs designed to meet the needs of both inspectors and specifications writers. Inspection and specifications are inseparable. The specifications not only are one of the principal elements of the contract documents, but serve as one of the principal tools of the inspector as well.

Allowances and Tolerances in Specifications

One of the factors that should be considered in the efforts to involve the specifications writer and the inspector in an effective working relationship is the inclusion in the specifications of tolerance limits, instead of the traditional methods of specifying absolutes. Some problems for which the inspectors have been blamed can actually be traced to the fact that the specifications either provided for no tolerance at all, or provided tolerances that were either unreasonable or unenforceable. An excellent example of properly specified tolerances can be seen in the American Concrete Institute *Standard Specifications for Tolerances for Concrete Construction and Materials (ACI 117-90 and Commentary (117R-90).* Tolerances for concrete construction expressed in metric (SI) units can be found in the U.S. Bureau of Reclamation *Metric Manual,* published by the U.S. Government Printing Office, Denver, CO. Where tolerances are specified, the inspector can be more secure in taking a stand with a contractor on an issue. The specifying of tolerances further eliminates many disputes with contractors who otherwise felt that rejections for nonconformance based upon the inspector's interpretation of absolute values was too subjective. As an absolute value is impossible to obtain consistently, an even greater burden is placed upon the inspector who is obligated to administer such a contract, for the inspector is then always being placed in the position of deciding "how close is close enough?"

All too often, such absolute values are not even well justified by the design computations, but may have been specified solely because the specifier lacked sufficient knowledge of actual construction materials, methods, and practices. Place the specifier in the field for a while and his or her attitude will most certainly change.

Even in a machine shop, where most people will admit that the demand for precision far exceeds that of construction in the field, the dimensions for machined parts are shown on the drawings, with tolerances shown for each dimension. Thus, even in precise work, it is recognized that an absolute dimension without an indicated tolerance is not only impractical, but virtually impossible. Let the construction industry take heed and reduce construction bid prices by applying reasonable tolerances, and at the same time minimize disputes with the contractor. Furthermore, by specifying tolerances during the design phase, the control is placed in the hands of the engineer or architect, where it rightfully belongs.

PROBLEMS

1. What are the three principal "parts" of the contract specifications document (or CSI "Project Manual" or "Contract Book" as some public agencies call the document)?
2. What are the three parts of a technical section under the CSI Format?
3. Prioritize the following documents in their traditional order of precedence:
   (a) _______ Technical Specifications
   (b) _______ Agreement
   (c) _______ General Conditions of the Contract
   (d) _______ Reference Documents
   (e) _______ Contract Drawings
   (f) _______ Supplementary General Conditions
4. Give three examples of unenforceable phrases if used in specifications.
5. If you were working with a set of project specifications prepared in CSI Format, under what section numbers would you find the following subjects?
   _______ Plumbing
   _______ Utility piping
   _______ Project meetings
   _______ Contractor submittals
   _______ Restoration of underground pipelines
   _______ Piling
   _______ Roof flashing and sheet metal
   _______ Sheet metal fabrications
   _______ Hardware
   _______ Butler buildings
   _______ Painting
6. Describe the function of the Special Provisions or Supplemental Specifications as used in a DOT project as opposed to the CSI Format.

7. Although CSI Format specifications are widely used, name one type of construction where they are seldom used.
8. Under CSI Format, explain the difference between the documents referred to as "General Conditions" and "General Requirements."
9. Under the CSI Format, the definition of *Contract Documents* does not include the Notice Inviting Bids, Instructions to Bidders, or the Bid itself. Name the principal exception to this rule, wherein the term *Contract Documents* may include the Notice Inviting Bids, Instructions to Bidders, and the Bid itself.
10. True or false? It is better to specify absolute dimensions than to indicate allowable tolerances in construction.
11. Which is more desirable in a specification: to require literal compliance with an absolute dimension or to specify a tolerance range for the inspector to work with?

7

USING THE SPECIFICATIONS IN CONTRACT ADMINISTRATION

GENERAL CONDITIONS OF THE CONSTRUCTION CONTRACT

The *General Conditions,* sometimes called the *General Provisions,* specify the manner and the procedures for implementing the provisions of the construction contract according to the accepted practices within the construction industry. These conditions are intended to govern and regulate the requirements of the formal contract or agreement. They do not serve as a waiver of any legal rights that either party to the contract may otherwise possess. The General Conditions of the Contract are intended to regulate the functions of either party only to the extent that his or her activities may affect the contractual rights of the other party or the proper execution of the work.

Although the General Conditions of most construction contracts vary somewhat from one set to another, depending upon the requirements of the agency that originated the document, most address similar issues, although not always in the same manner. The following examples will allow you to compare the content of three major national and international General Conditions documents. The following subjects addressed in the "Standard General Conditions of the Construction Contract" EJCDC[1] Document No. 1910-8 (1996 edition) may well be considered as somewhat typical of many engineering documents:

1. Definitions (terms used in the contract documents)
2. Preliminary matters (starting the project; preconstruction matters)
3. Contract documents, intent, amending, and reuse

[1]Engineers Joint Contract Documents Committee, c/o National Society of Professional Engineers, 1420 King Street, Alexandria, VA 22314.

4. Availability of lands, physical conditions, reference points (differing site conditions; soils reports)
5. Bonds and insurance
6. Contractor's responsibilities
7. Other work (related work at site; coordination)
8. Owner's responsibilities
9. Engineer's status during construction
10. Changes in the work (procedures)
11. Change of contract price (pricing provisions)
12. Change of contract times
13. Warranty and guarantee; tests and inspections; correction, removal, or acceptance of defective work
14. Payments to contractor and completion
15. Suspension of the work and termination
16. Dispute resolution
17. Miscellaneous

The EJCDC General Conditions are used mostly on engineering projects and by many public agencies, such as city and county governments, with supplements to adapt them to local state laws governing the conduct of public contracts in that state. In many instances they are used as a base document upon which to develop a proprietary document for a particular agency. The EJCDC documents were jointly prepared and endorsed by the following organizations:

American Consulting Engineers Council
American Society of Civil Engineers
National Society of Professional Engineers
Construction Specifications Institute
Associated General Contractors of America

Other standard sets of General Conditions have been developed by various segments of the construction industry, notably those of the American Institute of Architects for building construction. Although most published documents bear a superficial similarity to one another, some of the various documents do have serious differences.

The following subjects are addressed in the "General Conditions of the Contract for Construction" AIA[2] Document A201 (1987 edition), which is used almost universally on all architectural projects. If they are to be used on public building projects, however, they too would require a supplement to adapt to local state laws governing the conduct of public contracts in that state:

1. General Provisions (definitions of terms)
2. Owner (rights and obligations)

[2]American Institute of Architects, 1735 New York Ave., NW, Washington, DC 20006.

3. Contractor (obligations)
4. Administration of the contract (by the architect)
5. Subcontractors (owner's and architect's rights to control selection)
6. Construction by owner or by separate contractors (mutual responsibility)
7. Changes in the work (change process)
8. Time (delays; extensions; progress and completion)
9. Payments and completion (payment administration)
10. Protection of persons and property (by contractor)
11. Insurance and bonds
12. Uncovering and correction of work
13. Miscellaneous provisions (laws; notice; tests)
14. Termination or suspension of the contract
(Approved and endorsed by the Associated General Contractors of America.)

INTERNATIONAL CONSTRUCTION CONTRACTS

International projects have their own standards, usually FIDIC.[3] The documents of the "FIDIC Conditions of Contract for Works of Civil Engineering Construction" are used almost universally on international engineering projects financed by the international banking industry, the United Nations, and others. The scope of coverage of the FIDIC General Conditions includes:

1. Definition and interpretation
2. Engineer and engineer's representative
3. Assignment and subcontracting
4. Contract documents
5. General obligations
6. Labor
7. Materials, plant, and workmanship
8. Suspension

[3]International Federation of Consulting Engineers (FIDIC), 9 Carel van Bylandtlaan, The Hague, Netherlands (available in the United States from American Consulting Engineers Council, 1015 Fifteenth St., N.W., Washington, DC 20005).

FIDIC represents the interests of independent consulting engineers worldwide. It serves as the liaison for its membership with international banks and agencies, such as the International Bank for Reconstruction and Development, the Inter-American Development Bank, the African Development Bank, the Asian Development Bank, the United Nations Development Programme, the Food and Agriculture Organization of the United Nations, the World Health Organization, the United Nations Office of Technical Co-operation, the European Development Fund, the Kuwait Fund for Arab Economic Development, the Arab Fund for Economic and Social Development, and others. FIDIC documents are available in the United States from American Consulting Engineers Council, Washington, DC 20005.

9. Commencement and delays
10. Defects liability
11. Alterations, additions, and omissions
12. Procedure for claims
13. Contractor's equipment, temporary works, and materials
14. Measurement
15. Provisional sums
16. Nominated subcontractors
17. Certificates and payment
18. Remedies
19. Special risks
20. Release from performance
21. Settlement of disputes
22. Notices
23. Default of employer
24. Changes in cost and legislation
25. Currency and rates of exchange

Just as there are numerous differences in nomenclature between owners, architect/engineers, and contractors in various types of organizations, there are numerous differences between the terminology of international construction contracts in the United States and other countries.

For illustration, the following list compares terms of similar meaning from conventional American contracts and those from international contracts under FIDIC documents widely used outside the United States.

United States	International (FIDIC)
Addenda to Notice Inviting Bids	Appendix to Tender
Award of Contract	Letter of Acceptance
Beneficial Use	Taking-over
Bid	Tender
Bid Forms	Forms of Tender
Bid Schedule (unit price projects)	Bill of Quantities
Changes in the Work	Variations
Date of Notice to Proceed	Commencement Date
Differing Site Conditions	Adverse Physical Obstructions
Force Account	Daywork
Guarantee Period	Defects Liability Period
Owner	Employer
Partial Payment	Provisional Sum
Quantity changes in excess of 15 percent	Variations in excess of 15 percent
Retainage	Hold-back

Schedule	Programme
Schedule of Values	Breakdown of lump-sum items
Supplementary General Conditions	Conditions of Particular
Time-and-Materials work (sometimes called "force account")	Daywork

With the exception of those items that relate to matters of international currencies and assurances, it can be seen that all three documents—the EJCDC, the AIA, and the FIDIC—share many of the items of common concern to all people in the construction industry.

Standard forms of General Conditions have many advantages. Not only are they generally the result of collaboration with industry and government leaders, and thereby should represent a fair and equitable method of handling a construction contract, but also they have normally had the advantage of being thoroughly critiqued by members of both the architect/engineering profession and the legal profession. Furthermore, many have been tested in court and can be relied upon to provide similar protection to all who use them. Another, and frequently overlooked, advantage is that such documents have evolved into a form that has withstood the test of time and experience and have become familiar to the contractors who use them. After repeated usage, a contractor will clearly understand all of its terms, meaning, and implications. This is often reflected in a stability of bid prices, as the full effect of the document's provisions on the contractor has already been established, as far as costs are concerned.

It must be remembered, however, that the mere fact that a document *is* so "general" in nature means that certain types of specific data cannot be included; otherwise, it would not be applicable to all of the projects on which it is used. For example, a provision in the General Conditions for "liquidated damages" or for "bonds" will normally only cover the limitations, procedures, or other unchanging elements—but *not* the actual monetary amount. The specific dollar amounts must be referred to in the Supplementary General Conditions, a document that will be discussed later.

Some of the well-known standardized General Conditions documents in current usage are those of the following organizations:

AGC	Associated General Contractors of America
AIA	American Institute of Architects
CMAA	Construction Management Association of America
EDA	Economic Development Administration, Department of Commerce
EJCDC	Engineers' Joint Contract Documents Committee (ACEC, ASCE, NSPE, and endorsed by AGC and CSI)
EPA	Environmental Protection Agency
FHA	Farmers Home Administration, Department of Agriculture
FIDIC	Fédération Internationale des Ingénieurs-Conseils (International Federation of Consulting Engineers)
HUD	Housing and Urban Development
JPL	Jet Propulsion Laboratories

NASA	National Aeronautics and Space Administration
NAVFAC	Department of the Navy, Naval Facilities Engineering Command
SCS	Soil Conservation Service
USBR	U.S. Bureau of Reclamation
USCE	Corps of Engineers, U.S. Army
(Joint)	EJCDC Standard General Conditions of the Construction Contract and Standard Form of Agreement. Funding agency edition endorsed by U.S. Department of Agriculture, Rural Development Administration, Farmers Home Administration, and Rural Utilities Service

Obviously, there is no shortage of "standards," and the choice of which to use on any given project may be influenced by many things, including the specific owner requirements, source of construction funds, type of work to be constructed, whether the project is being built by a public or a private agency, and last but not least, whether a particular set of General Conditions contains provisions with which the owner and architect/engineer are willing to live.

From the multitude of standard General Conditions available, there would seem to be little valid reason to generate a new one, yet many architects and engineers and owners are doing just that when they type their own General Conditions to fit a particular job. Each time that a new set is written, there is always the danger that its provisions may contain subtle wording that may not afford the same contractual protection as that contained in the existing standard forms that have had their days in court—and survived.

General Conditions Portions of Standard Specifications

On some projects where the basic contract is written around a set of Standard Specifications, such as those used by the state Departments of Transportation, the General Conditions of the contract are usually contained in the early chapters of the Standard Specifications book. For example, a project for construction of a bridge under the Virginia Road and Bridge Specifications has its General Conditions specified in Division I, General Provisions, which includes Sections 101 through 110 of that document.

Similarly, the American Public Works Association, Southern California Chapter, in cooperation with the Associated General Contractors, published a book of Standard Specifications entitled *Standard Specifications for Public Works Construction,* usually simply referred to as the "Greenbook." (See "Non-DOT Standard Specifications" in Chapter 6.) Here, as in the standard state highway specifications, the General Conditions are actually Part I, General Provisions, Sections 1 through 9 of the Standard Specifications book. The principal difference between the "Greenbook" and the state DOT Standard Specifications lies in two areas: (1) the Greenbook contract provisions are designed to accommodate the broader type of construction often encountered in local public agency work, whereas the DOT highway and bridge specifications are designed primarily for state highways, bridges, and drainage facilities; and (2) in most cases the state highway Standard Specifications are designed for use by the state alone and therefore contain all of the special legal requirements that must be a part of state agency contracts in that state, whereas the Greenbook is written for use in Southern California, and attempts to include those legal requirements that apply to contracts by cities, counties, and other local agencies in California. In the State of Washington, the WSDOT Standard Specifications cover a broader scope of construction than would be ordinarily encountered in state highway and bridge construction. There, the purpose is to present a book of Standard Specifications that can be used by all levels of government, including state, counties, and cities. What is especially unique is that an APWA Supplement is provided that, if utilized, can adapt the General Provisions portions of the Standard Specifications to local agency use. Similar publications of APWA have been published in other regions of the United States.

Differing Site Conditions

Unforeseen Underground Conditions

Several of the provisions of the General Conditions deserve special discussion because of their importance to the inspector and to the contractor as a result of the possible interpretations of their provisions. One of the most misunderstood of all contract provisions, and the one that is frequently the cause of large contractor claims for additional work and change orders, is the provision for *differing site conditions.*

Subsurface and latent physical conditions at the site present a special problem. If they differ significantly from what is printed in the contract documents, the contractor may well be entitled to additional payment for any increased work involved. When this happens, it usually comes as a great surprise to the owner. Architects or engineers should explain the possibility of such claims to the owner before the construction contract is signed. They should also be particularly careful that they give no assurance as to the accuracy of any subsurface exploration, even when a special soils consultant was employed. Failure to advise the contractor of any available data regarding subsurface conditions may not only entitle the contractor to additional payment, but also may possibly be the cause of a significant delay in the project when underground conditions are discovered that are quite different from those shown on the plans.

One can no longer say that the responsibility for all such conditions and the delays that accompany them will always be that of the contractor. There are recent court decisions relieving the contractor from such responsibility even where the wording of the contract documents states that the contractor must be familiar with all conditions at the site that might affect the performance of the Work. In fairness to the contractor and to avoid risk of blame for causing delays on the job, the design firm or the owner should make available to the contractor all data used in design.

Federal Guidelines

A contracts policy in the federal government as it relates to unforeseen underground conditions is becoming a widely accepted standard of the industry and is highly recommended for all construction contracts, both public and private, because it provides for a fair and equitable approach to an otherwise difficult problem.

The provisions of federal contract forms call for the making of adjustments in time and/or price where unknown subsurface or latent conditions at the site are encountered. The purpose is to have the government accept certain risks and thus reduce large contingency amounts in contractors' bids to cover such unknown conditions.

The federal government includes provisions in its construction contracts that will grant a price increase and/or time extension to a contractor who has encountered subsurface latent conditions.

The wording of the "Differing Site Conditions" clause in federal contracts reads as follows:

(a) The Contractor shall promptly, and before such conditions are disturbed, notify the Contracting Officer in writing of: (1) Subsurface or latent physical conditions at the site differing materially from those indicated in this Contract, or (2) Unknown physical conditions at the site, or of an unusual nature, differing materially from those ordinarily encountered and generally recognized as inherent in work of the character provided for in this Contract. The Contracting Officer shall promptly investigate the conditions, and if he or she finds that such conditions do materially so differ and cause an increase or decrease in the Contractor's cost of, or the time required for, performance of any part of the Work under this Contract, whether or not changed as a result of such conditions, an equitable adjustment shall be made and the Contract modified in writing accordingly.

(b) No claim of the Contractor under this clause shall be allowed unless the Contractor has given the notice required in (a), above; provided, however, the time prescribed therefore may be extended by the government.

(c) No claim by the Contractor for an equitable adjustment hereunder shall be allowed if asserted after final payment under this Contract.

Types of Differing Site Conditions

It is common to refer to a differing site condition as being either a Type 1 or a Type 2 Differing Site Condition, based upon the definition given in the federal contract provisions and summarized here.

TYPE 1—DIFFERING SITE CONDITION
Subsurface or latent physical conditions at the site which differ materially from that indicated in the contract documents

TYPE 2—DIFFERING SITE CONDITION
Unknown unusual physical condition at the site which differs materially from conditions ordinarily encountered and recognized as inherent in this type of work

Bidder's Obligations

The premises upon which a differing site conditions clause in federal grant-assisted construction contract rests are as outlined in the following paragraphs.

1. Each bidder is not expected to perform an independent subsurface site investigation prior to submittal of its bid. However, bidders are generally advised in the contract documents to make a site inspection of their own.

2. The contract bid price is proportional to the degree of risk that the construction contractor must provide for in its competitive bid.
3. The most cost-effective construction is obtained by accepting certain risks for latent or subsurface site conditions.

It should be noted, however, that other clauses of the contract documents may impact upon the resolution of a change order that was originally initiated under a differing site conditions clause, such as a requirement to make a site inspection before bidding.

Sharing the Risk

One of the major risks on any construction project is the possibility that physical site conditions will differ from those expected by the parties to the contract. Regardless of whether the project involves new construction or modifications of an existing structure, it is impossible to predict accurately every physical condition that will be encountered in the field.

To a great extent, the level of detail and accuracy of the site information will depend upon the amount of time and money expended by the owner. On the other hand, the contractor has the opportunity, and usually the obligation, to inspect the site prior to bidding. Despite the good-faith efforts of both parties, it is almost certain that some unanticipated condition will be encountered that will affect the time or cost of construction, or both. This leads to the question of who should bear the risk of increased construction costs necessitated by site conditions that were not indicated in the contract documents.

While many owners were able to place the differing site conditions risk with the contractor, other owners began to question the wisdom of this approach. Careful contractors were forced to carry sizable contingencies in their bids as insurance against differing site conditions. This led some owners to the conclusion that the concept of contractual risk sharing would not only be more equitable, but more economical as well.

The pioneer in this risk-sharing concept has been the federal government. The "Differing Site Conditions" clause under the Federal Acquisition Regulations System (48 CFR 52.236-2) is the prototype for similar clauses in both the public and private sectors. In fact, the government's definition of a compensable differing site condition has become the standard throughout the construction industry.

Traditional Rule of Law

The traditional rule of law has been that the contractor has no implied contractual right to recover additional compensation for differing site conditions. The contractor's obligation is simply to complete the project, as designed, for the contract price. The rationale is that the contractor had the opportunity to inspect the site prior to submitting its bid. Therefore, the risk of differing site conditions is presumed to have been factored into the contractor's bid price.

The rather harsh effects of this doctrine gave rise to an exception created by the courts: To the extent that the owner includes specific site conditions in the contract documents, the owner is extending an implied warranty that the information is accurate.

If the contractor relies on the information in the contract documents during bid preparation, and that information proves to be inaccurate, the owner must bear the responsibility. The contractor is still obligated to complete the contract but may sue for the increased cost of performance on the theory of breach of contract.

Use of Disclaimers of the Accuracy of Site Information

Owners often attempt to include express disclaimers in the contract documents to counter the doctrine of implied warranty, as shown in the following example:

> Information regarding subsurface conditions and quantities is solely for the use of the owner. It is provided to bidders not as a representation of fact, but for general guidance only. Actual subsurface conditions encountered shall be at the sole risk of the successful bidder. Each bidder is responsible for thoroughly investigating the site and satisfying itself of the accuracy of the estimated quantities in the Bid Schedule. After all bids have been submitted, the bidder shall not assert that there was a misunderstanding concerning quantities of Work or the nature of the Work to be performed.

Another example of such a disclaimer is that used by the federal government in fixed-price contracts where physical data such as test borings, hydrographic, or weather conditions data will be furnished or made available to the bidders (48 CFR 52.236-4):

> Data and information furnished or referred to below is for the contractor's information. The Government shall not be responsible for any interpretation or conclusions drawn from the data or information by the contractor.

Generally, such disclaimers are enforceable unless the owner intentionally or negligently misrepresents site conditions or withholds site information. It should be noted, however, that the use of such exculpatory clauses will be strictly construed by the courts against the owner. Some courts have refused to enforce site data disclaimers on the grounds that investigation by the contractor was physically impossible due to lack of time between bid solicitation and bid submittal. In general, however, such disclaimers are considered as enforceable, and a wary contractor should proceed on that assumption [*Jahncke Service, Inc. v. Department of Transportation,* 322 S.E.2d 505 (Ga. App. 1984)].

An example can be seen in that decision handed down by a Georgia court. It enforced a public owner's disclaimer of the adequacy of designated borrow pits and disclaimer of the accuracy of accompanying boring logs. Jahncke Service, Inc. had been awarded a unit-price contract by the Georgia Department of Transportation to construct 10.6 km (6.61 miles) of highway embankment. The contract documents designated several nearby borrow pits where the department had obtained the right to remove gravel and other fill. Soil boring reports for each pit were also available. The contract documents disclaimed the department's responsibility for borrow materials in the following terms:

> The Department, in making this borrow pit available to contractors, assumes no responsibility if the Contractor relies on this information. . . . The quantity of material shown on the plans as available in the borrow areas is not guaranteed. . . . The obligation is upon the Contractor, before making its Bid or Proposal, to make its own investigation.

During construction, Jahncke discovered that the designated borrow pits did not provide the quantity or quality of materials that Jahncke anticipated from the boring reports. Forced to obtain materials from other sources, Jahncke submitted a claim for an increase in the unit price per cubic yard.

The department relied upon the contract disclaimers in denying the claim. The Court of Appeals of Georgia agreed that as a matter of law, the contract provisions precluded Jahncke from recovering additional compensation for insufficient material or inaccurate reports. The court stated: "It is undisputed that Jahncke was on notice as to the possibility of errors or discrepancies in the boring report and as to the necessity, before submitting a bid, of making an independent investigation rather than contenting themselves with relying on the boring report."

Materials and Equipment

One of the most common occurrences during construction is the constant search by the contractor to obtain products that cost less than those actually specified, and offer them to the architect/engineer as substitutes (sometimes without offering a share of the savings to the owner). The ever-present desire to use cheaper materials that frequently have not had the test of time to show that they will perform as well as a specified product frequently leads to claims against the architect or engineer for negligence if it is determined later that the product did not perform as required. Great care must be used in the approval of new, substitute products as well as in the application of some established ones. The architect or engineer has the duty to see that the products furnished in compliance with his or her drawings and specifications are actually suitable for the particular uses intended. Reliance on producers' sales literature is hazardous at best. There are several court decisions in which the architect or engineer has been held liable for failure to have a new material tested, or an established item tested for a new application prior to approving it. Thus, it is easy to understand his or her occasional reluctance to try new products.

It may even be desirable to require that the manufacturer of such new products furnish guarantees that extend beyond the usual time. The refusal of a producer to provide such guarantees may be sufficient reason for rejecting the product. In any case, the authority for acceptance of a product offered as an "or equal" item is reserved to the architect or engineer of record—not to the Resident Project Representative or other inspector. Consideration of a product as an "equal" should be deferred until after execution of the contract, and *not considered during the bidding phase* of a project.

The Contractor and Subcontractors

Almost all of the construction contract General Conditions are based upon having the Resident Project Representative deal solely with the general contractor, not directly with the subcontractors, material suppliers, or fabricators. The General Conditions usually state that the general contractor is fully responsible for all the acts and omissions of the subcontractors, and nothing in the General Conditions is intended to create any contractual relationship between any subcontractor and the owner or design firm or any obligation to assure that the contractor has paid the subcontractors or material suppliers.

The fact that only the *general* contractor is recognized should end the frequent disputes of subcontractors that revolve around definitions of the scope of their portion of the work (usually, the result of the failure of the general contractor to provide its subcontractors with a complete set of specifications and drawings). The general contractor may complain that it is the design architect's or engineer's fault because they failed to include certain items in a specification section that would be performed by a specialty subcontractor, and thus the contractor believes that additional funds should be paid to cover the added charges to the general contractor by the specialty subcontractors. Obviously, the answer is simple from a contractual standpoint—only one contract was let; the total scope of the work was specified; and it is not the responsibility of the architect or engineer to determine how the successful bidder plans to subcontract the work. Furthermore, the scope of any one class of work can often vary significantly even from one county to the next, because of differences in trade union contracts and the resultant jurisdictional agreements. One contract means one job; it is the general contractor's responsibility to contract properly with his or her subcontractors to assure a clear understanding of the scope of each such subcontract.

Shop Drawings and Samples

The Function of Shop Drawings

The shop drawing is the connecting link between design and construction. Because of the increasing complexity of today's construction, shop drawings in recent years have become one of the largest sources of professional liability claims against the designer. Unreasonable delay in processing shop drawings and ambiguous wording in the shop drawing approval stamps are the principal sources of trouble. Most specifications require that the contractor refrain from ordering material until the results of the review of the shop drawing submittal from the design organization have been received. Any delay in processing of shop drawings affects the contractor's scheduling and, in turn, may result in extra cost to the owner.

Normally, the contractor is obligated to submit a preliminary schedule of the submittals of shop drawings and other submittals required under the contract. This schedule should be reviewed by the design firm or the owner's engineering staff and finalized prior to construction. Shop drawing submittal procedures are one of the topics for discussion at the *preconstruction conference* described in Chapters 10 and 12. Careful attention to these preliminary matters can avoid misunderstandings at a later date, and an agreed-upon procedure should be set out on paper and copies circulated to all affected parties.

Shop drawings are drawings submitted to the owner's architect/engineer by a contractor or subcontractor. Shop drawings usually show in detail the proposed fabrication or assembly of project components. They are also used to indicate the installation, form, and fit of materials or equipment being incorporated into the project.

Shop drawings are needed because it is impossible for the plans and specifications to spell out every detail of every aspect of the work. This is particularly true of large construction projects.

While the plans and specifications usually define the overall nature of the project, the means and methods of construction are expected to be determined by the contractor. Owners and their architect/engineers expect this expertise from the contractor. Shop drawings also provide a way for contractors to propose and architect/engineers to approve a particular method of accomplishing a special requirement.

Approval of Shop Drawings

One of the most important and misunderstood facts about shop drawings is that a shop drawing approval does not normally authorize changes from the contract provisions [*Appeal of Whitney Brothers Plumbing & Heating, Inc.*, ASBCA No. 16876, 72-1 BCA 9448 (1972)]; these must properly be accomplished by a *change order.*

The review and approval of shop drawings involve some considerable risk to architect/engineers and owners. In an effort to maintain flexibility in determining the acceptability of the work, as well as some leverage over the contractor, architect/engineers frequently use shop drawing approval language that is evasive or ambiguous at best.

This language, which is usually found on the shop drawing approval stamps applied to each drawing, contains typical "approval" language, such as:

> Review is only to check for compliance with the design concept of the project and general compliance with the contract documents. Approval does not indicate the waiver of any contract requirement. Changes in the work may be authorized only by separate written change order.

The Design Professionals Insurance Company of Monterey, California, one of the major underwriters of errors and omissions insurance for architect/engineers, in its book *The Contract Guide,* advocates:

> . . . the review and approval of contractor submittals, such as shop drawings, product data, samples, and other data, as required by the engineer or architect, but only for the limited purpose of checking for conformance with the design concept and the information expressed in the contract documents. The review should not include review of the accuracy or completeness of details, such as quantities, dimensions, weights or gages, fabrication processes, construction means or methods, coordination of the work with other trades or construction safety precautions, all of which are the sole responsibility of the contractor. The engineer's or architect's review should be conducted with reasonable promptness, while allowing sufficient time in their judgement to permit adequate review. Review of a specific item must not indicate that the engineer or architect has reviewed the entire assembly of which the item is a component. The engineer or architect should not be held responsible for any deviations from the contract documents not brought to their attention in writing by the contractor, nor should they be required to review partial submittals or those for which submittals of corrected items have not been received.

However, in a case heard by the General Services Administration Board of Contract Appeals in 1985, the board ruled that "when a contractor clearly calls out a deviation from the original specifications, the government will be bound by its shop drawing approval regardless of disclaimers in the stamped language" [*Appeal of Montgomery Ross Fisher and H. A. Lewis, a Joint Venture,* GSBCA No. 7318 (May 14, 1985)].

Both the AIA in Articles 3.12.8 and 4.27 of General Conditions Document A201 (1987 edition) and the EJCDC in Article 6.26 of General Conditions Document 1910-8 (1996 edition) refer to the obligation of the architect or engineer to review *and approve* shop drawings. FIDIC documents do not use the term "shop drawings," but refer to such documents under the category of "Permanent Works Designed by the Contractor"; there, in Sub-Clause 7.2 of the "Conditions of the Contract" (1987 edition), they require such documents to be submitted to the engineer "for approval."

You should require a schedule of submittals from the contractor. As a general rule, do not review shop drawings or other submittals concerning the proposed implementation of means, methods, procedures, sequences, or techniques, or other temporary aspects of the construction process. Those are the responsibility of the contractor, and review of those submittals could subject you to responsibility not normally undertaken by an engineer or architect. If you receive uncalled-for submittals from the contractor, they should be stamped "Not Required for Review" and returned at once.

Misuse of Shop Drawings

A shop drawing must not be used as a change order, and any variation from the design drawings and specifications must be the result of a formal change order. Otherwise, it is not authorized [*Appeal of Community Science Technology Corp., Inc.,* ASBCA No. 20244, 77-1 BCA 12,352 (1977)].

In a recent example, a reservoir was to have a precast, prestressed (pretensioned) concrete roof provided under a contract with a fabricator who specialized in furnishing such work as a complete package; that is, they provided the design (within the criteria set by the architect/engineer); the fabrication of all the prestressed, precast concrete structural members; and the erection at the site of all such members into a complete roof system. On shop drawings transmitted during the execution of this contract, a small but significant design detail was changed from that shown in the original design concept that was approved by the engineer and the local building and safety department. The contractor contended that approval of shop drawings with the design change on them meant that they should build according to the shop drawings instead of the contract drawings. However, the contract provided that any change from the contract drawings must be accompanied by an authorized change order, and therefore the contractor was required to conform to the original detail as shown on the contract drawings. Again, only a *change order* should be used to authorize a deviation from the contract provisions, and a change order must normally be signed by the owner, as it is the owner who is a party to the contract, not an outside design firm. It cannot be done legally simply by showing changes on a shop drawing unless the architect/engineer creates an informal change order out of the shop drawing by adding the proper authorization and signature of the owner to the drawing sheets. This, of course, would be somewhat irregular and certainly undesirable from the record-keeping viewpoint.

Disapproving or Stopping the Work

Some General Conditions allow the Resident Project Representative, as a representative of an outside design firm, the right to stop the work. This is a very sensitive area and can lead to serious legal consequences if it can be shown that the action was unjustified and subjected the contractor to added cost. However, upon receipt of information from the architect or engineer that the work is defective, the *owner* may order the contractor to stop the work on these grounds. In any case, the more risky right to stop the work should be left to the owner, as a party to the construction contract, not to an outside design firm or their Resident Project Representative.

In some cases where the owner is relatively "unsophisticated," as the term relates to construction, there is a tendency to allow an architect/engineer or its field representative to exercise greater control over the project. It should be kept in mind at all times, however, that certain additional risks accompany such added responsibility.

One prime exception to the foregoing "stop the work" discussion is the case where the work is being carried on in an unsafe manner. Under these conditions, moral standards or the law may impose a duty on both the Resident Project Representative and the contractor for the benefit of employees and third parties to stop such work as a means of lessening the risk of death or serious injury that could result if such conditions were allowed to continue (see the discussion of Imminent Hazards in Chapter 9).

Any disapproval or rejection of the work should be communicated to the contractor in writing, stating the reasons for the disapproval. This should be done as early as possible after rejecting the work. In addition, the Resident Project Representative normally has the power to require special testing or inspection of all work that has been covered up without his or her consent, including work that has been fabricated, installed, or completed. In its Standard General Conditions of the Construction Contract, EJCDC Document 1910-8 (1996 edition) provides for the owner's right to stop the work in Article 13.10, as follows:

OWNER MAY STOP THE WORK

13.10. If the Work is *defective,* or CONTRACTOR fails to supply sufficient skilled workers or suitable materials or equipment, or fails to furnish or perform the Work in such a way that the completed Work will conform to the Contract Documents, OWNER may order CONTRACTOR to stop the Work, or any portion thereof, until the cause for such order has been eliminated: however, this right of OWNER to stop the Work shall not give rise to any duty on the part of OWNER to exercise this right for the benefit of CONTRACTOR or any surety or other party.

The unanticipated legal exposure resulting from giving the power to stop the work to the engineer resulted in a change to the National Society of Professional Engineers (NSPE) contract documents in 1967. Under Article 13.10 of the Standard

General Conditions of the Construction Contract of the EJCDC,[4] only the owner has the right to stop the work, and then only if that work is defective. It was believed that the right to reject defective work was a sufficient weapon for the engineer, and the severe and more risky right to stop other work should be left to the owner. Where the owner does not have sufficient technical sophistication, great reliance is placed on the engineer or architect to "do it all," but in so doing, the engineer or architect should be aware of the additional legal exposure assumed.

Similarly, in the General Conditions of the Contract for Construction, AIA Document A201 (1987 edition), Article 2.3.1., provides for the owner's right to stop the work. The principal difference between the EJCDC provisions and the AIA provisions for owner's right to stop the work appears to lie in the requirement in the AIA documents that a stop work order from the owner to the engineer must be delivered in writing and signed personally by either the owner or an authorized agent, who must also be designated in writing.

The FIDIC, in its Conditions of Contract for Works of Civil Engineering Construction (fourth edition 1987), places no such limitations on the authority of the engineer. It addresses the subject of work stoppage under the heading "Suspension of Work" in Article 40.1, where it states:

> **SUSPENSION OF WORK 40.1**
> The Contractor shall, on instructions of the Engineer, suspend the progress of the Works or any part thereof for such time and in such manner as the Engineer may consider necessary and shall, during such suspension, properly protect and secure the Works or such part thereof so far as necessary in the opinion of the Engineer unless such suspension is (a) otherwise provided for in the Contract, or (b) necessary by reason of some default of or breach of contract by the Contractor for which he is responsible, or (c) necessary by reason of climatic conditions on the Site, or (d) necessary for the proper execution of the Works or for the safety of the Works, or any part thereof (save to the extent that such necessity arises from any act or default by the Engineer or the Employer or from any of the risks defined in Sub-Clause 20.4), Sub-Clause 40.2 shall apply.

It can be noted in the FIDIC handling of work suspension (stopping the work) that the engineer may be obligated to "protect and secure" the work during the period of such suspension or stoppage, an obligation not imposed under either EJCDC or AIA documents. It further permits the engineer to stop the work, which in the United States could expose the engineer to the risk of tort liability under the principle that a party who has the right to stop the work is in *de facto* control of the work; thus if someone were injured on the job site, argument could be made that the engineer was under an obligation to prevent the injury by stopping the work beforehand to correct whatever defect resulted in the injury.

For further discussion of suspension or termination of the work, see Chapter 15.

---

[4] *EJCDC Standard Forms of Agreement,* Publication 1910-8 (1985), the National Society of Professional Engineers, 1420 King Street, Alexandria, VA 22314.

Supplementary General Conditions

As the name implies, the Supplementary General Conditions are simply an extension of the General Conditions. It is in this document that the special legal requirements of the contract are expanded to include provisions that apply solely to the project at hand. In some cases, the titles of articles within the Supplementary General Conditions will duplicate titles already mentioned in the General Conditions. This is neither repetitious nor necessarily a superseding provision. In most cases, both such paragraphs still apply. However, the provisions in the General Conditions may contain only procedural and responsibility clauses, whereas the same subject in the Supplementary General Conditions will add specific requirements that apply only to this job, such as amounts of liquidated damages or the amounts of bonds, or amounts of insurance required.

The Supplementary General Conditions portion of the contract documents may appear under several titles, without actually changing the nature of the document. Often, under older formats, this portion of the specifications was known as "Special Conditions" or on some of the newer formats, Supplementary General *Provisions* instead of *Conditions.*

In addition to items that are expansions of Articles already specified in the General Conditions, there are numerous other subjects that may, in all likelihood, be encountered in the Supplementary General Conditions. A sample of the contents of one such document follows:

1. Scope (of the entire project)
2. Supplementary Definitions (not covered in the General Conditions)
3. Legal Address of the Architect/Engineer and the Owner (needed for service of legal documents)
4. Amounts of Bonds (actual dollar or percentage values)
5. Amount of Liquidated Damages (actual dollar value)
6. Permits and Inspection Costs (who pays what)
7. Contract Drawings (complete list, by number and title, of all drawings that are made a part of the contract)
8. Applicable Laws and Regulations (specific requirements for this job)
9. Insurance (amount of coverage; additional insurance not specified in the General Conditions, and amount of its coverage)

Technical Provisions of the Specifications

It should be noted that two of the three parts into which the specifications document is usually divided have "general" clauses in them. They are Part II, Conditions of the Contract, and Part III, Technical Provisions. To avoid confusion, some distinction should be noted between these two portions of the specifications.

First, the "general" provisions of the *Conditions of the Contract* relate to the contractual relationships and legal obligations of the parties to the contract. However, the "general" provisions of the *technical* portion of the specifications should relate to those requirements of a technical nature that apply generally to the work of the entire project rather than to work of one trade, for example. Thus, the General Conditions of the Contract refer to legal and contractual relationships, whereas the General Requirements Section of Part III, Technical Provisions, relates to construction details, project features, procedural requirements for handling the work, and similar project-related functions.

Under the CSI Format, Division I of the 16-Division Format was reserved for this purpose, and the subjects generally included (if applicable to the project) would usually be either those shown in the following list, or similar subjects that the architect/engineer deems necessary to perform the work of the project properly (see also Figures 6.2 and 6.3):

GENERAL REQUIREMENTS—DIVISION I

01010. Summary of Work
- Work by others
- Items provided by owner
- Work included in this contract
- Work to be performed later

01025. Measurement and payment

01030. Alternatives

01200. Project meetings
- Preconstruction conferences
- Progress meetings
- Job site administration

01300. Submittals
- Construction schedules
- Network analyses
- Progress reports
- Survey data
- Shop drawings, product data, and samples
- Operation and maintenance data
- Layout data
- Schedule of values
- Construction photographs

01400. Quality Control
- Testing laboratory services
- Inspection services

01500. Temporary facilities and controls
- Temporary utilities (power, lighting, water, phones, sanitary)
- Construction elevators and hoists
- Guards and barricades
- Shoring, falsework, bracing, scaffolding, and staging

Access roads
Control of dust, noise, water, vapors, pollutants
Traffic control; parking; storage of materials
Temporary field offices

01600. Material and Equipment
Quality transportation and handling
Storage and protection

01700. Project closeout
Cleaning up
Project record documents
Touch up and repair
Operational testing and validation
Maintenance and guarantee
Bonding requirements during guarantee period

These and similar provisions are generally representative of the types of subject matter that are expected to be contained in Division I of the Technical Provisions where the CSI Format is being used. A comparison of these subjects with the subject area covered under the General Conditions of the Contract, Part II of the specifications, should give some idea as to the accepted grouping of the various subject matters.

Provisions for Temporary Facilities

During the preparation of the specifications, it is desirable to provide input data that are necessary for effective contract administration. These data should include provisions for certain temporary facilities and services so that they can be required of the contractor by a specific date.

It should be emphasized that it is the contractor's basic responsibility to provide all plant and equipment that is adequate for the performance of the work of the contract, within the time specified. All such plant and equipment must be kept in satisfactory operating condition and must be capable of performing the required work safely and efficiently. It is sometimes desirable to have all such items subject to the inspection and approval of the Resident Project Representative at any time during the duration of the contract.

A number of specific issues should be covered in the specifications to facilitate these requirements.[5] The following is a suggested list of items that may be appropriately covered where the item is applicable to the project being constructed:

- Temporary electrical services
  (Include: construction lighting, wiring, and circuit separation)

[5]As a part of the specifications provisions for temporary facilities, the specifications should also provide for some of these items to be included under a separate line item in the bid sheet, or as a part of a "mobilization" line item. In this manner, failure to provide such facilities on a timely basis can be reflected in the contractor's entitlement to the first progress payment. (See Chapter 17 for further details regarding Payment for Mobilization Costs.)

- Fire protection
  (Include: connections to contractor's water supply system)
- Temporary utility services
  (Include: water supply development and connections; later removal of such connections; power supply and connections; free local telephone service for both the contractor's and the owner's or architect/engineer's field offices; type of telephone service required, and cost of toll calls; time of installation of such temporary facilities)
- Sanitation
  (Include: toilet facilities and disposal of waste materials)
- Site access and storage provisions
  (Include: highway limitations, marine transportation provisions including pier and landing facilities and small boat-launching facilities, and contractor's work and storage area limitations)
- Environmental controls
  (Include: explosives and blasting, dust abatement, chemical use and disposal, and misplaced or discharged materials into waterways)
- Cultural resources
  (Include: historical, architectural, archeological, or other cultural resources endangered by the project; the owner's right to stop the work in case of a cultural resource "find")
- Field office facilities
  (Include: type of structure or facility; office equipment and furnishings to be provided; utility services to be provided; date of installation and completion of field office facility; and cleanup services required)

Addenda to the Specifications

Addenda to the specifications (or in singular form, an *addendum* to a particular set of specifications) are documents setting forth the changes, modifications, corrections, or additions to the contract documents that have been issued after the project has been advertised for bids, but *before* the time of opening bids—sufficiently in advance of the bid opening date, one hopes, to allow the bidder time to make the necessary changes in his or her bid.

The addenda may be specified as "Addendum to the Specifications" or as an "Addendum to the Notice Inviting Bids." Each has the same legal effect. Many public agencies prefer the latter (Figure 7.1). Normally, the addenda must be delivered to each party who has obtained a set of specifications in such manner as to provide the owner with written assurance of completed delivery before the opening of bids. This may be accomplished by sending a copy to each person who has obtained a set of plans and specifications, via certified mail with a return receipt requested. The return receipt is the confirmation of receipt by the bidder, and it should be carefully

CITY OF PALM SPRINGS
ENGINEERING DIVISION

ADDENDUM NO. 1

To all prospective bidders under Specifications for Re-roofing the Police Department Building, City Project No. 93-52, for which bids are to be received by the City of Palm Springs at the office of the Purchasing Manager at 3200 East Tahquitz Canyon Way, Palm Springs, California 92262 until 4:00 pm on Tuesday 21 March 1995.

I The existing three bid schedules in the Specifications and Drawings for this contract have been revised. Bid Schedules "A,""B," and "C" have been changed and revised Bid Schedules have been included as a part of this Addendum No. 1.

II The Specifications as originally issued, along with revised Schedules "A,""B," and "C," shall be used in submitting bids, and acknowledgement of receipt of this Addendum No. 1 shall be entered on Page 1 of the Bid. Failure to provide such acknowledgement shall render the bid as non-responsive and subject to rejection.

BY ORDER OF THE CITY OF PALM SPRINGS

13 March 1995

By Robert J. Rocket, PE
City Engineer
Civil Engineer C 28209

FIGURE 7.1 Specifications Addendum.

filed. A further safeguard to assure that all parties who bid the job are using the same edition of the documents is to use one of the following methods of assurance:

1. Require that the bidder sign an acknowledgment for the receipt of each addendum issued. Then, at bid-opening time, all such acknowledgments are checked before considering the bid as admissible or responsive.
2. Require that the bidder simply submit copies of all addenda along with the bidding documents at the time of opening bids. This is a fairly common procedure by many public agencies, which frequently require that the bid forms not be removed from the specification document, and that the entire book be submitted intact with the bid.

One of the first things to be done upon receipt of addenda is to check the specifications and drawings carefully, and mark all corrections, changes, modifications, or additions to the original documents. The next step is to cross-check to see if any of the data that were changed involve interfacing with other sections of the specifications or any other drawings. In the issuance of addenda, the architect or engineer seldom if ever provides cross-references. The Resident Project Representative's copy must be checked for this and the set of contract documents should be marked to reflect all such changes by addenda.

It should be remembered that addenda can be issued only *during the bidding period;* any changes that are made after the opening of bids should be issued as change orders during the construction phase. Thus the Resident Project Representative will have access to all such changes by addenda long before reporting to the project site.

Of prime importance is the fact that an addendum to the specifications (or to the *Notice Inviting Bids,* as it is sometimes called) will always take precedence over any portion of the plans or specifications that is in conflict with it. The Resident Project Representative should always check all addenda before requiring compliance with provisions of the original specifications or drawings.

Standard Specifications

By definition, a set of *Standard Specifications* is a preprinted set of specifications, usually comprising both a set of General Conditions and complete technical specifications for all types of construction and materials that the originating agency expects normally to cover in its kind of work. When adopted by a public agency or by a design firm working on a project for a public agency, the total content of the Standard Specifications becomes a part of the contract documents, subject only to changes set forth in a separate project-related document called the *Special Provisions* or *Supplemental Specifications,* which adapts the rather general treatment of the Standard Specifications to the specific needs of a particular project. In this manner, where alternatives are offered in the Standard Specifications, the Special Provisions or Supplemental Specifications serve to indicate which of the available choices apply to this specific project. Under this type of contract, contractually anything not modified by the terms of the Special Provisions or Supplemental Specifications is required to comply with all applicable provisions of the Standard Specifications.

In direct contrast, if a project does not cite the Standard Specifications as the principal contract document, but merely references certain sections from it, then nothing in the Standard Specifications will apply to that project except those items specifically referenced in the specifications. To assure proper application, the resident engineer or inspector must be very careful with regard to citations from Standard Specifications.

Particular attention should be paid to the exact numerical designation of a citation as well. For example, if a citation reads

> per Section 223.02(d) of the Virginia Road and Bridge Specifications

such a specific reference would preclude the use of anything that is not in that particular subsection (d); in this case it would mean that concrete must be cured solely by the use of the "liquid membrane seal" method, which is specified in detail. However, if the citation reads

> per Section 223 [or 223.02] of the Virginia Road and Bridge Specifications

the contractor would be free to select any of four specified acceptable methods of curing concrete; it would be the contractor's option.

Under California Department of Transportation Specifications, the coverage of each Section is broader; Section 90 of the document includes not only portland cement, but also aggregates, curing materials, admixtures, and similar items. The effect of making a reference as general as

> per Section 90 of the Caltrans Standard Specifications

in a contract where the Standard Specifications was the principal document would be to require the contractor's compliance with all provisions of that section, including a choice of each alternative provided. If the engineer wanted to limit the use of curing methods, for example, the Special Provisions or Supplemental Specifications would have to include a qualifying statement to that effect.

The use of a set of Standard Specifications without an accompanying set of Special Provisions or Supplemental Specifications is like asking to read a copy of a book, and being handed a dictionary instead with the comment, "All the words are here; just read the ones that apply."

Master Specifications (Guide Specifications)

A *Master Specification* or *Guide Specification* is another matter entirely. It is equivalent to a Standard Specification only in that it is a preprinted set of Specification Provisions. It is never used on a job in the preprinted form, however. Each time it is to be used, the Master Specification must be physically modified to meet the specific job requirements. This precludes any need to issue a set of Special Provisions or to issue sections of specifications that do not apply to the project at hand. Masterspecs are in-house tools to enable the architect/engineer or other agency to produce more effectively project specifications that reflect a fixed corporate or agency policy, and that may readily be updated to reflect current changes in construction methods, materials, and laws.

The Masterspec concept is based upon the same principle as that used in the old Corps of Engineers' Guide Specifications. Each subject area is separated into separate sections and a specification is prepared leaving blank areas or a selection of choices to be made to adapt the section to any specific project. After marking the guide or Masterspec, it may be typed as a part of the project specification.

One of the principal advantages of this system is the fact that old job specifications are not used to prepare specifications for a new project. Instead, the Masterspec is utilized each time as the basic document from which to begin. In this manner, too, updated material changes, legal requirements, ASTM references, and similar items subject to constant change can readily be incorporated in the Masterspec as the changes are noted. Thus, all subsequent project specifications will be up to date, without the commonly observed outdated phrases, materials that are no longer available, and similar faults inherent in the system of utilizing a previous project specification as a base to write a new one.

Special Material and Product Standards

If every time an item were specified for a project the specifications would have to contain all of the provisions that were necessary to assure that the product met all physical, chemical, geometrical, or performance standards required, every project specification would have to be from 10 to 100 times bulkier than it is now, and some would look like a set of encyclopedia. Worse yet, every architect/engineer without the benefit of coordination would have enough subtle differences in the description of a product as compared with another architect or engineer's description of the same product that the manufacturers would be solely in the business of manufacturing "custom" materials for every different project. Even if this could be accomplished, construction costs would skyrocket.

As a means of providing the uniformity necessary, various nonprofit associations as well as government agencies and manufacturers have established voluntary standards that are actually Standard Specifications for separate individual products. Thus, by referring to the published data for each of these products, an architect or engineer can design each project subject to the specified product limitations, with full assurance that such products are not only marketed, but also carefully regulated by each manufacturer to assure compliance with the previously established standards.

The agencies that issue such standards are sometimes governmental, sometimes industry trade associations, and sometimes independent standards associations whose only function is the preparation of such industry standards with the voluntary cooperation of industry, of course. In each case, the standards have been established as a coordinated effort between the manufacturer, architect/engineer, academic community, and other influences, as applicable.

Such standards become a part of a construction contract only if specifically called out in the specifications or drawings—and then only to the degree referred to. If a specification calls for a particular product by its ASTM designation, but includes something that was not a part of that ASTM standard, the product must conform to the cited standard *subject to the modifying provision*. Thus, it would actually be a "special" product requiring the manufacturer to make a "custom" item with appropriate increase in cost and delay in delivery schedule.

Such standards may be loosely divided into two basic classifications: (1) government standards and (2) nongovernment standards. Of the first category, the following are most commonly used.

Government Standards

Federal Specifications. Federal specifications describe essential and technical requirements for items, materials, or services that are normally bought by the federal government. They are also referred to extensively in specifications for nonfederal projects when commercial standards are not available for a particular item. They are

178 NUMERIC LIST OF FEDERAL SPECIFICATIONS & COMMERCIAL ITEM DESCRIPTIONS

DOCUMENT NUMBER	QPL	TITLE	FSC	PREP	DATE	PRICE
RR-W-670D		Wringer, Mop	7920	FSS	17 Oct 72	
RR-W-1101A		Waste Receptacle, Swinging Doors	7240	FSS	31 Oct 72	
RR-W-001588		Waste Receptacle, Wall Or Post Mounted	7240	FSS	20 Mar 70	
RR-W-1817A		Warning Device, Highway, Triangular, Reflective	9905	FSS	30 Jul 75	
SS-A-281B(1)		Aggregate; (For) Portland-cement-concrete	5610	ME	25 Jan 57	
SS-A-666D		Asphalt, Petroleum (Built-up Roofing, Waterproofing, And Dampproofing)	5610	FSS	7 May 68	
NOTICE 1					24 Jul 68	
SS-A-671C		Asphalt, Liquid; Slow-curing, Medium Curing, And Rapid Curing	5610	FSS	8 Sep 66	
SS-A-694D		Asphalt Roof Coating (Brushing And Spraying Consistency)	5610	FSS	9 Mar 73	
SS-A-701B		Asphalt, Petroleum (Primer, Roofing, And Weatherproofing)	5610	FSS	15 Jul 74	
SS-A-706D		Asphalt, Petroleum' road And Pavement Construction (Asphalt Cement)	5610	YD	25 Sep 78	
SS-B-656B		Brick, Building, Common (Clay Or Shale)	5620	ME	18 Feb 66	
SS-B-663B		Brick, Building, Concrete	5620	FSS	9 Jun 65	
SS-B-668B		Brick, Facing, Clay, Or Shale	5620	FSS	20 May 74	
SS-B-671C		Brick, Paving	5620	FSS	2 Jun 65	
SS-B-681B		Brick, Building, Sand-lime	5620	FSS	22 Dec 65	
SS-B-755A(1)		Building Board, Asbestos Cement' flat And Corrugated	5640	FSS	21 May 68	
SS-C-153C		Cement, Bituminous, Plastic	5610	FSS	13 Dec 74	
SS-C-160A(2)		Cements, Insulation Thermal	5640	FSS	11 Jan 79	
SS-C-161A		Cement' keene's	5610	FSS	23 Jan 74	
SS-C-255		Chalk, Carpenters' And Railroad	7510	FSS	30 Aug 49	
SS-C-00255A		Chalk, Carpenters' And Railroad	7510	FSS	10 Jan 69	
SS-C-266F		Chalk, Marking, White And Colored	7510	FSS	20 Aug 73	
SS-C-450A		Cloth, Impregnated (Woven Cotton Cloth, Asphalt Impregnated; Coal Tar Impregnated)	5650	FSS	10 Apr 67	
SS-C-466E		Cloth, Thread, And Tape; Asbestos	5640	SH	2 Jul 64	
INT AMD 2					22 Sep 76	
SS-C-540B		Coal Tar (Cutback) Roof Coating, Brushing Consistency	5610	FSS	11 Jun 75	
SS-C-621B		Concrete Masonry Units, Hollow (And Solid, Prefaced And Unglazed)	5620	FSS	19 Jan 68	
INT AMD 2					18 Jun 70	
SS-C-635B		Crayon Assortment, Drawing, Colored	7510	FSS	7 Dec 66	
SS-C-646B(2)		Crayon, Marking, Lumber	7510	FSS	31 Oct 63	
SS-C-661A		Crayon, Marking	7510	FSS	11 Jan 74	
SS-C-1783		Cloth, Asbestos	5640	FSS	29 Mar 73	
SS-C-1960/GEN		Cement And Pozzolan (General Requirements For)	5610	YD	2 Dec 75	
SS-C-1960/1A		Cement, Masonry	5610	YD	24 May 77	
[illegible]		[illegible]	5610	YD	28 Aug 78	
SS-C-1960/3B		Cement, Portland	5610	YD	28 Aug 78	
[illegible]		[illegible]	5610	YD	24 May 77	
SS-C-1960/5A		Pozzolan, For Use In Portland Cement Concrete	5610	YD	28 Aug 78	
SS-F-001032		Floor Covering, Asphaltic Felt (Bitumimous Type Surface)	7220	FSS	19 Oct 66	
SS-G-659A		Graphite, Dry (Lubricating)	9620	FSS	1 Mar 67	
SS-J-570B		Joint Compounds And Tape, Wallboard (For Gypsum Wallboard Construction)	5640	FSS	9 Aug 77	
SS-L-30D		Lath, And Board Products, Gypsum	5640	FSS	27 Feb 74	

FIGURE 7.2 Federal Specifications Index.

generally characterized by the unique letter–number type of designations used, such as "SS-C-1960/3B" for portland cement concrete in Figure 7.2.

Military Specifications (Department of Defense). Military specifications, or *Mil-Specs* as they are often called, specify products that are usually unique to the needs of the military; however, in some rare cases such as certain electrical devices, they may be the only source of an appropriate material specification. Their use is generally discouraged in civil projects because of the difficulty in obtaining copies of the standards as well as the restricted availability of manufacturers who produce to these standards. Mil-Specs are characterized by designations such as "MIL-R-0039016A" for an electrical relay (Figure 7.3).

UBC Standards

Although technically not a governmental agency, the UBC Standards, wherever referred to, are actually backed up by local ordinances, thus carrying the force of law. All cities and counties that have adopted the Uniform Building Code include the UBC

ALPHABETICAL LISTING

TITLE		DOCUMENT NUMBER	FSC	PREP	DATE			CUSTODIAN		
Valves, Aircraft Hydraulic Shuttle	Q	MIL-V-5530B (4)	•1650	71	24	Mar	70	AV	AS	71
Valves, Aircraft Hydraulic Unloading	Q	MIL-V-5519C (1)	1650	71	05	Aug	64			71
Valves, Aircraft Power Brake		MIL-V-5525C	1630	AS	21	Oct	59	AV	AS	11
Valves, Aircraft, Hydraulic Pressure Relief, Type Ii Systems (Asg)	Q	MIL-V-8813	•1650	AS	20	Nov	57		AS	71
Valves, Aircraft, Hydraulic Thermal Expansion Relief (Supersedes AN-V-28)	Q	MIL-V-5527A	1650	AS	14	May	51		AS	71
Valves, Angle, Shut-off, Packed, Receiver, REFRIGERANT-12		MIL-V-22854A (1)	4130	GL	30	Sep	68	GL	YD	82
Valves, Astern (For Shipboard Use)	L	MIL-V-22682A (2)	4820	SH	15	Aug	63		SH	
Valves, Ball	L	MIL-V-23611	4820	YD	19	Mar	63		YD	
Valves, Ball, Naval Shipboard, For Air, Nitrogen, Helium Or Hydraulic Service (Sizes 2-1/2 Inches I.p.s And Below)	Q	MIL-V-22687B (1)	4820	SH	16	May	66		SH	
Valves, Blocking	L	MIL-V-21517D (3)	1440	OS	17	Oct	63		OS	
Valves, Blow Off, Boiler	L	MIL-V-18406A	4820	YD	11	Jan	68		YD	
Valves, Boiler Blow, Shipboard Use	L	MIL-V-17737D	4820	SH	01	Feb	71		SH	
Valves, Check		MIL-V-18436C	4820	YD	30	Apr	70	ME	YD	82
Valves, Check Swing, Cast-iron And Steel		MIL-V-10386B (3)	4820	ME	17	Feb	67	ME	YD	82
Valves, Check, Controllable, Hydraulic, Aircraft, Type Ii Systems (Asg)	Q	MIL-V-19067A	1650	AS	30	Apr	57		AS	71
Valves, Check, Oxygen, High Pressure (Asg)	Q	MIL-V-5027D	1660	AS	12	Mar	68	AV	AS	71
Valves, Combined Vent-check For Submarine Mbt Blow Lines	QL	MIL-V-23953A (2)	4820	SH	12	Feb	70		SH	
Valves, Compressor Service, REFRIGERANT-12		MIL-V-22862B	4130	GL	30	Jul	68	GL	YD	82
Valves, Control, Air Diaphragm-operated	L	MIL-V-18030C (1)	4820	SH	26	Jan	70			
Valves, Cylinder, Gas, Carbon Dioxide Fire Extinguisher		QPL-17360-13	4210	SH	20	Nov	68			
Valves, Diaphragm, Stop	L	MIL-V-82026	4820	YD	13	Jul	64		YD	
Valves, Expansion, Thermostatic, REFRIGERANT-12		MIL-V-23450	4130	GL	10	Jun	71	GL	YD	82
Valves, Fuel Selector		MIL-V-5018A (2)	4530	AS	31	Oct	63	MO	AS	84
Valves, Fuel Shutoff Solenoid Operated, 28 Volt, D.c. (Asg)		MIL-V-8610A (1)	2990	82	05	Nov	71		AS	82
Valves, Fuel Shutoff, Electric Motor Operated		MIL-V-8608A (2)	2915	11	09	Nov	66	AV	AS	11
Valves, Gate, Bronze, 300 Psi	L	MIL-V-18827A	4820	YD	15	Dec	71		YD	
Valves, Gate, Cast Or Forged Steel And Alloy Steel, Outside Screw And Yoke (Basically Commercial Valves)	Q	MIL-V-18110D	4820	SH	24	Jul	61	MO	SH	82
Valves, Gate, Globe And Angle, Steel		MIL-V-18434B	4820	YD	28	May	71		YD	
Valves, Gate, Rising Stem, Double Acting Aluminum		MIL-V-58039B	4820	ME	20	Jun	66	GL	MC	
Valves, Globe, Angle And Y, Flanged Bonnet, Manually Operated (Sizes 1-1/2 Inches And Below)		MIL-V-22094B (1)	4820	SH	3	Oct	61	MO	SH	82
Valves, Globe, Angle, And Y, Cast Or Forged, Steel, And Alloy Steel, Outside Screw And Yoke (Sizes 2-1/2 Inches And Above) (Basically Commercial Valves)	Q	MIL-V-22052C (3)	4820	SH	24	Aug	64		SH	
Valves, Globe, Angle, Quick Change Cartridge Trim, High Pressure (H.p.) Hydraulic And Pneumatic (Sizes 1/8 - 1 1/4 Inches)	QL	MIL-V-24109A	4820	SH	31	Aug	71		SH	

FIGURE 7.3 Index of Mil-Spec Standards.

Standards as a part of their requirements by virtue of the fact that they are covered by reference in the Uniform Building Code. The majority of UBC Standards are, in fact, other commercial standards that have been adopted as UBC Standards and renumbered. The UBC designations are characterized by the following type designation: "UBC Standard 26-1-91" for Portland Cement and Blended Hydraulic Cements. This standard is based upon ASTM Designation C 150-94 and C 595-81a.

Nongovernmental Standards

ASTM (American Society for Testing and Materials). These are by far the most recognized of all American standards. They not only cover materials specifications, but contain testing requirements and, in rare cases, some performance standards as well. The listings are characterized by designations such as "ASTM Designation C 150-94," where the "94" denotes the year of the particular edition or revision of that particular standard (Figure 7.4). All ASTM standards are divided systematically into groupings with separate letter prefixes that enable a user to identify the type of material referred to from the number designation alone. The following is a complete listing of all current letter prefixes and the material categories to which they refer:

- **A** Ferrous metals
- **B** Nonferrous metals
- **C** Cementitious, ceramic, concrete, and masonry materials

C 133 C 261

C 133-91 Test Methods for Cold Crushing Strength and Modulus of Rupture of Refractories, **15.01**
C 134-84 Test Methods for Size and Bulk Density of Refractory Brick and Insulating Firebrick, **15.01**
C 135-86(1992) Test Method for True Specific Gravity of Refractory Materials by Water Immersion, **15.01**
C 136-93 Test Method for Sieve Analysis of Fine and Coarse Aggregates, **04.02**
C 137 Discontinued
C 138-92 Test Method for Unit Weight, Yield, and Air Content (Gravimetric) of Concrete, **04.02**
C 139-73(1989)$^{\epsilon 1}$ Specification for Concrete Masonry Units for Construction of Catch Basins and Manholes, **04.05**
C 140-91 Method of Sampling and Testing Concrete Masonry Units, **04.05**
C 141-85(1989) Specification for Hydraulic Hydrated Lime for Structural Purposes, **04.01**
C 142-78(1990) Test Method for Clay Lumps and Friable Particles in Aggregates, **04.02**
C 143-90a Test Method for Slump of Hydraulic Cement Concrete, **04.02**
C 144-93 Specification for Aggregate for Masonry Mortar, **04.05**
C 145 Discontinued 1992—Replaced by C 90: Specification for Load-Bearing Concrete Masonry Units
C 146-80(1989)$^{\epsilon 1}$ Test Methods for Chemical Analysis of Glass Sand, **15.02**
C 147-86(1991)$^{\epsilon 1}$ Method for Internal Pressure Test on Glass Containers, **15.02**
C 148-91 Test Methods for Polarascopic Examination of Glass Containers, **15.02**
C 149-86(1991)$^{\epsilon 1}$ Test Method for Thermal Shock Resistance of Glass Containers, **15.02**
C 150-94 Specification for Portland Cement, **04.01, 04.02**
C 151-93a Test Method for Autoclave Expansion of Portland Cement, **04.01**
C 152 Redesignated E 152
C 153 Discontinued—Combined with C 64
C 154 Discontinued
C 155-88(1992) Classification of Insulating Firebrick, **15.01**
C 156-93 Test Method for Water Retention by Concrete Curing Materials, **04.02**
C 157-93 Test Method for Length Change of Hardened Hydraulic Cement Mortar and Concrete, **04.02**
C 158-84(1989)$^{\epsilon 1}$ Method for Flexural Testing of Glass (Determination of Modulus of Rupture), **15.02**
C 159-91 Specification for Vitrified Clay Filter Block, **04.05**
C 160 Redesignated E 160

C 191-92 Test Method for Time of Setting of Hydraulic Cement by Vicat Needle, **04.01**
C 192-90a Test Method of Making and Curing Concrete Test Specimens in the Laboratory, **04.02**
C 193 Discontinued
C 194 Discontinued
C 195-90 Specification for Mineral Fiber Thermal Insulating Cement, **04.06**
C 196-93 Specification for Expanded or Exfoliated Vermiculite Thermal Insulating Cement, **04.06**
C 197 Discontinued
C 198-91 Test Method for Cold Bonding Strength of Refractory Mortar, **15.01**
C 199-84(1994)$^{\epsilon 1}$ Test Method for Pier Test for Refractory Mortar, **15.01**
C 200 Discontinued—Replaced by C 700
C 201-86$^{\epsilon 1}$ Test Method for Thermal Conductivity of Refractories, **15.01**
C 202-93 Test Method for Thermal Conductivity of Refractory Brick, **15.01**
C 203-92 Test Methods for Breaking Load and Flexural Properties of Block-Type Thermal Insulation, **04.06**
C 204-94 Test Method for Fineness of Hydraulic Cement by Air Permeability Apparatus, **04.01**
C 205 Discontinued—Replaced by C 595
C 206-84(1992)$^{\epsilon 1}$ Specification for Finishing Hydrated Lime, **04.01**
C 207-91(1992)$^{\epsilon 1}$ Specification for Hydrated Lime for Masonry Purposes, **04.01**
C 208-94 Specification for Cellulosic Fiber Insulating Board, **04.06**
C 209-92 Test Methods for Cellulosic Fiber Insulating Board, **04.06**
C 210-85(1990) Test Methods for Reheat Change of Insulating Firebrick, **15.01**
C 211 Discontinued—Replaced by C 700
C 212-93 Specification for Structural Clay Facing Tile, **04.05**
C 213 Discontinued—Combined with C 64
C 214 Discontinued—Replaced by E 96
C 215-91 Test Method for Fundamental Transverse, Longitudinal and Torsional Frequencies of Concrete Specimens, **04.02**
C 216-92d Specification for Facing Brick (Solid Masonry Units Made from Clay or Shale), **04.05**
C 217-85(1990) Test Method for Weather Resistance of Natural Slate, **04.07**
C 218 Discontinued
C 219-94 Terminology Relating to Hydraulic Cement, **04.01**
C 220-91 Specification for Flat Asbestos-Cement Sheets, **04.05**
C 221-91 Specification for Corrugated Asbestos-Cement Sheets, **04.05**
C 222-91 Specification for Asbestos-Cement Roofing Shingles, **04.05**
C 223-91 Specification for Asbestos-Cement Siding, **04.05**

FIGURE 7.4 Typical Page from ASTM Index of Standards. *(Copyright ASTM International. Reprinted with permission.)*

- **D** Miscellaneous materials
- **E** Miscellaneous subjects
- **F** Materials for specific applications
- **G** Corrosion, deterioration, and degradation of materials
- **ES** Emergency standards

ANSI (American National Standards Institute). This is perhaps the second most commonly known commercial and industrial standards organization. It is often mistaken for a governmental entity because of its name; however, it is a voluntary, nonprofit organization without any governmental connections.

ANSI is a coordinating organization for a federated national standards system. The ANSI federation consists of 900 companies, large and small, and some 200 trade, technical, professional, labor, and consumer organizations. In cooperation with its councils, boards, and committees, ANSI coordinates the efforts of many organizations in the United States that are developing standards.

ANSI does not develop any standards of its own, but does provide means for determining the need for them and ensures that organizations competent to fill these needs undertake the standards-development work. Most of the standards listed by ANSI are developed by trade, technical, professional, consumer, and labor organizations.

Coal
preparation plants, ANSI/NFPA 120
pulverizers, ANSI/ASME PTC4.2
Coated abrasives—*see* Abrasives
Coating and dipping, ANSI/NFPA 34
Coatings
organic, ANSI/NFPA 35; ANSI/UL 1332
steel water pipelines, ANSI/AWWA C203
underbody, surface vehicles, ANSI/SAE J671
viscosity, ANSI/ASTM D3236
COBOL (programming language), ANSI X3.23, X3.23a
Codes
bar, print quality, ANSI X3.182
carriage positioning characters, ANSI X3.78
communities in U.S., ANSI X3.47
counties in U.S., ANSI X3.31
dates, ANSI X3.30
geographic point locations, ANSI T1.205, X3.61
hydrologic units, ANSI X3.145
input/output to microprocessors, ANSI X3.95
ISO registration, ANSI X3.83
keyboard arrangements, ANSI X3.114
states in U.S., ANSI X3.38
time, ANSI X3.51
time of day, ANSI X3.43
videotex/teletext, ANSI X3.110
weights and measures, ANSI X3.50
Codes, ASCII
character sets, ANSI X3.4, X3.41
controls, ANSI X3.64
graphic representation, ANSI X3.32
Coding, color
identification, ANSI/EIA 359-A, 359-A-1
physical hazards marking, ANSI Z535.1
semiconductor devices, ANSI/EIA 236

Communication systems
electric power stations, ANSI/IEEE 487
inductive coordination, ANSI/IEEE 776
interconnection circuitry, ANSI/NEMA SB3
optical fiber, ANSI Z136.2
transmitters, ANSI/IEEE 377
Communications, fire services, ANSI/NFPA 1221
Compactors, surface vehicles, ANSI/SAE J1017, J1472
Compactors, waste, ANSI Z245.2
Components, surface-mount, ANSI/EIA 481-1
NECQ system, ANSI/EIA 601
Composites
insulators, ANSI/IEEE 987
slabs, ANSI/ASCE 3
structures (aerospace), ANSI/SAE ARP 1643A
welding rods, ANSI/AWS A5.21
Compressors, ANSI/ARI 510
air, motor-operated, ANSI/UL 1450
air systems, ANSI/ASME B19.1
centrifugal, ANSI/API 617
displacement, ANSI/ASME PTC9
performance test code, ANSI/ASME PTC10
process industries, ANSI/ASME B19.3
refrigerant, ANSI/ARI 520, 530; ANSI/UL 984
trucks and buses, ANSI/SAE J1340
Computer automated measurement and control—*see* CAMAC
Computer graphics
glossary of terminology, ANSI/IEEE 610.6
graphical kernel system, Ada binding, ANSI X3.124.3
graphical kernel system, FORTRAN binding, ANSI X3.124.1
graphical kernel system, functional description, ANSI X3.124

Computers (aerospace) ECS program, ANSI/SAE ARP 1623A
flight management systems, ANSI/SAE ARP 1570
performance, ANSI/SAE AS 8002
COMTRADE, ANSI/IEEE C37.111
Concrete
canal linings, ANSI/ASAE S289.1
curing, ANSI/ACI 308
heavyweight, selecting proportions, ANSI/ACI 2 11.1
lightweight, selecting proportions, ANSI/ACI 211.2
no-slump, selecting proportions, ANSI/ACI 211.3
reinforced, ANSI/ACI 315, 318, 318M
structural, ANSI/ACI 301, 318.1, 318.IM
Concrete construction
columns, ANSI/CRSI A38.1
forms for joints, ANSI/CRSI A48.1
forms for joists, ANSI/CRSI A48.2
masonry, ANSI/ACI 531
nuclear, ANSI/ACI 349
safety requirements, ANSI A10.9, A10.9a
tolerances, ANSI/ACI 117
Concrete pipe—*see* Pipe, concrete
Condensate removal devices, ANSI/ASME PTC39.1
Condensers, refrigerant, ANSI/ARI 490; ANSI/ASHRAE 22, 64
Conductors
aluminum, wire connectors, ANSI/UL 486B
copper, wire connectors and soldering lugs, ANSI/UL 486A
overhead, ANSI/IEEE 738
self-damping measurements, ANSI/IEEE 563
underground, ANSI/UL 486D
wiring terminals, ANSI/UL 486E

FIGURE 7.5 ANSI Index of Standards. *(Courtesy of ANSI. Reprinted with permission. Copies of ANSI standards may be purchased from the American National Standards Institute [ANSI], 25 West 43rd Street, New York, NY 10036, [212] 642-4900, http://webstore.ansi.org.)*

Each of these standards is then submitted to ANSI for recognition as a national consensus standard.

The system of standards identification used by ANSI prior to 1979 is no longer in use, and currently all ANSI standards are identified by the sponsor's own numbering system, prefixed by "ANSI" and the letters identifying the sponsoring agency (Figure 7.5).

AWWA (American Water Works Association). This is not as well known as ASTM or ANSI, except in the public works sector, where it has been the standard of the water industry. Its standards are broad, and many specify a single fabricated item, or an entire project feature, such as the designation "AWWA D100" [full designation is ANSI/AWWA D100-79(AWS-D5.2-79)] for the construction of welded steel elevated water tanks, standpipes, and reservoirs. AWWA publishes its data in several forms, including reference books, handbooks, manuals, standards, and periodicals and pamphlets, and some of its standards have also been adopted by ANSI, as in the example cited previously. As with the other standards organizations, a system of letter prefixes has been established for the orderly grouping of subject matter for easier recovery of data:

A Source
B Treatment
C Distribution

D Storage
E Pumping

ACI (American Concrete Institute). This possibly represents the most respected of concrete standards worldwide. Most of the ACI provisions have been adopted into all other codes and regulations, with minor changes. ACI is a nonprofit technical society. It is not industry supported [its nearest industry-supported counterpart is the Portland Cement Association (PCA)]. ACI publications are produced by standing committees, each of which is identified by a committee number. All publications are identified by a numbering system in which the ACI committee number forms the publication number, followed by the revision date. Thus committee 318 is the code committee, and each new revision of the ACI Building Code carries the same number 318, followed by the latest revision date.

There are numerous other standards that have not been mentioned, but the methods of identifying their publications are similar to the organizations already mentioned. The following is a partial list of other organizations that publish standards:

AASHTO	American Association of State Highway and Transportation Officials
AISC	American Institute of Steel Construction
AISI	American Iron and Steel Institute
AITC	American Institute of Timber Construction
APA	American Plywood Association
ASHRAE	American Society of Heating, Refrigerating, and Air Conditioning Engineers
ASME	American Society of Mechanical Engineers
AWPA	American Wood Preservers Association
AWPI	American Wood Preservers Institute
AWS	American Welding Society
CRSI	Concrete Reinforcing Steel Institute
CSA	CSA International (Canada)
IES	Illuminating Engineering Society
ISO	International Standards Organization
NEMA	National Electrical Manufacturers Association
NFPA	National Fire Protection Association
SSPC	Steel Structures Painting Council
UL	Underwriters' Laboratories, Inc.
WIC	Woodwork Institute of California
WRI	Wire Reinforcement Institute, Inc.

Access to Special Standards by Resident Project Representative and Contractor

All too often the specifier cites publications and standards that are not in the normal field library of the Resident Project Representative or the contractor. Although there are companies specializing in the sale of copies of such documents in many major cities that serve the needs of the contractor, often the Resident Project Representative of the architect/engineer or owner is the only party in the field who has no access to such information, and thus is placed at a considerable disadvantage in the field. It is important to remember that if it is important enough to specify that an item must conform to a certain set of standards, it should be important enough to see that the field office obtains a copy of this standard—otherwise, why even bother with inspection? There are certain popular standards that may be excluded from this, as they are so widely used and well understood that only a manufacturer's certificate of conformance with the cited standard should suffice.

Some firms solve the problem in an ideal fashion. A volume of photocopies of all the cited standards in a specific project is compiled and furnished to the Resident Project Representative at the beginning of the job. In this manner, the information is always accessible in the field when needed.

Building Codes, Regulations, Ordinances, and Permits

Building codes have been adopted by most cities, counties, and states. They are adopted by each governmental entity by ordinance or other means at their disposal to impart the force of law behind them. The codes carefully regulate design, materials, and methods of construction, and compliance with all applicable code provisions is mandatory. Most of these codes are based in whole or in part on the various national codes that are sponsored by different national groups. Changes, where made in the parent code, have generally been to accommodate local or regional needs and conditions and to make portions of the code more stringent than may have been provided for in the code as it was originally written.

There are, of course, many codes that are not based upon such national codes, but have been specifically written for a particular locality. Many large cities do this.

Although the Uniform Building Code was the basis for the majority of codes in use in the western states, there are several prominent codes in use throughout the United States that are equally important in the areas that they serve. Some of the more prominent codes are listed below:

1. *International Building Code:* compiled by the International Code Committee (ICC), comprised of ICBO, SBCCI, and BOCA.
2. *National Building Code:* compiled by the American Insurance Association; adopted in various localities across the United States.
3. *Uniform Building Code:* compiled by the International Conference of Building Officials (ICBO); widely used in the western states.
4. *Basic Building Code:* compiled by the Building Officials and Code Administrators International (BOCA); used mainly in the eastern and northcentral states.
5. *Southern Standard Building Code:* compiled by the Southern Building Code Congress International (SBCCI); used in most southern and southeastern states.

6. *National Electric Code:* compiled by the National Fire Protection Association (NFPA); widely adopted in all parts of the United States.
7. *National Plumbing Code:* compiled by the American Public Health Association and the American Society of Mechanical Engineers; used widely in all parts of the United States.
8. *Uniform Plumbing Code:* compiled by the International Association of Plumbing and Mechanical Officials (IAPMO); widely used in the western states.
9. *ICBO Plumbing Code:* compiled by the International Conference of Building Officials (ICBO); used principally in the western states.
10. *ICBO Uniform Mechanical Code:* compiled by the International Conference of Building Officials (ICBO; mostly HVAC and gas piping); used principally in the western states.

Of particular importance to the inspector as well as the contractor is the edition date on the code. Even though a new code edition has been published, there can be no assurance that a particular jurisdiction will adopt it at any certain time, if at all. A case in point is a city that for years used an old edition of the Uniform Building Code, which it kept updated by adopting a series of city ordinances annually to meet the requirements of the local building official. Thus, in fact, it was a special code, and only by carefully studying the basic old code edition plus all ordinances adopted afterward could an inspector be aware of the conditions that affected his or her project in that jurisdiction. That particular city has since adopted a newer version of the Uniform Building Code, and the cycle must begin all over again.

In addition to the provisions of any applicable codes, the contractor is obligated to conform to the provisions of all permits issued by public agencies having jurisdiction. Many of these permits are several pages long and resemble a small specification.

The International Building Code

Beginning with the 2000 edition, the International Building Code was born. It is the result of the efforts of the International Code Committee (ICC), which is a joint effort by three existing code-writing agencies: the ICBO, SBCCI, and BOCA.[6] Upon the release of the International Building Code in 2000, the participating agencies pledged to discontinue updates of their respective regional codes and to concentrate on continued development of the new International Building Code.

BOCA primarily represented the Northeast and Bermuda; SBCCI primarily represented the southeastern region, and included use in the Grand Cayman Islands; and ICBO represented the western and central states, as well as the U.S. territories in the Caribbean and Western Pacific along with Puerto Rico, Guam, the U.S. Virgin Islands, American Samoa, and Saudi Arabia.

The issuance of a new code carries with it no force of law; it is up to the local jurisdictions to adopt the IBC.

---

[6]International Conference of Building Officials (ICBO); Southern Standard Building Code (SBCCI); and the Basic Building Code (BOCA).

Projects Subject to Control by More Than One Agency

Often a project will involve property or facilities that are under the jurisdiction of another public agency. If a pipeline project, for example, crossed an interstate freeway route, a flood control right-of-way, or the pipeline of another public agency or utility, the contractor could well be facing the prospect of having not only the regularly assigned resident inspector on the job, but also a battery of other inspectors who represent each of the other affected agency properties. In each such case, a permit may be required from each affected agency, and the contractor would be bound not only by the terms of the contract documents for the project, but also by the terms of each such permit as long as the work was in an area within the jurisdiction of the issuing agency. Whenever the pipeline mentioned crosses any of these other improvements, the requirements of the agency owning such improvement will usually govern over the terms of the project specifications, unless its requirements are less stringent than those of the project specifications, in which case the affected agency may agree to the project specification requirements in lieu of its own.

The basic authority in such cases still lies with the Resident Project Representative. Inspectors representing any affected jurisdictional agencies should exercise their proper authority by working through the Resident Project Representative, and should not be instructed to communicate their requirements directly to the contractor.

Types of Drawings Comprising the Construction Contract

The drawings or plans that were prepared especially for the project, generally referred to as the *contract drawings,* are for the express purpose of delineating the architect's or engineer's intentions concerning the project that they have conceived and designed. The drawings normally show the arrangement, dimensions, geometry, construction details, materials, and other information necessary for estimating and building the project.

Occasionally, *standard drawings* of public agencies are defined as being a part of the contract documents. These drawings usually portray the repetitious details of certain types of construction that may be required by the local public agency for all similar work in their jurisdiction. A typical example is the design of drop inlet structures and other drainage structures that have been standardized in each community throughout the years. Instead of redrawing the same details over and over, the architect or engineer simply refers to a certain standard drawing of the jurisdictional agency involved.

Shop Drawings

Shop Drawings are those details and sketches prepared by the contractor or the material suppliers or fabricators that are necessary to assure the fabricator that the basic concept is acceptable before beginning costly fabrication. Shop drawings frequently contain information that is not related to the design concept, or information that is relative only to the fabrication process or construction techniques in the field, all of which are outside the scope of the duties and responsibilities of the architect or engineer. In approving shop drawings, the architect or engineer only indicates that the items conform to the design *concept* of the project and compliance with the plans and specifications prepared (see "Shop Drawings and Samples" in this chapter). The contractor remains wholly responsible for dimensions to be confirmed in the field, for information that pertains solely to the fabrication process or to techniques of construction, and for coordination of the work of all the trades.

Change orders are issued to accompany a written agreement to modify, add to, or otherwise alter the work from that originally set forth in the contract drawings at the time of opening bids. A change order is normally the only legal means available to change the contract provisions after the award of the contract, and normally requires the signature of the owner.

Record Drawings

Some confusion seems to accompany the use of the terms *"record" drawings* (Chapter 3) and *"as-built" drawings.* Generally speaking, *record drawings* are a marked set of prints prepared by the contractor or the Resident Project Representative in the field. Record drawings are contract drawing prints upon which the contractor or inspector records all variations between the work as it was reported by the contractor as having been actually constructed and the work as it was shown in the original contract drawings as they existed at the time the contract was awarded. All change orders should be reflected in appropriate marks on the record drawings. The term *as-built drawings* is unpopular because of some of the legal difficulties that have resulted in attempting to have the architect or engineer certify that a set of drawings truly represented the project "as built." If, some years later, an underground pipeline is struck because it was not at the location or depth indicated in the certified as-built drawings, the architect or engineer could be the defendant in a civil action pressed by the one-time client.

Order of Precedence of the Contract Documents

Where a discrepancy exists between the various parts of the contract documents, legal precedent has established some degree of order. However, many General Conditions cite the specific order anyway so that a dispute may be settled without the need for interpretation or arbitration.

Although, if specified in the contract documents, a listing of order will govern, the general policy follows, in order of decreasing authority:

1. Agreement
2. Specifications
3. Drawings

For further discussion of this subject, see "Order of Precedence of Contract Documents" in Chapter 20.

PROBLEMS

1. Is it still possible, under contracts involving separate inspections by other agencies having jurisdiction, to maintain the one-to-one relationship with the contractor?
2. What happens when the terms of the Supplementary General Conditions are in conflict with the General Conditions?
3. Can the Supplementary General Conditions be used to modify the Agreement form?
4. Which of the following listed documents is the preferred location for specifying the amount of liquidated damages? Which document normally is used to specify the terms and conditions of the liquidated damages?

   Notice Inviting Bids
   Instructions to Bidders
   Bid or Proposal
   Agreement
   General Conditions of the Contract
   Supplementary General Conditions
   Division 1 General Requirements
5. True or false? Changes on shop drawings are meant to supersede the original contract drawings.
6. Should an existing soils report be included as part of the contract documents?
7. Name the international (FIDIC) equivalent to the following terms commonly used in U.S. construction contracts:

   Differing Site Conditions
   Bid or Proposal
   Changes in the Work
   Date of Notice to Proceed
   Addenda
   Guarantee Period
   Bid Schedule (unit-price projects)
   Schedule
   Supplementary General Conditions
8. May changes to the documents be done by Addenda after opening bids, but prior to Award?
9. Who should be the only party with the authority to "stop the work"?

8

CONSTRUCTION LAWS AND LABOR RELATIONS

Compliance with Laws and Regulations

As with most endeavors, the performance of construction contracts is regulated by law. The difference between construction and other enterprises seems to lie in the fact that there are more jurisdictional agencies involved, and thus more laws to contend with.

Simply stated, it can be said in general that these laws fall into four major categories:

1. Contract law—those laws and regulations that affect the making of contracts, both public and private.
2. Laws governing the execution of the work being performed under the contract—including the issuance and conformance to the conditions of the various permits, regulations, ordinances, and other requirements of the many jurisdictional agencies that are frequently involved.
3. Laws that relate to the settling of differences and disputes that may develop out of the performance of the contract.
4. Licensing laws that govern not only the business practices but also the personal qualifications standards of the various people involved in the construction process.

The latter include the licensing of architects and professional engineers in every state in the United States, as well as the licensing of contractors in many states, the licensing of construction managers in a few states, and the licensing of inspectors in some states. In addition to these licenses, all of which require the demonstration of proficiency by some type of examination, there are the local business licenses, permits to do business in certain areas, and sales tax permits that authorize the collection of sales tax by businesses.

Some of the more common laws that are encountered on most projects are the following:

Americans with Disabilities Act (ADA) of 1990

Davis-Bacon Act requirements (for federal or federally funded projects)

OSHA and state safety requirements for industrial applications

OSHA and state safety requirements for construction

State Labor Code requirements

U.S. Department of Labor requirements, as applicable

State housing laws

Local building codes and ordinances

Sales and use tax regulations

Air pollution control laws

Noise abatement ordinances

Business licenses to conduct business in *each* locality

Mechanic's lien laws of the state

Unemployment insurance code requirements

Worker's compensation laws

U.S. Army Corps of Engineers and Coast Guard regulations for work in navigable waterways

Subletting and subcontracting laws

Licensing laws for architects, engineers, surveyors, contractors, and inspectors

Permits by special local agencies for construction, including building permits, grading permits, encroachment permits, street work permits, police permits for interrupting traffic, excavation permits, Environmental Protection Agency permits, special hauling permits, Department of Agriculture permits, and many others

Civil Rights Act of 1964 (federally assisted programs)

Civil Rights Act of 1991

All local city and county codes and ordinances

There are, of course, many others; but this list serves as a means of calling attention to the fact that a project cannot be built without regard for the regulations of the many jurisdictional agencies affected by the work. These include federal, state, county, city, special districts, including the many federal and state bureaus that have a legal interest in the effects of the proposed work on the area they have a legislative mandate to control, and all others who have a legitimate interest in the work that affects facilities or improvements over which they have legal jurisdiction and responsibility.

Public versus Private Contracts

As a means of protecting the public interest in projects involving the expenditure of public funds, or the public administration of projects built with private funds but intended for public use, the majority of all public contracts are required to conform to the general laws governing the execution of public contracts in that state.

Although every contractor has a legal obligation to observe the law in the conduct of business, many state and federal regulations are required to be spelled out in the project documents, even though the failure to repeat their terms and conditions would not relieve the contractor of the responsibility to conform to their requirements.

Some of the significant differences between the requirements for public contracts and private projects can be noted by observing the following restrictions that apply to public contracts in most jurisdictions:

1. The Project must be publicly advertised.
2. Bids must be accompanied by a bid bond, usually 5 or 10 percent.
3. The Notice Inviting Bids must normally contain a list of prevailing wage rates for all crafts to be used in the work. If federal funds are involved, an additional listing of the federal wage rates must be published.
4. Insurance policies and bonds covering public liability and property damage are required.
5. In some jurisdictions, a list of all subcontractors who will perform work on the project must be listed and filed with the general contractor's bid.
6. Wherever a brand name product is specified, the specifications must give the names of the brand name products, *plus* the words "or equal."
7. A performance bond and a payment bond (for labor and materials) must be provided in the amounts specified by law.
8. Award must be to lowest responsive, responsible bidder.

Some states can be classified as *code states* because they regulate virtually everything by specific code provisions. In principle, only Louisiana's legal system was not based upon English common law. However, some states that were supposedly under common law principles have evolved into code states. California is a notable example. As it applies to construction contracts, public works is most heavily regulated in that state. It is fruitless to list all of the applicable code requirements in a book such as this, for as the author learned in a law course many years ago: "Learn where to find the law; do not attempt to memorize it, for otherwise, with one sweeping move, the legislature could repeal your entire education." When asked to draft front-end documents for a set of construction specifications, it is wise to seek a review of the documents by competent legal counsel; however, be certain that the attorney is a specialist in contract law, especially as it applies to construction.

Limitations of Authority of a Public Agency

While the contract entered into between the contractor and the public agency responsible for a given public works project will generally define the relationship between them, it is also important for the parties to be aware of the limitations of the public agency's authority to enter into contracts in order to perform certain acts. As stated previously, a governmental agency has no power to enter into an act that it is not properly authorized to perform. Under certain circumstances, that fact can have a major impact on a contractor.

A public agency has only the ability to do what it has been granted the authority to do. Whatever a public agency wishes to do by way of a public works construction contract, it must be authorized to do under the laws of that particular jurisdiction. A public entity cannot act in excess of its jurisdiction, and any action taken by a public agency without the legal power to do so is without legal force and effect. In other words, a public agency can do only what the law allows or prescribes. The contractor, by knowing the particular laws that apply when dealing with a public agency, can use that knowledge to its advantage in order to force the public agency to do what is required by law.

The way that a limitation of authority can ultimately affect a contract is shown in the case of *Zottman v. The City and County of San Francisco* [20 Cal. 96 (1862)]. In that case, the city of San Francisco properly entered into a valid contract with a contractor to make improvements to a certain town square. After the contract was entered into, the scope of the contract was changed by the city, and the contractor was ordered to do certain additional work. The contractor was paid for the contract work, but the city refused to pay for the extra work. The contractor then brought legal action against the city for payment. The court denied payment because the contract for extra work was not let to the lowest bidder after proper notification as required by law. The court also said that the acts of the city officials in demanding that the work be done did not ratify the contract for the extra work, as the city had no authority to make such a contract in the first place. While the court acknowledged that denying payment to the contractor that performed the work might be a hardship, the court found that the contractor was aware of the law and would suffer only what should have been anticipated. The city was allowed to retain the work without paying for it. According to later decisions, had the contractor received any payments for any portion of the extra work, the city would have been entitled to demand that the contractor return such progress payments to the city. The risks that a contractor takes in working with a public agency outside the law or in violation of the law can easily be seen.

Traffic Requirements during Construction

Often the work on a project involves a degree of restriction on the local traffic in the project area. The contractor is obligated to get a permit from the local agency empowered to enforce traffic regulations whenever it becomes necessary to close a street or an intersection, or to restrict its traffic. The terms and conditions of this permit will become a part of the contract provisions, and must be enforced. The inspector should take careful note of the specific restrictions, which are sometimes spelled out in the project specifications as well. Frequently, as in cases involving streets that handle rush-hour traffic, the traffic requirements may vary depending upon the time of day. The inspector should watch in particular for requirements that a flagger be provided and used by the contractor to direct traffic through restricted traffic lanes, and to be certain that the conduct of the flagger is in full accordance with the traffic requirements of the jurisdiction having control. Whenever there is only one lane of traffic open on a street that must provide two-way travel, a flagger or a detour route will generally be required. In general, the minimum number of lanes allowed during construction affecting a city street is one 3-m (10-ft) traffic lane *in each direction*. The inspector should watch for requirements that excavations be continually fenced or covered, as might be the case in the vicinity of schools; or that not less than a specified length of pipe trench be allowed open at any one time ahead and behind a pipelaying operation; or that the contractor be required to provide temporary "bridges" over excavations that might otherwise prevent safe access to businesses and residences in the vicinity of the work.

Code Enforcement Agency Requirements

Whenever privately owned buildings are to be constructed, and frequently in the case of most public buildings or public works projects, the building code enforcement agency will be involved. In some jurisdictions, a public project built by another agency within the same jurisdiction enjoys immunity from the regulations imposed by another agency of the same government. Thus, a project constructed by that agency might apply its own rules to govern design and construction. Generally, however, it will be found that even the structures of agencies not obligated to conform to a sister agency's requirements will design in accordance with its code requirements and, in some cases, will even submit to its inspection requirements. If a permit fee is levied, however, it would simply be a paper transaction between departments.

Wherever the code enforcement agency has jurisdiction over a project under the Uniform Building Code, the provisions of Section 305(a) specify the conditions under which a project must be under the continuous inspection of a "special inspector" approved by the building official. In some jurisdictions, such approval is granted only after passing a written examination administered by the local department of building and safety. Where such programs are provided, even an architect or a registered engineer may not be allowed to perform the inspections unless the building department's examination has been passed. In other jurisdictions, no formal program exists, and the process of "approval" of the inspector by the building official is an informal one, usually simply upon recommendation of the architect or registered engineer of record on the project.

A complete listing of all specialty areas that require the employment of a special inspector is found in Section 305 of the Uniform Building Code, and in some other codes as well.

Work within or Adjacent to Navigable Waterways

It may not be frequent that an inspector will encounter a project subject to these requirements, but every now and then the inspector may be called upon to provide inspection of offshore construction as illustrated in Figure 8.1 or a project at a waterfront, such as where barge-mounted pile drivers or similar equipment may be used. On such work, some items that might not otherwise be thought of can be encountered.

For one thing, the contract documents should provide that the contractor must furnish, at the request of the design firm, the owner, or any inspector assigned to the work, suitable transportation from all points on shore designated by the design firm or owner to and from all offshore construction. If these provisions are not in the specifications, they should be arranged for with the contractor at the preconstruction meeting.

As for the transportation of the contractor's personnel and equipment to offshore sites, it is the contractor's own responsibility to obtain boat-launching facilities, or to make arrangements with local water carriers. The contractor is similarly obligated

FIGURE 8.1 Example of Offshore Construction Subject to Corps of Engineers' Jurisdiction.

to avoid the creation of navigation hazards or interference in any way with navigation routes except upon special permit from the agencies having jurisdiction.

If a contractor, during the course of the work, throws overboard, sinks, or misplaces anything overboard that could be interpreted as a hazard to navigation, the contractor must be required to recover such items as soon as possible. The contractor's first requirement after such an incident is to give immediate notice to the inspector, along with a description and location of such obstructions. If necessary, the contractor may also be required to place a buoy or other mark at the location until recovery. It is an item that must be conducted at the contractor's sole cost and expense unless specifically provided otherwise in the contract. Liability for the removal of a *vessel* wrecked or sunk without fault or negligence of the contractor is limited to that provided in Sections 15, 19, and 20 of the Rivers and Harbors Act of March 3, 1899, which is still very much alive and enforced.

All work being constructed in or involving the use of "navigable waterways" is subject to the orders and regulations of the Department of the Army, Corps of Engineers, and to the U.S. Coast Guard as they apply to the construction operations affecting property or improvements within the jurisdiction of these agencies. The principal jurisdiction on all construction matters lies with the Corps of Engineers, whereas the Coast Guard is concerned primarily with the movement of vessels. The term *navigable waters* is much broader now than in previous years, and is administratively defined in *Permits for Activities in Navigable Waters or Ocean Waters* on page 31324 of the *Federal Register* of 25 July 1975.

In addition, the Federal Water Pollution Control Act amendments of 1972 (Public Law 92-500) require a Corps of Engineers permit, under Section 404 of the act, for the discharge of any more than one cubic yard of dredged or fill material into any navigable waters. A word of caution—the list of waterways defined as navigable under this act is different from those defined in the Rivers and Harbors Act of 1899. Stripped of its regulatory language, the *404 Permit,* as it is commonly referred to in the industry, states that whenever a party deposits more than 1 cubic yard of anything in, on, or encroaching on any portion of the historic high water line of any "navigable waterway," an Army Corps of Engineers 404 Permit is required. This, as interpreted by the District Engineer of one Corps of Engineers District, includes portions of dry land that carried flood flows and previous streambeds even where, after a flood, a stream followed a new watercourse. Take into account, also, that the lead time in obtaining a 404 Permit is at the very least several months. Generally, it is preferable to plan construction operations to avoid such encroachment entirely.

An owner or architect/engineer should take into account the potential delays in the work that could result from expecting the contractor to obtain this permit. Additionally, federal laws may change the definition of *navigable waters* and *wetlands.* As political administrations change, the interpretation of the existing laws is also likely to change. Therefore, the author strongly advises that the Regulatory Functions Branch of the controlling District Office of the Corps of Engineers always be consulted to determine if, in fact, a permit is required, since it is tasked with enforcing these laws and will ultimately decide whether an area is a "navigable waterway" or a "wetland."

Fair Subcontracting Laws

For work on public projects in some areas of the United States, laws have been enacted to protect subcontractors and the public against the practice of *bid shopping* or *bid peddling* by general contractors. Such laws were prompted by the realization that projects that had been subject to bid shopping practices often resulted in poor-quality material and workmanship to the detriment of the public, and because it deprived the public of the full benefits of fair competition among prime contractors and subcontractors, and in addition it led to insolvencies, loss of wages to employees, and other evils.

Generally, such laws provide that any bidder on a public project must list all subcontractors that the bidder intends to use on the work, whose work, labor, or services exceed a specified percentage of the prime contractor's total bid.

Under the provisions of such *fair subcontracting* laws, if the general contractor fails to list a subcontractor for any portion of the work, the general contractor must perform all such work itself.

It is the Resident Project Representative's responsibility when construction begins to check carefully the identity of all subcontractors on projects subject to such laws to ensure that their provisions have not been violated. This may have to be monitored at intermittent intervals to assure compliance at all times.

The Hazardous Waste Problem

Hazardous waste management practices are influenced and in many cases controlled by a number of environmental laws and regulations. As the direct result of damage to life and the environment through mismanagement of hazardous wastes, in 1976 Congress enacted Subtitle C of the Resource Conservation and Recovery Act (RCRA), which imposed controls on the management of hazardous wastes.

Each state is encouraged to develop its own program, following the EPA's guidelines. The major provisions under RCRA for controlling hazardous waste are:

1. Definition of hazardous waste
2. A system to track hazardous waste from its generation to its point of final disposal
3. Standards for generators and transporters of hazardous waste
4. Permit requirements for facilities that treat, store, or dispose of hazardous waste
5. Requirements for state hazardous waste programs

The subject of hazardous wastes has become a sensitive construction issue, as owners are coming under fire to provide full disclosure of hazardous materials that a contractor may come into contact with in the course of construction of a project and to specify the means and methods for disposal of such wastes.

Contract wordings are varied, as it is a relatively new, untested area of risk for a construction contractor and project owner. In California, one of the states that did develop its own program within the guidelines of the EPA, any public works project of a local public entity that involves digging trenches or other excavations that extend deeper than 1.2 m (4 ft) below the surface *must* contain the following clause in the contract documents:

(a) In any public works contract which involves digging of trenches or other excavations that extend deeper than 1.2 m (4 ft) below the surface, the Contractor shall promptly, and before the following conditions are disturbed, notify the public entity, in writing, of any:

1. Material that the Contractor believes may be material that is hazardous waste, as defined in Section 25117 of the California Health and Safety Code, that is required to be removed to a Class I, Class II, or Class III disposal site in accordance with the provisions of existing law.
2. Subsurface or latent physical conditions at the site differing from those indicated.
3. Unknown physical conditions at the site of any unusual nature, different materially from those ordinarily encountered and generally recognized as inherent in work of the character provided for in the Contract.

(b) The public entity shall promptly investigate the conditions, and if it finds that the conditions do materially so differ, or do involve hazardous waste, and cause a decrease or increase in the Contractor's cost of, or the time required for, performance of any part of the Work shall issue a Change Order under the procedures described in the Contract.

(c) That, in the event that a dispute arises between the public entity and the Contractor whether the conditions materially differ, or involve hazardous waste, or cause a decrease or increase in the Contractor's cost of, or the time required for, performance of any part of the Work, the Contractor shall not be excused from any scheduled completion date provided for by the Contract, but shall proceed with all work to be performed under the Contract. The Contractor shall retain any and all rights provided either by Contract or by law which pertain to the resolution of disputes and protests between the contracting parties.

In addition, the California Department of Transportation (Caltrans) prescribes a construction hazardous waste contingency plan (Figure 8.2) for the orderly handling of unknown wastes discovered during construction. Although the foregoing applies only to California public works, similar requirements exist in other jurisdictions.

Federal Labor Laws

Labor–Management Relations Laws

The history of federal labor laws goes back to 1890, when the Sherman Anti-Trust Act was enacted. This provided the statutory beginnings for labor management legal policy. Although it is debatable whether the act was ever intended to apply to labor unions, a Supreme Court decision in 1908 ruled that the labor unions were indeed covered by the act.

During the intervening years, Congress passed other labor-related laws that more clearly outline federal labor policy. Today's labor policy is the outcome of the

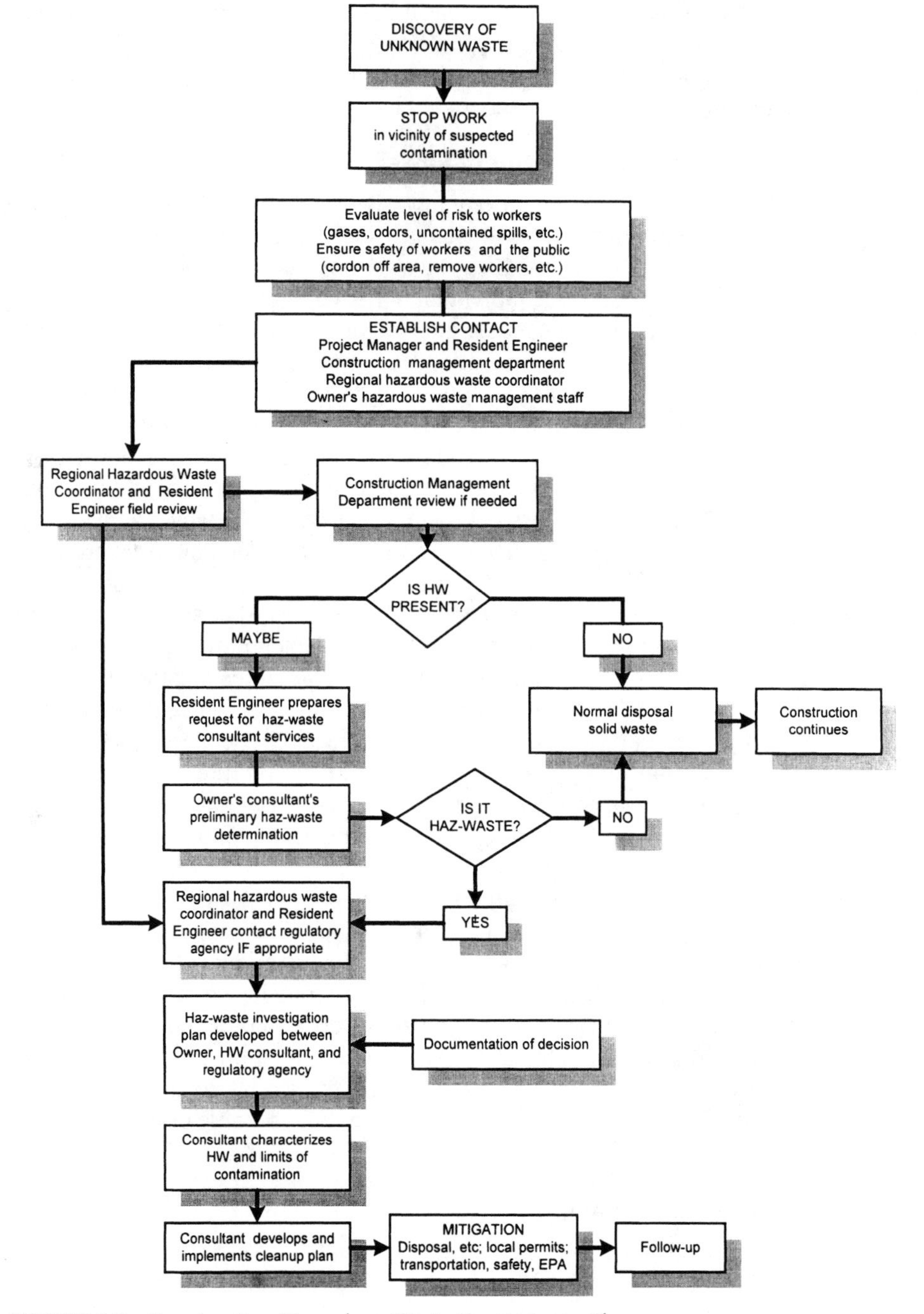

FIGURE 8.2 Construction Hazardous Waste Contingency Plan.

combined provisions of the Norris-LaGuardia Act of 1932 (Anti-Injunction Act), the National Labor Relations Act of 1935 (Wagner Act), the Labor Management Relations Act of 1947 (Taft-Hartley Act), and the Labor Management Reporting and Disclosure Act of 1959 (Landrum-Griffin Act).

Equal Employment Opportunity Laws

In recent years, many of the federal labor laws have been aimed at eliminating discrimination in employment for any cause, such as age, sex, race, religion, or nationality. The first of these was the *Civil Rights Act of 1964,* in which Congress confirmed and established certain basic individual rights with regard to voting; access to public accommodations, public facilities, and public education; participation in federally assisted programs; and opportunities for employment. Administration is handled through the Equal Employment Opportunity Commission (EEOC) that was created by the act. Under this act it is unlawful for any employer to refuse to hire or to discharge any individual or otherwise discriminate against him or her with regard to conditions of employment because of race, color, religion, sex, or national origin.

In 1965, the president issued *Executive Order 11246,* which applies to contracts and subcontracts exceeding $10,000 on federal and federally funded construction projects. Under this executive order, the contractors not only are prohibited from discrimination but also must take positive action to see that applicants are employed and that employees are treated during their employment without discrimination. Originally limited to discrimination based upon race, creed, color, or national origin, it was supplemented in 1968 by *Executive Order 11375,* which prohibits discrimination in employment based upon sex in all federal and federally funded contracts. Any contractor who fails to conform may be barred from future contracts involving federal funds, the present contract can be canceled or suspended, and the OFCC has the additional power to withhold progress payments from contractors who appear to be in violation.

In 1967, the *Age Discrimination in Employment Act* was passed. This act prohibits arbitrary age discrimination in employment. The act protects persons 40 to 65 years old from age discrimination by all employers of 25 or more persons in an industry affecting interstate commerce. Employment agencies and labor organizations are also covered. The prohibitions against age discrimination do not apply when age is a valid occupational qualification, when differentiation is based upon reasonable factors other than age, when differentiation is caused by the terms of a valid seniority system or employee benefit plan, or when the discharge or discipline of the individual is for good cause.

Americans with Disabilities Act

The *Americans with Disabilities Act* (ADA) of 1990 is a federal ban on disability-based employment discrimination (both physical and mental disabilities) covering all but the smallest employers. Insofar as the Act applies to employers, it applies to all employers with 15 or more employees. It is enforced by the Equal Employment Opportunity Commission (EEOC). Enforcement is through use of the Civil Rights Act of 1964 powers, remedies, and enforcement procedures.

Basically, the ADA makes it unlawful to discriminate against a qualified individual with a disability. A "disability" is defined in the Act, along with exclusions.

Because of the publicity accorded the ADA, it is highly likely that its provisions will be vigorously enforced, along with a significant increase in litigation.

Wage and Hour Laws

The *Davis-Bacon Act* of 1931, as subsequently amended, is a federal law that determines the minimum wage rates and fringe benefits that must be paid to all workers on all federal and federally assisted projects. The law applies to all projects over $2000, and it states that the wages of the workers must not be less than the wage rates specified in the schedule of prevailing rates of wages as determined by the Secretary of Labor for similar work on similar projects in the vicinity in which the work is to be performed. Under its terms, the contractor is required to pay once a week to all workers employed directly on the site of the work, at wages no lower than those prescribed. The federal minimum wage rates are currently published every Friday in the *Federal Register,* and all changes are reflected in the various trades and areas of the United States.

As passed in 1934, the *Copeland Act* (Anti-Kickback Law) forbids an employer to deprive any employee on a federal or federally assisted construction job of any portion of the compensation to which that employee is entitled under federal law. Other than deductions provided by law, the employer may not require kickbacks from its employees. Violation may be punished by fine, imprisonment, or both. The Copeland Act applies to all projects on which the Davis-Bacon prevailing wage law applies.

The Fair Labor Standards Act, usually known as the *Wage and Hour Law,* was enacted by Congress in 1938 and has since been amended several times. The act contains provisions governing minimum wage, maximum hours, overtime pay, equal pay, and child labor standards. An employer who violates the wage and hour requirements is liable to his employees for double the unpaid minimum wages or overtime compensation, plus associated court costs and attorneys' fees. Willful violation is a criminal act and may be prosecuted as such. Several classes of employees are exempted from coverage under the act, such as executive, administrative, and professional employees who meet certain tests established for exemption.

The Fair Labor Standards Act, as amended by the *Equal Pay Act of 1963,* provides that an employer may not discriminate on the basis of sex by paying employees of one sex at rates lower than he or she pays employees of the other sex for doing equal work on jobs requiring comparable skill, effort, and responsibility and performed under similar working conditions.

The basic minimum age for employment covered by the act is 16 years, except for occupations declared to be hazardous by the Secretary of Labor, to which an 18-year age minimum applies. Construction, as such, is not designated as hazardous, although some of its specific work assignments are designated as such by name.

The *Contract Work Hours and Safety Standards Act* passed by Congress in 1962 (Work Hours Act of 1962), which has since been amended, applies to federal construction projects and to federally assisted projects. It does not apply if the federal assistance involved is in the form of a loan guarantee or insurance. Its main requirement is that workers be paid not less than 1 times the basic rate for all hours worked in excess of 8 hours per day *or* 40 hours per week.

In addition to the wage and hour laws of the federal government, each state normally determines the minimum wages and benefits that must be paid to all workers on projects paid for with public funds. With few exceptions, such state wage rates apply to all public projects both with and without federal funds.

In a project funded in part by both the federal government and the state government, in case of conflict between the federal wage rates and the state wage rates on any individual trade or craft, the contractor is obligated to pay the higher of the two on each such craft or trade where the two wage rates differ.

The National Apprenticeship Act

In 1937, Congress passed the National Apprenticeship Act. One of its prime objectives was to promote cooperation between management and organized labor in the development of apprenticeship programs. Traditionally, apprenticeship programs have been a joint effort between union-shop contractors and the AFL-CIO building trades unions. However, in 1971, the Bureau of Apprenticeship and Training approved national apprenticeship standards for the employees of open-shop contractors. This was the first time a unilateral apprenticeship program was approved on a national basis and placed under the direct supervision of employers only.

Ethnic Minorities and Women in Construction

Affirmative action to increase participation of women and ethnic minorities in the construction force came about with the Public Works Employment Act of 1977, which made available $4 billion of federal subsidies for construction projects through local governments, but required that at least 10 percent of the work be awarded to ethnic minority companies, subcontractors, subcraftsmen, or suppliers. When bidding on a project involving such funds, the contractor must be aware of what is required to comply with the Equal Employment Opportunity (EEO) and Minority Business Enterprise (MBE) provisions. In many circumstances, a woman-owner business qualifies as an MBE.

While a history and full explanation of the effect of affirmative action programs such as these on construction is beyond the scope of this book, it is sufficient to say that bidding on many projects depends upon complete compliance with all such requirements. Particular attention should be paid to the sources of funding being used on any project to ensure that the contractor is aware of any such program requirements. If they are applicable, a contractor must scrupulously comply with every requirement in order to prepare a responsive bid and to stand a chance of being awarded the contract. Remember the Golden Rule: "He who has the gold make the rules." Thus, the source of construction funds is an important issue to the bidder.

Worker's Compensation and Employer Liability Insurance

The contractor's Comprehensive General Liability Insurance policy normally excludes coverage of any liability incurred under state Worker's Compensation laws.

The laws of every state require employers to be responsible for the payment of compensation benefits to employees who sustain job-related illnesses or injuries. Contractors normally finance such payments through the purchase of Worker's Compensation Insurance. Under Worker Compensation schemes, the insurance carrier then pays benefits to an injured employee at rates set by state law.

Very often Worker's Compensation Insurance is the most expensive type of insurance that a contractor must secure. There are numerous reasons for the high cost: the hazardous nature of construction work in general and the many job-related accidents that occur, long duration of payments, the statewide nature of Worker's Compensation programs, the pro-claimants on the Worker's Compensation boards, and the ease and simplicity of filing a claim.

While some states require contractors to purchase Worker's Compensation Insurance from state-administered insurance trust funds, contractors in most other states are free to purchase such insurance from private insurance carriers. There are significant variations in the administration and levels of benefits paid out from state to state that can pose a problem for contractors doing business in more than one state. When a contractor *does* have operations in more than one state, private carriers often provide "all states" endorsement to cover the contractor for every state listed in the policy. To ensure continued Worker's Compensation Insurance coverage, the contractor must supply updated information to the carrier, as required.

The costs of premiums for Worker's Compensation Insurance vary greatly, as they are geared in part to the contractor's safety record. If a contractor has a good record of job site safety, it will pay very small premiums in comparison to those paid by a contractor with a consistent loss record. The opportunity for a safety-minded contractor with a good job site safety record to save money on reduced premiums is significant. When purchasing such insurance, a contractor should seek a policy from a carrier that will provide comprehensive loss-prevention services as well as standard claim services.

Labor Relations

The involvement of the Resident Project Representative in the labor relations for a project is seldom extensive. In most cases, the Resident Project Representative plays the part of an impartial observer, able to enjoy the advantage of being close to the work and feeling the undercurrents operating there. Because of this closeness, the Resident Project Representative can be invaluable to the owner and the design firm as a barometer of on-site labor conditions. If employed on a construction management contract, this insight may be of even greater value, as the construction manager may be directly involved in the labor relations processes on the project.

One of the most important rules for the inspector to follow is to remain in an observer category only. No opinions should be expressed at the site with anyone except the project manager, and then only while in confidence. Neither should the Resident Project Representative or other inspector take sides on any labor issue, or even express any sympathy for either side of a potential controversy.

All direct relations of the resident inspector on the job are to be conducted through the foremen and superintendents only, with the one possible exception of the emergency handling of an "Imminent Hazard," defined in Chapter 9.

The resident inspector, although not expected to take part in labor negotiations, may be called upon to confirm that the contractors and their labor forces are in conformance with both the contract provisions and the labor laws of the state and federal governments.

Construction Unions

The labor union movement is deeply involved in the construction industry. There is no denying that its contribution has been an important one. It has had a stabilizing influence on what is a potentially unstable industry, although this is viewed as having both its good and its bad points. The unions do provide direct access to a pool of skilled and experienced labor from which a contractor can draw, as his or her needs require, and through the medium of negotiated labor contracts, fixed wage levels have been established, thus serving the dual purpose of upholding the living standard for the labor force as well as providing added stability to the bids of the contractor by eliminating the labor rates as a competitive bid item. The unions also help by maintaining discipline among their membership as well as setting achievement standards for the different skill levels.

Contractor–Employee Relationships

Since the advent of the large labor union in the construction industry, the old personal relationships that formerly existed between an employer and employees are all but gone. Loyalties of the building tradespeople are now largely to the union, which handles all of the employee's business relationships on a scale well beyond the confines of a single employer's shop. Thus, the rates of pay, holidays, overtime, and other employment conditions are not negotiated with the employee himself or herself, but with the union business agents. Thus, each employee is bound by the terms of the resultant labor contract just as the contractor is.

As a result of such relationships, the employees are not generally "company minded," but feel that they owe their allegiance to their respective unions instead. Even their social lives are intertwined with their unions. In a manner of speaking, it could be said that the tradespeople are employees of a single large employer—their union—who contracts with various contractors to provide their services. The paternalistic attitude of the unions to their members through the retirement benefits and similar allowances seems further to bear this out.

Collective Bargaining in Labor Relations

Under the provisions of the National Labor Relations Act, both management and labor are required to bargain in good faith. This does not necessarily mean that concessions must be made or that the two sides even agree, for that matter—just that they bargain in good faith.

Lack of good faith on the part of a contractor could be interpreted from ignoring a bargaining request, failure to appoint a bargaining representative with power to reach an agreement, attempts to deal directly with employees during negotiations, refusal to consider proposals, failure to respond with counterproposals, antiunion activities, or refusal to sign an agreement.

Although some contractors bargain with unions independently by dealing directly with them, the most prevalent form of labor negotiations in the construction industry is for contractor associations to bargain with the unions for all of the members of the association. In this manner, the contractor has greater bargaining power than as an individual. Local associations of general contractors generally bargain with the locals of the basic trades: the carpenters, cement masons, laborers, operating engineers, and construction teamsters. In some areas, iron workers are also included. The basic trades may bargain either as individual locals or as a group of locals affiliated with the same international union, or even through groups of locals of different unions such as building trades councils. The resulting labor agreements are generally referred to as *Master Labor Agreements*. The resulting agreements apply only within the jurisdiction of the union locals involved, and then only to the extent of the particular trades involved.

Administration of the Union Contract

The matter of union–contractor labor relations is not closed simply because a labor agreement has been reached, however. There is still the matter of administering the labor contract and assuring that all its provisions are being met. Any project is likely to have some disputes or disagreements either between the union and the contractor or between two unions representing different crafts. The labor agreements typically contain procedures for the settlement of such disputes. When a dispute occurs that cannot be resolved by a conference of the steward, business agent, superintendent, and any other party directly involved, the grievance procedure set forth in the agreement is followed. This procedure generally forces the matter up to progressively higher echelons of the contractor and the union, during which time no work stoppage is supposed to occur. If the matter cannot be resolved, arbitration of the matter may or may not be provided for in the labor agreement. Generally, the unions have resisted the concept of binding arbitration; however, there is a greater tendency now to provide for arbitration as a means of resolving or settling contract disputes without needing to resort to work stoppages.

Prejob Labor Agreements

On some projects where special employment conditions exist, and where the project is large enough to justify the procedure, a prejob conference is held with the local labor officials to establish standard conditions for field operations for the life of a particular project. The intended purpose is to establish a meeting of minds between the contractor and the unions involved regarding job conditions of employment, work rules, or jurisdictional responsibilities. In a project located in a remote area, for example, the contractor is interested in running the job in as economical a manner as possible, whereas the unions want fair labor standards to be maintained. The locals having jurisdiction must be checked to see that they have enough people available to do the job. If not, arrangements must be made to bring workers in from the outside, either by the union local or by the contractor. Although some contractors are tempted to resist prejob labor conferences on the grounds that the unions will be tempted to make exorbitant demands, most available evidence seems to indicate that the unions are more reasonable at this stage because they do not feel that the contractor is trying to hide anything. On one project in which the author was involved, such cooperation with the local building trades council allowed an open-shop condition to exist at a federal project, involving both union subcontractors and nonunion subcontractors on the same project, thus preserving the jobs of the members of an otherwise economically depressed community. Through an agreement involving mutual understanding of the various problems involved, no union member of the community with the skills necessary for this particular project was deprived of a livelihood as a result of the agreement, and thus the purposes of the owner, the contractor, and the labor force were realized.

Open-Shop Contracting

Open-shop contracting, so called because it is unhampered by union agreements or representation, has suffered more than its share of troubles in its lifetime. In recent years, however, it appears to have made rapid and possibly lasting strides to where it is now established as a fact of life in the labor market. There is a strong movement by many smaller contractors and subcontractors to embrace the open-shop concept, and it is estimated that close to 40 percent of the annual construction volume is now done by nonunion contractors. In recent years, even some of the larger contractors have gone nonunion.

Contrary to the first impression usually received, open-shop contractors are not necessarily anti-union. Although they do usually pay somewhat less than the union scale, the average nonunion worker is provided with full-time employment instead of working by the job. Thus, the income comes closer to being a guaranteed annual income, and as such often exceeds that of union counterparts. The fringe benefits of most of the companies are similar to those paid by union contractors. This does not mean that none of an open shop's employees are union members. Many such shops have a mixture of union and nonunion help. One of the basic rights an open-shop contractor stands for is the right of the contractor to decide on the size of his work crews and to what job a worker may be assigned. Similarly, they are free to use prefabricated materials and are not subject to jurisdictional disputes, featherbedding, forced overtime, and work slowdowns. Workers are paid according to their work and performance. If, however, an open-shop contractor bids a federal job subject to the Davis-Bacon Act, the employees must be paid the same minimum wage rates published in the Federal Wage Rate Determination, and the bids will have to be computed accordingly. Thus, competing in some markets for a project may be difficult. If a federal job is won, however, there is nothing in the federal law to prohibit an open-shop contractor from being awarded a contract.

PROBLEMS

1. If a contractor entered in contract with a public agency to construct a pipeline from points A to B, then the public agency later decides to construct a pump station at one end of the pipeline, may this be done with a change order or must it be separately advertised for bid under separate contract?
2. If in the project in question 1 the public agency, instead of building the pump station, decides to reroute the pipeline, but still run it between points A and B, may this be done with a change order?
3. What is the usual criterion that governs selection of the successful bidder in a public works contract?
4. True or false? Within those jurisdictions where public contracts are subject to "Fair Subcontracting Laws" to prevent bid shopping, each bidder is required to list its subcontractors as a part of its bid.
5. On a public contract covered under state laws regulating the payment of prevailing wages to all workers, what will be the effect if federal grant funds are involved in the project?
6. True or false? A public agency can do only that which the law allows or prescribes.
7. Executive Order 11375 prohibits discrimination in employment based upon sex on all federal and federally funded contracts. If a contractor fails to conform, what are the remedies open to the agency?
8. What are the subjects of the two federal laws that affect offshore construction or navigable waterways?
9. On most public works projects, if products are referred to by brand name, what additional phrase is required?

9

CONSTRUCTION SAFETY

As mentioned in Chapter 4, the contractors have the prime responsibility for construction safety; however, there are certain areas of concern for construction safety that the Resident Project Representative should not ignore. The degree of the inspector's involvement is to some extent influenced by the specific terms of the construction contract, but some recent court decisions point out the need for concern by the inspector on the job.

In the Illinois case of *Miller v. DeWitt* (37 Ill. 2d 273, 226 N.E. 2d 630), where a steel roof had to be shored up while construction took place beneath it, the roof fell and injured a worker. In this case, the court stated: "As a general rule it has been said that the general duty to 'supervise the work' merely creates a duty to see that the building when constructed meets the plans and specifications contracted for." Thus, the court said, under ordinary circumstances, the architect would not be regarded as a person in charge of the work. But in the *DeWitt* case, the courts added that despite the argument of the architects that the shoring was "a method or technique of construction over which they had no control, we believe that under the terms of the contracts the architects had the right to interfere and even stop the work if the contractor began to shore in an unsafe and hazardous manner."

In the New York State case of *Clinton v. Boehm* (124 N.Y.S. 789), it was ruled that the architect owed no duty to the workers to supervise the contractor's methods to assure the workers' safety, and in 1960 the rule was again upheld [*Olsen v. Chase Manhattan Bank,* 175 N.E.2d 350]; however, this rule was seriously challenged in Arkansas in 1960 when the court said that the "supervising" architect who saw that an excavation wall was badly shored had a duty to the workers to *stop the work* to make repairs and that failure to do so made the architect liable for the deaths of three workers in a cave-in [*Erhart v. Hummonds,* 334 S.W. 2d 869].

In southern California, an unsupported wall of a 1.2-m (4-ft) deep trench caved in, resulting in the death of a laborer who was in the trench. The project was a large residential development involving separate construction contracts not only for the various types of structures being built, but also for underground pipelines for water and sewer. An inspector was assigned to the entire project to monitor the progress of several of the different contracts then under construction. One of these projects involved a large underground sewer main, and it was on this project that the disaster occurred. It was reported that the inspector had observed an unshored trench prior to the accident and knew that workers were in the unshored trench. The widow of the deceased worker filed suit, and in the resulting decision, the court held that the excavation contractor had contributed to the accident through negligence, but also held that the engineer's office must share the responsibility. In the view of the court, the engineer had an inspector on the job during construction and although the inspector saw the contractor's employee "descend into the trench, he voiced no objection" [*Widman v. Rossmoor Sanitation, Inc.,* 1965–70, 97 Cal. Rptr. 52].

Both the Arkansas and the California cases seem to support the proposition that an architect/engineer who has knowledge of a safety problem has a duty to the workers to prevent harm to them. Without the knowledge of a safety hazard, it appears that the architect/engineer has no such duty to them. Thus, if a design or construction management firm has a contract for construction management or continuous inspection, as might be the case where a Resident Project Representative is employed, there would seem to be no way of pretending no knowledge of such conditions.

Thus, judging by the preceding examples of the inspector's involvement simply due to his or her presence on the project, it can be seen that the inspector has an important involvement in construction safety hazards that pose a threat to life or health. It is of particular note that the case just mentioned involved an inspector who was providing only *part-time* inspection on that particular project. It is from cases such as the one just mentioned that policies have been adopted by many agencies, both public and private, that spell out the obligations of an inspector with regard to personal knowledge of safety hazards on the job.

OSHA and Construction Safety

In 1970 Congress found that personal injuries and illnesses arising out of work situations imposed a substantial burden upon, and were a hindrance to, interstate commerce in terms of lost production, wage loss, medical expenses, and disability compensation payments. Consequently Congress passed the Williams-Steiger Occupational Safety and Health Act of 1970 (OSHA) [Title 29 USC 451, et seq.]. With the passing of the act, the federal government imposed nationwide safety standards on the construction industry. Under the act, each of the states was allowed to pass its own version of OSHA, as long as the state's plan was at least as strict as the federal standards. OSHA imposes strict employee safety and health standards to protect covered employees and enforces the same provisions for inspections, investigations, recordkeeping requirements, and enforcement procedures. Under OSHA, logs of accidents as well as supplementary information and inspection may be involved.

If a state exercises its right under the act to enact a safety plan at least equivalent to the federal OSHA regulations, it retains the right to be the sole safety enforcement agency within that jurisdiction. If a state does not come up with such a plan, construction in that state will be subject to inspection by federal safety inspection agencies as well as state inspection agencies.

The act established the Occupational Safety and Health Administration, U.S. Department of Labor, Washington, DC 20210. Regional offices are found in various cities throughout the country. OSHA is responsible for the establishment of safety and health standards and for the rules and regulations to implement them. Such rules and regulations are published in the *Federal Register* and can be obtained from local OSHA offices or federal bookstores. *OSHA Safety and Health Standards,* Code of Federal Regulations, Title 29, Part 1910, contains regulations relating to the safety features to be included by the agency or architect/engineer in the design of any project. *Construction Safety and Health Regulations,* Code of Federal Regulations, Part 1926, pertains specifically to construction work.

It was OSHA's intent in preparing Part 1926 to place in one volume all of the rules and regulations applicable specifically to construction work. It is suggested that each of the contractor's superintendents, foremen, or other supervisors have a copy for reference. To the extent that Part 1926 contains standards that are incorporated by reference, copies of applicable referenced material should be made available to the supervisors for reference as well.

In order to ensure compliance with applicable regulations, contractors should assemble copies of the pertinent regulations that are incorporated by reference, as they relate to those regulations that apply to the specific project under construction. OSHA documents can be inspected at OSHA in Washington or at any of the regional or field offices.

The act deals with all working conditions and includes the following broad categories:

General Safety and Health Provisions
Occupational Health and Environmental Controls
Personal Protective and Life Saving Equipment
Fire Protection and Prevention
Signs, Signals, Barricades
Materials Handling, Storage, Use, and Disposal
Tools—Hand and Power
Welding and Cutting
Electrical
Ladders and Scaffolding
Floors and Wall Openings and Stairways
Cranes, Derricks, Hoists, Elevators, and Conveyors
Motor Vehicles, Mechanized Equipment, and Marine Operations Excavations, Trenching, and Shoring
Concrete, Concrete Forms, and Shoring
Steel Erection

Tunnels and Shafts, Caissons, Cofferdams, and Compressed Air

Demolition

Blasting and Use of Explosives

Power Transmission and Distribution

Rollover Protective Structures; Overhead Protection

Recording and Reporting Work Injury Frequency and Severity Data and Accident Cost

These categories are further divided into sections listing specific requirements, including the posting of certain notices that projects are covered by the law and submittal of certain data within the time limits on standardized forms.

General Responsibility for Construction Safety

Construction sites can be considered as being one of the most hazardous types of working environments in the nation. In order to assure an accident-free environment, an accident prevention program aggressively supported by the management of the responsible organizations is essential.

The foremost area of critical concern is the relationship among the general contractor and its subcontractors. Regardless of the legal responsibility of the general contracting firm or its counterpart for a project, each subcontractor also has the legal obligation to assure the health and safety of its own employees. Each subcontractor, as well as the general contractor, will be held liable by OSHA if they allow their employees to be exposed to hazards that could cause serious physical harm or death. Of course, there will be those subcontractors who (1) do not require their workers to follow safety and health regulations; (2) do not provide personal protective equipment; and (3) permit the use of unsafe equipment. In general, these individuals consciously choose to ignore recognized safety and health standards. Prime contractors and construction managers should be acutely aware of this problem and should include a safety and health requirement in their own contracts with subcontractors.

Owner Participation in the Safety Program

Under conventional contracts there are two basic philosophies with regard to owner or engineer participation in project safety programs. The first, which is followed by many local public agencies and private firms on advice of their legal counsel, is for the owner's or engineer's personnel to avoid direct involvement in the contractor's safety program, with few exceptions. The other approach, and one which is widely followed by federal and state agencies and utility companies, is to take an active part in the approval and monitoring of the contractor's safety program.

From a risk avoidance viewpoint, the first seems preferable. Under OSHA, the contractor bears full responsibility for all safety on or around the site. Under this concept, except for chance observation of a safety hazard by the Resident Project Representative or inspector, no direct action would normally be taken by the owner's or engineer's personnel. If this position is chosen, the following guidelines should be observed:

1. Do *not* review or participate in the development of the contractor's safety program
2. Do *not* review the contractor's safety performance, lest you incur a "duty of care."
3. However, if in the normal course of business you should happen to encounter a serious safety hazard, appropriate action must be taken.

Some attorneys disagree on these principles. Many public agency attorneys argue that safety performance is part of the "means and methods" responsibility of the contractor and is a legal responsibility of the contractor only under OSHA. However, if reference is made in the specifications or other contract documents to safety obligations of the contractor, it becomes not only a legal responsibility but a contractual one as well. Under these conditions, the inspectors would be contractually obligated to assure compliance, thus incurring additional risk to the owner or engineer.

Typical Federal, State, and Utility Company Approach

For the benefit of those whose organizations choose to participate in project safety programs, such as state and federal agencies or utility companies, it should be understood that under this concept, the agency or utility company assumes shared responsibility for the hazards on the construction site, and thus increases its risk of loss significantly. This risk may even extend to potential tort liability. However, it is well understood that federal and state agencies and utility companies, as well as a few other major local public agencies, choose to participate in the contractor's safety program under the premise that their exposure to risk is somehow reduced because of increased control at the site.

Generally, wherever a party is in "control" it can be interpreted literally, and it can be presumed that whoever is in control had the means to have prevented any accident that may have occurred. Under this type of involvement, there is a widely held view that the party in control is subject to tort liability for any accidents at the site.

If, on the other hand, safety involvement is viewed to be in the owner's or engineer's interest, it is incumbent upon that organization to develop a viable approach to the safety planning and review of the contractor's safety performance. This requires a team of well-trained, knowledgeable, and experienced safety professionals in the employ of the agency.

Safety Responsibility under Construction Management and Turnkey Contracts

Although the Resident Project Representative is often spared the more detailed responsibilities of the general contractor and subcontractors in assuring safe working conditions at a construction site, there are two types of contracts for construction in which the Resident Project Representative may well be as involved in matters involving on-site construction safety as a general contractor under a conventional contract:

1. Professional Construction Management (PCM) contracts, depending on the specific contract provisions and scope of responsibility at the site.
2. Design–build and turnkey construction contracts.

Professional Construction Management Contracts

Under a Professional Construction Management (PCM) contract, the Resident Project Representative's responsibilities are extremely difficult to define, as they may vary widely from one firm to the next. There is no way of determining the exact limits of such responsibility under a PCM contract, as the specific scope and terms governing such arrangements vary significantly (see Chapter 1 for further details on PCM responsibilities). In many cases, however, the PCM firm can incur obligations and responsibilities that were previously reserved to the general contractor.

Where multiple prime contracts are being managed, as in fast-track construction, the PCM functions are very similar to those of a general contractor and the PCM might do well to include the following items in each of the various separate construction contracts:

1. All unsafe or unhealthy conditions observed should be reported to the Resident Project Representative so that immediate corrective action can be initiated.
2. Compliance with all applicable safety and health laws, codes, ordinances, and regulations should be made mandatory.
3. Recognized safety and health work practices should be made mandatory.
4. Safety equipment should be inspected regularly and should be properly maintained.
5. The PCM or design–build/turnkey contractor should reserve the right to approve all conditions and practices.
6. In his or her own best interests, the PCM or design–build/turnkey contractor should require that an ongoing accident-prevention program be implemented at the work site by each contractor and subcontractor.
7. Each contractor and subcontractor should be required to properly maintain all necessary personal protective equipment.

8. There should be a continuing communication channel open between the PCM or design–build/turnkey contractor and each of their prime or subcontractors concerning safety and health matters.
9. Periodic on-site inspections, including follow-up procedures, should be conducted by the PCM or design–build/turnkey contractor to guarantee continuance of acceptable accident prevention procedures by each contractor and subcontractor.

To establish a meaningful and successful accident prevention program, it is necessary to accept the premise that the PCM or design–build/turnkey contractor management has the legal and moral responsibility to ensure a safe and healthful workplace. The top executive, whether the president, manager, or owner, must outline policies, stimulate thinking, and exhibit personal concern and interest in such a program before expecting others to follow and cooperate. This same top management should show this personal interest and give positive evidence of a sincere commitment by personally informing all employees that accident prevention is good business, and everyone is expected to be an active participant. In larger organizations, the administration of such a safety program is a full-time job, and is generally handled by a person designated as a Safety Engineer, Hazard Control Engineer, Safety and Health Director, or a similar title. The duties of the safety professional should be administrative and advisory, and implementation should be through the regular line management staff.

The safety engineer should advise the operating management staff on safety and health matters of interest or importance to them. Although the safety engineer must have sufficient management backing to take action when the responsible operations management staff fails to respond, care must be exercised that this authority is not used to circumvent the authority of the other members of the management staff.

Elements of a Safety and Health Program

There are several key considerations to be taken into account in the development and implementation of an effective safety program. A brief summary follows:

1. Commitment by top management to the development of a feasible program
2. Establishment of a safety and health policy by top management
3. Provision for a reasonably safe and healthful environment
4. Provision for competent supervision
5. Delegation of adequate authority
6. Provision for training and education
7. Conduction of accident-prevention inspections
8. Investigation of accidents to determine cause (not blame)
9. Measurement of accident-prevention performance
10. Maintenance of proper documentation and records of construction accidents (Figures 9.1 and 9.2)
11. Provision of continuing support

RESIDENT PROJECT REPRESENTATIVE'S REPORT OF CONTRACTOR'S ACCIDENT

Date 21 JUNE 1999

DAY: S | M (X) | T | W | TH | F | S

Project: 5 MGD TREATMENT PLANT ADDITION

Unit:

Proj. No. IRV-100 Contract No. K-0433-21

WEATHER	Brite Sun ✓	Clear	Overcast	Rain	Snow
TEMP	To 32	32 50	50 70	70 85	85 up
WIND	Still ✓	Moderate	High	No.	

Contractor: ABC CONSTRUCTORS, INC.

Sub-Contractor: N.A.

Date of Accident: 6-21-99 Time: 09:20 (AM) PM Location:

Description of Accident: SITE CLEANUP WAS IN PROGRESS AFTER FORM REMOVAL. LABORER (VICTIM) WAS SWEEPING A LEDGE 9m ABOVE CONC. FLOOR. RECTANGULAR HOLE IN LEDGE, WHICH WAS PREVIOUSLY COVERED WITH PLYWOOD, WAS TEMP. UNCOVERED TO SWEEP DIRT INTO. LABORER BACKED INTO HOLE AND FELL 9 METERS TO HIS DEATH

Primary Cause: BY SWEEPING BACKWARDS WITH A PUSH BROOM LABORER WALKED BACKWARD AND FELL THRU HOLE TO HIS DEATH ON CONC SLAB, A DROP OF 9 METERS.

Contractor's Personnel or Equipment

Name of Injured Employee: JAMES L. MARTIN Age: 29

Occupation: LABORER Sex: M

Nature of Injury: SKULL FRACTURE

Degree of Injury: First Aid ☐ Doctor Visit ☐ Hospital ☐ Fatality ☒

Type of Equipment: NONE

Extent of Damage:

Other Persons or Property

Name of Injured Party: Age:

Address: City State

Nature of Injuries:

Data used are fictitious for illustration only

Name of Property Owner: Address:

Nature and Extent of Damages:

Was Use or Lack of Safety Equipment a Factor in This Accident: NO

If so, Explain:

What Safety Regulations Were Violated: NONE KNOWN. WALKWAY WAS PROPERLY PROTECTED WITH GUARD RAILS. FLOOR HOLE HAD BEEN PROPERLY COVERED

What Corrective Action Has Been Taken by the Contractor: USE OF GREATER CARE. WATCH WHERE THEY WALK. DO NOT WALK BACKWARDS

DISTRIBUTION:
1. Project Manager
2. Legal Staff
3. Engineer/Architect
4. Project File

Report by: [signature]

Title: CONSTRUCTION MANAGER

Fisk Form 8-13

FIGURE 9.1 Report by Resident Project Representative of Contractor's Accident. *(Fisk, Edward R., Construction Engineer's Complete Handbook of Forms, 1st edition, © 1993. Reprinted by permission of Pearson Education, Inc., Upper Saddle River, NJ.)*

FIGURE 9.2 Example of a Photograph Documenting a Fatal Accident to Accompany the Report Illustrated in Figure 9.1.

Design–Build/Turnkey Contracts

Generally, a design–build/turnkey contractor's responsibilities for safety are inseparable from those of a general contractor. In some states, the risks have been found to be even higher. Under a principle of law referred to as *strict liability,* design–build and turnkey contractors have been held liable for injury and damages where the injured party need prove only that an injury or loss was sustained. Fortunately, this principle has not yet found wide acceptance across the country.

Effect of Including Contractor's Safety Obligations in the Specifications

The Resident Project Representative is the person who is most directly involved in the administration of contract provisions and should have the responsibility of assuring that the contractor is in full compliance with all aspects of the contract, including applicable major safety requirements. The degree of control that the inspector may have over the contractor in requiring compliance with the OSHA construction safety requirements depends at least partly on the following conditions:

1. If no mention is made in the contract documents, whether on the drawings or in the specifications themselves, safety obligations of the contractor are primarily a legal obligation between himself/herself and the state or federal agency administering the provisions of OSHA. Although an inspector would seem to be obligated to call attention to observed deficiencies that constitute a serious hazard, and to notify the contractor that they should be remedied, the contractor's failure to respond can only be handled by the inspector through the service of a written notice to the contractor, with copies to the organization administering the construction contract and by filing an official notice to the local OSHA enforcement agency, which is administered at state level in many areas. Otherwise, the contractor's failure to comply is difficult to control, as the inspector normally possesses no special powers over the work.
2. If, however, the contractor's compliance with the safety requirements of OSHA is specified in the contract specifications, the inspector's subsequent demand that a contractor comply with certain OSHA provisions takes on a different light. In this case, the safety requirements, in addition to being the legal obligation of the contractor, have become a *contractual* one as well. Thus, the contractor's failure to comply can be interpreted as a breach of contract, and the design firm may recommend that the owner withhold payments for that portion of the work until the contractor complies. This does not take the place of the official notice mentioned in the preceding paragraph, but merely provides an additional recourse to the design firm and the owner beyond the steps already mentioned. Furthermore, under these conditions, assurance that the contractor is living up to safety obligations is now a part of the inspector's responsibility because it is written into the specifications and must therefore be considered as one of the inspector's field administrative responsibilities. This also functions as an allocation of a portion of the safety hazard risk to the inspector's employer.

Wherever the construction safety provisions are written as a part of the terms of the construction contract, the inspector in the administration of his or her part of the contract is required to see that the contractor properly provides for the safety of the workers. Under no circumstances should the contractor be instructed orally or in writing as to *how* to correct a deficiency. The unsafe condition should simply be *identified* and the specific regulation, if it is known, should be cited.

As mentioned previously, it is quite probable that the inclusion of safety requirements in the construction contract will incur additional responsibility on the part of the architect/engineer to assure that proper safety precautions have been taken. It must also be recognized that there may be some additional liability to the architect/engineer in case of a job-related injury involving the failure of a contractor to observe safety requirements. However, such a provision is the only means readily at the disposal of the architect/engineer to assure performance of the contractor's safety obligations, and the risk of loss to the architect/engineer could be even greater in the absence of such controls if a fatal or crippling accident did occur, as one may be certain that the architect/engineer, the owner, and the Resident Project Representative would all be named in any resulting litigation.

Applicability of State and Federal OSHA Provisions to a Project

As a way of setting the groundwork, it should first be mentioned that the federal OSHA provisions are in two volumes. The first book, *OSHA Safety and Health Standards,* Code of Federal Regulations, Title 29, Part 1910, deals with safety features that are intended to be included in the design of the project. This is the responsibility of the designer to include on the plans as a part of the project design. The second book is the one that relates to the construction phase of the work and generally concerns the temporary hazards and conditions that exist as a direct result of the construction activities. This volume is entitled *Construction Safety and Health Regulations,* Code of Federal Regulations, Title 29, Part 1926. In addition, under the federal safety program, each state has the right to enact a safety code that is at least equivalent to the federal OSHA provisions and, by so doing, retains the right to be the sole safety enforcement agency within its jurisdictional borders. If a state does not choose to exercise this option, construction in that state will be subject to inspection by both federal and state safety inspection agencies. If a state elects to upgrade its safety code to meet the OSHA requirements, it has a three-year period to accomplish this. Under these conditions, local safety enforcement will be by state agency only, both during and after enactment.

Special Applications

Although ordinarily it would be assumed that a state safety program would not have jurisdiction within the confines of a federal reservation, and that only the federal OSHA program would govern there, an interesting state government interpretation has been made on this subject, as described in the following paragraph.

A federal military reservation was the site of a construction project that was planning to base its safety requirements on federal OSHA requirements administered by federal safety inspectors. The state announced jurisdiction, based upon the fact that its safety provisions were a part of its labor code, which was enacted to protect workers in their various occupations. It was further stated that none of the construction workers on the military base were federal employees, and as such were all subject to the provisions of the State Labor Code. This appears valid, as all other provisions of the State Labor Code apply to the contractor's employees, and the fact that a particular construction project takes them into federal property does not strip them of the protection afforded by the labor code.

It appears, then, that a project on a federal reservation, if built by nonfederal employees from off the reservation, will be subject to local state safety regulations—possibly in addition to federal OSHA if the state involved has failed to meet, or has not participated in, the OSHA upgrading program.

Procedural Guidelines

In carrying out the owner's and the design firm's responsibilities of assuring safety compliance as a contract requirement, the following guidelines are suggested where the owner feels that inspector monitoring of the contractor's safety program is desirable:

1. IMMINENT HAZARD (Figure 9.3): a condition that if not corrected would probably result in an accident causing severe or permanently disabling injury or death.

   PROCEDURE: When an imminent hazard condition is known to exist, or when a contractor either delays in correcting or permits repeated occurrences of a hazardous condition, the Resident Project Representative should immediately order the contractor to suspend the operations affected and not permit work to resume on these operations until the condition has been

FIGURE 9.3 Photographic Documentation of an Imminent Hazard.

corrected. The hazard should be photographed, and the project manager of the design firm and the owner and the state or federal agency having jurisdiction over construction safety should be notified of the hazardous condition and of the action taken. In addition, a letter giving all the details should be prepared, covering all the events leading up to the suspension, and this letter should be submitted to the project manager.

2. DANGEROUS CONDITION (Figure 9.4): a condition that does not present an immediate danger to workers, but if not corrected could result in a disabling injury and possibly death, or could develop into an imminent hazard as just described.

   PROCEDURE: When a dangerous condition is known to exist, the resident inspector should notify the contractor in writing of the condition and allow a reasonable period of time for correcting the condition. If the

FIGURE 9.4 Photographic Documentation of a Dangerous Condition.

resident inspector is not certain of the remedial measures proposed or taken by the contractor, the services of a construction safety engineer should be requested. If the contractor does not correct the dangerous condition, or if the condition is deteriorating into an imminent hazard, the design firm should consider recommending that the owner suspend the affected operations.

3. MINOR OR NONSERIOUS CONDITION: conditions that could result in minor or less serious injuries, or that are small in nature, but that may still be classified as a threat to health.

   PROCEDURE: When a minor or nonserious condition is known to exist, the Resident Project Representative should advise the contractor of the condition and of the necessity of eliminating it. If the contractor fails to correct the problem or permits its repeated occurrence on subsequent operations, the design firm or owner should be notified.

The construction safety activities of both the contractor and all project personnel must be documented in the inspector's diary. It is important for inspectors to realize that their duties include only responsibility for seeing that the contractor complies with the project safety requirements through the use of normal administrative procedures. The legal enforcing agency is the federal or state OSHA officers. The inspector should keep the name and telephone number of the local safety compliance officer handy—it is often a most effective compliance tool.

It should be noted that the mention of suspension of portions of the work in the foregoing procedural guidelines *relates to the immediate area* of the hazardous condition only. Nothing described here is intended to suggest that the inspector, the design firm, or the owner would be justified in closing down an entire project or even a significant portion of a project for such local conditions.

References to the capability of payment retention by the owner upon the recommendation of the design firm on contracts where safety provisions have been made contractual requirements as well as legal obligations are presented solely as a matter of interest to the inspector. It must be kept in mind that only the project manager has the authority to approve or recommend that part or all of a contractor's monthly progress payment be withheld. However, recommendations of the Resident Project Representative may bear a heavy influence on the decision of the project manager in such cases.

Shoring and Bracing

The federal OSHA Part 1926 *Construction Safety and Health Regulations* requires that all trenches and earth embankments over 1.5 m (5 ft) deep be adequately protected against caving in by a system of sheeting, shoring, and bracing or by sloping the sides of the trench or other excavation to an acceptable angle.

Trench and excavation shoring is one of the critical safety hazards referred to in the OSHA Construction Safety and Health Regulations. Numerous fatalities have resulted from failure of the contractor to provide adequately for worker safety under these conditions. The inspector should take particular note of the fact that all trenches or other excavations on any project, public or private, require sheeting, shoring, or bracing if they are 1.5 m (5 ft) deep or deeper (1.2 m [4 ft] in some states). Details of safety codes vary somewhat from state to state, but there is a trend toward greater uniformity, and the safety codes of each jurisdiction should be carefully checked prior to beginning work in another state to confirm the specific limitations and regulations that will control.

It is quite probable that all of the excavation cave-ins of record could have been prevented if there had been proper engineering design and inspection of the support system. The additional cost of engineering may well be compensated by lower construction costs when contingency and liability costs are minimized through reduced risk of construction hazards.

As a result of research conducted by the Associated General Contractors of America (AGC) and reported in the American Society of Civil Engineers *Journal of the Construction Division*, some revealing facts were discovered regarding excavation cave-ins:[1]

1. At least 100 fatalities occur each year from cave-ins. At least 11 times as many receive disabling injuries.
2. The majority of these cave-ins are in shallow excavations, primarily sewer trenches.
3. No part of the country is immune to cave-ins.
4. Every type of soil is susceptible to cave-ins.
5. Most cave-ins occur in unsupported excavations.
6. Major factors influencing trench failure are the presence of construction equipment near the edge of an excavation and adverse climatic conditions.
7. Usually, engineers do not specify shoring requirements prior to bidding.
8. Approximately 50 percent of the contractors surveyed would prefer that engineers specify the shoring requirements prior to bidding.
9. Virtually no one is designing support systems for shallow excavation based on soil properties and site conditions. "Traditional" or "standardized" methods are the present governing procedures.
10. Engineers are not investigating cave-ins to determine the causes by soils investigation and engineering analysis.
11. Any time an excavation fails, there should be an engineering investigation to determine the cause of the accident. This information should be published to aid other engineers in preventing future failures.

---

[1]L. J. Thompson and R. J. Tannenbaum, "Survey of Construction-Related Trench Cave-ins," *Journal of the Construction Division*, Vol. 103, September 1977, p. 511. Proceedings of the American Society of Civil Engineers. Reprinted with permission.

The Competent Person

As a part of the federal OSHA standard safety requirements for earthworks, in 1989 a new category of "Competent Person" was established (29 CFR 1926 Subpart P). The new regulation affects construction in all states. The program affects all underground construction, with primary emphasis on trench safety.

The definition of a "Competent Person" under the OSHA Standards is "one who is capable of identifying existing and predictable hazards in the surroundings, or working conditions which are unsanitary, hazardous, or dangerous to employees, and one who is authorized to take prompt corrective measures to eliminate them."

In order to be qualified as a Competent Person under OSHA, a person must have had specific training in, and be knowledgeable about, soils analysis, the use of protective systems, and the applicable requirements of the OSHA standards.

The Competent Person having such training and knowledge must be capable of identifying existing and predictable hazards in excavation work and have the authority to take prompt measures to abate these hazards. Thus, a backhoe operator who might otherwise meet the qualification requirements for designation as a Competent Person is not qualified if he or she lacks the authority to take prompt corrective measures to eliminate existing or potential hazards.

The designated "competent person" should be an employee of the contractor who can act as his or her employer's designee for choosing a protective system from the legal options available. It further becomes the responsibility of the Resident Project Representative to verify that a Competent Person has been designated, and that the designated person meets OSHA requirements for the assignment.

Safety Requirements in Construction Contracts

Many public agencies include safety standards as a part of the construction contract documents, which then become a contractual obligation as well as a legal one, as explained previously. Many state highway departments include a safety code in their construction contracts. Several federal agencies, including the U.S. Army Corps of Engineers, the Naval Facilities Engineering Command, and the U.S. Bureau of Reclamation, include health and safety standards in their construction contracts. OSHA provides that such federal agencies may continue to provide their own safety inspection and enforcement; however, this does not preclude the state from requiring compliance through its own safety enforcement officer. Thus it can be seen that the inclusion of safety requirements as a part of the contractual obligations of the contractor is a growing practice. Upon beginning a project the inspector should make a careful study of the plans and specifications to determine whether or not such safety provisions are included as a contractual requirement, thus placing an additional burden of responsibility upon the inspector to assure compliance and to take appropriate administrative action in case of default by the contractor.

PROBLEMS

1. What party to a construction contract normally has primary responsibility for construction safety unless otherwise covered under the contract: the owner, the engineer/architect, or the contractor?
2. What are the three hazard classifications discussed in the text?
3. The recommended procedure to deal with an unshored trench hazard is initially to:
   (a) Stop the work
   (b) Order the workers out of the trench
   (c) Call OSHA
   (d) Do nothing (it is the contractor's problem)
4. Does the Resident Project Representative have a duty to document accidents by the contractor or its personnel? If so, what kind of documentation should be used?
5. Many contracts require the engineer/architect to review the contractor's safety program and to monitor the contractor's safety performance. Discuss the merits or demerits of that approach.
6. Federal and state contracts generally require their agencies to take part in the project safety program and accept a portion of the responsibility for project safety. What are the risks, benefits, and obligations of the owner under this arrangement?
7. On a Professional Construction Management contract, can the Construction Manager be cited by OSHA for safety violations or is the contractor the only one that can be cited by OSHA?
8. Who sets minimum standards for occupational safety and health in construction?
9. If safety requirements are not specified in the contract, is the contractor relieved of its safety obligations on the contract?
10. If not contractually obligated to accept safety responsibility, can the Resident Project Representative or inspector still be at risk?

10

MEETINGS AND NEGOTIATIONS

The meeting is the communications center of every organization. It is a decision-making body where the best-prepared member's ideas are accepted because such a member is more direct, thinks more quickly, and knows how to use effective communications methods. As in any other organized group, the meeting displays a blending of a number of individual talents. When the objectives of the meeting are to formulate in-house decisions relating to project design, construction, or administration, the concept of performing individual roles in the meeting format can lead to better decisions and prevent later problems. In negotiations, the meeting *team* concept is preferable, where all of the talents of each individual member of the negotiation team are put together to function as a formula for success.

Any dispute or conflict can be negotiated successfully if the parties are willing to engage in good-faith discussions on mutually defined issues in an attempt to reach an acceptable settlement. Each party must be willing to compromise and accept less than its total demands. Negotiation involves a series of compromises. A person must have a thorough understanding of the basic elements of the bargaining process to negotiate effectively. Likewise, a person must adequately prepare for and competently conduct negotiations to avoid poor performance, fragmentation, mistakes, and miscalculations. The outcome of a negotiation should be satisfactory to both parties and should promote good relationships. The key to a successful negotiation is to satisfy the needs of all parties: Everyone should gain something.

Types of Meetings in Construction

Most construction projects involve numerous types of meetings. Some are in-house meetings of the owner, the architect/engineer, or the contractor; others are meetings between representatives of these organizations as they meet the challenges of the various problems that are thrust upon them regularly throughout the construction process. Generally, in-house meetings are organization, coordination, or decision-making meetings, whereas meetings between different organizations may also involve negotiations. Some of the many different purposes for holding meetings for construction are summarized in the lists that follow. Meetings of an in-house nature are generally attended only by members of the same firm, and their principal functions are fact finding, decision making, and coordination. Some examples are:

1. Project conceptual design meetings (in-house)
2. Project presentation meetings (to potential client)
3. Project "kickoff" meetings (in-house project team)
4. Project manager's organization or coordination meeting
5. Project budget meetings for production
6. Project design meetings (technical coordination)
7. Project constructability meetings (design/construction interface)
8. Planning and scheduling meetings for project workload
9. Value engineering meetings
10. Prenegotiation meetings (in-house preparation)

As a project enters into the construction phase, the owner or the owner's architect/engineer may also become involved in a series of meetings with the contractor. Such meetings may be for the purpose of communication, coordination, resolution of difficulties, or to negotiate prices or other terms. Such meetings may include:

1. Prebid conference (information—not recommended for public works projects because of risk exposure)
2. Preconstruction conference (communication and coordination)
3. Requests for substitutions of materials or products
4. Change Orders and Extra Work (negotiations)
5. Unforseen underground work (negotiations)
6. Corrective work required (negotiations)
7. Scheduling changes (coordination; negotiation)
8. Mediation of construction difficulties
9. Protests and disputes (negotiations)
10. Difficulties caused by conflicts in plans and specs
11. Contractor-sponsored Value Engineering Proposal costs
12. Punch list meetings
13. Project acceptance and transfer of responsibility
14. Postproject review meeting

Although there may be numerous reasons for holding meetings, these lists represent some of the more frequent reasons. Some of the meetings listed are design-phase meetings and are not considered to be within the intended scope of this book. Several of the others, however, will be discussed further.

Basically, it should always be remembered that depending on the purpose of the meetings and the types of people who will be in attendance, the philosophy of conduct at some meetings may be required to be quite different from what it is at others. Most in-house meetings are not negotiation sessions; therefore, each individual will probably be called on to contribute ideas. In that context, one should be prepared to offer ideas without regard for whether the viewpoint is in keeping with the consensus. In meetings held with a potential client, however, as in the making of a presentation, or meetings between the contractor and the owner or architect/engineer that have been arranged for the purpose of negotiating, it is important for each group to maintain an appearance of unity. This can be done effectively only after careful preparation, rehearsal, and previously established position policy.

Who Should Attend

Often, the success or failure of a presentation or negotiation meeting is determined before the first word has been spoken. Careful selection of those who will be in attendance is of vital importance to the success of the meeting objective.

The types of persons and the specific parties represented at each type of meeting will vary somewhat depending upon the purpose of the meeting and the type of organization represented. However, it is vital that certain key persons be present and, for maximum effectiveness, the attendees should be limited to just those key persons.

In a few cases, notably where disputes or potential claims are involved, one of the parties may bring an attorney to the meeting. If this should happen, the meeting should be delayed long enough to enable the other party's attorney to be present also. Never conduct a meeting with anyone when his or her attorney is present without contacting your own attorney to receive instructions as to the course of action to take.

Meeting Resources

In addition to the conventional meeting format, meeting documentation can be enhanced through the use of a PC computer running a construction management software program designed primarily for owner-engineer applications. Unfortunately, most programs on the market are directed to the contractor's needs and may be of little use to the architect/engineer.

Under such a program, the user can define multiple tracks of similar meetings such as prebid, preconstruction, weekly project, and closeout meetings. In addition, tools are available for a virtual meeting in which the participants meet electronically to discuss the topics. Such a virtual meeting can free all of the participants from their schedule constraints or physical location by allowing them to participate when they are available. The agenda, meeting discussions, reader comments, and any attached information can be made available to all users having an appropriate security designation. See Chapter 5 for further discussion.

Handling Yourself at a Meeting

One of the first and most basic rules of conduct at a meeting is to become familiar with the meeting agenda. It is important to learn in advance as much as possible about the subject of the meeting if you expect to turn it to your advantage. Some guidelines follow:

1. Determine who called the meeting.
2. Find out the reason for the meeting.
3. Understand the background of the subject.
4. Establish your own position on the subject.
5. Determine your goals or objectives before the meeting.

The most important point brought out by these guidelines is that of establishing your own position on the subject. Then, at the meeting, direct all of your efforts toward reaching that objective during the meeting, while carefully observing the responses of the others in attendance.

If you receive mixed reactions to a proposal, a vote may be premature, so while confidently holding your position, you may suggest the need for a recess or deferred action on the issue until a consultant can be heard, or that more information is required and should be reviewed before a final decision is arrived at. Then, slowly, the subject may be eased out by withdrawing your push and pointing to the fact that the question may be easy to resolve at the next meeting or after a meeting recess.

After the recess is a good time to begin a little lobbying. Each apparent dissenter may then be approached individually to put the point across. There may be some compromises needed, but if they fit your objectives, they may be well worth a reasonable compromise. After mustering sufficient support, the issue is ready to be reintroduced to the floor for a general vote.

A delayed meeting has other advantages. It allows preparation time to strengthen your position after your opponent has tipped his or her hand. No amount of advance preparation can offer the advantages of being able to study the issues and plan your position after your opponent has made full disclosure of his or her position. If you succeed in obtaining the delay, it can place you on the offensive, where you have a distinct tactical advantage.

Importance of Your Image

Appearance is of considerable importance at meetings and negotiation sessions. The best general advice is to dress conservatively and maintain an appearance of authority. At the meeting, speak to others in a confident, objective way and never give the appearance of begging or patronizing your opponent.

Seating Advantage

Seating position is seldom spoken of as having an influence on the outcome of a meeting. Yet its subtle psychological advantages are felt if not actually observed. Unless place markers are provided, it is to your considerable advantage to position yourself very carefully at the conference table.

Many people attempt to put themselves in as unobtrusive a position as possible. This approach can work to your disadvantage, as it will minimize your effectiveness at the meeting. If you have a point to get across, you will never make it that way. The position that you occupy at the meeting table can also have a significant psychological effect upon whether your ideas will get attention or whether you come out ahead on a negotiation session. A few brief suggestions are offered that might allow you some advantage at the conference table.

1. Sit at the opposite end of the meeting room table from, or on the side closest to, the leader.
2. If, however, the leader is the type who frequently consults the person next to him or her, sit next to the meeting leader.

Some seating positions to avoid at a meeting include any seat where visual contact with the leader is obstructed. If you happen to be late in arriving, this may not be possible to avoid, but that is another reason for arriving on time. If the Resident Project Representative attends a meeting accompanied by the project manager, where the meeting is presided over by higher management, it is advisable to seat yourselves in such a way that will not place either party between the leader and the other party. If you do, and the leader directs a question your way, one party can simply field the question and leave the other party invisible. It is advisable for the Resident Project Representative and the project manager to seat themselves at opposite sides of the table, facing each other. In this manner, either party could add to the other's statement, and, while doing so, all those present would have their eyes upon both members of your team. High visibility is important, for with high visibility goes attention, and with attention goes authority.

Determine the Opponent's Motivation

Part of your group's success at a meeting will be based upon your ability to correctly assess the opponent's motivation at the meeting. The opponents' drive toward their objective can offer clues as to their probable reaction to any point brought to a vote.

Methods and Techniques

The first and basic rule to turn the meeting to your team's own advantage is to take a positive approach. Ideas should always be presented in a systematic and orderly fashion. It is not enough merely to present the facts; they must also follow a logical course toward the point that you are attempting to put across. Often, the problem in presentation meetings where the architect/engineer is attempting to present a proposal for a project to the owner is that each member will respond individually as the opportunity seems to arise. It is of vital importance in this type of meeting to prepare carefully ahead of time, coordinate the activities of each of the people who are involved, and rehearse the presentation just as though it were a theatrical production. Many a disaster has been precluded by this simple expedient.

1. *Leader encouragement.* In most organizations, group members need encouragement to feel free to disagree with the boss or a group leader. The subordinates in the group must feel free to disagree if they are to contribute

the best of their thinking. The leader should encourage free expression of minority viewpoints. Although group members holding minority views are more likely to be on the defensive and more hesitant in voicing their opinions, to introduce balance into the situation, the group leader must do all that he or she can to protect individuals who are attacked and to create opportunities for them to clarify their views.

2. *Diversity of viewpoints.* Attempt to structure the group so that there are different viewpoints. Diverse input will tend to point out nonobvious risks, drawbacks, and advantages that might not have been considered in a more homogeneous group.
3. *Legitimized disagreement and skepticism.* Silence is usually interpreted as consent in a meeting. It should be explained that questions, reservations, and objections should be brought before the group and that feelings of loyalty to the group should not be allowed to obstruct expression of doubts. Genuine, personal loyalty to the group that leads one to go along with a bad policy should be discouraged. Voicing of objections and doubts should not be held back for fears about "rocking the boat" or reluctance to "blow the whistle." Each member of the meeting group should take on the additional role of a critical evaluator and should be encouraged by the group leader and other members to air all reservations.
4. *Idea generation versus idea evaluation.* A major barrier to effective decision making is the tendency to evaluate suggested solutions as soon as they appear instead of waiting until all suggestions are in. Early evaluation may inhibit the expression of opinions, and it tends to restrict freedom of thinking and prevents others from profiting from different ideas. Early evaluation can be particularly destructive to ideas that are different, new, or lacking in support. The group leader should encourage initial emphasis on problem solving at the expense of early concentration on solutions.
5. *Advantages and disadvantages of each solution.* The group should try to explore the merits and demerits of each alternative. The process of listing the sides of a question forces discussion to move from one side of the issue to the other. As a result, the positive and negative aspects of each strategy are brought out into the open and may become the foundation for a new idea with all its merits and few of its weaknesses.
6. *New approaches and new people.* In many cases, thinking about the problem by oneself or discussing it with another person can result in refreshing new perspectives. Any belief that one should be able to generate correct answers to complex problems and issues the first time that they are dealt with should be done away with. In fact, the norm, during the design phase at least, should be to "think about it again" and "think about it in a new way." This implies remembering the answer derived by one approach, putting it aside for a while, then coming back to the problem afresh. Also, it may be helpful if, in the intervening time, each of the group participants consults a trusted colleague, who is not a member of the group, to bounce it off him or her for a reaction. Ideally, these colleagues should be someone different in expertise

and background from the rest of the group members, so that they can offer critical, independent, and perhaps fresh ideas that can be reported back to the group. These recommendations, although desirable in decision-making meetings, such as design-phase meetings in an engineering organization, are quite unsuited to the method of handling the implementation meetings involving the decision-making process during construction when time is of the essence and a quick response is necessary.

PRECONSTRUCTION CONFERENCE[1]

The optimum time for the *preconstruction conference,* also known as a *construction coordination conference,* is after all of the subcontracts have been awarded but prior to the beginning of actual construction. This will permit the subjects discussed at the conference to form a background of understanding of the intended operational plan for each of the members of the construction team.

It is essential that all key members of the construction team be represented at this meeting. The presence of the owner will enable him or her to better appreciate the potential operational problems encountered by the project team, will aid the total construction team by providing greater insight on specific owner needs, and will help the architect/engineer to secure and translate team cooperation into a good quality job that is consistent with the scheduled time and costs.

The preconstruction conference is a logical method by which the problems of economic waste and disruptive construction problems can be discussed and possibly prevented. It is designed to benefit all concerned by recognizing the responsibilities for the various tasks *before* the project is begun. The benefits include the following:

1. Recognition and elimination of delays and disagreements.
2. Establishment of agreements that curb increases in construction costs.
3. Predisposition of gray-area responsibilities that, if left unassigned, can cause later disputes.
4. Unification of management requirements and the establishment of clear understanding of these requirements.

Definitions

The *preconstruction conference,* or *construction coordination conference,* is a meeting of the principal parties involved with the planning and execution of the construction project, and should include:

- The owner or authorized representative
- The architect/engineer and Resident Project Representative

[1]Adapted from *Construction Coordination Conference,* Copyright © 1974 by the American Subcontractors Association, the Associated General Contractors of America, and the Special Contractors, Inc. All rights reserved. Used with permission.

- The general and all prime contractors and their superintendents
- The subcontractors and their superintendents
- Key suppliers
- Public agency representatives, as necessary

Full attendance and participation by all key team members can be assured by including a requirement in the conditions of the contract that all contractors and subcontractors attend this meeting.

Purpose

The primary purpose of the conference is to establish acceptable ground rules for all parties concerned, and to assure that each contractor understands the complete job requirements and coordinates the work to produce a completed job in a minimum amount of time, with maximum economic gain, and in harmony with the owner, architect/engineer, prime contractors, and all subcontractors.

Time for the Conference

The preconstruction conference should be scheduled to permit sufficient time to cover the total agenda. This could involve from one to several days. In any case, whatever time is spent should be considered as *preventive* rather than corrective.

Topics for Discussion

The topics of the discussion will depend on the nature, size, and complexity of the project. It is necessary, however, to assign priorities to the tasks. Although each job is different, there are certain factors that are common to all types of construction. As an aid to the development of an agenda, or as a handy means of listing all of the subjects discussed and determinations reached, it may be helpful to utilize a preprinted form for this purpose, such as that illustrated in Figure 10.1 or in Figure 12.13.

Agenda for a Typical Preconstruction Conference

1. *Progress payments.* When, how, and to whom are payments to be made? State in exact terms so that no questions remain about requirements and responsibilities. The subject of retention and final payment should also be discussed at this time.
2. *Form of payment requests.* Identify the form that is required to apply for progress payments. Also, are suppliers' and subcontractors' lien waivers required? Can supplier and subcontractor waivers be one payment behind? (See also Chapter 17 regarding partial payments to the contractor.)
3. *Payroll reports.* Specify the requirements, if any, for payroll reports, as required on projects subject to Davis-Bacon Act provisions.
4. *Shop drawing and sample submittal requirements.* These include the form and procedure for the submittal of shop drawings and samples, identity of

PRECONSTRUCTION CONFERENCE

Checklist of Subjects to Be Considered for Agenda

Construction Coordination (Preconstruction) Conference

- Participants
- Time required for conference
- Use of an agenda—Topics for discussion
- Minutes of the meeting
- Acknowledgment of correction of minutes

Identification of Key Personnel of All Parties

- Names and 24-hour telephone numbers of Contractor, Engineer, and Owner
- Define authority and responsibility of key personnel
- Designate sole (one-on-one) contact for administration of contract

Authority and Responsibilities

- Methods of construction (methods may be challenged)
- Rejection of work by inspector
- Work performed during absence of inspector (unacceptable)
- Work performed during absence of Contractor's superintendent (unacceptable)
- Stopping the work (right reserved to Owner)
- Safety at the site
- Issuance of Field Orders from Engineer/Owner
- Authority of the inspector

Conformance with Plans and Specifications

- Call attention to areas of special concern
- Answer Contractor inquiries
- Clarification of specification provisions

Contract Administration

- Notice to Proceed
- Time of the contract
- Liquidated damages
- Record drawings (procedure; responsibility for)
- Mobilization (identify scope of)
- Contractor submittal procedure (through one-on-one contact)
- Surveys and staking
- Bid allocation of lump-sum bids (schedule of values) not valid for pricing of extra work
- Environmental requirements (cleanup; sanitary; dust; blasting; chemicals; etc.)
- Change orders and extra work procedures
- Unforeseen underground condition procedures
- Type 1 and Type 2 Differing Site Condition procedures
- Coordination of on-site utility work
- Closeout procedures
- Progress payment and retainage procedures

FIGURE 10.1 Checklist of Subjects to Be Considered for Preconstruction Conference Agenda.

Materials and Equipment

- Substitutions of "or equal" items
- Long-lead procurement (prepurchase) items
- Assignment of procurement contract to Contractor
- Owner-furnished materials and equipment
- Storage and protection
- Concealed shipping damages

Contractor's Schedule

- Owner-Engineer rights to approve
- Submittal requirements
- Owner-Engineer can set milestone dates

Change Orders and Extra Work

- Who has authority to issue
- Effect on time and cost
- Field Order vs Change Order
- Cumulative Change Orders

Subcontractor and Suppliers

- Contractual relationship
- Submittal requirements
- Owner/Engineer: no direct contact with subcontractors and suppliers
- Fair subcontracting act requirements

Coordination with Other Agencies and Contractors

- Interface requirements
- Testing and validation of systems
- Highway departments and railroad companies
- Code enforcement agencies
- Other governmental regulatory agencies

Handling of Disputes, Protests, and Claims

- Must exhaust all contractual means
- Resolution by the Engineer or Owner

Labor Requirements

- Davis-Bacon Act
- Documentation and audit requirements
- Federal, state, and local requirements

Rights-of-Way and Easements

- Permanent easements for project
- Temporary easements for construction
- Dumping sites and storage areas
- Access to the site by heavy equipment

FIGURE 10.1 (continued)

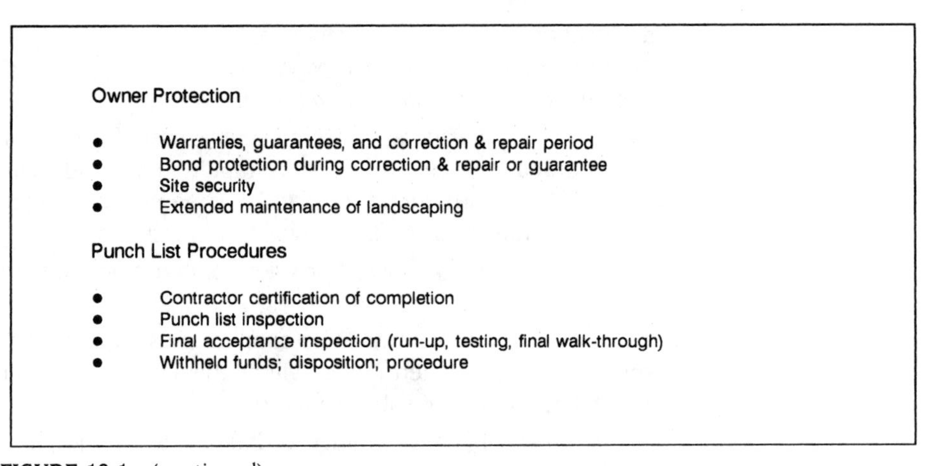

Owner Protection

- Warranties, guarantees, and correction & repair period
- Bond protection during correction & repair or guarantee
- Site security
- Extended maintenance of landscaping

Punch List Procedures

- Contractor certification of completion
- Punch list inspection
- Final acceptance inspection (run-up, testing, final walk-through)
- Withheld funds; disposition; procedure

FIGURE 10.1 (continued)

parties authorized to receive submittals, where submittals are to be delivered, number of copies required of each submittal, turnaround time required to return submittals by the architect/engineer, precedence of contract drawings over shop drawings, type of action indicated by the architect/engineer, and the limits of responsibility for review of submittals.

5. *Requirements for interference and/or composite drawings.* Who initiates them, and what will be the order of progression of these drawings? What is the impact on construction time if composite drawings are required?
6. *Insurance requirements; permits required.* Identify the time for providing insurance; who obtains and pays for permits?
7. *Job progress scheduling.* A preconstruction conference affords an opportunity for the essential involvement of subcontractors in the development and correlation of the individual schedules that make up the construction schedule for the project. Many large projects are conducted on an overall, or "master," schedule, using CPM or the PERT systems. Such scheduling is useless unless it is understood and followed by all of the parties in the construction team.
8. *Temporary facilities and controls.* These are the utility services that are essential to the construction process but do not form a part of the finished project. (See also Chapter 15 regarding temporary facilities provided by the contractor.) Under the CSI Format, these requirements are spelled out in Division 1, General Requirements. Under other specifications formats these may not be as well defined. This is often a gray area, requiring clear definitions of responsibility at the time of the conference. Some of the questions that must be answered on this subject are:
   (a) Who provides the services?
   (b) Who maintains them?
   (c) Who pays for these services?

   (d) If shared cost, in what proportion?
   (e) What are the *contractual* responsibilities, if any, under OSHA?

9. *Storage facilities and staging area.*
10. *Job site security during nonworking hours.* Losses from vandalism and theft at unguarded construction sites are rapidly mounting sources of expense to the entire construction industry. An agreement sharing the costs of better security measures might be worthy of discussion.
11. *Cleanup and trash removal.* Consider containers, scraps, sanitary wastes, and so on.
12. *Available hoisting facilities.* Who supplies hoisting facilities? If the general contractor supplies them, what will be the arrangement to make the hoist available to individual subcontractors?
13. *Change orders.* Because change orders are the subject of more disputes than any other single aspect of construction, they should be discussed in complete detail (see Chapter 19). Typical items for discussion relating to change orders are:
   (a) Percentages for overhead and profit to be applied to change orders. What costs will or will not be included in the change order price?
   (b) Length of time that a change order proposal price is to be considered firm.
   (c) Identify individuals who are authorized to approve change orders.
   (d) Procedures to be followed when submitting initiator change orders or change order proposals.
   (e) Change order forms that must be used.
   (f) Time extension requests made by subcontractors due to changes in drawings or specifications.
   (g) Amount of detail required of subcontractors when submitting change order proposals or initiator change orders. Will a complete breakdown of all costs be required? Brief description. Descriptive drawings.
   (h) Overtime resulting from change orders; consideration of decreased productivity.
   (i) *When* materials or equipment are to be removed because of a change, which party owns the removed material or equipment, and who removes it from the site of the job.
   (j) Responsibility for preparation of Record Drawings brought about by change orders.
14. *Warranty requirements.*
15. *Employment practices.*
16. *Listing and identification of all tiers of subcontractors.*
17. *Punch lists.* Establish timely punch list items and avoid an excessive number of punch lists.
18. *Record drawings* and final document submittals.
19. *Final payment* and retainage.

The establishment of meaningful communications between the parties involved on a construction project is essential for its successful development. Only with such a basis of understanding can the necessary planning and work proceed without conflict or costly disputes. The preconstruction conference, or construction coordination conference, as some call it, is a logical means toward this end.

A suggested checklist of subjects to be considered for inclusion in a preconstruction conference agenda is shown in Figure 10.1. Obviously, no agenda will contain all of the listed items, but with the list as a reference while compiling the actual agenda, there is less chance that an item will be omitted.

PRINCIPLES OF NEGOTIATION

Generally, negotiation may be considered as the art of arriving at a common understanding through bargaining on the essentials of the contract, such as specifications, prices, time of performance, and terms. A negotiator, when fully aware of his or her bargaining strength, can know where to be firm and where to make concessions on prices or terms. The negotiation occurs when both parties with differing viewpoints and objectives attempt to reach a mutually satisfying agreement. The negotiation process involves:

1. Presentation of each party's position
2. Analysis and evaluation of the other party's position
3. Adjustment of one's own position to as many of the other party's views as are reasonable

If one party, after thoroughly analyzing the position of the other party, acknowledges the fairness of the other party's position and that it is in the best interest of both parties, this concurrence represents an equitable agreement.

It should be understood from the beginning, however, that both sides have not only the right but also the obligation to obtain the best deal that they can for their own firm. Generally, the best negotiation strategy is based upon being reasonable within the rules of the game. This involves objectivity in negotiations and an awareness that the negotiator's prime purpose is to obtain the best possible deal for the employer. We talk of knowing the rules of the game, yet it must be appreciated that complete disclosure is not necessarily desirable nor practical (nor are you likely to encounter much of it). In short, each party is there to *win*. Nevertheless, a reputation for honesty is certainly useful, as it improves the image of the negotiator and therefore the chance for winning. Remember, negotiation is an adversary action, and complete disclosure of one's position may be counterproductive.

The person chosen to negotiate on behalf of the owner, architect/engineer, or contractor organization must be a strong individual who is capable of controlling the members of the negotiating team. The chief negotiator must also be acquainted with the broad details of the issue, the work to be done, and the method by which the cost information was developed. The negotiator must have freedom of action so that the various factors involved in any negotiation may be considered. It is equally important for the negotiator to be capable of organizing the team into a harmonious group, planning the objective, and explaining the objective to the negotiating team members so that they will be able to coordinate their efforts effectively.

Contractor's Position on Change Orders and Extra Work

If the contractor is in a sole source position, knows it, and is sure to be assigned the work eventually, an attempt to hold the line at the negotiation table may be delayed. However, if in an unsound or weak position, the contractor may assign the most alert and aggressive people as representatives in negotiations to take advantage of any opportunities that might arise during the course of the negotiations. Some organizations attempt to obtain a psychological advantage by a show of force, such as having several members of their top management or their attorneys represent them at the negotiations. Sometimes this can be to an organization's disadvantage as well, for it emphasizes the importance of the issue to it, and thus places the other party at a tactical advantage.

The Philosophy of Team Playing

Both sides in the negotiations will generally select participants in the negotiation process on the basis of special positions or skills that they possess. In many cases, these specialists may have authority and positions within their organization that are superior to those of the person selected as the principal negotiator. Although the project manager, and sometimes the Resident Project Representative, may be a part of the negotiation team, it is important to understand that the authority of the principal negotiator must take precedence during negotiations, and that negotiation is not intended to be a discussion among individual specialists.

At the beginning, an in-house meeting of the negotiation team should be held, at which time the subject of team conduct and communication at the bargaining table should be discussed. Each member of the negotiation team must know what he or she can and cannot do during negotiations. The negotiator must not assume that all members of the team will know how to act, but must instruct them beforehand. Project managers and other team members who are accustomed to leadership roles in their own organizational environment often find it quite difficult to play secondary or supporting roles during negotiations. It is essential that each team member be constantly reminded that the principal negotiator is the team leader and spokesperson, and that this negotiator is the only one who actually negotiates with the other party.

Sometimes a team member will forget this and become overeager, resulting in an active discussion with the other party. The principal negotiator should not hesitate to call for a recess if it is felt that there is a developing loss of control over the team members. Even if another team member feels that the principal negotiator failed to take advantage of an opportunity that the other party inadvertently provided, that team member should remain silent during the session. A few missed opportunities are preferable to an undisciplined team.

The principal negotiator should be careful about asking for a recess, however, as it can also become a disadvantage. It can serve as a red flag to the other party;

they will sense your position on the issue at hand. One strategy is to delay calling for a recess until another subject is brought up, or as some prefer, call for an occasional recess at random times, just as a break, so as not to give any indication to the other party that any specific issue is involved.

Basic Negotiation Policy

There are two basic methods of approaching a negotiation. First, you may consider the package as a whole, an overall method; second, you may elect to resolve each of the elements of the package separately, in sequence, as the negotiations progress.

Inherent difficulties of the sequential method of negotiation are the facts that it is necessary to reach separate agreements on each of the items before progressing to the next item, and to agree upon the order in which each item will be considered. Another problem is that there is no later opportunity for making trade-offs of one item for another.

The basic principles of negotiation do not require that an agreement be reached on individual items, but that compromises may be made by each party to reach an overall determination. Sometimes a contractor's main interest may fall into an area that the architect/engineer is willing to concede. Therefore, negotiations should not be conducted on the basis of reaching a firm agreement on each issue but, rather, on the basis of discussing each issue in sequence and arriving at a general meeting of the minds. The final solution will then represent not an agreement on individual items, but a resolution of all of the points of disagreement within the total contract.

One advantage of the overall method of negotiating is that it does not require a specific agreement on an issue before moving on to the next subject. Instead, a general meeting of the minds or an implied acceptance of the issue is understood and each party may then move on to the next item of the negotiation. Under this principle, it is understood that the implied agreement is binding only in the context of the entire negotiation.

Occasionally a contractor, for example, may attempt to bypass the negotiator and negotiate directly with the management of the architect/engineer or the owner. This is a practice that must be discouraged, as a precedent will have been set that will have the effect of undermining the negotiator's effectiveness. If successful in this ploy, the contractor may subsequently insist upon conducting all future negotiations that way. The solution is for the management to reiterate the position taken by the negotiator. Another method, which was used by a contractor on one of the author's projects, was to drive toward the home office of the architect/engineer after failure to win a point in the field, with the intent of negotiating with the principal. After telephoning the engineer's office to announce the contractor's visit so that the principal could be properly prepared with the position taken in the field, it just so happened that on the contractor's arrival, the principal of the firm was always tied up in conference. After several hours of waiting in the outer office, the contractor would give up and leave. After about three or four incidents of this type, the contractor was resigned to the fact that the matters would have to be negotiated with the Resident Project Manager, as originally provided for in the contract.

Negotiation Guidelines

As a means of obtaining a tactical advantage over the opponent in any negotiation, the following principles are suggested:

1. Keep the objective in mind.
2. Adjust your end to suit your means.
3. Exploit the line of least resistance.
4. Take an approach that offers alternative objectives.
5. Keep your plan adaptable to changing circumstances.
6. Do not put your weight behind an approach while your opponent is on guard.
7. Do not renew an attack along the same lines or in the same form after it has failed once.

In the early stages of a negotiation, an architect/engineer, an owner, or a contractor will find it to his or her advantage to plan the overall strategy as an offensive one. There is a choice, however, of various tactics to use. The strategist may elect not to reveal its position and maneuver the other party from one position to the other until its objective has been reached; or, more often, after an initial fact-finding session, the strategist may reveal its minimum position as a counterproposal. In this case, the counterproposal must be a realistic one. Another tactic is to reveal the minimum figure, then immediately offer the objective. This provides the user of this method with very little bargaining room, however, and may be considered as a sign of weakness by the other party.

Techniques of Negotiation[2]

Numerous times during the conduct of a construction project, the project manager and the Resident Project Representative will be called on to become involved in some form of implementation meetings and negotiations with the contractor. At any meeting where a decision must be made that will influence the financial position of either the contractor or the owner, it is quite likely that the meeting will involve some form of negotiation, even if the symptoms were not recognized.

Negotiation is an art. Let no one tell you otherwise. However, some of the basic rules and guidelines of the art can be acquired by observing and learning from the experts.

It is said that in the marketplaces of the Middle East, there often comes a point in the haggling when the buyer starts walking away to signal that there is to be no deal. This action in itself does not impress the seller, but tradition holds that when the customer has walked 30 paces, he or she will not return. So if the merchant is willing to offer a better price, he or she waits until the customer has gone 28 paces

[2]Adapted from G. W. Harrison and B. H. Satter, "Negotiating at 30 Paces," *Management Review,* April 1980, pp. 51–54 (New York, AMACOM, a Division of American Management Associations).

or so and then runs after him. An amusing game to an observer, perhaps, but similar games are played for much higher stakes around the construction conference table. Price is usually the key issue for both the contractor and the owner.

Without going into the psychological analysis of the art of negotiating, we should take a practical look at some of the ploys that surface on both sides. The *30 paces* ploy is a last-ditch maneuver that should never be used on minor obstacles to reaching an agreement. It is equivalent to saying "do it my way or else," and it is dangerous if cooperation is needed at a later stage in the negotiations.

Another approach is the *commissar's technique,* which is said to be somewhat typical of the "iron curtain" countries. Under this ploy, a team of people of impressive credentials and corporate titles negotiates a deal. But if, on reflection, the team is not fully satisfied, more negotiators will come in the next day, saying that they are superiors to the first group and that the first group did not have the authority to make a final commitment (sounds like some car salesmen I have heard). Then the process may be repeated. Obviously, it is not likely to work a third time, as the credibility of the offending negotiating team probably has been stretched as thin as it can get.

What is the point? It is that you should know the exact authority of the negotiator at the table, even if it is the chairman of the board, for even that person may have to obtain approval of the board of directors before he or she can make a binding commitment. It is also wise to make sure that the issues to be discussed have already passed through the appropriate management committees, lawyers, auditors, or any others whose judgments or approvals are necessary before final consideration.

Who, and how many, will attend the negotiating session is also important. Too many people present on either side makes it difficult to get a consensus. Although the collective thinking phenomenon is unlikely to be present at the average negotiation session, as all of the position policies of each side should have been well established in prenegotiation meetings, the adversary relationship that may be present between the two groups may complicate efforts to reach a consensus. In addition, always remember that a "handshake" deal may be a stronger weapon for one side than for the other. Think out in advance how an agreement should be arrived at and make sure that you are in control on all of the issues that are important to you.

Bargaining Strategy

The substance of good strategy is something that must be tailored to the specific circumstances and facts. However, there are certain basic practices that can lead to the formulation of strategies. The most important step is to determine your objective by discussing it with your negotiating team, and compare that with what you perceive as the objective of the other party. Seek out common ground that can be settled without disagreement, and isolate key issues. The negotiator must recognize the weaknesses of his or her own arguments as well as emphasize the strong points.

Negotiating Tips

The following is a list of negotiating tips from *Presenting the Claim to the Contracting Officer and Appeals Board,* Government Contracts Claims Course Manual (Federal Publications, Inc.):

1. Strive to determine the real objectives of the other party.
2. Do not let personality differences frustrate the progress of the negotiations.
3. Avoid being too dogmatic or inflexible.
4. List and discuss your side's objectives with all members of your team.
5. Be prepared when you begin negotiation.
6. Recognize the consequences to your side if negotiations should fail.
7. Many successful negotiations are easier in a better environment.
8. Maintain a written record of the negotiations on a daily basis.
9. Continually verify the information being presented and received during the course of the negotiations.
10. Never walk out on a negotiation unless you are prepared to terminate it.
11. Patience is a virtue, but progress is a necessity.
12. Be prepared to make reasonable concessions.
13. Be prepared to turn a disadvantage into an advantage.
14. Quit when you are ahead. You don't always have to knock something off an offer. If the offer is good, take it.

Once a settlement has been reached, the owner or architect/engineer should prepare a written memorandum of agreement that clearly spells out the terms of the settlement. This should be done as soon as a basic agreement is reached; send a copy to the other party requesting that they sign it, indicating confirmation of the agreement. Details may be worked out in subsequent meetings and need not be included in the basic agreement letter.

Negotiations that have broken down should also be detailed in writing, since statements of fact constituting an admission can be used in court. Also, interim agreements reached during negotiations can serve as a basis for final award in court.

Psychology of Negotiation

The story is told of a shrewd country lawyer in the South who always asked the big-city negotiating lawyers who came down South, "When are you taking the plane back to New York?" If the answer was "this afternoon," he knew they were eager for an agreement. But if they were booked into the motel for a week, he knew that he was in for a siege. Finding out the other party's timetable for making a deal is always one of the objectives of a good negotiator, as it can have a decided effect on the outcome. Never rush anyone who is not ready to make a quick decision.

A knowledge of *real values* of the work to be accomplished is essential to negotiate intelligently, although the owner and the contractor may choose not to use the same basis of comparison. The determination of the real value of work does not always dictate the price, but it does keep both sides in the same ballpark. Getting someone to come down in the asking price is a big accomplishment only if the final price is not $50,000 or $100,000 more than the work was worth.

A tale has been told about the boy who told his father that he had decided to sell his dog for $1 million. The amused father gave his permission, and a few days later his son told him that he actually had sold his dog for $1 million. The surprised father asked the boy if he had been paid in cash. "No," the son said, "I traded him for two $500,000 cats."

Some negotiators, unfortunately, fail to compute the cash equivalent of what is being offered in payment for the work in question. Anything other than cash must be carefully analyzed to determine its real worth. This includes bonds, stocks, and other instruments.

Strategic Ploys

Throwing up obstructions or creating distractions with minor issues is sometimes a deliberate strategy in negotiations, but can be a dangerous one. Negotiations should be directed at the settling of major issues, without getting bogged down in technicalities that the lawyers and the accountants can deal with after a broad agreement has been reached. As much as possible, both the contractor and the owner should concentrate on the prospects of making a deal, not breaking one. A good negotiator always labels a participant as a "deal maker" or "deal breaker," depending on how the outcome would affect the negotiator's own position.

It was a wise person who advised never to bargain for the last possible penny but always "leave a little something on the table." The other side must have the opportunity to save face, both at the negotiating session and back at the home office. It is not a good idea to alienate everyone through overzealous haggling, especially if those present at the session must also stay on to run the operation. Develop a sense of how far the other party can go, and when to back off a little.

Setting the Pace

Contractors and owners, or their architect/engineer representatives, frequently engage in gamesmanship of who will make the first move in the negotiations, neither wanting to be typecast as the anxious one. Normally, such maneuvering produces no real advantages, and there are really no rules of protocol about such matters. From a practical standpoint, the initiator of a particular deal or step in the negotiation process is out to take the lead in supplying information, suggesting a price, and so on.

Whether a party to the negotiations establishes itself as a professional or a newcomer at the negotiation table strongly influences how the session is conducted. Here are some guidelines:

1. Make sure that you know the scope of the meeting, the issues to be discussed, and the amount of time available to discuss them.
2. Start off with some smaller issues to get the "feel" of the personalities involved.
3. Do not get into the position of waiting for the other shoe to drop. Know all of the issues before reaching a compromise.

4. Decide in advance whether it is to your advantage to reach a rough or a precise agreement, and if publicizing the conditions before a commitment would work for or against you.
5. Make sure that you are bargaining with someone who has the authority to sign for his or her organization.
6. Have a walk-away price in mind that is consistent with the realities of the circumstances.
7. Bear in mind that there is no one "best" negotiating strategy other than good common sense.
8. If the other side brings legal counsel, contact your side to see whether or not your attorney wishes to be present. In some cases, unexpected arrival of the other party's lawyer at a meeting was cause for delay of the meeting until the owner could be afforded similar representation.

Who Won?

It is often impossible to know who wins in a negotiation. The mere fact that a contractor came down in price, or an owner or architect/engineer increased the offer, proves nothing without knowing what they were really willing to settle for. Although one lawyer is reported to have said that if both sides say they are unhappy, it was probably a fair deal, the fact is that in any transaction based upon sound analysis and realistic negotiation, both sides should be satisfied. Often, in winning the point, a good negotiator will leave something "on the table" for the other party. A face-saving gesture, perhaps, but a valuable concept for maintaining a continued amicable relationship with the other party. Thus, each will come away from the bargaining table with a little something and may well be prompted to say, after conclusion of the negotiations, "I like doing business with you."

Problems

1. List at least five important types of in-house meetings occurring prior to the construction phase.
2. List at least five important types of meetings held during the construction phase of a project.
3. Should any of the parties bring an attorney to a construction meeting? If one does, what should the other party do?
4. What are the five types of information that you should determine prior to attending a meeting?
5. Is there any advantage to be obtained by choosing a seating position at a meeting, if the opportunity presents itself?
6. Name two recommendations for selection of good position at a conference room table.

7. Name the benefits of holding a preconstruction conference.
8. Name the principal parties that should be asked to attend the preconstruction conference.
9. Name three important principles of negotiation, with reference to your own preparation.
10. Name at least five principles that should guide you in obtaining and maintaining a tactical advantage over the opponent.

11

RISK ALLOCATION AND LIABILITY SHARING

In recent years, a great deal of lip service has been paid to the concept of risk allocation and liability sharing. It seems to be the inevitable result of the many losses suffered by owners and engineers alike in connection with the projects they build, due to a new emphasis on the resolution of construction disputes through litigation and arbitration. Many of the risks were there all the time, but in earlier years, the contractors were always expected to bear the responsibility for as many construction risks as the owners could pass off on them through indiscriminate use of exculpatory clauses for everything from risks of unforseen underground conditions to substandard designs and specifications.

Far from trying to operate in a risk-free environment, a contractor understands that risk is part of the business. All that is wanted is a fair reimbursement for taking such risks. However, the modern contractor is no longer content to sit idly by and take all of the risks while being locked into a guaranteed maximum price, but now stands ready to fight back. Contractors have come to realize that they have the means and often the right to recover the losses that are the result of the imposition of unfair contract conditions or administration.

Whereas part of the job of the Resident Project Representative is to minimize exposure of the owner and the architect/engineer to risk of claims losses, there are other risks that are rightfully within the responsibility area of the project manager to control. Some of the risks may be transferred to others by contract. However, it should be recognized that all risks are rightfully the owner's unless transferred or assumed by another party for fair compensation. The principal guideline in determining whether a risk should be transferred to another is whether the party assuming the risk has both the competence to assess the risk and the expertise necessary to control or minimize it. The choice must be made before the allocation of risk takes place. One such approach to the decision-making process is outlined in the flowchart in Figure 11.1.

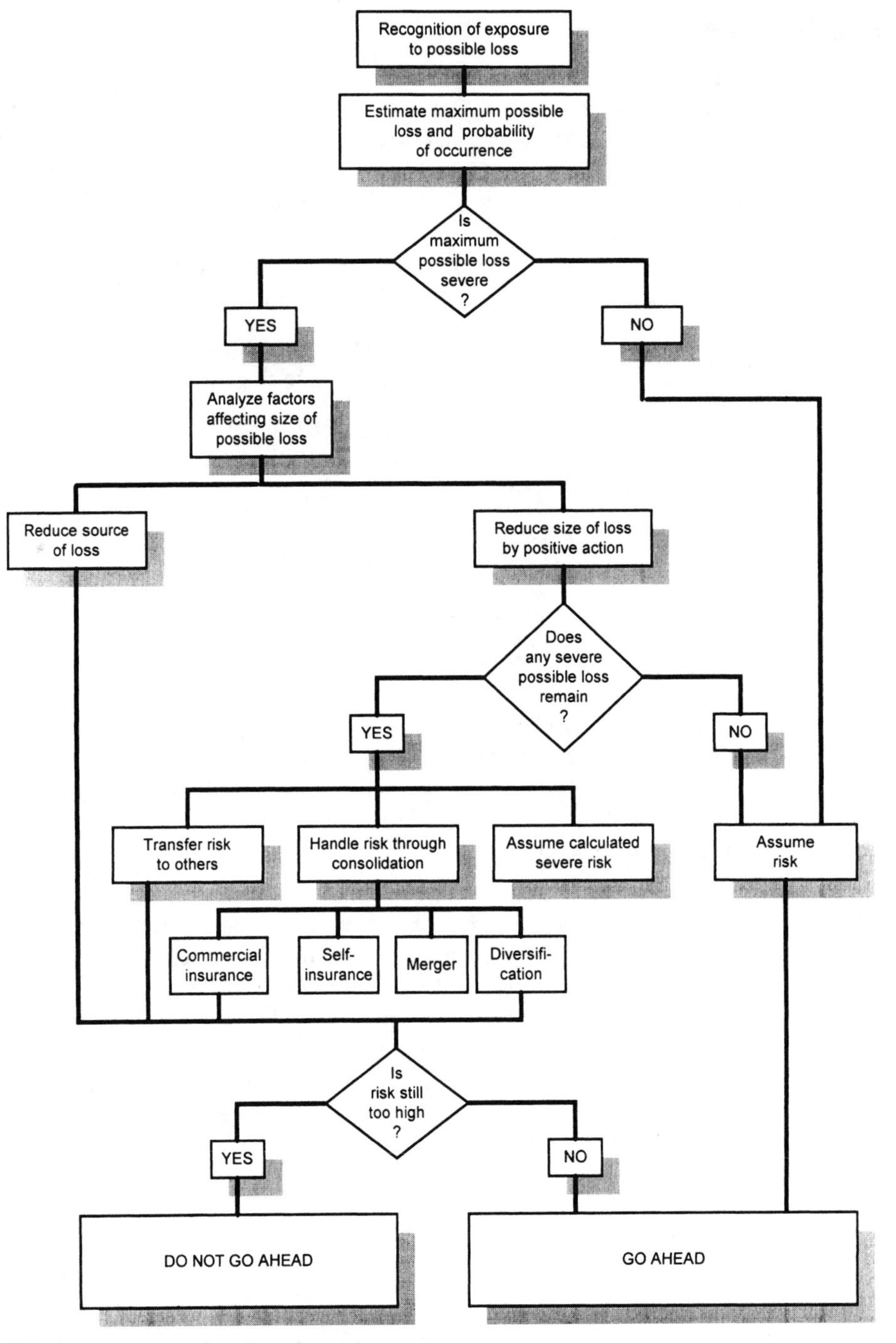

FIGURE 11.1 Logic Flowchart for Risk Decisions.

Risk Management

The first thing that must be recognized is that risks, especially in construction, do exist but are not necessarily fairly distributed. Some kinds of risks must be recognized as being inevitable in engineering and construction and therefore must be accepted philosophically and realistically as a part of the situations to be dealt with. Managing risks means minimizing, covering, and sharing of risks—not merely passing them off onto another party.

Although some risks can be avoided, risk management deals primarily with the following concepts:

1. Minimizing risks—regardless of whose risk it is
2. Equitable sharing of risks among the various project participants

The parties must be able to sit down together, prior to the start of the work, to come to a better understanding of the realities of the risk responsibility, assumption, and allocation. The parties must be prepared to discuss and to decide on the following issues:

1. What levels of risk are realistic to assume?
2. Who can best assume each risk?
3. What levels and kinds of risks are properly and most economically passed on to insurance carriers?

Risk exists wherever the future is unknown. Because the adverse effects of risk have plagued humankind since the beginning of time, individuals, groups, and societies have developed various ways for managing risk. Because no one knows the future exactly, everyone is a risk manager, not by choice but by sheer necessity.

Definition of Risk

Risk has been defined in various ways. There is no single "correct" definition. In order to emphasize the major objective of risk management, we will choose to define *risk* as the *variation in the possible outcomes that exist in nature in a given situation.* Another way to clarify this definition of risk is to distinguish between risk and probability. Risk is a property of an entire probability distribution, whereas there is a separate probability for each outcome.

Both risk and probability have their objective and subjective interpretations. The true state of things is different from the way it appears. Because a person acts on the basis of what is believed to be correct, it is important to recognize this distinction. To the extent that a person's estimates are incorrect, that person's decisions are based on false premises. Consequently, risk managers must constantly strive to improve their estimates. Even with perfect estimates, decision making about risk is a difficult task. Uncertainty is the doubt that a person has concerning his or her ability to predict which of the many possible outcomes will actually occur. In other words, it is a person's conscious awareness of the risk in a given situation.

Scope and Applicability to Construction

Risk, as such, is present in all situations and businesses. As applied to construction, we will use the categories detailed by the American Society of Civil Engineers at the Specialty Conference on Construction Risks and Liability Sharing in Scottsdale, Arizona, in January 1979, where the principal categories listed included:

1. Construction-related risks
2. Physical risks (subsurface conditions)
3. Contractual and legal risks
4. Performance risks
5. Economic risks
6. Political and public risks

As it may well be imagined, not all of these items are considered in the same order of priority by owners, architect/engineers, and contractors, and there are also considerable differences of viewpoint as to the percentage of each item that should be shared by the various parties.

As a means of defining the range or spectrum of risk and liability sharing in construction, consider that at one end of the spectrum, the contractor is assigned the entire risk or liability with no risk to the owner. This usually results in high cost; in effect, the contractor is acting as the insurer of the owner.

At the other end of the spectrum, the contractor is released from all risks or liability, and the burden is assumed by the owner. This usually results in a lower cost; in effect, the owner assumes a self-insured role.

Either end of the spectrum has both advantages and disadvantages. Owners must thoroughly assess their own situations to determine what allocation or distribution of risks would serve them best. For instance, the viewpoint of a large city, with an excellent engineering force of its own, might be that it is to its advantage to minimize contractor risk and assume more of the risk itself. An owner with a large amount of engineering resources and a thorough and expert design staff could afford to assume a larger portion of the risk to obtain the overall benefit of lower costs. Other factors that should influence the owner's assessment are the type of construction (does it include high-risk construction such as underground work?), the degree of detail, accuracy and/or completeness of the plans and specifications, in some cases the urgency of the project, and similar considerations.

Identification and Nature of Construction Risks

Although construction risks can be categorized in many ways, only four groupings are presented here: physical, capability, economic, and political and societal. In the process of identifying risks, only those that are created by the parties themselves in their attempts to transfer risks are included.

Contractual Allocation of Risk

In the absence of contractual provisions to the contrary, our legal system already allocates most construction risks between the designer, owner, and contractor. Therefore, when we speak of risk allocation, we really mean risk reallocation, risk spreading, or reaffirmation of the existing allocation of risk so that the risk stays where it otherwise would be.

There are two basic precepts or guidelines that should be recognized as criteria for the sharing of risks inherent in a construction project.

1. All risks are rightfully those of the owner unless and until contractually transferred to or assumed by the contractor or insurance underwriter for a fair compensation.
2. The principal guideline in determining whether a risk should be so transferred is whether the receiving party has both the competence to assess the risk fairly and the expertise necessary to control or minimize it. An additional guideline is the determination of whether the shift of the risk from the owner to another party will result in a savings to the owner and the public.

The principal means available for contractual allocation or reallocation of risk are the construction specifications for the construction contract and the owner/architect/engineer agreement for the design of the project. Under the format endorsed by the Engineers Joint Contract Documents Committee,[1] which has been approved and endorsed by the Associated General Contractors of America, such provisions would logically be spelled out in the General Conditions or the Supplementary General Conditions. In the case of a public works contract, the contractor does not have the opportunity to participate in the wording of the agreement between the parties. However, in private contracts risk allocation could become a valid bargaining consideration. The wording should in any case be prepared by competent legal counsel—avoiding the attitude adopted by some attorneys that their client is best served by exculpatory clauses that would seem to relieve their client of any responsibility for anything, including the negligence of their own personnel. The weakness of this attitude may become more evident during the settlement of some of the large claims that frequently follow such issues.

Exculpatory Clauses

An exculpatory clause is one that attempts, by specific language, to shift a risk or burden of risk from one party to another. As the impacts of such clauses can be very great, a contractor must be very careful to review the contract for such exculpatory

[1]Standard Forms of Agreement, Engineer, Owner, and Construction Related Documents prepared by Engineers Joint Contract Documents Committee (American Consulting Engineers Council, American Society of Civil Engineers, Construction Specifications Institute, and National Society of Professional Engineers).

clauses to determine what risks are being shifted to it and how the bid should be adjusted to reflect that risk.

A contract, for example, may contain an exculpatory clause that is intended to shift the responsibility for the engineer's errors and omissions to the contractor, such as the clause illustrated in the following example:

> If the Contractor, in the course of the work, becomes aware of any claimed errors or omissions in the Contract Documents, it shall immediately inform the Engineer. The Engineer will then promptly review the matter and if an error or omission is found, the Engineer will advise the Contractor accordingly. After discovery of an error or omission by the Contractor, any related work performed by the Contractor shall be done at its own risk unless otherwise authorized in writing by the Engineer.

Under such a provision, the owner may claim that work done by the contractor in accordance with incorrect plans or specifications will not be paid for. While it is the intent of the clause to prevent the contractor from knowingly exploiting any errors or omissions the contractor may have become aware of to the detriment of the owner, one can see how such a clause could also be easily used against the unwary contractor.

In the following example of an exculpatory clause, a similar situation is created:

> The Contractor shall give all notices required by law and shall comply with all laws, ordinances, rules, and regulations pertaining to the conduct of the work. The Contractor shall be liable for all violations of the law in connection with work provided by the Contractor.

Under the terms of the foregoing provision, the contractor can be held liable for violations of the law, *even though those violations were the result of building the project in accordance with the plans and specifications that violated the law.* It is clearly the responsibility of the design professional to prepare plans and specifications that comply with all applicable laws; this clause unfairly shifts the responsibility for such compliance over to the contractor.

Although most exculpatory clauses are viewed by both state and federal courts on the basis of what is equitable to the parties as well as the specific language of the clause, some courts on occasion disregard the equitable considerations and apply only the harsh terms of the clause. In a New York court decision, Kalisch-Jarcho, Inc. was a successful bidder on an $8 million heating and air-conditioning contract for the construction of a new police headquarters in New York City. The contract included a provision in which Kalisch agreed to make no claim for damages for delay in the performance of the contract occasioned by an act or omission of the City, and any such claim would be compensated by an extension of time. Kalisch sued the city for breach of contract, claiming damages for 28 months of delay. The trial court awarded Kalisch approximately $1 million, but the decision was reversed upon appeal. The court found that the exculpatory clause protected the city from a claim of damage as there was no evidence that the delay had been intentional [*Kalisch-Jarcho, Inc. v. City of New York,* Ct. App. N.Y. (March 29, 1983)].

Who Should Accept What Risks?

There is no fixed rule to help answer this question, but Figure 11.2 suggests a starting position for determining who should bear what risks. It should be recognized, however, that in some cases more than one party to the contract may share a common risk. In such cases, though, the risk may be shared in name only, as the specific risk carried by each party may differ materially in terms of the specific details of the risk they carry.

Types of Risks and Allocation of Those Risks[2]

Site access is obviously an early risk and one that the owner should retain. The contractor lacks the capacity to influence those in control of the site to render it available. However, permit requirements that relate to a contractor's capacity or safety control program can be rightfully assumed by the contractor.

Subsurface conditions of soils, geology, or groundwater can be transferred to the contractor, who is in a better position to assess the impact of these conditions on the project cost and time. However, as an essential party of the transfer process, the owner has the responsibility to undertake precontract exploration measures, and the designer has the responsibility to design for the conditions expected. The extent that this is not feasible should determine the degree to which the owner retains a portion of the risk under an "unforeseen conditions" clause.

Weather, except for extremely abnormal conditions, is a risk for the contractor to assume, as its impact on construction methods can be better assessed by the contractor.

Acts of God, such as flood or earthquake, are exposures that have no purpose in being transferred beyond the owner, except that the architect/engineer can assume the responsibility for designing to minimize their impact. However, to the extent that it can be occasioned by the contractor's operations, fire may be one shared with the owner.

Quantity variations are another form of risk frequently encountered. Within reasonable tolerances, quantities of work can be reasonably estimated and any variances assumed by the contractor for all quantities in excess of, for example, 15 to 25 percent. Where quantities are dependent upon subsurface or other lesser-known conditions, significant variations should be shared only to the extent that exploratory information is available. Quantity changes triggered by late changes in the owner's requirements, however, should be at the owner's risk. Some types of variation, such as tunneling overbreak, are contractor controlled and should be borne by the contractor.

Capability-related risks are the result of the different capacity and expertise that each of the parties brings to the construction project. The consequences of failure of

[2]This section follows J. Joseph Casey (President, Gordon H. Ball, Inc., Danville, CA), *Identification and Nature of Risks in Construction Projects: A Contractor's Perspective,* presented at the ASCE Specialty Conference on "Construction Risks and Liability Sharing," Scottsdale, AZ, January 24–26, 1979.

Type of Risk	Contractor	Owner	Engineer	Comments
Site Access		•		
Subsurface Conditions		•		a
Quantity Variations	•	•		b
Weather	•			c
Acts of God		•		
Financial Failure	•	•	•	
Subcontractor Failure	•			
Accidents at Site	•			
Defective Work	•			
Management Incompetence	•	•	•	
Inflation	•	•		d
Economic Disasters		•		
Funding		•		
Materials and Equipment	•			
Labor Problems	•			
Owner-Furnished Equipment		•		
Delays in the Work	•	•	•	e
Environmental Controls		•		
Codes and Regulations		•		
Safety at Site	•			
Public Disorder		•		
Union Strife	•			
Errors and Omissions			•	
Conflicts in Documents			•	
Defective Design			•	
Shop Drawings			•	

a—Can be transferred to the contractor; however, owner has obligation to undertake precontract exploration measures, and the designer has the responsibility to design for the conditions expected.

b—Contractor can be expected to assume risk up to 15 to 25 percent. Where quantities are dependent upon unforseen subsurface conditions, owner must assume the risk.

c—Normal weather for the time and location only. Unusual inclement weather that delays the work is the owner's responsibility.

d—Sharing of escalation risk should be limited to 12- to 18-month span.

e—Usually the contractor's risk; however, owner could incur some liability.

FIGURE 11.2 Construction Risk Allocation to Participants.

any party to measure up to these standards should be borne by the failing party. Unfortunately, this is not always the case. Too often the contractor who has the practical task of building the project carries the burden of the owner's or architect/engineer's failure. This, in turn, renders the contractor's performance task either unfeasible or feasible only at considerable extra cost.

Defective design is a risk usually associated with the architect or engineer. The tremendous expansion of construction has placed great burdens upon the design professions. Maintaining performance standards in the face of this is quite difficult, and occasionally, design or specification defects occur that create construction problems. Unfortunately, it is usually the owner and the contractor who suffer the consequences of such failures instead of the architect/engineer who created the problem in the first place. Design failures or constructability errors are becoming more and more apparent, and the architect/engineer should bear the true cost of such failures. Often, ill-advised use of performance specifications are provided as an escape from the responsibilities of design.

Subcontractor failure is a risk that is properly assumed by the contractor except where it arises from one of the other listed risks attributable to the owner or architect/engineer. The prime or general contractors are in the best position to assess the capacity of their subcontractors, and therefore it is they who should bear the risk of not assessing the risk properly.

Defective work of construction, to the extent that the problem is not caused by a design defect, should be the contractor's risk.

Accident exposures are inherent to the nature of the work and are best assessed by the contractors and their insurance and safety advisors. Furthermore, the contractors have the most control over site conditions that can increase or decrease accident exposure.

In the viewpoint of some, the recent trend toward "wrap-up" insurance coverage is a mistake. The safety record on a construction project is so heavily affected by the contractor's methods, site conditions, worker attitudes, and supervisor awareness that the owner will quite possibly obtain the opposite of what is sought for. Ultimately, the cost of insurance is the cost of the losses plus the cost of administering the compensation for these losses.

Managerial competence is a risk that must be shared by each party, as they each have their own set of managers. It is an ongoing challenge for each organization to assign personnel according to their respective competence levels.

Financial failure is a risk not frequently mentioned and can happen to any of the parties to a contract. Although infrequent, the order of magnitude of such failure should be considered. It is a shared risk, as the parties need to look at the financial resources of themselves, their partners in joint undertakings, and the other parties to the contract.

Inflation is one of the world's realities. Every owner is conscious of its impact on the viability of any project. It is important that the owner retain the true cost of a project. Government experts in finance have so far been unable to predict where the country will be a few years from now, so it is unfair to expect the contractor to do better than so-called government experts. The contractor's apprehensions will result in higher cost to the owner, or unwarranted optimism will result in the contractor's own financial harm. A default resulting from such a failure will result in even greater costs to the owner. The sharing of the escalation risk should therefore be limited to a short span of time, approximately 12 to 18 months, when union agreements usually expire and beyond which is pure speculation.

Economic disasters, as referred to herein, are periodic economic disasters of such magnitude that a contractor could not properly assess either their probability or their cost impact. An example might be OPEC decisions, nationwide strikes, devaluation, tax rate changes, and similar large-scale incidents. The owner should retain the risk of such disasters.

Funding is obviously a risk beyond the capacity of the contractor to control. Improper sources of these funds may occasion delays or create interest costs that are not anticipated and financing problems that to many contractors are unbearable. There is no moral justification for a competent contractor being driven out of business by delayed compensation for services rendered. This is especially true in the protracted negotiation of changes. All too often the owner plays the cash-flow game to lever dispute negotiations to the owner's advantage. Some large contractors with financial capability may be able to fund these delays, with great outlay of interest, but all too often, the smaller contractor cannot even survive.

Labor, materials, and equipment involve considerable risks. The availability and productivity of the resources necessary to construct the project are risks that it is proper for the contractor to assume. The expertise of the contractor should follow the assessment of cost and time required to obtain and apply these resources. This is the basic service that the owner is paying for.

Acceleration or suspension of the work is a risk properly retained by the owner, but is all too often pushed onto the contractor in the form of "constructive acceleration" or "constructive suspension." An objective appraisal of the facts underlying the situation and acceptance of responsibility where it belongs are necessary. It is important to realize that this applies to legitimate acceleration, however, and not to false claims of acceleration as described in Chapter 14 under the heading "Who Owns Float?"

Political and societal risk is an area of growing importance to any effort at risk allocation. It is an area in which political and social pressures from parties having little interest in a project but having a great impact on such a project greatly influence its outcome. This is an unclear area and deserves much careful thought as to how the risk should be allocated—in some cases it is clear, in others vague.

Environmental risks rightfully belong to the owner alone and should be retained by the owner except to the extent that they are influenced by construction methods determined by the contractor, or created by suppliers controlled by the contractor.

Regulations by government in the social area, such as safety and economic opportunity, are the rules under which the contractor rightfully must operate. Although there is additional risk in this less known and interpretive area, it is similar to the work rules established by union contract or agreements.

Public disorder and war are political catastrophes of such impact that their risk is best retained by the owner, lest it becomes necessary to pay an unusually high price for transferring the risk to another party.

Union strife and all that it entails are risks that are properly taken by the contractor. Unjustified work rules and similar problems are all risks that the contractor must assess and provide for.

Risk Distribution

To many contractors, risk management is the nature of their business. That is what they are paid to do. Management of risk first involves a "go/no-go" decision on risk assumption. To the extent that this process is complicated by unwarranted "risk dumping," the costs in time and money eventually find their way to the owner in the form of higher prices.

There have been some construction contracts where total physical risk was assigned to the contractor, including the risk of unforeseen (changed) conditions and of variations of quantities required to complete the work. Even under such extreme allocation of risk, the owner still retains a very substantial risk that the contractor may not comply with the terms of the contract and properly complete the work within the allotted time. Even if a claim is made under a performance bond, substantial damage may certainly have already occurred. Under the best of circumstances, from the owner's viewpoint, the owner will retain material risk. It is therefore important that the owner recognize the existence of this risk and not be lulled into the false sense of security that it has somehow been passed on to the contractor or the contractor's surety. Unreasonably burdening the contractor does not necessarily rid the owner of the risk. Default on the part of the contractor in whole or in part is also a very real risk that the owner can be left with.

Inclusion in a contract of the frequently used (or "misused") disclaimer provision relating to geological information furnished the bidders for underground construction may not be as effective in passing risk to the contractor as some might think, because such clauses are of questionable enforceability. Despite this, such exculpatory clauses continue to be used, and such use actually places the risk upon all of the parties to the contract, without at the same time providing any relief to anyone. Under such circumstances, a dispute will frequently arise if the geological formation does not coincide with what is found in the field. Such a dispute may not be resolved until many years have passed, and then only at great expense to all of the parties to the contract, with the resulting effect of creating a substantial increase in the cost of the project. The final result is that the owner may have to bear certain risks anyway, despite any contract clauses to the contrary. In view of this, it may be wiser for the owner to recognize these facts at the beginning and provide contractually for the owner's assumption of such risks.

The fact that the contractor carries a substantial burden of risk is beyond dispute. Unfortunately, it seems that the viewpoint of many who design and administer construction contracts is that the contractor should carry virtually all of the risk, whether provided in the contract or not, and it is precisely this attitude that has contributed to the inevitable litigation that will follow.

The Contractor's Viewpoint[3]

It is not sufficient to establish policy on how risk is to be allocated among the parties. Follow-up is required to ensure that these policies are actually being implemented.

---

[3]Adapted from Norman A. Nadel (President, MacLean Grove & Company, Inc., New York, NY), *Allocation of Risk—A Contractor's View,* presented at the ASCE Specialty Conference on "Construction Risks and Liability Sharing," Scottsdale, AZ, January 24–26, 1979.

The owner may determine that it is in its own interest to undertake extensive geological investigation for an underground project. However, this effort may be wasted if the frequently used disclaimer provision regarding geological information furnished to the bidders is included in the contract documents.

In the specifications for a recent subway construction contract that involved underground rock tunneling, there was a "changed condition" clause included. The borings generally indicated reasonably sound rock with ample cover over the roof of the tunnel. Despite this, the specifications writer saw fit to include the following paragraph:

> The contractor's procedures for tunnel excavation shall provide for such construction techniques, including but not limited to, reduced heading advance, multiple narrow drift excavation, forepoling, pumping of cement or grout to reduce water inflow, and any other techniques applicable to rock tunneling, that may be required due to the nature of the rock encountered.

Here, any benefit thought to be obtained from the use of the changed conditions clause by the owner was clearly lost when that paragraph was added.

Contractors sometimes find themselves at the mercy of contract administrators who lack the courage to implement the contract in accordance with its terms. For example, a contractor was engaged in the construction of a rock tunnel for a subway in which the borings indicated sound rock with substantial cover over the tunnel roof. In actuality, the rock cover disappeared and approximately 30 meters (100 ft) of the tunnel had to be excavated utilizing soft-ground techniques. Before encountering the area of reduced cover, the rate of progress was such that the excavation for the stretch of tunnel in question would have taken approximately 8 days. Actually, more than six months were required to redesign the support system, to secure the necessary materials, and to excavate the tunnel utilizing a much slower and costlier procedure. Despite these facts, the contract administrator refused to find that a changed condition existed. Instead, it was determined that "extra work" was being performed for which, under the terms of the contract, the contractor could be paid only for certain direct costs.

The contractor had been denied payment for the very great costs that resulted from the delay of the work. Under the provisions of the changed conditions clause of the contract, the contractor was very clearly entitled to payment for the costs of the delay. The contractor was left with no alternative but to seek relief in court, a process that is going to be very expensive and time consuming to both the contractor and the engineering firm and owner.

As an intelligent application of the risk allocation procedures, it would seem that the application of the following guidelines could be helpful to produce construction at the lowest possible cost:

1. If a risk is imposed upon a party, an opportunity for reward to that party should exist for properly dealing with the risk.
2. Allocation of the risk to the party who is in the best position to control it.
3. Allocation of the risk to the party in whose hands the efficiency of the system is best promoted.

4. Allocation of the risk to the party who is best able to undertake it financially.
5. Steps should be taken to assure that risks are actually allocated as intended.

Occasionally, it may not be apparent how a given risk should be allocated. In such a case, careful consideration of what may be the motivation of the parties involved may be productive.

How Are Risks Allocated?[4]

How are risks allocated, and how should they be allocated? The allocation is initially made by the owner's legal department or its specifications department, which prepares the contract forms that are offered to the bidders on a take-it-or-leave-it basis. To the extent that such persons or departments are sensitive to practical construction and contract administration problems, possibly some contractor organizations or engineers may be capable of exerting some influence upon them in the preparation of these documents and the resultant reallocation of risks. More coercive and effective (and more expensive) allocations are the kinds made through the courts.

In the traditional construction contracting practice, the owner would allocate almost all of the risks to the contractor, saying in effect: "You deal with all of the construction problems and all the third parties, and don't bother me." In this same tradition, the architect/engineer would design a structure in its finished condition, and if any thought was given to the construction problems that might be involved in building it, considerable care was taken not to express any opinions on these matters in the contract documents.

This one-sided attitude fostered two results:

1. Contractors added high contingencies to their bids to cover the costs of the risks.
2. Litigation of construction contract claims followed.

Broadly speaking, the owners lost; the courts reallocated many risks that the owners thought they had laid on the contractor, and, as a result, the owners paid for their risks *twice*—once in bidding contingencies and a second time in court. Meanwhile, the contractors were not profiting either. They were losing money on delays and disputes, and often just breaking even (if they were lucky) in court. Construction law, as a result, is rapidly becoming a very profitable field—for lawyers.

Risks Reserved to the Contractor

In addition to the types of risks referred to under "Types of Risks and Allocation of Those Risks," the following are typical of the risks reserved to the contractor:

1. The contractor should bear all risks over which the contractor can exercise reasonable control. These include all matters relating to selection of

[4]Following Thomas R. Kuesel (Senior Vice President, Parsons, Brinckerhoff, Quade, & Douglas, Inc., New York, NY), *Allocation of Risks,* presented at the ASCE Specialty Conference on "Construction Risks and Liability Sharing," Scottsdale, AZ, January 24–26, 1979.

construction methods, equipment, and prosecution of the work except as this control is affected by the action of third parties.

2. In the area of third-party effects, risks should be allocated to those best able to deal with the third party. This principle would assign to the owner the risks related to government agency regulations and to agreements with adjacent property owners. Risks associated with labor and subcontractor agreements and disputes should be assigned to the contractor.
3. Construction safety should be the responsibility of the contractor, although financial risk with regard to third parties is properly allocated to insurers (either the contractor's or the owner's).

Construction is a highly complex business. Guidelines, recommendations, regulations, contracts, and even legal rulings can only provide direction for judging a particular situation. Among the most difficult and important to define factors in evaluating and allocating risk are the reputations of the parties to the contract. Some owners and some architect/engineers have earned reputations such that reputable contractors will not bid on their projects. Others have reputations that even attract bidders who would pass up similar work in other jurisdictions. Conversely, some contractors have earned reputations that invite contract administration "by the book," whereas others enjoy the ability to secure many contract modifications by negotiation. The risk of an unfavorable reputation (or the benefit of a favorable one) is earned by all parties over a long period. It is not allocable, and it is not rapidly changed.

Minimizing Risks and Mitigating Losses[5]

The provisions and methods used in allocating risks should be clear and straightforward enough so that all of the parties know in advance what risks they have assumed, how they will be compensated, and that they can monitor the process. Otherwise, the owner may lose the benefit of the allocation and may even end up paying for the risk twice.

In the allocation of risks, it is important not to discourage designer innovation or production of ultraconservative, defensive designs. Designers cannot innovate if placed in a position where the amount of their fee does not cover their risks, unless the owner will protect them as a means of encouraging new concepts.

The concept of minimizing risks and mitigating losses can be implemented initially by the adoption of a set of management policy positions that are vital to the success of the program. Whereas any one policy item may in itself appear to be somewhat insignificant, collectively they can save a company a considerable amount of trouble.

---

[5]Adapted from Henry J. Jacoby (Chairman, Grow Tunneling Corporation, New York, NY), *Summary Session,* and David G. Hammond, *Minimizing Risks and Mitigating Losses,* presented at the ASCE Specialty Conference on "Construction Risks and Liability Sharing," Scottsdale, AZ, January 24–26, 1979.

The following principles are essential to an effective risk management program:

1. Thorough engineering, with competent advance planning to minimize the delays
2. Adequate subsurface exploration and interpretation of the data as it affects both design and construction
3. Full disclosure of all information available to the owner and the designer to the contractor
4. Permits and rights-of-way obtained in advance by the owner
5. Strong, competent management by all parties
6. On-site decision-making ability to minimize delays
7. Adequate procedures for handling disputes promptly, as the work progresses
8. Adequate financial security of all parties
9. Contractor participation during design phase, or for public agency contracts, contractor value-engineering clauses in the contract with shared savings provision
10. Good labor contracts and conditions to improve productivity

No attempt was made in the foregoing list to assess the effect of regulatory agencies and the interagency disputes that delay and increase the cost of projects. These are recognized as major risks and are considered beyond the ability of the owner, architect/engineer, or contractor to control or minimize except through the legislative or political process.

Contractor Participation in Value Engineering

Value engineering by the contractor is a subject that is viewed as controversial by some, but the concept has potential benefits as a means of minimizing risks and cost overruns. Value engineering by the contractor (see Chapter 16) involves contractor proposals for changing construction methods or designs as a means of reducing project construction costs. Generally, such concepts involve a cost-sharing provision on any savings realized. Many designers resist such potential intrusions of the contractor onto their hallowed ground. They often feel that their design approach was the proper one in the first place, and it is a reflection upon their design ability if the contractor dares to question it. Nevertheless, it should be recognized that for most projects, the identity of the contractor will not be known at the time of design. Obviously, what an architect/engineer might call for or permit to be done by a contractor whose qualifications were known and trusted might be quite different from that which could be allowed for a contractor who is known to the designer only as the "low bidder" with bonding capability, but otherwise possesses unknown virtues.

Management Structures

Of major importance in the minimizing of risks and mitigating of losses is the management structure used for prosecution of the project. This includes all management, supervisory, and working levels of the owner, architect/engineer, construction manager, and the contractor.

Long-Lead Procurement

Before making a final decision about prepurchase of materials or equipment by the owner, the impact should be considered of the added risks that are thus assumed by the owner, such as latent defects, damage in transit, storage costs, compatibility problems, maintenance, and others. There can be several valid reasons for owner prepurchase, but saving money on the purchase price is not generally one of them. The contractor must still add a percentage for handling and installation, and, after the project is ready for testing and operation, if the prepurchased equipment does not work properly and substitutions or replacements must be ordered, the owner will become liable for the added cost of project slowdown and delays. Thus, instead of liability reduction, liability has only increased.

There are certain cases, of course, where it may be necessary to prepurchase equipment, such as in the case of long-lead, factory-fabricated items. Here again, the owner must consider the risks and costs and the ability to furnish the equipment to the contractor when it is needed, what handling and erection costs will be, and whether there is warehousing and maintenance capability available. As an alternative, on long-lead purchases the owner may include an assignment of contract provision in the construction contract. In this manner, the cost of the prepurchase can be included in the project cost as an allowance, and the successful contractor contractually agrees to accept assignment of the prepurchase contract as soon as the construction contract has been executed. In this manner, the contractor makes payment directly to the manufacturer or fabricator of the item at the prescribed time, expedites delivery, and arranges for delivery, unloading, storage, maintenance, and installation just as would have been done had the contractor personally initiated the order in the first place. This reduces the owner's risk and places it in the hands of the contractor who generally has more experience and ability in the handling of such transactions. The cost will be reflected in the contractor's bid just as if the purchase had been a sole source procurement in the construction contract.

Permits and Rights-of-Way

Another way to minimize losses is for the owner to obtain permits and rights-of-way in advance of construction. Some permits can often be better obtained by the owner than by the contractor. In addition, there will be some savings to the owner by not having the contractor's markup added to the task of obtaining them and paying the fees. Some permits can be better handled by the contractor, and some judgment must be exercised in the allocation of these responsibilities. Permits for things such as haul routes, disposal sites, and similar items are better handled by the contractor, who has a better understanding of the plan of work.

It is generally agreed by all that whenever possible, it is better to have the owner obtain rights-of-way in advance of construction and make them available to the contractor upon notice to proceed. Although this is the ideal arrangement, often the complete rights-of-way are not available at the time that the construction should start. Here the owner must weigh the somewhat known risk of delaying the contractor while the contract is in force until the rights-of-way become available for the contractor to start, knowing full well that there may be some negative effect on the contractor's efficiency of operation and therefore on the project cost to both the owner and the contractor.

Disputes

Even in the area of disputes there is still some leeway for cost-saving measures. After a difference of opinion has been expressed, adequate machinery must be put into action for the resolution of such problems. The *second worst* way to handle claims is to ignore them; the *worst* way is to allow them to go to litigation. If handled promptly and vigorously, most disputes can be resolved without their being permitted to degenerate into large problems affecting not only the cost to the project, but also the progress of the work. It is important to both the efficient progress of the work and the lowest cost to both the owner and the contractor in the performance of "changed conditions" work that such changes be negotiated and settled as soon as possible.

Disclosure of Information

It has been common until recent times for owners to keep design or site information in their possession to themselves, providing the contractor only with so-called "factual information." Only recently has this been recognized as not in the best interest of the owner or the architect/engineer. On the Baltimore Subway Project, for example, complete disclosure was made to the contractor in the bidding documents not only of factual data, such as boring logs and cores, but also of the designer's interpretation of how the ground or rock was expected to act during construction. A complete geotechnical report including both factual data and design analysis was made available to the bidders in the form of a geotechnical report. It is considered probable that the furnishing of such complete information enabled and forced the owner and engineer to do a more complete subsurface exploration and thus avoided some surprises. The evidence to date appears to support the premise that lower bids may be anticipated from bidders when this approach is taken. Low bids on this project averaged 10 percent below the engineer's estimate.

In contrast, on the Chicago TARP Project, where the initial philosophy was that all risks were to be assumed by the contractor, the contract did not provide any provisions such as those for changed conditions and the geotechnical information provided to the contractor was factual only, for which the owner further disclaimed any responsibility. Initial bids taken on early contracts resulted in few bids being submitted, and those that were submitted were in amounts more than double the engineer's estimates. Subsequent modifications in the philosophy in the contract documents resulted in an improvement in that situation.

PROBLEMS

1. True or false? All risk belongs to the owner unless transferred by contract.
2. Describe the two primary concepts involved in risk management.
3. What are principal issues that should be considered in the allocation and assumption of risk responsibility?
4. Name the six categories of risk.
5. In a construction contract, which is the best party to bear the various risks?
6. What are the four guidelines for allocation of risk in construction contracts?
7. Name one alternative to long-lead purchase and issuance to the contractor of products as "owner-furnished equipment."
8. Name four types of construction risks that are properly allocated solely to the contractor.
9. Name four types of risks that are properly retained solely by the owner.
10. Name four types of risks that are properly allocated solely to the engineer or architect.

12

PRECONSTRUCTION OPERATIONS

Description of Approach

On a project involving a field staff of two or more persons, the organizational structure of the Resident Project Representative's field office is often determined by a design or construction management firm as the representative of the owner of the project. It should be clearly understood before work on the project begins that all assignments and limits of authority and responsibility are as delegated to the Resident Project Representative by his or her employer. On larger projects, it might be desirable to draft an organizational diagram or chart that clearly defines all levels of responsibility and authority on the proposed project. Such a chart can be of immeasurable help in expediting the work when new tasks must be done and the normal procedures are not applicable.

Preconstruction operations can generally be grouped into five phases:

1. Advertise and award phase (including prebid meeting and site visitation)
2. Development of quality control program or construction surveillance and inspection plan on CQC projects
3. Field office organization phase (planning)
4. Preconstruction conference
5. Establishment of a field office at the construction site for the administration and quality assurance of the work for the owner

Frequently, the Resident Project Representative is not involved in the project during the advertise and award phase at all; however, this phase of the work will be covered in this chapter for the benefit of those whose obligations do include this phase of the work. In cases where that obligation is not included, the Resident Project Representative's tasks may be limited to the performance of items 3, 4, and 5 only of the foregoing list.

Constructability Analysis

Often overlooked, except by some large engineering organizations such as the Army Corps of Engineers, a constructability analysis is an essential element in the plan to provide competent plans and specifications and avoid or, at the very least, minimize the owner's exposure to preventable claims.

Although the value of a constructability analysis was discovered by the rest of the construction industry rather belatedly, constructability is evolving as one of the most significant enhancement opportunities in the construction industry. Although practiced for many years by a few organizations, only recently has it come to focus as offering major benefits in construction cost and schedule. The Business Roundtable's Construction Industry Cost Effectiveness (CICE) Project identified constructability as an area with a major potential for improved cost-effectiveness.

There is no single method to achieve the desired results. The type of contract in which to implement constructability is on turnkey (design–build) projects; it is much more difficult to do on projects where design and construction are accomplished by distinct and separate contracts. Each discipline seems to resent the intrusion of the other onto what it considers as its "turf."

If constructability is to work, there must be a bridge between the traditional separation of design and construction. This is sometimes accomplished through the involvement of Professional Construction Management firms, which often possess these areas of expertise, and, when acting as direct representatives of the owner, have the clout to carry it through with the design organizations.

As proposed by the Business Roundtable, systems were examined that have proven to be the best for implementing constructability. Company or project size is no barrier to constructability, as it is equally valuable to both large and small organizations. However, some smaller companies have elected to combine some constructability functions.

One of the most effective means of achieving involvement of constructability specialists into the planning and design process is to bring in the construction discipline as a component of the project team, starting the first day. In order to be effective, the construction discipline must be accorded full participation in the planning design sessions.

One of the biggest obstacles to good constructability is the *review syndrome.* This occurs when construction personnel are excluded from the planning process and are invited only to "review" completed or partially completed work received from the designers. This prevents construction knowledge and expertise input in the early planning, when cost sensitivity is at a maximum and construction should be making its most important contributions.

When constructability is approached solely on a review basis, it inevitably becomes inefficient and ineffective. The designer becomes defensive because he or she has already committed publicly on the drawings and specifications being reviewed, and the reviewer is reluctant to comment lest he or she be perceived as being overly critical.

The most effective approach is to place the entire constructability team into active roles in an integrated planning and design development process. In that environment, the whole array of alternatives is discussed up front and jointly evaluated. One excellent technique is to hold a series of project constructability brainstorming sessions as the first project activity. Analyses and trade-offs can be made before the project design is finalized.

Advertise and Award Phase

During the advertising phase, the Resident Project Representative should review the contract documents carefully to make certain that all important field considerations have been provided for in the specifications. If, during the award phase, omissions are noted in the plans or specifications or if conditions are specified that may create conflicts in the field, there is still time to provide written notice to the design firm of any such omissions or conflicts so that it can issue an addendum to the specifications prior to bid opening time. Thus the problem can be corrected in time to eliminate the need of a change order during construction.

An example of such omission might be the failure of the design firm to specify that the contractor shall furnish a field office for the Resident Project Representative's use throughout the life of the project, unless the design firm or owner intends to provide such facilities. The omission of a requirement for a field office in the specifications was once the cause of the author spending a long, uncomfortable year at a project site in a prefabricated tool shed that the contractor was talked into providing as a field office at no additional cost. Anyone who has ever had this experience will realize its shortcomings. In rainy weather the choice was to close the window flaps and keep dry—in the dark, or to leave the flaps open and get plenty of light—and get wet. It was hot in the summer and freezing in the winter, and the dirt was often so thick that the floor was obscured.

Another thing to look for in the specifications is a requirement for a field telephone. If a free telephone is not specified, the contractor will not be obligated to furnish a field telephone, except at added cost to the owner. After all, if a telephone was not in the specifications, there is no valid reason for the contractor to have included the cost of a free one for the Resident Project Representative in the bid. If a pay phone is installed, the Resident Project Representative will have to keep a pocket full of change at all times to be capable of communicating with the home office. The ideal approach is to require that the contractor provide, maintain, and pay for, for the full term of the project, a field telephone in the Resident Project Representative's field office as well as in the contractor's field office. This is fair to the contractor as well as to the owner, for it allows a fixed telephone cost to be bid as part of the proposal. This telephone should also be specified to be connected to an established exchange for toll service and with all other phones that may be used by the contractor. This may sound like an unnecessary precaution, but consider the fix the author found himself in some time ago when the telephone in the field office was part of a private telephone system of the contractor and *could not be connected* to a regular public telephone service for local or toll calls! It is equally important to see that the telephones provided in the contractor's field office and in the Resident Project Representative's field office be on separate trunklines. A party-line or extension telephone, including key phones that allow selection of both the contractor's and the inspector's trunkline, is undesirable, as it allows no security of communications for either party.

An alternative, of course, might be a cell phone, but that has disadvantages if a land-line telephone is not also provided.

Naturally, the mention of toll service may scare the contractor unless some prearranged agreement has been reached concerning the use of the telephone for long-distance calls. It is recommended that the contractor permit the Resident Project Representative, the design firm, the owner, or any of their authorized representatives to use the telephone without cost for all calls that do not involve published toll charges. Calls that do involve toll charges should then be billed to the owner by the contractor *at the actual rate charged* by the telephone company.

There are many other points to watch for in the specifications; many of them involve technical matters. In all cases, if a description of the problem can be submitted to the design firm in time, and if they agree with the inspector's recommendations, this item can become a part of the contract requirements before bid-opening time.

Prior to bid-opening time, the Resident Project Representative should meet with the project manager to develop an agenda and a list of key subjects to be discussed at the preconstruction conference. In addition, the project manager should contact the owner to see if assistance will be wanted during the preparations for and during the bid opening.

During the bid-holding period following the opening of all bids, but prior to determination of the award of the construction contract, the Resident Project Representative should be available, if requested by the design firm or the owner, to assist in the evaluation of bid data, costs, and other contractually significant items. In this manner, the architect's or engineer's job of making recommendations to the owner for award of the contract can be made much simpler and easier.

Advertise and Award Scheduling

While usually considerable flexibility exists for scheduling the bid solicitation and award period, that is not true of public works contracts. In a city, for example, coordination must be maintained between the engineering department, the reproduction department, the city clerk's office, the advertising media, and in some cases the procurement department in order to meet a strict timetable (Figure 12.1).

Several activities must be scheduled to occur at a specified time to meet legal constraints. In California, for example, all projects must be advertised in a newspaper of general circulation (not a trade publication) for not less than two times and at least five days apart. This results in a minimum bid period of two weeks.

After opening bids, and before award, a council report must be prepared by the engineer for inclusion in the city clerk's agenda a week in advance of the regular city council meeting. Many cities have two city council meetings each month, such as the second and fourth Tuesday of each month. Award of the contract can only be done at a council meeting, so everything must run like clockwork. An example is shown in Figure 12.1 of one city's timetable. Miss one date and the project award is delayed at least two weeks.

CITY OF THOUSAND OAKS, CALIFORNIA
DEPARTMENT OF PUBLIC WORKS
ENGINEERING DIVISION

ADVERTISING AND AWARD SCHEDULE

Project Name_______________

C.I.P. No._______________

DATE	ACTIVITY	ELAPSED TIME (DAYS)		DAY
		MINIMUM	ACTUAL	
______	Engineering Review	0	______	Any
______	Obtain Bid Opening Dates from City	1	______	Any
______	Send Specs to print shop	2	______	Any
______	Send Notice Inviting Bids to Plan Rooms	7	______	Any
______	Plans available at counter	9	______	Any
______	1st Advertisement appears	20	______	Any
______	2nd Advertisement appears	27	______	Any
______	Schedule Prebid Meeting (optional)	31	______	Any
______	Open Bids	37	______	WED
______	City Clerk Agenda for Council Meeting	39	______	FRI
______	Prepare Council Report for project	*	______	
______	Award by City Council	50	______	TUE
______	Execute Agreement	65	______	Any

* Get on City Clerk's Agenda 2nd Friday before the next City Council Meeting
City Council meets on 1st and 3rd Tuesday of each month

Times shown are approximate minimum based on the day of the week and NO intervening holidays. "Any" refers to any weekday

FIGURE 12.1 Example of a Public Works Advertising and Award Schedule for a California City.

ISSUANCE OF BIDDING DOCUMENTS

At first, the subject of issuing bidding documents (plans and specifications and related materials) seems like a minor one. Yet if not properly handled, particularly on public works projects, serious claims or project delays can result. The following summarizes just a few of the more important details that must be handled during the bidding phase of a project:

1. After advertising for bids, keep an accurate log of all sets of contract documents issued, the firm name and legal address of each bidder, the name and position of the party picking up the documents for the firm, the date picked up, the amount of fee or deposit received and whether it was in the form of cash or check, the number of sets picked up by each bidder, and if the sets are numbered, the serial numbers of the sets delivered to each bidder. This log, in the form of a preprinted form, can be kept by a receptionist, and construction documents can be issued at the front desk. One recommended format is shown in Figure 12.2.
2. Whenever there are changes in the drawings that are issued as an addendum to the bidding documents, a copy of the addendum should be mailed to each holder of a set of contract documents. Where a single bidder has obtained more than one set of documents, a copy of the addendum should be sent for each set obtained. All such addenda should be sent to each bidder by certified mail, with a "return receipt requested," and should be addressed to each bidder's legal address (preferably *not* a post office box). Meanwhile, all as-yet-undistributed sets of documents should be brought up to date simply by including a copy of the addendum with each set so that all bidders picking up sets of documents after issuance of an addendum will receive theirs along with the initial issue.
3. At the time that a project is advertised for bidding and the estimated number of document sets have been reproduced, all project drawings should be sent out to have reproducible copies made so that there will be a permanent record of the set as originally issued, and in case of a sellout on all preprinted sets of documents, additional identical sets can still be reproduced, even though later changes may already have been posted on the original transparencies. In this manner, if later changes are made to the drawings and you run out of bid document sets, another set of drawings *in their original form* can still be printed.

It is of vital importance that all bidders receive *identical* sets of bidding documents. It is not proper to issue later drawings where changes covered in previous addenda have already been posted, as charges of bidding irregularity may well be made and sustained. If this were to happen, the first group of bidders would have received the old drawings with a copy of an addendum that instructs the bidders as to what changes are to be made—and the bidders would have to post them on their own drawings. This means that the addendum was subject to the bidder's interpretation. However, if a bidder picked up an updated set of drawings later, the later bidder would have received a set of documents upon which the architect/engineer had posted the corrections, and thus the addendum would have been subject to the architect/engineer's interpretation—an obvious irregularity. Not only that, but problems in drafting errors can cause difficulty, and often do. When the architect/engineer posts all corrections, as on the later sets of drawings referred to, a drafting error could easily be the cause of the later set of drawings being not even technically the same as the earlier set plus its addendum.

GLEASON, PEACOCK & FISK, INC.
SUBSIDIARY OF CONSTRUCTION CONSULTING GROUP, INC.
OLEN POINTE BREA, THREE POINTE DRIVE
SUITE 103 · BREA, CALIFORNIA, 92821-3845
(714) 256-1922 · FAX (714) 256-1920

RECORD OF BID DOCUMENTS

Project Manager: G. F. Albrecht   Phone ext.: 544

Project Name: Parker WWTP Expansion   Project No.: 3-704   Am't req'd per set $ 50.00   fee [x] deposit [ ]   Must sets be returned? No

DATE ISSUED	FIRM NAME & ADDRESS OF RECIPIENT (Legal Address of Firm · Street Address)	NAME OF INDIVIDUAL RECIPIENT TELEPHONE NO. OF FIRM (Include Area Codes)	FEE OR DEPOSIT					DOCUMENTS		ADDENDA ISSUED				
			AMOUNT	Check	Cash	DATE REC'D	DATE RET'D	NO. SETS	SET NUMBERS	1 / 6	2 / 7	3 / 8	4 / 9	5 / 10
3/3/80	LMN Contractors 1000 Argus St. Centerville, MO 63100	G. F. Edwards (314) 000-0000	50 –	x		3/3		1	001	✓	✓			
3/3/80	ABC Construction, Inc. 1000 N. Tustin Orange, CA 92600	L. M. Ciabottoni (714) 000-0000	150 –		x	3/3		3	002 003 004	✓	✓			
3/3/80	T & G Paving 2765 Columbine Ave. Suburbia, MO 63100	H. F. Jorgensen (816) 000-0000	50 –	x		3/2		1	005	✓	✓			
3/4/80	Concrete Specialties, Inc. 50 Samsone St. Centerville, MO 63100	H. Bedrosian (314) 000-0000	50 –	x		3/3		1	006	✓	✓			
3/5/80	Emerson & Farnsworth 1000 Any Street Dallas, TX	Geo. F. Carson (214) 000-0000	100 –			3/2		2	007 008	✓	✓			

CM 101

FIGURE 12.2 Example of a Record of Bid Documents. *(Fisk, Edward R.,* Construction Engineer's Complete Handbook of Forms, *1st edition, © 1993. Reprinted by permission of Pearson Education, Inc., Upper Saddle River, NJ.)*

Such irregularities can be sufficient cause to have all bids set aside and the entire project forced to be rebid. This cannot only lead to costly delays in getting a project started, but might also influence the amounts of each bid after each bidder has had the opportunity for full disclosure of the amount of each of the other bids. Frequently, this is not to the financial advantage of the owner.

An additional option is to have the bid documents prepared by the owner or architect/engineer posted and made available on the Web site so that all interested contractors can download the bid forms and view the drawings and specifications on line. Additionally, provisions might be made to allow the bidders to order either CD copies or paper prints.

Prequalification of Bidders

Some states provide for prequalification of bidder on public contracts. In fact, in Massachusetts a court held that "bidder prequalification is not mere formality; it is a cornerstone of the competitive bidding statute," and ruled that the statute must be construed strictly [*Modern Continental Construction Co. v. City of Lowell,* 465 N.E.2d 1173 (Mass. 1984)].

Several states employ prequalification requirements for public works bidding. These include California [Govt Code §14310 et seq (1980)]; Colorado [Colo. Rev. Stat. § 24-92-107 (1982)]; Delaware [Del. Code Ann. tit. 29, §6905 (1983)]; Hawaii [Hawaii Rev. Stat. § 103-25 (1976)]; Indiana [Ind. Code § 5-16-1-2 (1974)]; Maine [Me. Rev. Stat. Ann. tit. 5, §1747 (1979)]; Massachusetts [23 B.C.L.Rev. 1357 (Mass. 1982)]; Michigan [Mich. Comp. Laws Ann. §123.501 (1967)]; Nebraska [Neb. Rev. Stat. §73-102 (1981)]; Pennsylvania [Pa. Stat. Ann. tit. 71, §642 (1962)]; Virginia [Code of Va. §11-46 (1984 Supp.)]; and Wisconsin [Wis. Stat. Ann. §66.29 (1965)].

Bonds

Bid Bonds

When an owner decides that some security is needed to protect it against the financial disadvantages that may occur because a bidder later refused or was unable to sign an agreement to construct the work after submitting a bid, there are a couple of alternatives open. In any case, the bid security should amount to at least 5 percent and is often 10 percent of the contractor's bid price. Bid bonds, certified checks, cashier's checks, and sometimes negotiable securities may be accepted as bid security. The security assures the owner that if the contract is awarded to the successful bidder, that bidder will enter into contract with the owner. After the bids are opened and an award made, bid securities are returned to all bidders. If, however, the selected bidder cannot or will not enter into contract, the bid security is forfeited, and award is made to the next bidder in line on the same basis.

The bid bond is far preferable to other securities, as it is less of a burden upon the contractor. Bid bonds should preferably be in an amount of 10 percent of the bid price for the project.

A surety company makes only a nominal charge to a contractor for a bid bond, and although writing one does not commit the surety to write a performance and payment bond later, it is a good indication that it is willing to do so. Bid bonds should be of the "forfeiture" type that gives the owner a direct right of action under the bond.

Although optional on private contracts, all public works contracts require the contractor to post a bid bond as security with the bid. Under the terms of a bid bond, the surety company agrees to pay a stipulated sum of money to the owner, as a forfeit, if the bidder is awarded the contract and fails to enter into an agreement with the owner to construct the work.

Performance and Payment Bonds

After award of the contract, the contractor is required to provide performance and payment (labor and materials) bonds on all public works contracts. Although this is not common in private works, and is actually actively opposed by some groups, there is a current move toward the utilization of bonds on private works as a hedge against potential lien claims and failure to complete the work.

Under a performance bond, the surety has an obligation to the owner for any additional costs to complete the contract due to the contractor's failure to comply with its contract requirements. The most common reason for a contractor not completing a contract is insolvency. Therefore, sureties are interested in the contractor's financial condition as well as other qualifications before writing a performance bond. Generally, a contractor's ability to take a large contract is a function of bonding capacity, as the bonding company will only risk a slight increase over previous bond amounts. In this manner, the road from being a small contractor to becoming a large contractor doing $20 million projects is a long, slow, step-by-step process with the bonding company.

Under a payment bond, the surety guarantees the payment of all legitimate labor and materials bills that result from the performance of the contract. The surety has an obligation to the owner for the additional costs that are the result of failure of a contractor to pay the labor and materials bills due to the performance of work on the contract.

Combination payment and performance bonds, which include in one instrument the obligation for both the performance of the contract and the payment of laborers and material suppliers, have resulted in difficulties and delays in the handling of claims. It is recommended that where bonds are to be provided, they should be separate bonds issued by the surety as a "package," for which no additional charge is made. Under the two-bond system, the surety is enabled to make payment without awaiting a determination as to owner's priority. The customary amount of public works bonds the author has seen are in an amount of 100 percent on performance bonds and 50 percent on payment bonds. However, the Construction Industry Affairs Committee of Chicago, with membership spanning both the design profession and the contractor associations, recommends that both the payment and performance bonds be written in the amount of 100 percent of the contract price.

When prime contractors require a surety bond from their subcontractors, the prime contractor's position is similar to that of an owner. Prime contractors should be careful to obtain bonds from their subcontractors that are of the same form and not less than the guarantee that the prime is giving the owner under the owner's own bond.

Where awards can be made to prequalified contractors, such as on private work, surety bonds might be eliminated if the financial stability and record of performance of the contractor are known to be satisfactory. Performance bonds and payment bonds are not guarantees of trouble-free jobs but do protect the owner from additional costs due to the contractor's failure to complete a contract or to pay bills.

Time of Submittal of Bonds

Under the 5-step project initiation process described in Chapter 1, a bid bond must be submitted at the time of submitting bids (item 2 in Figure 12.3). Upon being awarded the contract, the successful bidder is obligated to enter into contract with the owner for construction of the work. If the awardee fails to enter into contract with the owner, the bid bond is forfeited, and the owner will then award to the next-lowest bidder.

On public works projects, after award of the contract, a bidder who enters into a contract with the owner must submit a performance bond and a payment bond to the owner. These should be submitted to the owner at the time of submitting the signed agreement form to the owner for its signature (item 4 in Figure 12.3). Upon receipt of the signed agreement from the successful bidder, the owner releases all remaining bidders' bonds and affixes its own signature to the agreement. Then, subsequent to execution of the agreement by both parties, a Notice to Proceed may be issued (item 5 in Figure 12.3).

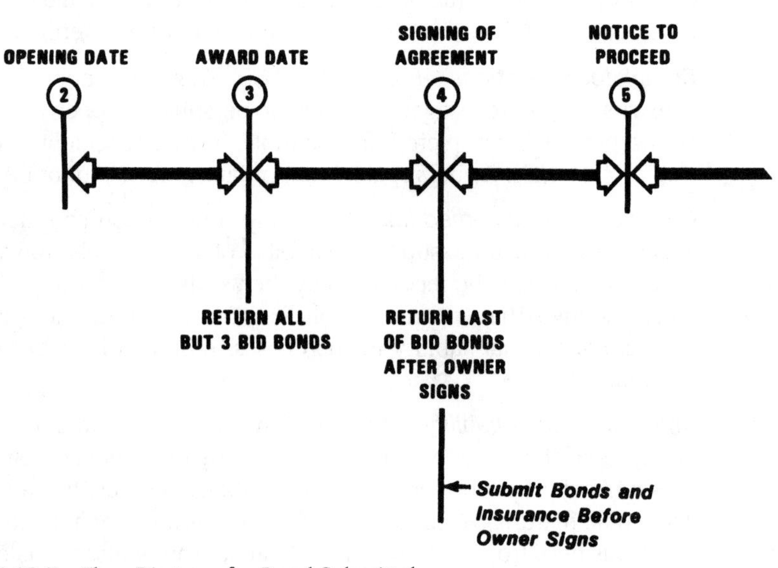

FIGURE 12.3 Flow Diagram for Bond Submittal.

Liability Forms of Insurance

Comprehensive General (Public) Liability Insurance

This type of insurance protects against legal liability to the public. There are many forms of liability insurance, but the one usually recommended for construction is the Broad Form Comprehensive Liability Policy (automobile included). Under this type, all forms of liability insurance are combined in one contract. Physical damage may also be included on all owned automobiles.

The following forms of liability insurance may or may not be included in the comprehensive form. If not, separate policies may be arranged.

Premises—Operations. This coverage protects against the legal liability for bodily injury to persons other than employees, and damage to the property of others that is not in the contractor's care, custody, or control.

Exclusions should be checked carefully. For example, explosion, collapse, and underground damage are normally excluded from coverage under the basic policy for most types of work. These are usually designated exclusions "x," "c," and "u," respectively. Although in some cases these exclusions cannot be removed, each project should be carefully examined for exposure to these hazards, and coverage secured where possible and necessary.

Personal injury. This protects against legal liability for claims arising from false arrest, libel, wrongful entry or eviction, and related wrongs against a person. A check of the policy will reveal exclusions.

Independent contractors—Protective liability. Coverage provides for the insured's legal liability that may arise from acts or operations of a subcontractor or its employees and damage to property of others if that property is not in the care, custody, or control of the contractor. Certificates of insurance should be secured from all subcontractors and the scope of their insurance coverage verified.

Completed operations—Product liability. This is an optional coverage that, subject to exclusions, protects against liability to persons or property of others that may arise after a project is completed; for example, from an accident due to faulty workmanship or materials. Actual replacement of faulty work cannot be covered.

Contractual or assumed liability. Many construction contracts include a clause in which the contractor assumes the liability of someone else toward third parties. Such clauses can usually be recognized by the words "hold harmless" or "indemnify." It is recommended that insurance policies of contractors and subcontractors provide blanket contractual liability in such cases. Assumed liabilities can then be covered automatically.

Umbrella excess liability. This insurance provides catastrophe coverage for claims in excess of the limits of liability afforded by other policies, and also for some hazards normally excluded in underlying liability policies. It may be subject to a large deductible feature, $10,000 to $25,000, and is a reasonably inexpensive way to protect a business from claims that could arise from a disastrous occurrence.

Automobile. All liability from existence or operation of any owned, hired, or nonowned vehicle may be included in this provision. This insurance should be on an automatic basis to provide coverage of newly added equipment. A special endorsement may be needed if employees or their families use company cars. A "use of other car" endorsement, naming each person so protected, would provide coverage for individuals using cars not owned by them or their employer.

Automobile medical payments. This may be added to a policy. In some states it is required.

Automobile physical damage. This covers damage to property of others and may also be endorsed to include comprehensive and collision coverage on owned vehicles. A high deductible of $250 to $500 or more can result in a considerable savings in premiums.

Property Forms of Insurance

Standard Builder's Risk Insurance

This type of coverage protects against physical damage to the insured property during the construction period resulting from any of the perils named in the policy. This coverage provides reimbursement based upon actual loss or damage rather than any legal liability that may be incurred. There are three principal methods used to establish amounts of coverage and to determine the premium.

1. *Completed Value.* This method is based upon the assumption that the value of a project increases at a constant rate during the course of construction. While the policy is written for the value of the completed project, the premium is based upon a reduced or average value. The dollar coverage provided is the actual value of completed work and stored materials at any given time. This form of builder's risk must be taken out at the start of construction. It is recommended that this method be used for the typical building project.

   During construction the contractor must notify the insurance carrier periodically of the increase in value of a project. Coverage and premiums are based upon the reported value. This method is advantageous where completed value is low during most of the construction period but increases very rapidly toward the end. However, failure to report an increase in value may result in lack of proper coverage.
2. *Automatic Builder's Risk.* This policy form gives a contractor temporary protection automatically, pending the issuance of a specific policy for each project.
3. *Ordinary Builder's Risk.* This seldom-used type of policy form is written for a fixed value. Coverage may be increased by endorsement at the request of the insured.

Within the framework of the preceding methods of writing a builder's policy, the following perils may be covered:

1. Fire and lightning.
2. Extended coverage—covers windstorm, hail, riot, civil commotion, nonowned aircraft, smoke, and explosion (other than that from boilers, machinery, or piping).
3. Vandalism and malicious mischief—excludes pilferage, burglary, larceny, theft, and damage to glass (other than glass block).
4. Additional perils—The standard builder's risk policy may be endorsed to provide for specific additional perils. These may include collapse (not caused by design error, faulty materials, or workmanship); landslide; groundwater; surface water (other than flood); sprinkler leakage; explosion or rupture of boilers, machinery, or piping; breakage of glass; pilferage; and theft.

It may be possible to obtain endorsements or separate policies to cover perils of flood and earthquake; however, such coverage can be difficult to obtain. For these perils it is recommended that the contractor request the project owner to secure this coverage in conjunction with its permanent insurance for the completed structure.

Some of these additional perils can be covered by adding an "all-risk" type of endorsement, or a multiple-peril builder's risk policy (as described in the next section) may be preferred. These additional coverages generally require a deductible clause.

Multiple-Peril (All-Risk) Builder's Risk Insurance

This is a nonstandard type of policy that provides similar but broader coverage than the standard builder's risk policy. Although the name "all risk" is widely used, all perils are *not* covered—it is a relative term denoting the broader-than-usual coverage. Generally, each insurance carrier writes its own form of multiple peril.

Rather than naming the perils insured, this type of policy insures against all risks of direct physical loss or damage to property from any external cause *except* those specifically excluded in the policy. Thus, coverage is determined by what is excluded, not included, and the policy must be checked closely to determine the coverage provided. Some forms will require a deductible clause for some of the perils covered.

This type of policy (written on a completed-value basis) is recommended over a standard builder's risk policy *if* sufficient care is taken at the outset to make sure that all desirable coverage is included. Protection can be tailored to an individual contractor's needs.

Submittal of Evidence of Insurance

The contract documents should require that evidence of specified insurance be submitted at the time of submitting the performance and payment bonds (item 4 in Figure 12.3). This should be in the form that will disclose the total policy coverage (Figure 12.4), not merely a statement certifying that coverage was obtained by the contractor. During the submittal phase, the project manager should maintain a Bond and Insurance Record such as that illustrated in Figure 12.5

MINIMUM INSURANCE REQUIREMENTS

TYPE OF COVERAGE		LIMITS OF LIABILITY	
Comprehensive General Liability			
	Bodily Injury	$300,000 Each Occurrence	
	Property Damage	$100,000 Each Occurrence	$100,000 Aggregate
Comprehensive Automobile Liability			
	Bodily Injury	$100,000 Each Person	$300,000 Each Occurrence
	Property Damage	$50,000 Each Occurrence	
Excess Liability		$1,000,000 Combined Single Limit	
Workers' Compensation		Statutory	
and **Employer's Liability**		$100,000	

Comprehensive General Liability — Including:

Coverage for damage caused by blasting, collapse, underground damage or explosion:

Independent Contractors:

Products, completed operations:

Contractual Public Liability and Property Damage covering liability assumed in the contract between the Insured and the owners/agents designated in the contract.

Comprehensive Automobile Liability — Including:
Coverage for all owned, hired or non-owned licensed automotive equipment. Coverage to comply with Motor Carrier Act of 1980.

Excess Liability — Including Employer's Liability

See Certificate On Reverse Side

REVISED 1-2-84

FIGURE 12.4a Examples of a Certificate of Insurance Showing Policy Coverage.

FORM NO 4500-0008

Certificate of Insurance

THIS CERTIFICATE IS ISSUED TO THE OWNERS/AGENTS LISTED BELOW DESIGNATED IN THE CONTRACT WITH THE INSURED

Texas Utilities Company
Its Subsidiaries:
Texas Utilities Services Inc.
Texas Utilities Mining Co.
Texas Utilities Fuel Co.
Basic Resources Inc.

Texas Utilities Electric Company
Its Divisions:
Dallas Power & Light Co.
Texas Electric Service Co.
Texas Power & Light Co.
Texas Utilities Generating Co.

NAME AND ADDRESS OF INSURED	INSURANCE COMPANIES AFFORDING COVERAGE
	COMPANY A
	B
	C
	D

TYPE OF WORK PERFORMED & LOCATION

COMPANY LETTER	TYPE OF INSURANCE	POLICY NO	POLICY EXP DATE	Limits of Liability in Thousands (000)	EACH OCCURRENCE	AGGREGATE
	COMPREHENSIVE GENERAL LIABILITY Including: ☐ EXPLOSION AND COLLAPSE ☐ UNDERGROUND DAMAGE ☐ PRODUCTS/COMPLETED OPERATIONS ☐ CONTRACTUAL INSURANCE ☐ BROAD FORM PROPERTY DAMAGE ☐ INDEPENDENT CONTRACTORS ☐ PERSONAL INJURY			BODILY INJURY	$	$
				PROPERTY DAMAGE or	$	$
				BODILY INJURY AND PROPERTY DAMAGE COMBINED	$	$
				PERSONAL INJURY	$	
	COMPREHENSIVE AUTOMOBILE LIABILITY Including: ☐ OWNED ☐ HIRED ☐ NON-OWNED ☐ MOTOR CARRIER ACT			BODILY INJURY EACH PERSON	$	
				EACH ACCIDENT	$	
				PROPERTY DAMAGE or	$	
				BODILY INJURY AND PROPERTY DAMAGE COMBINED	$	
	EXCESS LIABILITY Including: ☐ EMPLOYER'S LIABILITY			BODILY INJURY AND PROPERTY DAMAGE COMBINED	$	
	WORKERS' COMPENSATION and EMPLOYER'S LIABILITY Including: ☐ LONG SHOREMEN'S AND HARBOR WORKERS			STATUTORY		
				E L	$	(EACH ACCIDENT)
	OTHER					

The undersigned certifies that he is the representative of the above listed insurance companies, that he has authority to execute and issue this certificate to Certificate Holder, and, accordingly, does hereby certify on behalf of said insurance companies that policies of insurance listed above have been issued to the insured named above and are in force at this time. Notwithstanding any requirement, term or condition of any contract or other document with respect to which this certificate may be issued or may pertain, the insurance afforded by the policies described herein is subject to all the terms, exclusions and conditions of such policies. Copies of the policies shown will be furnished to the certificate holder upon request.

This Certificate does not amend, extend or alter the coverage afforded by the policies listed.

Cancellation: Should any of the above described policies be cancelled before the expiration date thereof, the issuing company will mail 30 days written notice to the below named certificate holder.

NAME & ADDRESS OF CERTIFICATE HOLDER

DATE ISSUED ____________

BY ____________
AUTHORIZED REPRESENTATIVE OF INSURANCE COMPANIES AFFORDING COVERAGE

____
ADDRESS

REVISED 1-2-84

FIGURE 12.4b Reverse Side of Certificate of Insurance.

BOND AND INSURANCE SUBMITTAL

PROJECT NO. 99-15 CONTRACT NO. NSY-15309

PROJECT 15 MGD WATER TREATMENT PLANT

OWNER CITY OF FAIRVIEW

ENGINEER R. E. BARNES, P.E.

CONTRACTOR ABC CONSTRUCTORS, INC.

INSTRUMENT	COMPANY	AMOUNT	EXP. DATE
PERFORMANCE BOND	COLONIAL INDEMNITY	100%	10-5-01
LABOR AND MATERIAL BOND	COLONIAL INDEMNITY	100%	10-5-01
MAINTENANCE AND GUARANTEE BOND	COLONIAL INDEMNITY	10%	10-5-01
PL AND PD INSURANCE	AETNA CASUALTY & SURETY	PER SECTION 8.02	10-5-01
WORKERS' COMPENSATION INS.	AETNA CASUALTY & SURETY	PER SECTION 6.05(3)	10-5-01
BUILDER'S RISK INSURANCE	NOT REQUIRED		
WRAP-UP INSURANCE			

Data used are fictitious for illustration only

DISTRIBUTION:
1. Owner
2. Project Engineer
3. Resident Project Representative
4. Project Files

Fisk Form 6-6

FIGURE 12.5 Example of a Bond and Insurance Submittal Record.

OPENING, ACCEPTANCE, AND DOCUMENTATION OF BIDS

In the administration of private contracts, where informality may be the preferred or at least accepted choice of the architect/engineer, no rigid procedures are necessary, nor, for that matter, is the issue of bid irregularity or responsiveness a subject of critical concern unless desired by the owner or architect/engineer.

The administration of a contract for a public agency, however, whether it be by the public agency itself or a private architect/engineering firm under contract to the agency, does invite a number of somewhat inflexible rules for the opening, acceptance, and documentation of all bids received.

Several primary matters of concern during this phase include:

1. Receipt of sealed bids at the designated time and place
2. Confirmation that all bids are responsive
3. Acceptance and logging of each bidder's name and the bid amount for all responsive bids
4. Summary of all line-item prices for all unit-price bids of responsive bidders

Bid Shopping or Bid Peddling

It is sometimes the custom in the construction industry, where there are no protective laws to shield a subcontractor from the unfair practice of bid shopping, for the prime contractor, after being notified of selection, to shop around to other potential subcontractors and offer to substitute them in the general contractor's bid if they will underbid the subcontractors originally used in determining the bid price. Where local laws prohibit this bid shopping, or "bid peddling" as it is sometimes called, the custom is varied so that the prime contractor may withhold submittal of its bid until the very last moment allowable before the closing of the bid acceptance period. The contractor, having previously been provided with a copy of the bid with penciled-in prices, will often spend the last few moments of time before bid opening in an attempt to negotiate last-minute price concessions from the subcontractors and suppliers, or even solicit new subcontractors or suppliers to replace those who originally quoted the job. Then, just before closing of the bid acceptance period, the contractor will ink-in the final prices, insert the bid in its envelope, seal it, and submit it. All details of the contractor's original bid will have been carefully prepared in the office before arriving at the bid opening, and only a minor amount of work is involved in any last-minute changes. All that needs to be done is to replace the originally penciled-in figures with the latest figures—in ink—and compute the new bottom-line price before submitting the bid.

Bids Must Be Responsive

A responsive bid is one that meets all of the requirements specified in the Notice Inviting Bids and the Instructions to Bidders. To be responsive, a bid must be submitted on time, including all required documents; all forms must be completely and properly filled out, signed by a responsible party, and dated; any required bid bond must be executed on the prescribed form and attached to the bid; on unit-price bids, all line items must be filled out; no changes or alterations of any of the documents are permitted; no changes or conditions may be attached to the bid; and all addenda must be properly acknowledged.

In addition, where required by the terms of the bid solicitation, the documents may be required to be filled out and left bound in the bound volume of specifications for submittal to merit consideration. Also, in some cases, instead of a simple written acknowledgment that all addenda have been received, a bidder may be asked to submit the actual copies of all addenda received. One reason for requiring that bids be executed in the bound set of specifications and the original addenda returned with the contractor's bid is that many public agency owners prefer to assemble a legal record bid set by drilling a hole through the corner of all bid documents and inserting a wire with a lead seal. Often, where this procedure is used, the owner will also require that one of the original agreements be executed on the documents bound in the same specifications copy. Normally, when this system is used, additional loose sets of bidding documents and agreement forms are provided for additional originals.

Acceptance and Recording of Bids

Prior to opening the bids, the architect/engineer or owner should prepare a checklist of all items that must be confirmed before a bid can be considered as responsive. The terms *responsive* and *responsible* should not be confused, however. Responsible, under existing law means a bidder who has demonstrated the attribute of trustworthiness, as well as quality, fitness, capacity, and experience to satisfactorily perform the contract (c.f. Calif. Public Contract Code § 1103) A bidder may be determined to be *nonresponsible* (sometimes referred to as "informal") after having been ruled as properly *responsive.*

The bid opening can be conducted most efficiently using two persons, each with a specific task to perform during the bid opening. One person should be responsible for opening and extracting all of the bids from their packages and recording the name of each bidder on a log sheet such as the Bid-Opening Report illustrated in Figure 12.6.

Each bid should then be carefully checked for being responsive. This is accomplished by taking each bid, in turn, and having one person read from the checklist while the second person checks the bid for conformance. After completing each review, an appropriate entry is made after the name of any bidder on the log whose bid has been found to be nonresponsive or informal.

Next, one of the two persons, usually the project manager, will read each bidder's name and total bid price aloud to the assembled bidders. The second person should then enter all bid totals opposite the names of the responsive bidders previously recorded in the Bid-Opening Report.

Finally, after all of the bids have been opened, the nonresponsive or informal bids rejected, and all responsive bids received, the project manager may, by inspection, determine the identity of the *apparent* low bidder, subject to later confirmation after a careful bid review and evaluation. It is important that any mention made at the time of the bid opening be referred to as only the *apparent* low bidder. A better system would be to provide copies of all tabulated bids, as received, and let the other

LAWRANCE, FISK & McFARLAND, INC.
CONSULTING ENGINEERS • SANTA BARBARA • ORANGE

BID OPENING REPORT

Bids were opened on 18 July 1990 at 10:00 a.m.

For Construction of 5 MGD Treatment Plant

Santa Barbara Sanitary District, CA

Pre-Bid Estimate $6,353,444.00

	CONTRACTOR:	BID AMOUNT:
1.	A.J. Pegno Construction Corporation, College Point, NY	$6,194,675.00
2.	Petracca & Sons, Incorporated, Jamaica, NY	$6,202,109.00
3.	Frank Mascall Construction Company, Flushing, NY	$7,431,639.00
4.		
5.		
6.		
7.		
8.		
9.		
10.		

Data used are fictitious for illustration only

Signed:

Date: 18 July 1990

THE ABOVE BID AMOUNTS HAVE NOT BEEN CHECKED.

THE BID TOTALS ARE SUBJECT TO CORRECTION AFTER THE BIDS HAVE BEEN COMPLETELY REVIEWED.

Purpose: Documentation of all bids received, including the name of each contractor and the amount of each bid.

Prepared by: Administrator at the bid opening.

Directed to: Project Manager; owner.

Copies to: Project Manager for engineering review; official files.

Comments: Bids must be carefully examined at the time of opening to determine if they are responsive prior to recording. Nonresponsive bids must be rejected without being entered into the record.

FIGURE 12.6 Bid-Opening Report for Recording Bids. *(Fisk, Edward R.,* Construction Engineer's Complete Handbook of Forms, *1st edition, © 1993. Reprinted by permission of Pearson Education, Inc., Upper Saddle River, NJ.)*

bidders draw their own conclusions, just in case the bid evaluation process turns up an error that could change the identity of the low bidder.

The issue of whether or not a bid is responsive can be potentially far more serious than it might seem to be at first. The author was involved in one case, involving a city water department, where failure of the apparent low bidder to sign one of the minor documents prior to submittal of the bid was challenged by the second-lowest bidder as being a nonresponsive bid that should thus be disqualified. Each threatened to sue the city if the contract was awarded to the other party, and, in the end, the city attorney ruled that all bids would be rejected and the project readvertised, even though this would create a delay in beginning the project, and even though it was understood that there was a strong possibility of receiving higher bids on the project the second time around. Under the circumstances, the city would most certainly have been involved in litigation if an award had been made to either one of the original two lowest bidders.

Summary of Bids for Evaluation

Following the close of the bid-opening session, the individual bids must be tabulated and evaluated, and each bidder must be checked for financial responsibility, licensing (where required), and integrity. Where unit prices form the basis of the bid, every individual line item must be tabulated for each bidder in preparation for a detailed, item-by-item evaluation, analysis, and comparison of bids. This information is normally entered onto a specially prepared form (spreadsheet) such as that illustrated in Figure 12.7, and all line-item amounts are entered opposite the corresponding engineer's estimate for that item. Additional information on the subject of bid evaluation is presented in Chapter 15.

Cost Breakdown of Lump-Sum Bids (Schedule of Values)

Traditionally, contractors are understandably reluctant to provide unit-price breakdowns of their lump-sum bids. However, the problem of determining the value of work performed each month for the purpose of making fair progress payments is quite difficult without some reasonable basis for allocating portions of the total contract price to the various individual tasks represented.

The contractor may not want to make a commitment on unit-price amounts for various reasons. One reason, of course, is the fear that in case of later negotiations for possible extra work, the contractor would be at a disadvantage at a bargaining session if unit-price estimates were offered on a project in which the contractor is entitled to a fixed price for doing the work, regardless of cost distribution.

It is generally necessary to specify that such a price allocation be made representing the major task items if it is to be obtained at all. Even then, it must be fully understood that its only purpose is to allow monthly evaluations of work performed for the determination of progress payment amounts, and must *not* be considered as a commitment to prices in case of later negotiations. To accomplish this through specifications provisions, the contract documents should include copies of price allocation forms (schedule of values) similar to the one illustrated in Figure 12.8, with the line items already determined by the architect/engineer. Thus the contractor need fill in only a price representative of an allocation of portions of the lump-sum contract price to the specific tasks listed.

E. R. FISK CONSTRUCTION
CONSULTING CONSTRUCTION ENGINEERS

BID OPENING
Date:
Time: 10:00 AM
Place: SACRAMENTO

SHEET 1 OF 1
Computed by R. B. BARNES

SUMMARY OF PROPOSALS RECEIVED
for: FAIRVIEW PROJECT
CL-14059

ITEM NO	DESCRIPTION	QUANTITY	UNIT	ENGINEER'S ESTIMATE		HUNTER, BARCLAY & THOMAS, INC. 333 MARKET ST. SAN FRANCISCO		COLLINS, INC. 4000 SAN PEDRO ST. OAKLAND, CA					
				UNIT PRICE	TOTAL	UNIT PRICE	TOTAL	UNIT PRICE	TOTAL	UNIT PRICE	TOTAL	UNIT PRICE	TOTAL
1	Furnishing & handling cement	[illegible]	bbl	[illegible]	[illegible]	4.00	8,000.00	4.50	9,000.00				
2	Furnishing & handling pozzolan	50	ton	[illegible]	[illegible]	40.00	2,000.00	50.00	2,500.00				
3	Aggregate base	9,900	ton	[illegible]	[illegible]	3.30	32,670.00	2.75	27,225.00				
4	Asphalt concrete	8,250	ton	[illegible]	[illegible]	7.00	57,750.00	7.50	61,875.00				
5	Asphalt dike	[illegible]	L.F.	[illegible]	[illegible]	0.60	4,800.00	0.75	6,000.00				
6	Radiographing welds	[illegible]	L.F.	[illegible]	[illegible]	10.00	10,000.00	18.00	18,000.00				
7	Site grading	L.S.	L.S.	--	[illegible]	L.S.	18,000.00	L.S.	50,400.00				
8	Modif. to existing drainage structures	L.S.	L.S.	--	[illegible]	L.S.	5,000.00	L.S.	8,100.00				
9	Conc. curbs, gutters, sidewalks, slabs	L.S.	L.S.	--	[illegible]	L.S.	17,000.00	L.S.	18,340.00				
10	Chain link fence	L.S.	L.S.	--	[illegible]	L.S.	7,000.00	L.S.	7,000.00				
11	Interim raw water supply line	L.S.	L.S.	--	[illegible]	L.S.	38,000.00	L.S.	38,100.00				
12	Water distribution system	L.S.	L.S.	--	[illegible]	L.S.	98,000.00	L.S.	104,800.00				
13	Natural gas distribution system	L.S.	L.S.	--	[illegible]	L.S.	8,000.00	L.S.	6,300.00				
14	Natural gas service	L.S.	L.S.	--	[illegible]	L.S.	3,000.00	L.S.	3,900.00				
15	Water tank and tower	L.S.	L.S.	--	[illegible]	L.S.	62,000.00	L.S.	67,100.00				
16	Sewage collection & disposal system	L.S.	L.S.	--	[illegible]	L.S.	70,000.00	L.S.	77,400.00				
17	Storm water drainage system	L.S.	L.S.	--	[illegible]	L.S.	47,000.00	L.S.	49,000.00				
18	Power transmission & distr. sys. (overhead)	L.S.	L.S.	--	[illegible]	L.S.	25,000.00	L.S.	30,200.00				
19	Power Communication Sys. (underground)	L.S.	L.S.	--	[illegible]	L.S.	53,000.00	L.S.	52,600.00				
20	Intercom system	L.S.	L.S.	--	[illegible]	L.S.	7,500.00	L.S.	6,800.00				
21	Fire detection & alarm system	L.S.	L.S.	--	[illegible]	L.S.	4,000.00	L.S.	4,200.00				
22	Guard station	L.S.	L.S.	--	[illegible]	L.S.	3,000.00	L.S.	12,400.00				
23	Loading ramp	L.S.	L.S.	--	[illegible]	L.S.	2,000.00	L.S.	2,400.00				
24	General maintenance warehouse	L.S.	L.S.	--	[illegible]	L.S.	129,000.00	L.S.	113,700.00				
25	Heavy equip. storage building	L.S.	L.S.	--	[illegible]	L.S.	39,000.00	L.S.	52,200.00				
26	Mobile equipment repair building	L.S.	L.S.	--	[illegible]	L.S.	131,000.00	L.S.	126,600.00				
27	Vehicle storage building	L.S.	L.S.	--	[illegible]	L.S.	85,000.00	L.S.	90,300.00				
28	General maintenance headquarter bldg.	L.S.	L.S.	--	[illegible]	L.S.	184,315.00	L.S.	198,627.00				
	TOTALS				[illegible]		$ 1,144,085.00		$ 1,156,027.00				
	AMOUNT OF PROPOSAL GUARANTEE					10% of Proposal Amount		10% of Proposal Amount					
	SURETY					The Aetna Casualty & Surety Co.		The Travelers Indemnity Co.					
	COMMENTS												

Data used are fictitious for illustration only

I hereby certify the above to be a true and correct summary ... Project Manager

FIGURE 12.7 Summary Sheet for Recording the Engineer's Estimate and All Bids Received. *(Fisk, Edward R.,* Construction Engineer's Complete Handbook of Forms, *1st edition, © 1993. Reprinted by permission of Pearson Education, Inc., Upper Saddle River, NJ.)*

Development of a Quality Control or Assurance Program

On many federal projects, one of the construction phase requirements is that the architect/engineer or its field representative develop a *construction surveillance and inspection plan* for submittal to the contracting officer (Figure 12.9). A prerequisite to the development of a surveillance and inspection plan is the development of a *quality requirements control system.* The purpose of this system is to develop a method of control for all specification requirements that are required to be done in writing, plus some other items that the architect/engineer may add. The *quality requirements control system* should contain a record of all quality control requirements in tabular form or other equivalent method.

In addition, such federal contracts often include the requirement that the architect/engineer's Resident Project Representative review the *construction contractor quality control* (CQC) plan. This involves a review of the contractor's CQC plan for clarity and completeness and a recommendation to the owner's contracting officer for acceptance, rejection, or revision.

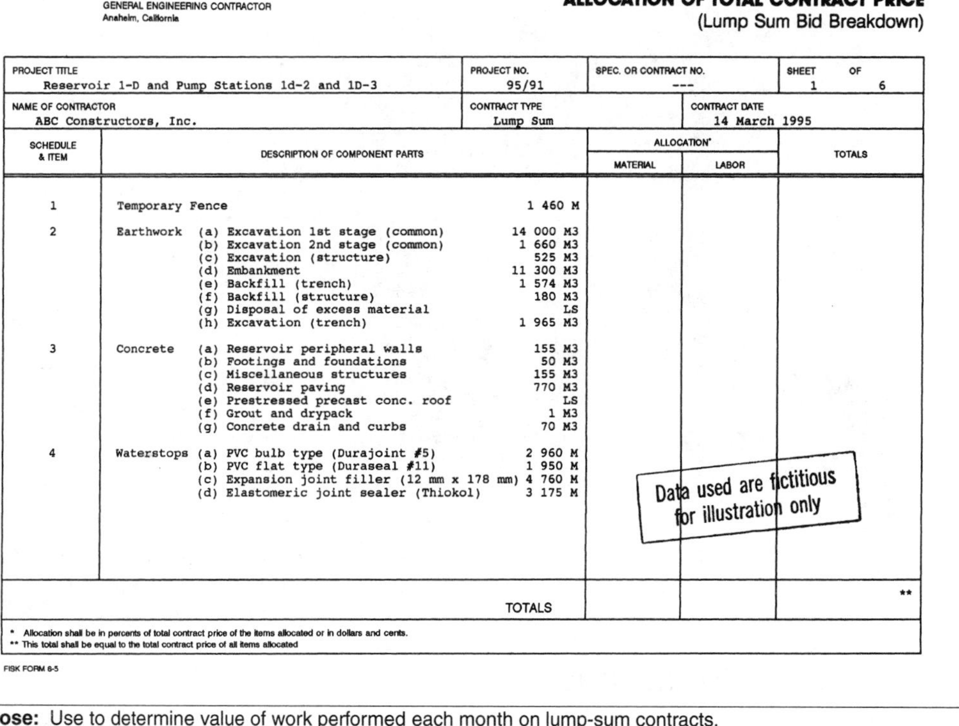

E.R.FISK CONSTRUCTION
GENERAL ENGINEERING CONTRACTOR
Anaheim, California

ALLOCATION OF TOTAL CONTRACT PRICE
(Lump Sum Bid Breakdown)

PROJECT TITLE	PROJECT NO.	SPEC. OR CONTRACT NO.	SHEET	OF
Reservoir 1-D and Pump Stations 1d-2 and 1D-3	95/91	---	1	6

NAME OF CONTRACTOR	CONTRACT TYPE	CONTRACT DATE
ABC Constructors, Inc.	Lump Sum	14 March 1995

SCHEDULE & ITEM	DESCRIPTION OF COMPONENT PARTS		ALLOCATION* MATERIAL	ALLOCATION* LABOR	TOTALS
1	Temporary Fence	1 460 M			
2	Earthwork (a) Excavation 1st stage (common)	14 000 M3			
	(b) Excavation 2nd stage (common)	1 660 M3			
	(c) Excavation (structure)	525 M3			
	(d) Embankment	11 300 M3			
	(e) Backfill (trench)	1 574 M3			
	(f) Backfill (structure)	180 M3			
	(g) Disposal of excess material	LS			
	(h) Excavation (trench)	1 965 M3			
3	Concrete (a) Reservoir peripheral walls	155 M3			
	(b) Footings and foundations	50 M3			
	(c) Miscellaneous structures	155 M3			
	(d) Reservoir paving	770 M3			
	(e) Prestressed precast conc. roof	LS			
	(f) Grout and drypack	1 M3			
	(g) Concrete drain and curbs	70 M3			
4	Waterstops (a) PVC bulb type (Durajoint #5)	2 960 M			
	(b) PVC flat type (Duraseal #11)	1 950 M			
	(c) Expansion joint filler (12 mm x 178 mm)	4 760 M			
	(d) Elastomeric joint sealer (Thiokol)	3 175 M			
	TOTALS				**

Data used are fictitious for illustration only

* Allocation shall be in percents of total contract price of the items allocated or in dollars and cents.
** This total shall be equal to the total contract price of all items allocated

FISK FORM 6-5

Purpose: Use to determine value of work performed each month on lump-sum contracts.

Prepared by: Contractor.

Directed to: Resident Project Representative; Project Manager.

Copies to: Project Manager; Resident Project Representative; contractor.

Comments: On lump-sum contracts it is difficult to determine value of work performed without a bid breakdown prepared by the contractor. These cannot be considered as firm prices, however, and cannot be used as a basis of determining the value of extra work.

FIGURE 12.8 Contractor's Allocation (Schedule of Values) of Lump-Sum Bid Price to Specified Construction Tasks for Computing Partial Payments. *(Fisk, Edward R.,* Construction Engineer's Complete Handbook of Forms, *1st edition, © 1993. Reprinted by permission of Pearson Education, Inc., Upper Saddle River, NJ.)*

Initially, to develop a successful quality control program, a complete definition of each and every task must be determined. This is done in coordination with the drawings and specifications and other controlling documents. Following completion of the quality control program, personnel assignments must be made, along with a definition of the duties and responsibilities of each person on the quality control team.

Prior to the time when the need actually arises, arrangements should be made for the selection of the testing laboratory that will provide testing services under contract and, if required, supplementary inspection personnel for specialized tasks of short-term duration.

FIGURE 12.9 Construction Surveillance and Inspection Plan.

Inspection and Testing Manual

Although not universally done on private or local public agency projects, a requirement of many federal agency construction services contracts calls for the preparation of a *quality control plan,* as mentioned earlier. Actually, this is a good idea to use on any job, and it not only eases the work burden during the job but also provides for a more orderly administration of the entire construction phase of the job. It does call for a great deal of effort at the beginning of a project, however, as it not only requires that all the policy matters that apply to the quality control functions be set out in writing, but it also requires the preparation of technical checklists for all work on the project as a means for reminding the inspector of each item that must be checked, as well as setting forth the standards for each such inspection. Normally, a quality control plan also sets out, in tabular form, a complete listing of all testing that is required on all materials and equipment for which tests are specified, and it further specifies the frequency of such tests (see again Figure 3.13). It can be said that a quality control plan is actually a separate "specification" for the testing and inspection phase of an entire project, and such a document can easily run to over a hundred pages of word processed text.

Workflow diagrams should be developed before the work on the project has begun. In this manner, preliminary planning efforts will not interfere with the orderly flow of the inspector's work after the contractor has mobilized at the site. The inspector's workflow diagrams are intended solely for the use of the field office staff and should include reminders to the Resident Project Representative as to when certain important events are to happen, as well as the assignment and responsibility for them. Generally, all these things can be shown in a simple chart form with a few explanatory sentences.

A significant part of the input data needed to prepare the workflow diagrams can be obtained from the contractor's CPM or bar charts. These charts are the keys to each significant construction event, as no other document can provide so much data in such a usable form. To be sure, adjustments will be required throughout the life of the job, but in general, the CPM or bar chart will set the stage for all construction events to follow.

Preparation of an Inspection and Testing Manual

The engineer or architect should take a systematic approach to the quality control and assurance functions that are required for the project. The actual task of developing such a document may rightly fall upon the shoulders of the Resident Project Representative, whose firsthand knowledge of construction can be of immeasurable value in the preparation of a workable inspection plan.

The Resident Project Representative should plan ahead by developing an outline of all the inspections that must be made, a checklist of points to look for, and a list of the types and frequencies of all tests that are required. An inspection plan might reasonably be expected to cover all or some of the following items:

1. Establishment of detailed inspection procedures
2. Outline of acceptance/rejection procedures
3. Preparation of a chart showing all tests required, when they are needed, the frequency of sampling and testing, the material being tested, and who is obligated to perform the tests
4. Establishment of who will be responsible for calling the laboratory for pickup of samples for testing, who will call for special inspectors when needed, and to whom such outside people will be directly responsible on the project
5. Identification of who must physically prepare samples for testing—the contractor or the inspector; determination of whether the contractor will provide a laborer to assist the inspector in obtaining samples and transporting samples for testing
6. Establishment of ground rules for acceptable timing of work operations after sampling and testing; mandatory scheduling to assure not only time to make samples and tests, but also to make corrections needed before work may be allowed to continue

A prerequisite to the development of a surveillance and inspection plan is the development of a quality requirements control system. The quality requirements should be shown in tabular form or other equivalent method. Typical examples of the data that must be presented in the form of tables or lists as a checking method throughout the job are:

1. Proofs of compliance (from the various specification sections)
2. Qualifications for soil-testing service
3. Tests for proposed soil materials
4. Reports required
5. Excavation methods (approval of)
6. Concrete testing and inspection service (approval of)

Prior to the time when the need actually arises, arrangements should be made for the selection of the testing laboratory that will provide testing services under contract and, if required, supplementary inspection personnel for specialized tasks of short-term duration.

A Testing Plan (Figure 12.9) should be developed to provide for the orderly administration of the construction phase of the work and as a means for reminding the inspector of each item that must be checked, as well as setting forth the standards for each such inspection. Normally, a quality control plan also sets out, in tabular form, a complete listing of all testing that is required on all materials and equipment for which tests are specified, and it further specifies the frequency of such tests. As a part of an inspection plan, it is desirable to include a flow diagram (Figure 12.10) for guidance of the Resident Project Representative, inspector, and contractor in administering the inspection of nonconforming work or test results.

Field Office Organization of the Owner or the Field Representative

Mission of the Field Engineer or Inspector

The mission of the Resident Project Representative or inspector as the design firm's or the owner's resident project representative is to assure compliance with the plans and specifications and to cause the end product to meet the needs of the ultimate user. In certain cases where public agencies are involved, the Resident Project Representative is further charged with the responsibility of checking on the compliance by the contractor with certain legislation, such as that pertaining to labor and the use of domestic materials.

The Planning Stage

In the early stages of design or in some cases prior to that time, the owner must decide who is going to operate its field office to monitor construction. There are many current practices, some of which are described in the following paragraphs.

As practiced in many cases, the owner may engage its own Resident Project Representative to monitor the construction phase of the project (under the responsible charge of an architect or a professional engineer), or the owner may hire an architect/engineer or a construction management firm to perform this function.

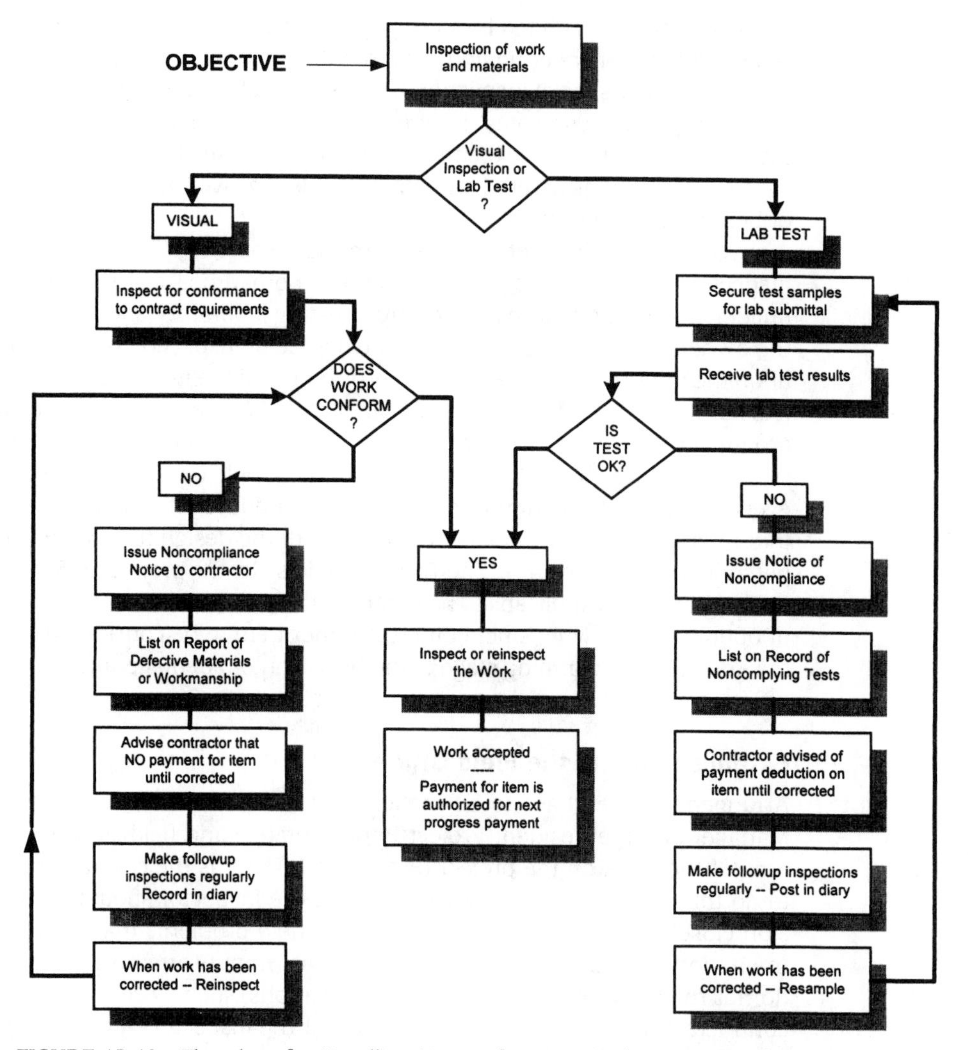

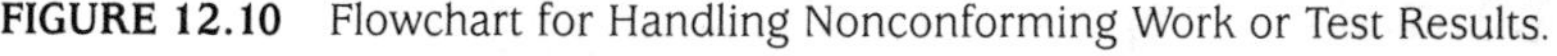

FIGURE 12.10 Flowchart for Handling Nonconforming Work or Test Results.

Another approach is to engage the services of an independent agency such as a testing laboratory or, on public projects, one governmental agency may solicit the help of another government agency for furnishing construction services. An example of this was the employment of the U.S. Army Corps of Engineers in the early postwar hospital construction program of the Veterans Administration, and later for Post Office construction.

Another practice is for the owner to enter into a professional construction management (PCM) contract wherein the construction manager oversees both design and construction and may actually contract with the construction contractor. Or, the owner may hire a single contractor to perform design, construction, and supervision of construction on either a lump-sum or a cost-plus-fixed-fee (CPFF) basis. The latter is a combined engineering and construction contract often called a design–build or turnkey job. For special inspection, the services of a testing laboratory may be hired for testing soils, aggregates, lumber, masonry units, steel, and concrete. Similarly, there are expert freelance technicians who inspect steel framework to assure soundness of connections (riveted, bolted, or welded) as well as the leveling and plumbing of beams and columns.

Having decided who is to perform the inspection, the extent and scope of the inspection services must be determined. Sometimes contracts provide that the builder will arrange for inspection by a testing laboratory and furnish test results to the owner or its architect/engineer. The degree of inspection required often depends upon the importance of the work or the function of the item in question. For instance, it is hardly worthwhile to take test cylinders on 2000-psi concrete used for sidewalks around a single-story residence, but it is absolutely essential that it be done on a prestressed concrete bridge girder. The inspector should be cautioned at this point, however, as the determination of the frequency and level of tests required is a decision that must be made by the engineering staff of the design firm or owner, not by field personnel unless specifically authorized to do so by the project manager of the owner or design firm, as applicable. The extent of the inspection is a major factor in the cost of operating the owner's field office. Another item to be considered by the owner in the construction planning stage is the provision of adequate funds to cover the field office costs of whatever agency is chosen to inspect the project.

Establishment of the Field Office

Arrangements will have to be made to set up the field office for the Construction Manager and the Resident Project Representative and field staff prior to the beginning of the work on the project (Figure 12.11). If it is the contractor's responsibility to do this under the project specifications, the inspector should monitor this provision closely, as there are often last-minute field problems that tend to delay implementation; also, the contractor may make last-minute attempts to offer facilities that do not meet the provisions of the specifications, such as setting up one-half of the contractor's field office trailer for the use of the inspector—even though the specifications may call for a separate structure.

The Resident Project Representative should make certain that the field office telephone has been ordered in time if it is the contractor's responsibility to provide such utilities to the inspector's field office. Without it, the Resident Project Representative will be out of communication with the design firm's or the owner's office during the early portion of the job. In addition, if the contractor is obligated under the provisions of the specifications to provide an electronic pager (radio "beeper") or other communications equipment, be sure that such equipment is provided in time to cover the critical early part of the job.

Finally, a requisition should be made and sent to the office of the design firm or the owner by the Resident Project Representative in sufficient time to enable them to make up a supply order on a routine basis, and to be able to obtain the office equipment that may be needed in the field office.

FIGURE 12.11 Large Field Office Complex (Foothill/Eastern Transportation Corridor, Construction Engineering Management Field Office, Orange County, California).

Field Office Responsibilities

In the establishment of a field office, the owner, or its design or construction management firm, is guided by many of the criteria that are considered by the contractor for setting up its own field office: the delegation of authority, the functions to be performed, the remoteness of the work site, and the extent and complexity of the job.

The delegation of authority from the design firm or the owner to the Resident Project Representative or from the contractor to its CQC representative can vary within wide limits. In some cases, it is merely to inspect the work; in others, the greater task of construction project administration is assigned. Costs of field office operation will vary widely with the extent of the work delegated. Usually personalities, experience, remoteness of the job site, and functions performed are serious considerations governing the delegation of authority. With respect to the Resident Project Representative's or CQC representative's *delegated* functions, the following factors must be considered:

1. What specific functions of the Resident Project Representative or CQC representative are specified in the contract?
2. How much testing will be performed by the Resident Project Representative or CQC representative and his or her staff, and how much will be contracted out to commercial testing laboratories?

3. How much of the materials specified will require preinspection at the factory before being shipped to the job site?
4. How much use will be made of special inspectors, such as for concrete, masonry, or structural steel and welding inspections?
5. Is the contract lump-sum, unit-price, or cost-plus-fixed-fee?
6. What is the authority of the architect/engineer to initiate, estimate, negotiate, and execute contract modifications?
7. What authority will be granted to the design or construction management firm by the owner to approve partial and final pay estimates?
8. How much of a check on the contractor's costs must be maintained by the Resident Project Representative, particularly on unit-price and cost-plus-fixed-fee contracts?
9. What responsibilities will the Resident Project Representative have, in the case of public agency contracts, for enforcement of such labor laws as the Davis-Bacon Act, the 8-hour law, and the Copeland Act or similar laws?

Other questions more specifically related to the *direct* responsibilities of the Resident Project Representative or CQC representative include the following:

1. To what degree must testing be conducted? In cases such as cement or steel, will mill certificates be accepted or will specific tests be required?
2. How much surveying must be supplied by the design firm or owner's field office?
3. Will record drawings be required to be prepared by the contractor or by the Resident Project Representative of the owner or design firm?
4. What reporting and other documentation requirements will be imposed on the Resident Project Representative or CQC representative and their field office staff?
5. What will be the responsibility of the Resident Project Representative or CQC representative with regard to intermediate and final acceptance of test results?
6. Will the Resident Project Representative or CQC representative be asked to participate in any portion of the review phase of shop drawings? If so, to what degree?
7. What, if any, safety responsibilities will be expected of the Resident Project Representative or CQC representative?

In the establishment of the Resident Project Representative's field office, the matter of staffing should be considered together with the assignment of individual responsibilities (Figure 12.12). The number of personnel needed to perform the administrative and quality control functions normally associated with the Resident Project Representative's field office may vary from one part-time construction administrator/inspector to a staff of 10 to 40 key people on an exceptionally large project. Most commonly, the small projects involve the use of one part-time construction coordinator who provides construction administration and general technical inspection,

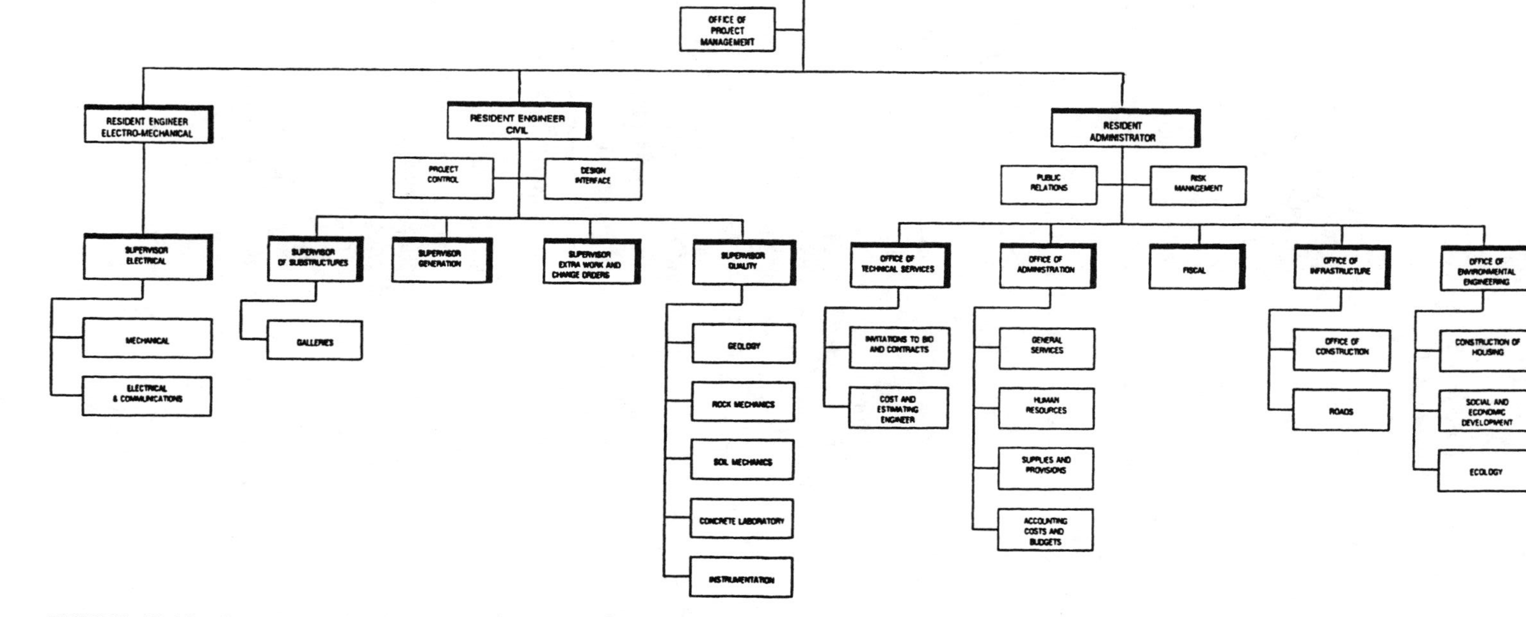

FIGURE 12.12 Representative Operational Diagram for a Large Project.

supplemented by special inspectors where necessary. On slightly larger projects, a full-time Resident Project Representative may be justified. Although the Resident Project Representative may be the only full-time member of the field office staff, special inspectors may be utilized where necessary. Beyond this level, the full-time Resident Project Representative is usually provided with a full-time clerk-typist to assist in the numerous routine office tasks involved. In addition, one or more full-time field inspectors may also be assigned to the project, supplemented by part-time special inspectors as necessary.

Projects of a size and complexity that require a larger or more specialized staff are generally under the direction of a resident engineer who is a registered professional engineer or architect as the Resident Project Representative (Figure 12.12).

Outline of Field Office Cost Items

Among the items contributing significantly to the cost of establishing and operating a construction field office for the owners or design firm are the following:

1. *Supervision.* Usually, on the larger projects, a full-time Resident Project Representative is assigned to each job site. The Resident Project Representative may work under the direct supervision of a project manager in an architect/engineer's office who often is in charge of several projects and many different contracts. The project manager *is* the architect/engineer as far as the project is concerned. A Resident Project Representative may be assigned to supervise the inspection of more than one project on an intermittent basis—usually for the same project manager. In addition to supervising the inspection forces, the Resident Project Representative maintains liaison with the contractor's project superintendent and sometimes with a representative of the owner, and is often asked to conduct on-the-job conferences.
2. *Operations.* This generally comprises the inspection (quality) group. It may be subdivided into functional areas such as the dam, the spillway, or the powerhouse; or on a contract basis such as the paving contract, the drainage facilities contract, the pump station contract, or similar tasks. Also to be considered are the various special inspectors such as the mechanical, electrical, control systems, welding and structural steel, masonry, concrete, soils, and paving inspectors. Laboratory operations may also be assigned to test soils, aggregates, concrete masonry units, sealants, asphalt, and other construction materials.

   Keeping a daily diary is the duty of each inspector (Chapter 4), and all activities must be documented thoroughly. The Resident Project Representative is also responsible for checking the contractor to assure that all insurance, bonding, and permits are in order. The operations tasks also include review of payment requests, measurements of work completed as a means of verifying payment requests (Chapter 17), the evaluation of requests by the contractor for change orders (Chapter 19), and the submittal of recommendations to the design firm or the owner regarding the recommendations of the Resident Project Representative for action to be taken on such requests. The Resident Project Representative, although not performing any construction surveys personally, should be involved to the extent of coordinating the survey tasks required for the project. The field office operations include estimating of costs of construction for partial and final payments, change order requests, and value engineering proposals. All transmittals of samples, shop drawings, and other material intended for the design firm or the owner should be routed through the Resident Project Representative, who should log all such submittals prior to forwarding to the design firm or the owner for action. The responsibility of the Resident Project Representative does *not* include approval of shop drawings, samples, approval of substitutions of materials, *nor* the interpretation of the intent of the plans and specifications—these should be reserved for the architect or engineer's project manager. On a very large project, a professional engineering staff may be assigned the job of field administration of the project, in which case, exceptions to the aforementioned responsibilities may be established by the design firm.

3. *Administration.* In addition to the administrative tasks included in the foregoing paragraph, the Resident Project Representative and the field office staff must perform general clerical services, communications services, limited personnel functions, mailing and shipping, ordering and maintaining of supplies and equipment, administration of transportation and travel of personnel, expense account submittals, minor purchasing, maintenance of company owned vehicles, and similar functions. In addition, on some governmental projects, particularly federal work, the Resident Project Representative field office may be required to check the contractor's payrolls for compliance with labor legislation, and assist in negotiations with craft unions and labor relations. Wherever owner-furnished equipment is to be installed by the contractor, the Resident Project Representative also may be called upon to expedite delivery and control, and provide for the storage and issuance to the contractor of such owner-furnished equipment.

The Preconstruction Conference

In communicating with the contractor and the contractor's personnel, it is important that the design firm or the owner and their Resident Project Representative make their positions very clear right from the start of the job. It is far better to get started on a basis of administering the contract firmly in accordance with the plans and specifications than it is to correct a difficult situation later in the job that is the result of a lax relationship with the contractor.

As a means of establishing the "ground rules" and calling the contractor's attention to the critical areas of construction, the preconstruction conference is an invaluable tool. Initially, it allows the key personnel of both sides to be introduced, and the responsibilities and authorities of each can be defined at that time. It also allows the parties to get a clear understanding of the procedures involved in contractor submittals, sampling and testing, construction surveys, inspections by outside agencies, payment requests, procedures for claims and disputes, unforeseen job conditions, change order requests, and similar items. The contractor can take this opportunity to raise questions about any of these items and clear up any misunderstandings. A checklist of items to cover is illustrated in Figure 10.1.

During the course of the preconstruction meeting, mention can be made of the contractor's responsibility to provide insurance documents as specified and all required bonds as well as to obtain (and usually pay for) all permits from building departments, street departments, police departments (for traffic control), flood control districts, environmental protection agencies, or other agencies having jurisdiction. The contractor should be reminded at the preconstruction meeting that all such documentation is required to be submitted before work can begin, and that the contract has a specified length of time after the signing of the contract or the issuance of a Notice to Proceed within which time the contractor is to begin the work.

Another item that should be brought up at the preconstruction conference is the schedule of job site and management meetings, the location and frequency of such meetings, and who should be in attendance.

Before closing the conference, the subject of the construction schedule should be raised. It may be necessary to remind the contractor that the initial project schedule submitted at the beginning of the project will require periodic adjustment, and that all such adjustments must meet with the approval of the design firm and the owner. It should also be emphasized that the design firm or the owner, through the Resident Project Representative, has the right to require the contractor to revise the work schedule, increase the work hours or personnel, or to make other adjustments that will assure completion of the project within the agreed-upon time schedule. Before making any such demands upon the contractor, however, be certain that the provisions of the contract will allow the architect or engineer to exercise such authority. Also, check both the time allowed to complete the job and whether the exact wording of the contract makes specific reference to "time" as "the essence of the contract." Failure to specify time as the essence of the contract may place the owner in a weak position for requiring the contractor to accelerate the work to complete the work "on time" [cf. *Kingery Construction Co. v. Scherbarth Welding, Inc.*, 185 N.W.2d 857 (1971)].

A careful record should be kept of all matters discussed at the preconstruction conference, and a copy of the record should be provided to all who attended the conference (Figure 12.13).

City of Thousand Oaks

RECORD OF PRECONSTRUCTION CONFERENCE

DATE 6 APRIL 1990

Page 1 of 4

NAME OF OWNER	ADDRESS (Including Zip Code and Telephone)
City of Thousand Oaks Utilities Department	2150 West Hillcrest Drive; Thousand Oaks (805) 497-8611
NAME OF ENGINEER	**ADDRESS (Including Zip Code and Telephone)**
Lawrance, Fisk & McFarland, Inc. Consulting Engineers	328 Garden Street, Santa Barbara (805) 564-2441
NAME OF CONTRACTOR (FIRM)	**ADDRESS (Including Zip Code and Telephone)**
XYZ Constructors, Inc.	100 Conejo Drive - Thousand Oaks (805) 000-0000

PROJECT TITLE	JOB NUMBER
Beatty Canyon Water Treatment Plant	89/90-00

SUBJECTS TO BE DISCUSSED

1. Identification of Official Representatives of Owner, Engineer, Contractor and Other Interested Agency:

OWNER: H. F. PAULSEN, GEN. MGR. ENGINEER: C. H. LAWRANCE, VICE PRES.

ADDRESS: 18100 SIERRA MADRE, AZUZA, CA ADDRESS: 328 GARDEN ST., SUITE 1 SANTA BARBARA, CA

CONTRACTOR: G. ALBRECHT, VICE PRES. OTHER AGENCY: Data used are fictitious

ADDRESS: ADDRESS: for illustration only

2. Responsibilies of the Engineer: (Engineer does not "Supervise" the Contractor's employees, equipment, or operations)

PER GEN. COND. ARTICLES 1.02, 1.03, AND 3.01. ALL COMMUNICATIONS TO CONTRACTOR WILL BE SUBMITTED AT CONST. SITE THRU RES. ENGR. ALL ORDERS IN WRITING. ONLY RES. ENGR. HAS AUTHORITY TO ISSUE ORDERS.

3. Responsibilities of Owner's Governing Body (Actual Contracting Organization.)

OWNER WILL WORK THRU ENGINEER ON ALL PROJECT MATTERS. OWNER'S APPROVAL REQUIRED ON ALL CHANGE ORDERS INVOLVING A CHANGE OF PROJECT TIME OR COST.

4. Responsibilities of the Contractor: (Review contract terms for specifics)

PER GEN. COND. ARTICLES 1.04, 1.05, AND 3.05 PRINCIPALLY. CONTR. MUST KEEP AUTH. SUPT. ON JOB AT ALL TIMES. MUST DO NOT LESS THAN 20% WITH OWN CREW

5. Responsibilies of Materials Testing Organization: UNDER DIRECT CONTRACT W/ OWNER, BUT ASSIGNED TO WORK WITH RES. ENGR. REQUIRES MIN. 24-HR NOTICE OF NEED

6. General Discussion of Contract:

A. Contractor Submittals: (Shop Drawings; Samples; Record Drawings; Schedules; Insurance & Bonds; etc.)

ALL SUBMITTALS PER SECTION 130 OF SPEC. ALL SUBMITTALS MUST BE MADE DIRECTLY TO RES. ENGR. AT SITE, WHO WILL FORWARD TO ENGR.

B. Notice to Proceed; Mobilization:

NOTICE OF AWARD DOES NOT AUTHORIZE SITE ACCESS. WORK TO START WITHIN 10 DA OF NOTICE TO PROCEED. MOBILIZATION PER SEC. 110; NO PRORATION

FIGURE 12.13a Sample Record of a Preconstruction Conference. *(Fisk, Edward R.,* Construction Engineer's Complete Handbook of Forms, *1st edition, © 1993. Reprinted by permission of Pearson Education, Inc., Upper Saddle River, NJ.)*

Page 2 of 4

C. Completion Time for Contract: (Does everyone understand contract requirements and methods of Computing?)

TOTAL 370 CALENDAR DAYS. TIME IS OF ESSENCE REF. SPEC. SEC. 101.04

D. Liquidated Damages

LIQUIDATED DAMAGES BOTH FOR FAILURE TO COMPLETE PROJ. ON SCHEDULE AND FOR FAILURE TO MEET SPECIFIED MILESTONE DATES FOR INTERFACE W/OTHER CONTRACTORS

E. Requests for Extension of Contract Time: SUBJECT TO TERMS OF GEN. COND. ARTICLE 3.11 (c). MUST BE IN WRITING. DELIVER TO RES. ENGR. WITHIN 15 DAYS OF DELAY BEGINNING

F. Procedures for Making Partial Payments: DISCUSS QUANTITIES WITH RES. ENGR. BEFORE SUB-MITTING ON 25TH OF MONTH. SUBMIT FORMAL PAY REQUEST ON FORMS PROVIDED.

G. Partial Utilization and Beneficial Use:

PORTIONS OF PROJECT SUBJECT TO PARTIAL UTILIZATION BEFORE PROJ. COMPLETION. ATTENTION IS DIRECTED TO PROVISIONS OF GEN. COND. ARTICLE 6.16.

H. Guarantees and Warranties on Completed Work: (Materials; Workmanship; etc.)

PER SPEC. NOTE SOME SECTIONS OF SPEC REQUIRE GUARANTEES IN EXCESS OF BASIC ONE YEAR. WARRANTIES BEGIN AFTER COMPLETION, EXCEPT PER ART. 6.16

I. Breakdown of Contract Costs: (Allocation of Total Contract Price)

CONTRACTOR MUST SUBMIT LUMP-SUM BID BREAKDOWN IN FORM PROVIDED. AGREED THIS IS NOT BASIS FOR COST OF EXTRA WORK. DUE WITHIN 10 DA. OF "PROCEED."

7. Contractor's Schedule:

A. Analyze Work Schedule in Sufficient Detail to Enable the Engineer to Plan His Operations: (Consideration must be given to needs of the Owner and the planned operations of other contractors.)

CPM SCHEDULE CONTROL REQ'D. WITH MONTHLY UPDATES TO RES. ENGR. PRELIM. SCHEDULE WITHIN 30 DA. OF AWARD. FINAL DUE WITHIN 30 DA. OF PROCEED.

B. Equipment to be Used by the Contractor: CONTRACTOR'S OPTION, SUBJECT TO ENGRS RIGHT TO REJECT UNSATISFACTORY EQUIP. OR REQUIRE DEMONSTRATION TO PROVE.

C. Contractor's Plans for Delivering and Storing Materials and Equipment at the Site: LONG-LEAD ITEMS MAY BE STORED AT SITE SUBJECT TO CONTR. CONSTRUCTING SUITABLE, HEATED SHELTER UNTIL USE. PMT OF 95 PERCENT INVOICE UNTIL INSTALLED.

D. Expediting of Long Lead Assigned Purchase Contracts: ALL LONG-LEAD PURCHASE EQUIPT HAS BEEN PREASSIGNED TO CONTR. WHO WILL EXPEDITE & RECEIVE.

8. Subcontracts: (Review and approval of proposed subcontractor work schedules) NO SUBSTITUTION OF SUB-CONTRACTORS AFTER BID OPENING EXCEPT AS ALLOWED BY LAW & SUBJ. TO ENGR & OWNER APPROVAL.

9. Status of Materials and Equipment Furnished by the Owner:

A. Delivery, Storage, and Issuance to the Contractor: NO OWNER-FURNISHED EQUIPMENT

FIGURE 12.13b (continued)

Page 3 of 4

8. Transportation and Shipping to Project Site:

NO HIGHWAY OR BRIDGE RESTRICTIONS

10. Change Orders and Extra Work: (Outline procedures to be followed and who has authority to issue)
NO CHANGES FROM PLANS & SPECS EXCEPT ON WRITTEN AUTHORITY OF RES. ENG. ALL SUCH CHANGES MUST ULTIMATELY BE SUPPORTED BY FORMAL CHANGE ORDER. EMERGENCY FIELD ORDERS OK, BUT MUST BE CONFIRMED.

11. Staking of Work: (Clearly define responsibilities of Engineer and Contractor. Line and Grade must be furnished by Engineer.)

ENGR. WILL PROVIDE LINE & GRADE ONLY (2 POINTS & ONE ELEVATION)

12. Project Inspection:
A. Functions of the Engineer, including Records and Reports: RES. ENGR. IS DESIGNATED AS SOLE ON-SITE REP. OF ENGR. & OWNER. AUTH. PER NSPE FORM 1910-1-A.

B. Responsibilities of Owner:
EXERCISE PROJECT CONTROL THRU ENGR.

C. Safety and Health Obligations of the Contractor: (Cite Engineer's procedures in handling an imminent hazard)
CONTR. HAS TOTAL RESPONSIBILITY. ENGR RESERVES RIGHT TO ORDER IMMEDIATE CORRECTION OF OBSERVED MAJOR HAZARDS, HOWEVER.

13. Final Acceptance of the Work: (Include requirements for tests and validation of systems and project cleanup)
CONTR. TO PROVIDE LIEN RELEASES & AFFIDAVIT OF PAYMENT. OWNER WILL HOLD RETAINAGE PLUS TWICE VALUE OF OUTSTANDING PUNCH LIST ITEMS.

14. Labor Requirements:
A. Equal Employment Opportunity Requirements:

PER SPEC REQUIREMENTS — SUPPL. GEN. CONDITIONS

B. Davis-Bacon Act: PROJECT SUBJECT TO FEDERAL WAGE REQUIREMENTS. IF STATE MINIMUMS ARE DIFFERENT, CONTR MUST PAY HIGHER RATES.

C. Other Federal Requirements:

CORPS OF ENGINEERS 404 PERMIT FOR DUMPING OR STOCKPILING

D. State and Local Requirements: STATE HIGHWAY ENCROACHMENT PERMIT REQ'D. STREET DEPT EXCAVATION PERMIT; TRAFFIC CONTROL; BLDG PERMIT

E. Permit Responsibilities: AS LISTED THEREIN. ALL REQUIRE CLEANUP, NIGHT COVERING OF OPEN TRENCHES & TRAFFIC PER ANSI D6.1-1968

F. Union Agreements:

NO SPECIAL AGREEMENTS

15. Insurance Requirements: PER. GEN. COND. ARTICLES 6.05 AND 6.06, ALSO SUPPL. GEN. COND. ARTICLE 8.02. MUST SUBMIT TO OWNER AT TIME OF AGREEMENT SIGNING

FIGURE 12.13c (continued)

Page 4 of 4

16. Rights-of-Way and Easements:

A. Explain any Portion of Project Site Not Available to Contractor: CONTR. NOT PERMITTED TO MOVE ONTO PRIVATE PROPERTY UNTIL COPIES OF EASEMENTS ARE OBTAINED

B. Contractors Responsibilities During Work Covered by Contract: NO TEMP. CONST. EASEMENTS HAVE BEEN OBTAINED. IT IS CONTR. RESPONSIBILITY TO ARRANGE THESE & PAY.

17. Coordination with Utilities, Highway Departments, and Railroads: CONTR. MUST NOTIFY AFFECTED UTILITIES WHEN WORKING NEAR THEIR FACILITIES. FOLLOW UTIL. INSTRUCTIONS

18. Placement of Project Signs and Posters: PER SPECIFICATION & EPA STANDARD. SIGN MUST BE ERECTED WITHIN 30 DA AFTER NOTICE TO PROCEED

19. Handling of Disputes, Protests, and Claims: FIRST STEP: RES. ENGR; SECOND STEP: PROJECT ENGINEER. REF. GEN. COND. ARTICLES 3.06 AND 3.07.

NOTED AND CONCURRED WITH, But understood not to be a modification of any existing contracts or agreements:

(Contractor Representative)	(Contractor Representative)
(Engineer Representative)	(Engineer Representative)
(Owner Representative)	(Owner Representative)

OTHERS IN ATTENDANCE at Preconstruction Conference	ORGANIZATION Represented
	ABC CONSTRUCTORS, INC.
	PARKSIDE ENGRG CORP
	LAWRANCE, FISK & McFARLAND, INC.

DISTRIBUTION:
1. Contractor
2. Project Engineer
3. Resident Project Representative
4. Owner
5. Project File

FIGURE 12.13d (continued)

Study Plans and Specifications

Prior to the beginning of any actual work on the project, and preferably prior to mobilization by the contractor, the Resident Project Representative or CQC representative *and all other inspectors* assigned to the project should obtain a complete set of project plans and specifications, *plus all addenda* and all key reference books or standards that are cited in the specifications. The plans, specifications, addenda, and references should be carefully studied, and the inspector's copy of the plans and specifications should be marked with a highlighter pen or other marker to identify all key inspection provisions. The placing of index tabs at the beginning of each section of the specifications is also a good idea, as it will facilitate rapid reference to each section when needed. If the construction documents review can be conducted in the office of the architect/engineer, it is even better, as the inspector will be in a better position to get factual answers to specific questions from the project manager and the design engineers and architects themselves. In addition, this one-to-one level of communication allows the supervisor of each design discipline to meet the project field representative and to develop a working relationship. At the same time, contacts can be made with the mechanical and electrical department heads to determine their schedules for requesting special inspections of those portions of the work for which they are personally responsible.

Key Dates

The Resident Project Representative or CQC representative should note on the field calendar every date that has special significance on the project, whether it be for tests, special inspections, payment request due dates, delivery dates, or other important milestones. All such data should be obtained either from the specifications (if listed there) or from the contractor's CPM or bar chart schedule.

Listing of Emergency Information

It is quite important to maintain an up-to-date list of all key personnel of the owner, the engineer, and the contractor to be contacted in case of a project emergency (Figure 12.14). Each person listed should provide telephone numbers where he or she may be contacted at any time, day or night, as well as on weekends and holidays.

Review the list from time to time to confirm and update names and telephone numbers. In addition, the Resident Project Representative should keep a list of emergency and utility services and the corresponding telephone numbers.

Distribution of the key personnel list should be restricted to those key personnel on the list and management personnel of their respective organizations. No general posting of the list should be made.

City of Thousand Oaks

UTILITIES AND SERVICES INFORMATION

PROJECT: La Crescenta Reservoir and Pumping Plant PROJECT NO. 89/90-00

OWNER: City of Thousand Oaks

CONTRACTOR: XYZ Constructors, Inc., Thousand Oaks

ENGINEER: Lawrance, Fisk & McFarland, Inc. 928 Garden St., Suite 1, Santa Barbara CA

KEY PERSONNEL

	Business	Residence
OWNER REPRESENTATIVE		
C. P. Owens, Project Manager	(805) 000-6391	(805) 000-3110
CONTRACTOR'S SUPT/PROJECT MANAGER		
Everet K. Jefferson	(805) 000-2632	(805) 000-9431
PROJECT MANAGER/ENGINEER		
Charles M. Parker	(805) 000-9141	(805) 000-6152
RESIDENT PROJECT REPRESENTATIVE		
R. E. Barnes	(805) 000-4426	(805) 000-3716

PROJECT UTILITIES AND SERVICES

OSHA REPRESENTATIVE	Cal-OSHA Consultation Service	(213) 786-3870
	Cal-OSHA (Enforcement)	(213) 736-3071
GAS	Southern California Gas Co.	
	Excavation Pipeline Locations	1-800-422-4133
POWER	Southern California Edison Co.	
	Santa Barbara	(805) 000-0000
TELEPHONE	Pacific Telephone & Telegraph Co.	
	Underground Service Alert	1-800-422-4133
WATER		
SANITATION		
STREETS & HIGHWAYS	City of Thousand Oaks	
	Department of Public Works	(805) 000-0000
FIRE PROTECTION	Thousand Oaks Fire Department	
	Thousand Oaks	9-1-1
POLICE PROTECTION	Thousand Oaks Police Department	
	Thousand Oaks	9-1-1

All key personnel should have a copy of this information, and a copy should be posted in each field office in a prominent location.

FIGURE 12.14 Example of a Project Emergency Services Information Sheet. *(Fisk, Edward R.,* Construction Engineer's Complete Handbook of Forms, *1st edition, © 1993. Reprinted by permission of Pearson Education, Inc., Upper Saddle River, NJ.)*

Agency Permits

At the beginning of the job, the Resident Project Representative should see that all required permits have been obtained and should prevent any construction work from proceeding whenever a required permit controlling such work has not been obtained. It should be noted that sometimes the terms of a permit may be quite lengthy and the permit conditions may read like a specification in itself. Such terms and conditions of a permit must be considered as binding upon the contractor and will normally take precedence over the terms of the project specification in case of conflict.

Starting a Project

The beginning of a construction project normally starts with the award of the construction contract. This may be accomplished in the minutes of a city council or county board of supervisors action, by letter, or by issuance of a preprinted *Notice of Award* form (Figures 12.15 and 12.16). The giving of a Notice of Award is similar in its legal effect to the issuance of a letter of intent, as it obligates the owner to sign the construction contract if the contractor does what is required of it within the time specified. The Notice of Award *does not authorize the start of construction,* because no work is supposed to start until after the owner/contractor agreement has been signed by both parties. Under the terms of the EJCDC General Conditions, the contract time will begin running on the 30th day after the owner has signed and delivered a fully executed agreement, but it may start sooner if a formal Notice to Proceed is issued. Under the AIA General Conditions, if no Notice to Proceed is issued, the contract time will begin as of the date of signing of the agreement. Under FIDIC Conditions of Contract, the Work must begin "as soon as is reasonably possible after receipt of a notice" (to proceed).

The issuance of a *Notice to Proceed* (Figures 12.17 and 12.18) formalizes the date that the project is to begin and sets the stage for computation of the total project construction time. This will greatly facilitate the establishment of an accurate count of construction time for the computation of liquidated damages. It is considerably more reliable than relying solely upon the Notice of Award, as can be seen by the variation in terms within the contract provisions of the two major societies who offer standardized General Conditions to be used in the project specifications. Many public agencies allow a 10-day period after the issuance of a formal Notice to Proceed for the contractor to begin work at the site.

A construction contract can but should not designate a specific commencement date because of the uncertainty of when the work can begin. Before the contractor can begin it must be given access to the site. This can require that easements be obtained, public approvals be given, and funds be obtained.

As a result, the commitment to begin work is usually expressed in terms of a number of days after access to the site. The owner also uses this approach to avoid responsibility for delay in its ability to grant access to the site.

Some owners attempt to establish a date to proceed by basing the starting date upon the date of execution of the agreement. This can be unfair to the contractor, who

City of Palm Springs

NOTICE OF AWARD

Dated MARCH 22 1991

To: Walsh Construction of Illinois dba Walsh Pacific Construction
(Bidder)

Address: 791 Foam Street, Suite 200
Monterey, CA 93940

Project: 1991 EXPANSION OF PALM SPRINGS CONVENTION CENTER

City's Project No. 90-38

Contract for 1991 Palm Springs Convention Center Expansion

You are hereby notified that your Bid dated MARCH 19 1991 for the above Contract has been considered. You are the apparent successful bidder and have been awarded a contract for the above-named project.

The Bid Price of your contract is $ Five million, Six hundred, seventy-eight thousand dollars ($5,678,000.00) Dollars

Three copies of each of the proposed Contract Documents (except drawings) accompany this Notice of Award. Three sets of the Drawings will be delivered separately or otherwise made available to you immediately.

You must comply with the following conditions precedent within 10 days of the date of this Notice of Award; that is, by: APRIL 2, 1991.

1. You must deliver to the City ~~three~~ four fully executed counterparts of the Agreement, including all the Contract Documents. This includes the ~~triplicate~~ duplicate sets of Drawings. Each of the Contract Documents must bear your signature on the cover page.
2. You must deliver with the executed Agreement, the Payment and Performance Bonds and the Insurance Certificate as specified in the Instructions to Bidders, the General Conditions (Article 5), and the Supplementary General Conditions.

Failure to comply with these conditions within the time specified will entitle the City to consider your Bid abandoned, to annul this Notice of Award, and to declare your Bid Security forfeited.

Within 10 days after you comply with the foregoing conditions, the City will return to you one fully signed counterpart of the Agreement with the Contract Documents attached.

CITY OF PALM SPRINGS

By Harold E. Good, CPPO

Title Purchasing Manager

Copy to Architect by Certified Mail

______ Return receipt Requested

Data used are fictitious for illustration only

Palm Springs Project 90-38
Convention Center Expansion
A00670-022391
© 1990 by E. R. Fisk

NOTICE OF AWARD
CONTRACT ADMINISTRATION
FORMS - PAGE 1

FIGURE 12.15 Example of a Notice of Award *(Fisk, Edward R.,* Construction Engineer's Complete Handbook of Forms, *1st edition, © 1993. Reprinted by permission of Pearson Education, Inc., Upper Saddle River, NJ.)*

NOTICE OF AWARD

Dated ____________________, 19____

TO: ______________________________
(BIDDER)

ADDRESS: ______________________________

____

____

PROJECT ______________________________

OWNER's CONTRACT NO. ____________________

CONTRACT FOR ______________________________

____
(Insert name of Contract as it appears in the Bidding Documents)

You are notified that your Bid dated ____________________, 19____ for the above Contract has been considered. You are the apparent Successful Bidder and have been awarded a contract for__________

____

____
(Indicate total Work, alternates or sections or Work awarded)

The Contract Price of your contract is______________________________

______________________________Dollars ($).

[Insert appropriate data in re Unit Prices. Change language for Cost-Plus contracts.]

____ copies of each of the proposed Contract Documents (except Drawings) accompany this Notice of Award.
____ sets of the Drawings will be delivered separately or otherwise made available to you immediately.

You must comply with the following conditions precedent within fifteen days of the date of this Notice of Award, that is by

____________________, 199____.

1. You must deliver to the OWNER ___ fully executed counterparts of the Agreement including all the Contract Documents. This includes the triplicate sets of Drawings. Each of the Contract Documents must bear your signature on (the cover) (every) page (pages ________).
2. You must deliver with the executed Agreement the Contract Security (Bonds) as specified in the Instructions to Bidders (paragraph 18), General Conditions (paragraph 5.1) and Supplementary Conditions (paragraph SC-5.1).

EJCDC No. 1910-22 (1990 Edition)
Prepared by the Engineers Joint Contract Documents Committee and endorsed by The Associated General Contractors of America.

FIGURE 12.16a Example of a Notice of Award Keyed to the Documents of the Engineer's Joint Contract Documents Committee (ACEC, ASCE, CSI, NSPE). *(Copyright © 1990 by National Society of Professional Engineers.)*

3. (List other conditions precedents).

____

Failure to comply with these conditions within the time specified will entitle OWNER to consider your bid in default, to annul this Notice of Award and to declare your Bid Security forfeited.

Within ten days after you comply with the above conditions, OWNER will return to you one fully signed counterpart of the Agreement with the Contract Documents attached.

(OWNER)

By:

(AUTHORIZED SIGNATURE)

(TITLE)

ACCEPTANCE OF AWARD

(CONTRACTOR)

By:

(AUTHORIZED SIGNATURE)

(TITLE)

(DATE)

COPY to ENGINEER
(Use Certified Mail,
Return Receipt Requested)

FIGURE 12.16b Reverse Side of EJCDC Notice of Award.

E. R. FISK & ASSOCIATES
P.O. Box 6448 • Orange, CA 92613-6448

NOTICE TO PROCEED

To

XYZ CONTRACTORS, INC.
5892 First Street, Suite 104
Santa Barbara, 93000

Project Caldwell Reservoir No. 1134-57

Construction Contract No. CN07-8976-4366

Type of Contract Unit Price

Amount of Contract $423,975.00

You are hereby notified to commence work on the referenced contract on or before 23 April 1990, and shall fully complete all of the work of said contract within 365 consecutive calendar days thereafter. Your completion date is therefore 23 April 1991.

The contract provides for an assessment of the sum of $ 500.00 as liquidated damages for each consecutive calendar day after the above established contract completion date that the work remains incomplete.

Dated this 9th day of April 1990.

Data used are fictitious for illustration only

By [signature]

Title Project Manager

ACCEPTANCE OF NOTICE

Receipt of the foregoing Notice to Proceed is hereby acknowledged

By XYZ Contractors, Inc.

this 11th day of April 19 1990.

By [signature] J. P. Campbell

Title President, XYZ Contractors, Inc.

Wiley-Fisk Fo. 6-1

FIGURE 12.17 Example of a Notice to Proceed *(Fisk, Edward R.,* Construction Engineer's Complete Handbook of Forms, *1st edition, © 1993. Reprinted by permission of Pearson Education, Inc., Upper Saddle River, NJ.)*

NOTICE TO PROCEED

Dated ____________________, 19 ______

TO: ______________________________________
(CONTRACTOR)

ADDRESS: ______________________________________

____

PROJECT ______________________________________

OWNER's CONTRACT NO. ______________________________

CONTRACT FOR ______________________________________

____
(Insert name of Contract as it appears in the Bidding Documents)

You are notified that the Contract Times under the above contract will commence to run on __________, 19 ____ . By that date, you are to start performing your obligations under the Contract Documents. In accordance with Article 3 of the Agreement the dates of Substantial Completion and completion and readiness for final payment are ______________, 19 ____ and __________, 19 ____ .

Before you may start any Work at the site, paragraph 2.7 of the General Conditions provides that you and Owner must each deliver to the other (with copies to ENGINEER and other identified additional insureds) certificates of insurance which each is required to purchase and maintain in accordance with the Contract Documents.

Also before you may start any Work at the site, you must
(add other requirements)

____

____

____

____
(OWNER)

By: ______________________________________
(AUTHORIZED SIGNATURE)

____
(TITLE)

ACCEPTANCE OF AWARD

By: ______________________________________
(CONTRACTOR)

____
(AUTHORIZED SIGNATURE)

____
(TITLE)

____
(DATE)

Copy to ENGINEER
(Use Certified Mail,
Return Receipt Requested)

EJCDC No. 1910-23 (1990 Edition)
Prepared by the Engineers Joint Contract Documents Committee and endorsed by The Associated General Contractors of America.

FIGURE 12.18 Example of a Notice to Proceed Keyed to the Documents of the Engineer's Joint Contract Documents Committee (ACEC, ASCE, CSI, NSPE). *(Copyright © 1990 by National Society of Professional Engineers.)*

may not be given access to the site until long after the date on the agreement. Often, the contractor will not receive a signed copy of the Agreement until one or more weeks after execution by the owner, thus using construction time that should have been available for the contractor. If the contractor runs into a period of liquidated damages, it will more assuredly file a claim to recover the time lost by failure to receive confirmation of the execution of the agreement. EJCDC documents provide that the contract time will begin to run on the 30th day after the effective date of the Agreement, or if a Notice to Proceed is given, on the day indicated in the Notice to Proceed. There are many complications possible; however, all of the risks could be avoided if the owner fixed a starting date in the Notice to Proceed or upon receipt of the Notice to Proceed.

Problems

In each of the first three questions, insert the letter from the following list of multiple-choice answers that represents the proper time to accomplish the indicated task.

- **a.** during the bid phase
- **b.** after Agreement, but before the Notice to Proceed
- **c.** prior to advertising for bids
- **d.** after bid opening, but before Award of Contract
- **e.** after Notice to Proceed
- **f.** after award, but before owner signs the Agreement

1. A prebid conference should be held when? ____________
2. A preconstruction conference should be held when? ____________
3. When should the bidder submit the payment bond, the performance bond, and the certificate of insurance? ____________
4. What is the proper method of responding to bidder inquiries during the bid phase of a project?
5. True or false? A builder's risk "all-risk" policy of construction insurance covers all perils without additional endorsement.
6. What is a nonresponsive bid?
7. Explain the nature and purpose of a price allocation form, or schedule of values.
8. Is it the responsibility of the Resident Project Representative to review shop drawings, evaluate sample submittals, approve substitute materials, or interpret the intent of the contract plans and specifications?
9. What is the significance of a Notice of Award? What does it do? Is access to the site authorized thereby?
10. Is it an acceptable contract administration procedure to permit access to the site or issue a Notice to Proceed after Award but prior to execution of the formal Agreement?
11. On a project that supports a full-time Resident Project Representative at the site, with the project manager located in the home office, to whom should all contractor submittals be submitted first?

12. Should an inspector always try to get the contractor to give the owner the highest quality materials and workmanship, regardless of the quality level called for in the specifications?
13. If the specifications are silent on the subject, but a contractor asks you, as the Resident Project Representative or inspector, how to perform the work, which of the following is the recommended procedure?
    (a) Be ready to help by telling the contractor your opinions, because as a member of the construction team, it is your responsibility.
    (b) Politely decline because construction means and methods are the sole prerogative of the contractor.
    (c) Direct the contractor how to perform the work because you know that if he asked you, it meant that he did not know how, and the project may otherwise suffer delays or failures.
14. What types of bonds are required to be provided by a bidder/contractor on a public works contract?

13

PLANNING FOR CONSTRUCTION

Construction planning and scheduling should not be considered as one of the deep mysteries of life, but rather as an application of common sense, a logical analysis of a construction project together with all of its parts, and a thorough knowledge of construction methods, materials, and practices.

The process of planning is simply an application of the thought process that must be entered into before the actual scheduling can begin. The planning process should include answers to the following *preliminary* questions:

1. *Long-lead purchases.* Are there any items that will require purchase orders to be placed long in advance of the time that the item is needed on the job because of material shortages, fabrication time, or similar delay factors?
2. *Utility interruptions.* Is there any part of the project that will involve an outage of utility services such as water, power, gas, or other essential services? If so, has the utility owner been contacted to determine the maximum length of the outage, the time of day that outages will be permitted, the calendar dates during which outages will be permitted, or similar restrictive controls?
3. *Temporary utilities.* Will temporary utility lines be required to be built to bypass the construction area, and will temporary roads be required to provide detour routes for street traffic?
4. *Temporary construction utility service.* Who provides temporary construction utility service, and from where must it be obtained?
5. *Labor.* Have representatives of labor been contacted in the area of construction to establish the jurisdictional responsibilities of the various trades to be used in the work, as well as to determine the union work rules in the affected area?
6. *Work and storage areas.* Have provisions been made for contractor's work and storage areas?

7. *Traffic requirements.* Have local traffic control regulations been investigated? Will construction equipment be allowed to operate on public streets, will street closures in the construction area be permitted, and will special traffic control and flaggers be required to direct traffic around construction?
8. *Temporary access.* Will temporary access, including temporary bridges, be required to provide continued access to residences and places of business during the construction period? Will temporary access require a permit for crossing railroad right of way?
9. *Other contractors.* Will other contractors be working in the same area, thus requiring schedule coordination with them to complete the work of this contract?
10. *Interdependency of tasks.* Are some of the tasks in this project dependent upon the completion schedule of another contractor or utility owner before they can be started?
11. *Environmental controls.* Will special environmental controls be required; if so, which ones? For example, when and for how long barge operations must be suspended at a rock quarry for salmon spawning.
12. *Special regulations.* Are there special regulations, such as FAA requirements for work at airports, that may affect the construction scheduling or construction time?
13. *Special construction equipment.* Will special equipment be required for construction? If so, is it off-road equipment that will require special haul routes? What are the load limits and bridge clearances for roads in the area?
14. *Time for construction.* Is the time allowed to complete the project adequate for the location and the seasons, or will it require increased crew sizes or premium time? For example, is the time between annual regatta festivals adequate to rebuild a riverfront park?

Each of these items is an important consideration in the planning stage of a project. After determination of the various limitations and constraints on the conduct of the work, the task of planning the actual construction effort can begin. Many large contracts can be divided into separate stages of construction. In some cases there may be a single general contractor responsible for the construction of all the stages, which may simply have been planned to accommodate a fixed order or sequence of construction. In other cases, it may be possible that portions of a construction project may begin even before the entire work of design has been completed. This is generally accomplished by the owner entering into phased construction contracts ("fast-track"), in which each stage of the work is let for construction as soon as the design effort on that particular phase of the work has been completed (Figure 13.1).

As mentioned previously, this approach has the added advantage that it can result in completion of a project at an earlier date than would be possible if it were necessary to wait until all design work had been completed. It may also offer a distinct financial advantage to the owner in many cases by allowing the early bidding of completed portions of the work to avoid the added costs that could result from the cost

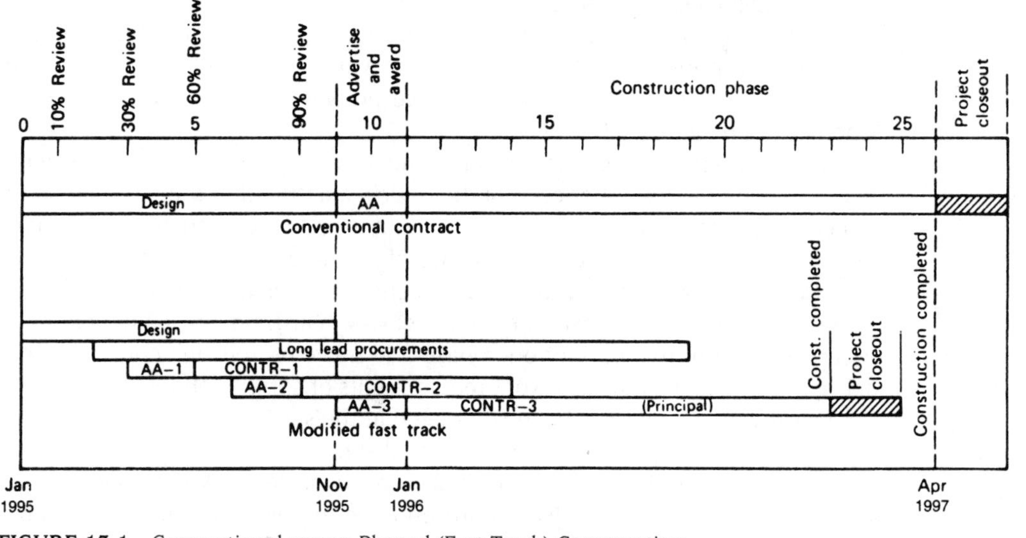

FIGURE 13.1 Conventional versus Phased (Fast-Track) Construction.

inflation spiral if the work were to be delayed. The fast-track method of construction is often a money saver for the owner despite added field administration costs, as the earlier construction completion date is often money in the owner's pocket.

Phased construction (fast-track) has its disadvantages, too, as it creates considerably more work for the project manager and the Resident Project Representative, and may require a larger field staff to administer. One of the difficulties that may have to be overcome is created by the fact that under a fast-track system of construction, numerous other prime contractors are on the job. Because of this, the added responsibility for coordination of each of these various *prime* contractors falls upon the shoulders of the contract administrator and the Resident Project Representative, as there would be no general contractor with this responsibility. Similarly, interfaces between the various scopes of the work of each of the separate prime contracts should have been very carefully worked out in the contract documents. Furthermore, if the project in which these various separate contractors are involved concerns separate elements that must ultimately become a part of a single operational system, the question of responsibility for the proper functioning of the overall system may pose a serious problem for the owner. Finally, there may be severe schedule impacts if there are any changes in design, especially those that affect something already built.

Generally, under a system of fast-track contracts, it is most desirable to operate under a design–build/turnkey or a construction management contract, wherein coordination responsibilities can be handled by the construction manager and his or her project team. In such cases, the task of administration of the construction will be carried by an on-site project manager, and the resident inspector will normally be performing the functions of a quality control supervisor.

Another approach to the problem of responsibility over fast-track contract operations that has not yet come into common use is one in which all but one of the prime contracts awarded contain an assignment clause that allows one of the other prime contractors to function as a lead contractor. In this capacity, the lead contractor will have responsibility for coordinating the work of all the contractors on the project and assuring that all work will interface properly without gaps or overlaps of responsibility (Figure 13.2).

The planning problem is further complicated by the fact that consideration must be given to the work of each of the trades to assure that there will be no interference by others whose work takes them into the same construction area; nor will the work of one trade be held up because of the failure of another to complete its work. Costs of mobilization must also be considered, as it would certainly be both uneconomical and time consuming to move heavy earthmoving equipment to a site for construction of the work of one phase and move off the site, only to be moved back again to execute the earthwork requirements of another phase of the work if all of the earthwork could have been completed at the same time.

Obviously, to get a clear picture of all the interrelated problems at a construction site, some kind of a graphic picture must be drawn to enable the contractor and the Resident Project Representative to get a clear understanding not only of the sequence of events, but also of their times of beginning and anticipated completion.

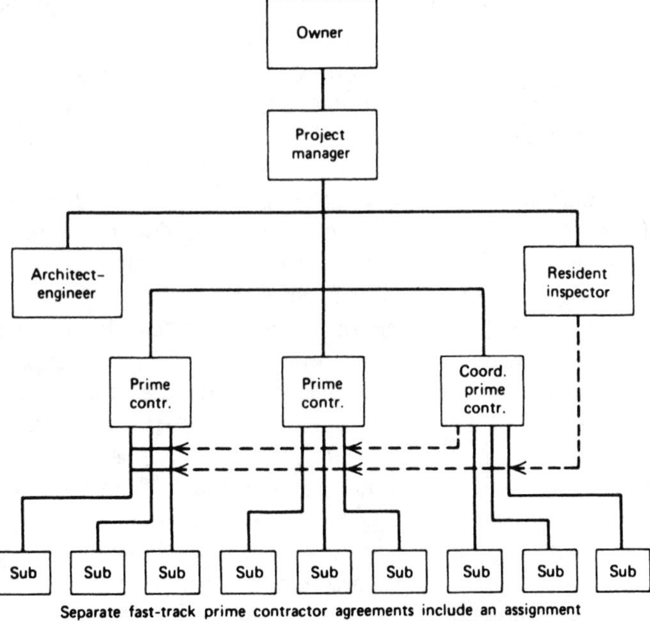
Owner
Project manager
Architect-engineer
Resident inspector
Prime contr.
Prime contr.
Coord. prime contr.
Sub
Sub
Sub
Sub
Sub
Sub
Sub
Sub
Sub
Separate fast-track prime contractor agreements include an assignment clause to allow principal (coordinating) contractor, when he is brought onto the project, to have authority for coordination and interface responsibility. This is NOT AN ASSIGNMENT OF CONTRACT.
Principal (coordinating) contractor's agreement will contain a clause requiring him to provide overall coordination between his own forces and those of all fast-track prime contractors as well as maintain interface control between the work of all tasks.

FIGURE 13.2 Phased (Fast-Track) Construction with Assigned Coordination.

Construction Schedules as Related to Building Costs

Unrealistic completion dates, coupled with exorbitant liquidated damages charges against the contractor for failure to meet such requirements, impose severe financial burdens and undeterminable risks on the contractor. Furthermore, the owner, as a direct result, pays more than the building is worth, with no guarantee that the job will be completed by the stipulated deadline.

Some of the reasons are stated here. To cope with an unrealistic completion schedule, a contractor must do everything possible to attract adequate labor from a current short supply of workers, and must add to the bid sufficient money to cover the daily costs of liquidated damages for the days that are expected to run over the completion date. In addition, the contractor must figure in the premium costs of overtime work rates, which run from one and one-half to two times the normal rate.

Another result will be evident in the disparity of bids received, as no two bidders are likely to evaluate the unknown and uncontrollable factors in the same manner. The essence of proper bidding is an accurate estimate of determinable costs within the power and judgment of the bidder. Unrealistic completion schedules and liquidated damages charges also force the bidder to guess as to the availability of materials and equipment, over which the contractor has no control. Thus, another added cost must be considered.

In addition, disputes and seemingly endless paperwork seem to result from the inevitable claims for extensions of time due to causes outside the scope of the contract.

Scheduling Methods

There are relatively few basic scheduling systems in use today, although numerous variants of each are in use. Generally, scheduling methods can be classed in four major categories:

1. Bar charts
2. Velocity charts (S-curves)
3. Line-of-balance charts
4. Network diagrams

Bar charts seem to have been with us since the beginning of time, and they are still in extensive use today. They are still an extremely useful tool and may often be seen accompanying a network diagram. The bar chart as it is used today is somewhat similar to the charts discussed by one writer in the nineteenth century involving a work versus time graphical representation. It was Henry L. Gantt and Frederick W. Taylor, however, who popularized its use early in the twentieth century. Their Gantt charts are the basis of today's bar graphs or bar charts. Although their work was originally aimed at production scheduling, it was readily accepted for planning construction and recording its progress. One of its principal advantages is that it is readily understood by all levels of management, supervision, and laypersons.

The velocity diagram used as a management tool predated network techniques; however, the method is not too well known in the United States. The velocity diagram is similar in many respects to the well-known S-curve, which is customarily used in the United States for project cost control and progress reporting (Figure 13.3). The velocity diagram as a scheduling tool shows the relationship between time and output of a construction project in a straightforward and simple way. It rep-

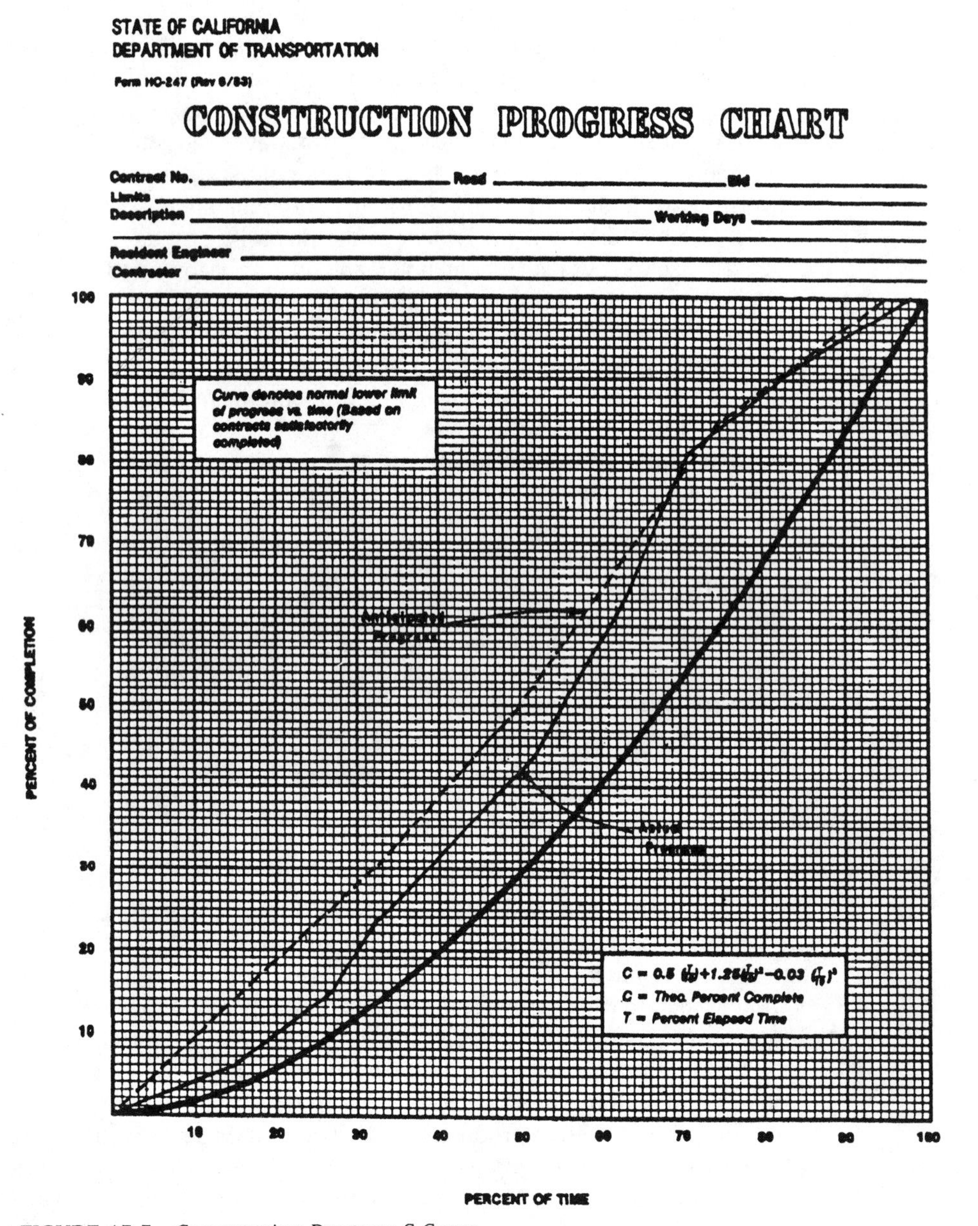

FIGURE 13.3 Construction Progress S-Curve.

resents the determined route of a construction process, or the construction volume on one axis and time on the other. The construction advance rate or *production velocity* is indicated by the slope of the production line. As with the traditional S-curve, the user can determine at a glance whether a project is proceeding on schedule or whether it is behind but catching up or falling further behind.

For that reason, the velocity diagram lends itself efficiently to projects that are linear in nature. On a project such as a pumping station or water treatment plant, there are no rational means of gauging work progress except as a function of cost. However, in a linear project, progress can be related to units of length. Examples of projects ideally suited to velocity diagram scheduling control are linear construction sites such as pipelines, canals, and highways. Although adaptable as a time/progress *reporting* tool for projects involving vertical linearity, it is not as well suited to the *control* of a vertical project such as a high-rise building because of the complex dependencies.

Line of balance is a specialized technique for repetitive work. It was derived from the manufacturing industry and has been found to be effective in planning construction work that is truly repetitive. Examples of successful applications include planning and construction of identical floors in high-rise construction and large housing developments.

Line of balance has been found to be difficult to use on projects that require a large number of trades or operations to construct each identical unit. The problems result not from the technique itself but from the difficulty of showing all of the information on one chart, especially when using the technique to monitor progress. When used to plan, it can be an effective means of relating resources, activity durations, and the general pace of work on the site.

The technique is not widely used in the United States, however, and most contractors in this country utilize either bar (Gantt) charts or network scheduling techniques. Network diagrams represent the development of systems that not only record the graphical work-versus-time relationships of each phase of the work, but also enable the user to see the interrelation and dependencies that control the project. Network planning did not come into being until the middle of the 1950s, when the U.S. Navy Special Projects Office set out with its consultants to devise a new method of planning for special weapons systems. The result was called the *Program Evaluation Review Technique,* now universally called PERT. It was an *event-oriented* system and was designed primarily as a project monitoring system. The attempt to apply PERT to the construction industry was not too successful, primarily because of event orientation, the use of three time estimates instead of one, and the technique of starting at the end and working toward the beginning. Consequently, further research was done by others in the construction field, and the result was the *critical path method,* known as CPM. This method was *activity oriented,* used single time estimates, and usually started at the beginning and worked to the end of the project.

In actual practice, the CPM and PERT techniques are very similar in principle, except for the emphasis on probability in PERT. Network planning is the basis for both systems, and although in contemporary usage a computer is generally used, they can be done by manual methods as well.

By comparing the bar chart methods with network diagramming methods, the user can see the advantages of one system over the other. With a network diagram, it can be shown to be obvious that work item E could not proceed until work item D had been completed, whereas a bar chart could only indicate the scheduled date that task E was to be performed, and if task D was late, there was no way of determining from the bar chart that task E would be delayed also because of the interrelationship between the two tasks. In the preparation of a bar chart, the scheduler is almost necessarily influenced by the desired completion dates, often working backward from the completion dates. The resultant mixture of planning and scheduling is often no better than wishful thinking.

By comparison with the bar chart and network diagram systems, the velocity chart appears to share few of the advantages of either system, yet, in turn, offers some distinctive advantages of its own. Whereas both the bar chart and the network diagrams identify specific project tasks or activities and provide the user with progress information relative to the individual activities, and whereas the network systems go even further by providing additional control by means of established activity dependencies, the velocity chart is only capable of tracking the project as a whole. What it *can* offer that neither of the other systems can, however, is the ability to display the *rate* of completion and a graphic indication of whether an off-schedule project is catching up, falling behind, or accelerating its scheduled completion date. Some engineers and contractors find it an extremely useful tool for smaller projects and seem to prefer it to either bar chart scheduling or network scheduling.

BAR CHARTS

If a bar graph is carefully prepared, the scheduler goes through the same preliminary thinking process that the network planner does. However, a bar graph does not show or record the interrelations or dependencies that control the progress of the project, and thus cannot be used alone in scheduling a project. At a later date, even the originator of a bar graph may find it difficult to explain the plan using the bar graph. In the example in Figure 13.4, a simplified bar chart shows the stages of construction of a small one-story office building. Suppose that after this 10-month schedule has been prepared, the owner asks for a 6-month schedule instead. By using the same time for each activity, the bar chart can be changed as shown in Figure 13.5. Although this may look correct at first, it is not based upon logical planning; it is merely the juggling of the original bar graph. In the case of *Mega Construction Co., Inc. v. United States* [29 Fed.Cl. 396 (1993)], involving the construction of a post office building in Canoga Park, California, the Court of Federal Claims denied a claim because the contractor's bar charts failed to establish the interrelationship between disrupted tasks and other activities on the schedule's critical path. In the words of the court, the "Plaintiff's bar chart depicted its version of the numerous work items. However, it failed to prove that the claimed delays occurred along the critical path, because it does not indicate the interdependence of any one or more of the work items that were on the critical path while the project was ongoing, but offered no credible evidence of the interdependence of the project's activities."

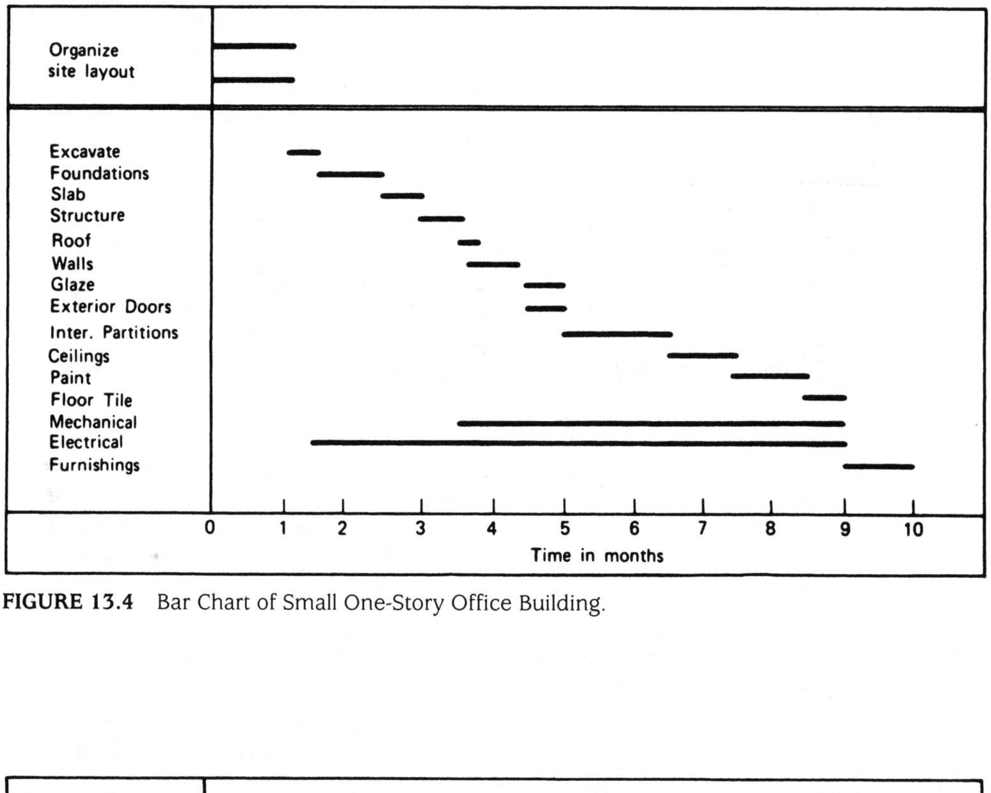

FIGURE 13.4 Bar Chart of Small One-Story Office Building.

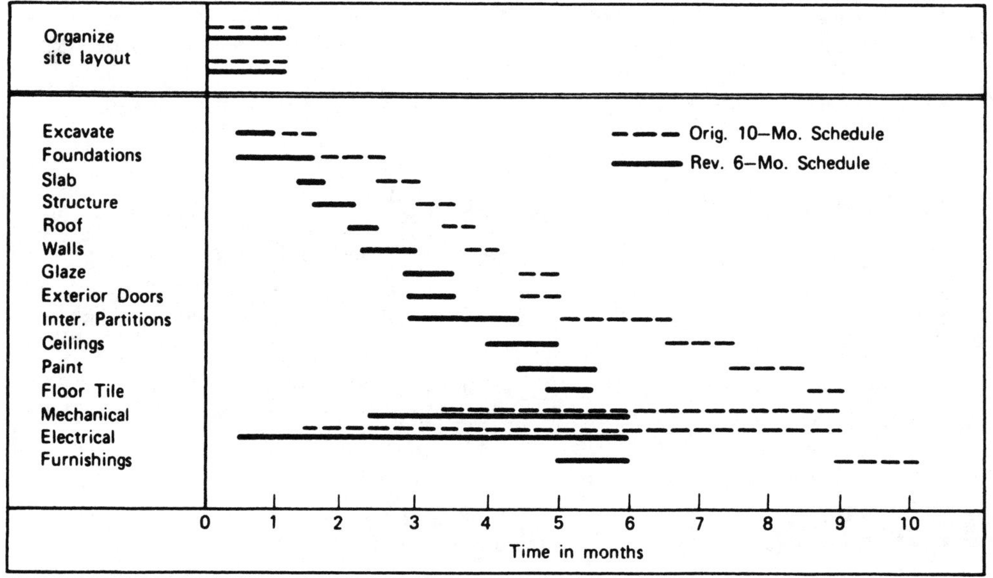

FIGURE 13.5 Revised Bar Chart of a Small One-Story Office Building.

In Figures 13.4 and 13.5, note that the general contractor's work is broken down in some detail, while the mechanical and electrical work are each shown as a continuous line, starting early and ending late. In conformance with the bar chart, the general contractor often pushes the subcontractor to staff the project as early as possible with as many mechanics as possible, although the subcontractors would like to come on the project as late as possible with as few mechanics as possible. The general contractor often complains that the subcontractor is delaying the job through lack of interest in the progress of the work. At the same time, the subcontractor is complaining that the general contractor is not turning over work areas and that the subcontractor will have to go into a crash effort to save the schedule. As in most matters, the truth is probably somewhere between both extremes. Network diagrams offer the means to resolve many of these differences with specific information rather than with generalities.

The bar chart is often actively used early in the project, but seems to be nowhere to be found later in the project. One can assume that the reasons for this may be that somewhere before the construction phase the design firm and owner are all trying to visualize the project schedule to set realistic completion dates and, once this is accomplished, lose interest in the specifics of the schedule. Most specifications require the submittal of a schedule in bar graph form by the contractor soon after the award of the contract. When the project begins to take shape in the field, the early bar charts become as useful as last year's calendar, because the bar graph does not lend itself to planning revisions.

In one form, the bar chart persists, and rightfully so. As a means of communicating job progress information to nontechnically trained people, or even to construction experts whose need to know is limited to progress data only, the bar chart excels as a means of showing such data in a clear and concise manner. Such charts record the progress in each of the major elements of construction as a solid bar along a corresponding time scale, and are generally updated monthly (Figure 13.6).

S-Curve Scheduling or Velocity Diagrams[1]

Although it was stated earlier in the chapter that a velocity diagram (Figure 13.7) is essentially a form of the popular S-curve, it is how the diagram is used as a scheduling tool that makes it unique, and thus deserving of its own identity. As with the S-curve, the construction advance rate is indicated by the slope of the line. During the actual construction process, the Resident Project Representative or project manager must compare the scheduled construction velocity with that of the actual construction velocity (rate of progress). From the updated velocity diagram, surplus or deficit can be recognized easily. Therefore, decisions such as starting construction at a second place on the project can be made, in case the actual velocity is lower than the scheduled velocity, and compliance with the time deadline has first priority.

[1] J. Dressier, "Construction Management in West Germany," *Journal of the Construction Division, ASCE*, Vol. 106, No. C04, Proceedings Paper 15878, December 1980, pp. 477–481.

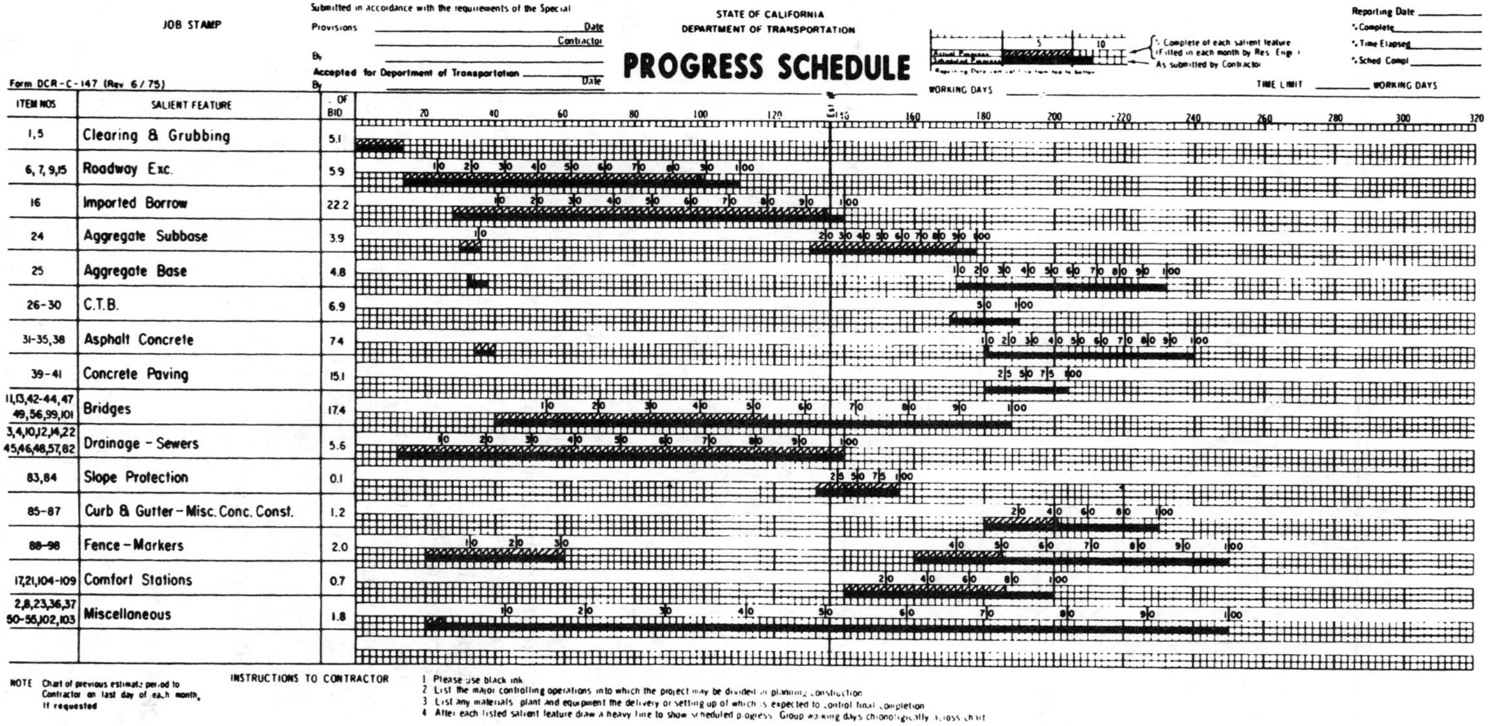

JOB STAMP

Submitted in accordance with the requirements of the Special Provisions ______ Date ______ Contractor

By ______

Accepted for Department of Transportation ______ Date

By ______

STATE OF CALIFORNIA
DEPARTMENT OF TRANSPORTATION

PROGRESS SCHEDULE

% Complete of each salient feature (Filled in each month by Res. Engr.)
As submitted by Contractor

Reporting Date ______
% Complete ______
% Time Elapsed ______
% Sched Compl ______

WORKING DAYS ______ TIME LIMIT ______ WORKING DAYS

Form DCR-C-147 (Rev 6/75)

ITEM NOS	SALIENT FEATURE	% OF BID
1,5	Clearing & Grubbing	5.1
6,7,9,15	Roadway Exc.	5.9
16	Imported Borrow	22.2
24	Aggregate Subbase	3.9
25	Aggregate Base	4.8
26-30	C.T.B.	6.9
31-35,38	Asphalt Concrete	7.4
39-41	Concrete Paving	15.1
11,13,42-44,47 49,56,99,101	Bridges	17.4
3,4,10,12,14,22 45,46,48,57,82	Drainage - Sewers	5.6
83,84	Slope Protection	0.1
85-87	Curb & Gutter - Misc. Conc. Const.	1.2
88-98	Fence - Markers	2.0
17,21,104-109	Comfort Stations	0.7
2,8,23,36,37 50-55,102,103	Miscellaneous	1.8

Working days scale: 20 40 60 80 100 120 140 160 180 200 220 240 260 280 300 320

NOTE: Chart of previous estimate period to Contractor on last day of each month, if requested

INSTRUCTIONS TO CONTRACTOR

1. Please use black ink.
2. List the major controlling operations into which the project may be divided in planning construction.
3. List any materials, plant and equipment the delivery or setting up of which is expected to control final completion.
4. After each listed salient feature draw a heavy line to show scheduled progress. Group working days chronologically across chart.

FIGURE 13.6 Bar Chart Progress Schedule for Unit-Price Project.

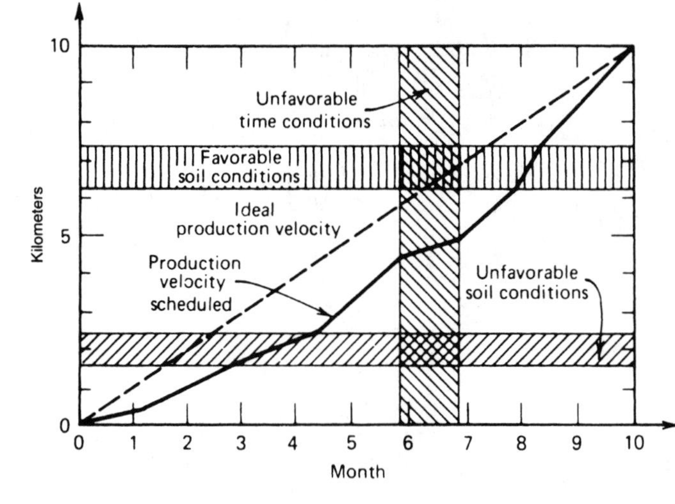

FIGURE 13.7 Velocity Diagram as Time-Scheduling Tool.

A simplified illustration of a velocity diagram applied to a pipeline project is illustrated in Figure 13.7. Here, if a pipeline must be built from station A to station B, a distance of 10 kilometers (6.2 miles) in 10 months (100 percent), the ideal production velocity would be 1 kilometer per month (dashed line in Figure 13.7). However, as there will be changing soil or other environmental conditions, and one will also encounter changing weather conditions, the most probable schedule must be developed according to the solid line.

Figure 13.7, of course, represents a greatly oversimplified example. However, the velocity diagram has proved its efficiency for complicated operations as well, in particular when schedules depend on minimum time lags between parallel operations, or minimum distances relating to space, and so forth. In that context, the velocity diagram was used as a management tool for the first time during the planning and construction of the St. Gotthard Tunnel in Switzerland about 100 years ago.

As a reporting device, instead of a tool for *control,* the popular S-curve is frequently used as a management reporting tool to show comparisons of anticipated progress with actual progress, even where bar charts or network diagrams are being used to schedule the work in the field. An example was shown in Figure 13.3, where an S-curve was used by the California Department of Transportation as a monthly progress reporting device. Another variation of the S-curve is based upon cost instead of time and is used as a cost control tool (see Figure 15.1).

Line-of-Balance Charts

Line-of-balance charts (Figure 13.8) are used to plan the construction of a number of similar items. The technique is used to analyze the application of labor and plant resources to assure that each resource can progress from one item to the next in an orderly way, completing its own work on all the items without being delayed in waiting for the preceding work to be completed. Thus, the technique is

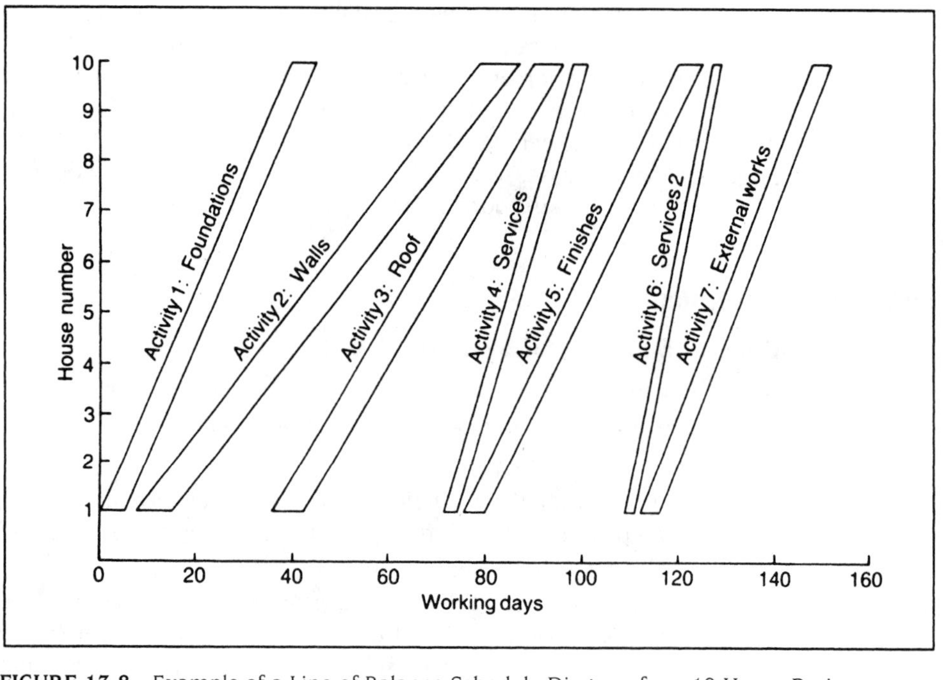

FIGURE 13.8 Example of a Line-of-Balance Schedule Diagram for a 10-House Project.

based upon the concept of keeping all of the resources in balance, each following the other productively.

The main objective of using line-of-balance is to make optimum use of all resources. To accomplish this, it is important that the activities and resources are related closely (i.e., the job is planned on a trade basis).

The purpose of a line-of-balance analysis is to balance the rate of progress of the activities, and to schedule the activities to eliminate interference. This is done by:

1. Adjusting the rate of production for each activity so that this approximates to a common rate of production for all activities
2. Delaying the start of those activities that (even after adjustment) proceed faster than the activity immediately preceding them, to maintain at least the minimum buffer specified at all times

In the classical, factory-based line-of-balance analysis, it is assumed that individual resources make equal contributions to progress, regardless of the number of resources used. For example, if a task takes 10 person-hours, one person would complete the task in 10 hours, whereas 10 people would complete the task in 1 hour. Thus, the rate of progress of each activity may be adjusted quite finely; consequently, the activities may be made to work at almost the same rate of progress.

In construction, such an approach is unrealistic. Tradesmen rarely work as individuals or in large groups. Years of experience and practice have established the most effective size for a group.

In construction, estimates of activity duration are only approximate. In repetitive construction where the activities and trades follow in succession, any delay in the planned completion of an activity will result in the following trades having to wait unproductively for its completion. Therefore, it is considered as wise to plan a short delay or buffer between each activity. The estimate of buffer times is related to the project manager's assessment of the reliability of the estimate of activity durations. Where the reliability is poor, large buffers must be used.

Network Diagrams

General Summary of Systems in Use

Network diagrams, so called because of the net effect of the interconnecting lines used to indicate dependencies and interrelationships, can be divided into two basic categories.

1. *CPM* or *critical path method,* which was originally developed specifically for the planning of construction from PPS (Project Planning and Scheduling), which was originated by du Pont.
2. *PERT* or *Program Evaluation Review Technique,* which was originated by the Special Projects Office of the Navy Bureau of Ordnance, and with few exceptions is used almost exclusively in military, aerospace, and project management work. Adaptations of the basic technique have been developed for use in construction.

Occasionally, modifications are used that utilize some of the separate features of the previously mentioned systems. Contractors for construction of ground support and launching facilities for missiles in space may encounter PERT in construction; otherwise, it appears that little will be seen of it in construction in an unmodified form.

Critical Path Method

The workhorse of network scheduling methods used in construction is the popular critical path method (CPM), which is used almost universally wherever network scheduling methods are called for. The CPM system shows the order and interdependence of activities and the sequence in which the work is to be accomplished. The basic concept of the network analysis diagram is to show the start of each given activity and its dependence upon the completion of preceding activities, and how its completion also restricts other activities to follow. The CPM network provides for the construction phase only, using either manual or automated methods, involving the following activities:

1. Long-lead purchases and deliveries of critical materials
2. Fabrication, installation, and testing of critical equipment
3. Submittal and approval of material samples and shop drawings
4. All activities that affect progress on the job
5. Required dates of completion for all activities

If a project is of such size that a single network cannot be easily shown on a single sheet, a summary network diagram will probably be provided. The summary sheet is usually a network of from 50 to 150 activities and is based upon the detailed diagrams of all the remaining tasks. The mathematical analysis of the CPM network diagram includes the following information for each activity:

1. Preceding and following event numbers
2. Activity description
3. Estimated duration of activities
4. Earliest start dates (calendar dates)
5. Earliest finish dates (calendar dates)
6. Scheduled or actual start date
7. Scheduled or actual finish date
8. Latest start date (calendar date)
9. Latest finish date (calendar date)
10. Monetary value of activity
11. Responsibility for activity (prime; subs; suppliers; etc.)
12. Labor required
13. Percentage of activity completed as of each report
14. Contractor's earnings based upon portion of work completed
15. Bid item of which the activity is a part

In addition to the tabulation of activities, the CPM computer printout should also include an identification of activities that are planned for expediting by the use of overtime or double shifts to be worked, including possible Saturday, Sunday, or holiday work. It should also provide an on-site manpower loading schedule and a description of the major items of construction equipment planned for operations on the project. Where portions of the work are to be paid for under unit costs, the estimated number of units in an activity that was used in developing the total activity cost should be shown.

The computer printout generally sorts certain classes of frequently used data into groups or "sorts." Generally, the data on a CPM will be grouped into the following sorts:

1. By the preceding event or activity number from the lowest to the highest, and then in order of the following event number.
2. By the amount of slack, then in the order of earliest allowable start dates.
3. In order of the latest allowable start dates, then in order of preceding event numbers, then in order of succeeding event numbers.

The data provided must be timely, must be responsive to the needs of management at all levels, and must be fully capable of providing a sound basis for management decisions. The computer printouts, in which automated methods are used, will include the following:

1. Sorts by early and late start, criticality, responsibility, and building area.
2. Allocation of material and labor costs to each work item.

3. Generation of cash-flow projections and contractor payment request verifications.
4. Master schedule, design schedule, preconstruction schedule, and occupancy schedule.
5. Separate reports, including summary and bar charts, contractor payments, purchase orders, shop drawings, and samples.

PERT Management Control Systems

PERT can be used by private architect/engineers in the project management of numerous projects. PERT provides for the design phase of a project using both manual and automated procedures as a support for the following functions:

1. Planning
2. Organizing
3. Scheduling
4. Budgeting
5. Reporting of design and construction progress
6. Reporting of design and construction expenditures
7. Accounting
8. Documentation
9. Identification of variances and problems
10. Decision making
11. Decision implementation

Learning to Use a Network Diagram

Network diagrams may seem unduly complicated to read at first, and if you happened to be called upon to try to develop one, they would indeed *be* complicated. Fortunately, the Resident Project Representative's responsibility in network diagraming is limited to reading, understanding, and using the diagrams to good advantage on the project. The skill of learning to read and understand them is not that complicated and can be acquired by anyone willing to learn a few of its basic principles. These are explained in more depth in Chapter 14.

It is essential that the Resident Project Representative should become reasonably familiar with the principles of network diagraming so as to perform effectively the associated duties and responsibilities involved with those projects that use them. The inspector not only must understand the network diagrams themselves, but also should be capable of reading and interpreting the computer printouts of all CPM data received by the field office. Architects, engineers, and Resident Project Representatives, as owner's agents, should check the content of contractor schedules to ensure they know the status of the project. Updated schedules should not just be received and filed, as that would be wasteful of the owner's money. Admittedly, the system may only be encountered on larger projects, but a basic knowledge of the general principles of network diagraming should be a part of every Resident Project Representative's education.

Specifying CPM for a Project

There are a number of sources for guide specifications for preparing a meaningful CPM specification, but the one favored by the author is a slightly modified version of the guide specification published by the Associated General Contractors of America in its book *Using CPM in Construction.* The principal modifications made by the author were to require time-scaled network diagrams, provide columns of both free float and total float on the report form, and specify frequency of printouts not exceeding four weeks apart. In addition, the author ties in the submittal of the schedule to the owner or engineer with the contractor's initial mobilization payment to assure that the contractor will abide by the requirements to submit the proposed schedule in time. A sample specification meeting these requirements can be seen in Figure 13.9.

Computerized Progress Payments

Progress payments are a logical extension of the project scheduling environment. If the project can be logically scheduled, priced, and tracked on a regular basis, an effective, accurate payment schedule can be easily developed for that project. That, of course, depends upon who controls the payments and the nature of the contract agreement regarding progress payments. The simplest payment structure is full payment upon completion. This system is generally limited to short-term projects, however. Another payment structure, usually seen in commercial construction, is the regular periodic payment of a fixed percentage of the total project cost or lump-sum value. By far the most common payment structure is regular payments based upon the percentage of work completed within the payment period based upon unit-price values or a *schedule of values.*

As long as the project stays on schedule and within budget, there should be no substantial difference between costs and revenues. However, in many instances, changes in both time and costs occur with surprising regularity. In such events the project will immediately display a variance between costs and revenues. The knowledgeable contractor will submit requests for payment that are out of conformance with the baseline project schedule and show supporting documentation for the cost difference. In little time, the cost of accounting begins to escalate. If done by hand, over a long construction period, the potential for error becomes staggering. Here the PC and its associated software can more adequately fill the need for progress payment control.

Some progress payment software has been developed in response to the observed need for a quicker and more efficient means of producing accurate progress pay estimates for large jobs. In the past, the usual method of producing pay estimates involved (1) collecting data from field personnel on percentage of work completed, materials used from storage, and change orders for each payment period, and (2) hand calculating and typing of the revised estimate. Most of this work is easily adapted to computers, assuring numerical accuracy and allowing for easy revisions of project data. An interactive program package can reduce the time needed to

SECTION 01310

CPM CONSTRUCTION SCHEDULE

1.01 GENERAL

A. The project management scheduling tool "Critical Path Method," a network scheduling system commonly called CPM, shall be employed by the Contractor for planning, and scheduling, of all work required under the Contract Documents. All schedule reports shall be in the form of computer printouts. The Contractor may elect to use bar charts (Gantt Charts) as an onsite scheduling tool; provided, that all such bar charts shall be generated from the approved CPM network.

1.02 QUALIFICATIONS

A. A statement of computerized CPM capability shall be submitted in writing prior to the award of the Contract and shall verify that either the Contractor's organization has in-house capability qualified to prepare and use the CPM scheduling technique, or that the Contractor employs a CPM consultant who is so qualified.

B. In-house capability shall be verified by description of construction projects to which the Contractor or its consultant has successfully applied computerized CPM scheduling techniques and shall include at least 2 projects valued at not less than half the expected value of this project, and at least one project which all project activities and procurement were controlled throughout the duration of the project by means of a computerized, periodic, systematic review of the CPM schedule.

1.03 SUBMITTAL PROCEDURES

A. Submittal of all schedules and schedule reports shall be as specified in Section 01300 Contractor Submittals

1.04 CHANGE ORDERS

A. Upon issuance of a Change Order or Work Directive Change, the approved change shall be reflected in the next submittal of the Revised Construction Schedule by the Contractor.

1.05 CPM STANDARDS

A. Definition: CPM scheduling, as required by this Section, shall be interpreted to be generally as outlined in the Associated General Contractors publication, "The Use of CPM in Construction," except that either arrow or precedence diagramming format is acceptable.

FIGURE 13.9 Sample Guide Specification for CPM Scheduling Based upon AGC Recommendations.

B. Construction Schedules: Construction schedules shall include a computer-generated graphic network and computerized, construction schedule reports, as described below.

C. Networks: The CPM scheduling network shall be in the form of a time-scaled arrow or precedence diagram, shall be of the customary activity-on-arrow or activity-on-node type, and may be divided into a number of separate pages with suitable notation relating the interface points among the pages. Individual pages shall not exceed 900 mm by 1525 mm (36-inch by 60-inch). Notation on each activity arrow shall include a brief work description and a duration estimate (see below). Precedence diagrams in box-node format are not acceptable.

D. All construction activities and procurement shall be indicated in a time-scaled format and a calendar scale shall be shown on all sheets along the entire sheet length. Each activity arrow or node shall be plotted so that the beginning and completion dates and free float time of said activity can be determined graphically by comparison with the calendar scale. All activities shall be shown using the symbols that clearly distinguish between critical path activities, non-critical activities, and free float for each non-critical activity. All non-critical path activities, shall show estimated performances time and free float time in scaled form.

E. Duration Estimates: The duration estimate indicated for each activity shall be computed in working days, converted to calendar days, and shown on the construction schedule in calendar days, and shall represent the single best estimate considering the scope of the Work and resources planned for the activity. Except for certain non-labor activities, such as curing concrete, paint drying, procurement, or delivering of materials, activity durations shall not exceed 10 working days (14 calendar days) nor be shown as less than one working day unless otherwise accepted by the Engineer.

F. The requirement for activity durations not in excess of 10 days shall apply to all schedule submittals excepting the Proposed Construction Schedule required to be submitted prior to the Notice to Proceed.

G. Schedule Reports: Schedule Reports shall be prepared from the Initial Construction Schedule and from all subsequent Revised Construction Schedules, and shall include the following minimum data for each activity:

1. Activity Numbers or i-j Numbers.

2. Estimated Activity Duration.

FIGURE 13.9 (Continued)

3. Activity Description (including procurement items)

4. Early Start Date (Calendar Dated).

5. Early Finish Date (Calendar Dated).

6. Late Start Date (Calendar Dated).

7. Late Finish Date (Calendar Dated).

8. Status (Whether Critical).

9. Total Float for Each Activity.

10. Free Float for Each Activity.

H. Project Information: Each Schedule Report shall be prefaced with the following summary data:

1. Project Name.

2. Contractor Name.

3. Type of Tabulation (Initial or Revised; if revised, show revision number or date).

4. Project Duration.

5. Project Scheduled Completion Date.

6. The date of commencement of the Work as stated in the Notice to Proceed.

7. If an updated (revised) schedule, cite the new project completion date and current project status.

1.06 CONSTRUCTION SCHEDULE MONITORING

A. At not less than monthly semi-monthly weekly intervals, and when specifically requested by the Engineer, the Contractor shall submit to the Engineer a computer printout of a Revised Schedule Report for those activities that remain to be completed.

B. Each Revised Construction Schedule, including an updated network diagram, if required, shall be submitted in the form, sequence, and in the number of copies requested for the Initial Construction Schedule.

FIGURE 13.9 (Continued)

C. Where it is elsewhere provided in these Specifications that payments will be allowed for materials delivered to the site but not yet incorporated in the Work, separate pay items shall be established for such materials and for the furnishing and the installation of such items. Costs of such materials delivered to the site but not yet incorporated into the Work shall not be included in the cost value of the installation of such materials but shall be covered under a separate cost value report.

- END OF SECTION -

FIGURE 13.9 (Continued)

produce a completed estimate from a week to less than 2 days. Along with reducing turnaround time, there can be substantial reductions in labor and other costs required to produce progress payment estimates.

The basic tool of progress payment control is the accounting process. Simply put, you try to pay only once for something and then pay for it only after you get it. There are as many "fair" ways to accomplish this as there are parties to the contract. The ground rules for progress payments should be written into the contract; in this way, everyone's needs will be served.

Lump-Sum Projects

On lump-sum contracts, there are two methods available to utilize a PC computer in determining the amount of payment due to the contractor each month:

1. Progress payment program designed for unit-price contracts (PROGPAY, QuatroPro, or Excel) but adapted to a lump-sum project by utilizing the values shown on a schedule of values created for payment purposes.
2. Cost loading the network schedule by assigning a cost value to each work activity. Then, if the schedule is being followed faithfully, payment will be computed based upon the number of work items completed each month. (A schedule of values is not used under this plan.)

Unit-Price Projects

On unit-price projects there is really only one option, namely, the use of a small program expressly created for the purpose of computing progress payments (PROGPAY, QuatroPro, or Excel) based upon the number of units completed of each item on the bid sheet. Cost loading of a CPM for a unit-price job is counterproductive, as it leads to a considerable increase in payment administration costs because of the disparity between the pay-line items as shown in the unit-price bid sheet and the separate work activities shown in the CPM network, neither of which can be effectively correlated. In short, you will be comparing apples with oranges.

Selection of PC Scheduling Software

A word of caution if you are planning to buy a CPM program for your personal computer: It seems that a large number of the earlier scheduling programs on the market were created by nonengineer computer programmers to meet design-oriented scheduling needs and who failed to recognize the practical needs of the construction engineers. Some of these programmers appeared to show an elementary knowledge of some facets of the construction industry—a knowledge that was sorely lacking in others.

Several of the programs found by the author are either PERT systems or combinations of PERT/CPM, with far too much of the PERT influence. In the author's opinion, PERT systems have no place as an operational tool for construction scheduling in the field office. PERT was conceived as a project-management scheduling technique for situations in which neither time nor cost could be accurately estimated, and completion times therefore had to be based upon probability. In most construction projects, however, both time and cost can be reliably estimated, and thus the more definitive approach provided by CPM is practical.

It is therefore important to understand that PERT systems involve a "probability approach" to the problems of planning and control of projects and are best suited to reporting on works in which major uncertainties exist. Although there are some uncertainties in any construction project, the cost and time required for each operation involved can be reasonably estimated, and all operations may then be reviewed by CPM in accordance with anticipated conditions and hazards that may be encountered on the site.

PERT, then, is of little value in a construction field office. In fact, if you are obligated to construct an as-built schedule such as that used in claims preparation, such a statistical approach is totally useless. What is actually needed is a program that will allow the user to enter only *i-j* numbers, activity identities, activity durations, and dependencies or job logic and in return receive a report format from the computer that will identify early and late start dates, early and late finish dates, actual start and actual finish dates, total float, and free float. Any of the various CPM formats, whether they be activity-on-arrow, node or precedence diagraming, or activity-on-node, are all quite suitable for field engineering use, as long as they are basically CPM programs, *not* PERT programs.

Before buying software for a PC to use for schedule review, changes, or "what if?" scheduling investigations, it would be wise to check carefully to see that the programs offered are properly suited to construction scheduling. It should be a true CPM system, specifically designed for construction by construction engineers. Far too many of the programs currently offered are designed for project management, design management, or a manufacturing plant, not for actual construction scheduling.

There seem to be many people sitting in ivory towers, who have never even gotten their boots muddy, trying to second-guess what the field engineer needs. The sooner the programmers get around to actually working with construction engineers to solve construction problems instead of just the design engineers and office-trained project managers, the sooner the problems will disappear. PERT has its place, but not in construction.

Typical CPM Software Available

Several CPM scheduling programs that truly reflect the needs of the owner's construction engineer are available in the software marketplace; most available software, unfortunately, does not. Some even attempt to emphasize PERT in their construction scheduling program. The good ones, however, get right down to the business of scheduling by CPM using either arrow or precedence diagram techniques, or both.

Although there is a large movement toward the generation of network graphics by computer, principally to keep the cost down, many of the large companies whose business depends entirely on the preparation of construction schedules for contractors still resort to the old tried-and-true hand-drawn networks. There is good reason for the preservation of the hand-drawn networks: They are much easier to follow, as the draftsperson who produced them has the capability of expanding wherever necessary to present the information clearly, while the computer on a complex project creates a document that requires considerable effort and visual skills to follow. The difficulty rests primarily in the area of reading dependency lines that are often too close together and cross other activities, making them very difficult to follow without a magnifying glass and a colored pencil to trace the lines (Figure 13.10). The plotting capabilities of the computers seem to be limited to one or more of three precedence diagraming formats, and none of them seem to have solved those problems.

Such programs are available to operate on most of the major microcomputers utilizing MS-DOS, CP/M, and UNIX operating systems, as well as a few others. For microcomputers, the greatest selection available, of course, is written for the various IBM-compatible computers.

Many of the available scheduling programs offered by the software manufacturers refer to all network scheduling diagrams as PERT charts, when in fact they may actually be CPM charts that have been misnamed by the programmer or software supplier because of an unfamiliarity with network scheduling terminology. The only way you can tell if a program is going to provide the desired information is to see it and try it. Some initial confusion seems to result from the similar-sounding terms used by the programmers or software suppliers and the construction engineers. To the programmer and the software supplier, CP/M is a computer operating system having nothing to do with CPM scheduling; to the construction engineer, CPM is a particular type of network scheduling system, whether done manually or by computer.

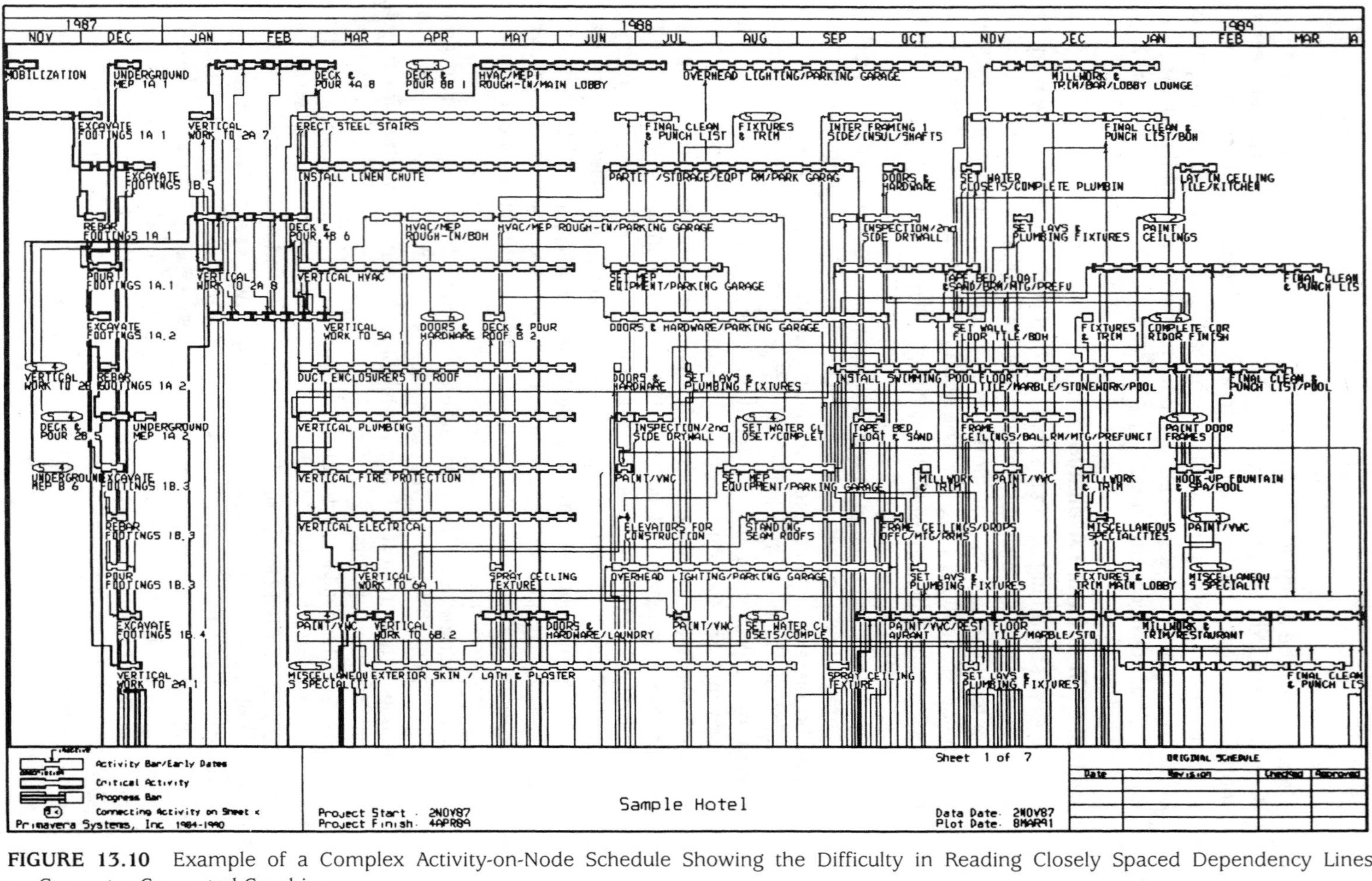

FIGURE 13.10 Example of a Complex Activity-on-Node Schedule Showing the Difficulty in Reading Closely Spaced Dependency Lines on Computer-Generated Graphics.

PROBLEMS

1. Phased construction is also known by what other name?
2. Name four types of scheduling methods.
3. Line-of-balance scheduling is best adapted to what type of application?
4. What is the principal advantage of the traditional S-curve?
5. What is the primary disadvantage in the use of bar charts for scheduling?
6. Name the types of network scheduling systems in use and state the proper type of work for each system.
7. In computer reports, what is a "sort"?
8. Identify which system of network scheduling is best adapted to each of the following types of work:
   (a) Construction scheduling
   (b) Research and development
   (c) Project management
   (d) Fast-track construction
9. It is well known that computerized CPM can be used for effective scheduling of time. Name two other applications that are possible using CPM.

14

CPM SCHEDULING FOR CONSTRUCTION

Construction projects are complex, and a large job will literally involve thousands of separate operations. If a project is to be completed within the time called for in the contract, the work must be very carefully planned and scheduled in advance. If all the tasks would simply follow each other in consecutive order, the job of scheduling would be much easier. Unfortunately, the problem is not that simple. Each operation within the project has its own time requirement, and often it cannot start until certain other operations have been completed. There are many other tasks, however, that can be carried on simultaneously because they are entirely independent of one another. Thus a typical project involves many tasks that are interdependent as well as many other tasks that are totally independent of one another, and when interrelated in a project they create a tangled web of time and sequence relationships. When all these tasks are superimposed, it becomes obvious that project planning and scheduling is a very complicated and difficult management function.

It is not the intent of this book to go into the subject of CPM scheduling in sufficient depth to enable the resident engineer or inspector to be able to set up a project schedule by CPM, but certainly deep enough so that he or she should be capable of reading, reviewing, and understanding one. While the contractor uses the schedule to manage the work, one of the most important purposes for the schedule is to allow both the contractor and the Resident Project Representative to monitor its progress. It is this monitoring of the work that gives rise to the concept of Management By Exception. As the project progresses, if an activity falls behind the schedule, it is that activity, and that activity alone, that gets detailed management attention, the presumption being that the other activities are progressing correctly and need no additional attention. This allows a more efficient use of managerial manpower than giving equal attention to all activities, regardless of whether they are on schedule or not. It is often one of the Resident Project Representative's responsibilities to make regular evaluations of the contractor's construction schedules to determine whether they are meeting their schedule requirements and will complete the work within the agreed time. The project representative should know the subject well enough to know what to look for, and be fully capable of recognizing the difference between a logical and an illogical chart.

The traditional basis in past years for scheduling construction work had always been the bar and S-charts. There is no question but that these are still useful tools for showing the established schedule of operations and recording its progress. However, the bar chart falls somewhat short of being an adequate tool for project planning, and the resulting construction schedule is based more upon the contractor's experience and intuition than on any rational analysis of the work to be performed. Its major weakness lies in the fact that it does not show the interrelationships and interdependencies that exist among the various phases of the work. Also, there is no way to determine which operations actually control the overall time progress of the project. In fact, numerous court decisions would indicate that a CPM schedule is the *only* way to adequately see the logic of the project and the impact of everyone's actions dealing with delays, acceleration, and change orders. The S-curve chart or velocity diagram as a scheduling tool is better suited to linear projects on smaller contracts.

CPM: What It Is and What It Does

The CPM is essentially a project management system that covers the construction phase of a project and allows the user to aid the decision-making process by guiding the contractor in selecting the best way to expedite the job and by providing a prediction of future labor requirements as well as equipment needs. Completing a project successfully is highly unlikely when decision making consists of the contractor's superintendent waking up in the morning and asking "What shall we do today?". A simple schedule, well thought out, is more valuable than a complex one that no one understands, and hence no one follows. It should be prepared by someone who understands scheduling and job logic rather than by someone who is just entering keystrokes on a computer keyboard into a high-priced scheduling software program. The schedule is used to effectively coordinate work activities and fosters cooperation among all the project participants.

Project planning is the first step in a CPM procedure. This step consists of the following:

1. Identify the elementary work items needed to complete the job.
2. Estimate time durations for each activity.
3. Determine what tasks must be completed before each work item can be started.
4. Establish the logical order in which these work items must be done.
5. Prepare a graphic display in the form of a network diagram.

The next step is the scheduling phase, and it requires an estimate of the time required to accomplish each of the work items previously identified. With the use of the network diagram, computations are made to provide information concerning the time schedule characteristics of each work item and the total time necessary to complete the project.

Although these comments seem to suggest that CPM must follow a definite step-by-step order, this is not the case in actual usage. For example, the five planning steps often proceed concurrently. For the purposes of this discussion, it will be assumed that they are treated separately in the order listed.

The computations previously mentioned are actually only simple additions and subtractions. Although the actual computation is simple and very easy, there are usually so many of them required on an actual project that the process becomes very tedious. For this reason, many contractors use computers to produce their CPM schedules. Furthermore, with the use of the personal computer, the schedule may be updated even on a weekly basis without undue hardship. With the addition of some programmed logic in a computerized network system, complex scheduling problems can be worked out rapidly and optimum solutions can be reached. A task that would be impossible within the allowable time if it had to be done by hand methods can be accomplished in minutes by computer after developing the input data required. For more sophisticated schedules, some computer management firms can provide additional levels of service for network scheduling to owners, contractors, architects, and engineers.

For a full understanding of the method, some understanding of the calculations that are necessary and an understanding of the terminology used are required. In addition, there are many applications of CPM in which manually developed data are adequate and quite usual. Therefore, the following explanations are based upon manual procedures. They are followed by some examples showing how the data developed are printed out when computer methods are used, along with some elementary instructions as to the use of the computer printouts by the Resident Project Representative.

Basic Procedure in Setting Up a CPM Schedule

Normally, the network scheduling is started right after the award of the project. Because the prime purpose of the system is to produce a coordinated project plan, the principal subcontractors must also be entered into the planning stage. Normally, the general contractor sets the general timing for the project; the individual subcontractors then review their portions of the work, and the needed alterations are made.

The basic procedure used by the planning group is to "talk" the project through first. This way the project is subject to careful, detailed, advance planning. This planning alone justifies the time spent on CPM. Usually, the network diagram is then constructed in a rough form and the job is broken up into basic elements; then the sequential order of construction operations is discussed. It is often helpful to list the major operations of the project and use them as a means of developing the preliminary diagram.

Project Planning

The first phase of CPM is that of planning. The project must first be broken down into time-consuming activities. An *activity* in CPM is defined as any single identifiable work step in the total project. The extent to which the project is subdivided into activities depends upon the number of practical considerations; however, the following factors must be taken into account:

1. Different areas of responsibility, such as subcontracted work, that are distinctly separate from the work being done directly by the prime contractor.
2. Different categories of work as distinguished by craft or crew requirements.
3. Different categories of work as distinguished by equipment requirements.
4. Different categories of work as distinguished by materials such as concrete, timber, or steel.
5. Distinct and identifiable subdivisions of work such as walls, slabs, beams and columns.
6. Location of the work within the project that necessitates different times or different crews to perform.
7. Owner's breakdown for bidding or payment purposes.
8. Contractor's breakdown for estimating or cost accounting purposes.
9. Outage schedules or limiting times that existing utility services may be interrupted to construct the project.

The activities chosen may represent relatively large segments of the project or may be limited to only small steps. For example a concrete slab may be a single activity or it may be broken into separate steps necessary to construct it, such as erection of forms, placing of steel, placing of concrete, finishing, curing, and stripping of forms or headers.

As the separate activities are identified and defined, the sequence relationships between them must be determined. These relationships are referred to as *job logic* and consist of the necessary time and order of construction operations. The three logic relationships are predecessor, successor, and concurrent. A *predecessor* activity is one that is followed by another activity. A *successor* activity is one that is preceded by another activity. *Concurrent* activities are those activities that can proceed at the same time. When the time sequence of activities is being considered, *logic constraints* must also be considered; these are the practical limitations that can influence the start or finish of certain activities. For example an activity that involves the placing of reinforcing steel obviously cannot start until the steel is on the site. Therefore, the start of the activity of placing reinforcing steel is "restrained" by the time required to prepare and approve the necessary shop drawings, fabricate the steel, and deliver it to the job. It is quite common to treat logic constraints much the same as activities and to represent them on the network diagram.

Fundamentals of CPM

The first step in understanding the critical path method (CPM) is to learn the meaning of the terms used, the symbols involved, and the rules of network scheduling. The following paragraphs are presented to define the principal terms and summarize the most important of the rules for network planning and scheduling.

Activities

After the activities have been identified and their logic established, it is time to construct the job graphically in the form of a network diagram. If a computer is to be used to develop the network, it is still best to start by hand-drawing a network diagram. Starting with the computer keyboard is similar to installing the electrical system in a building without a wiring diagram. This can be a costly way of finding errors.

The basic symbol for an activity is an arrow or a bar. It is a general practice to think of the arrow as moving from left to right, and that "time" also passes from left to right on the diagram. A basic relationship is that of the event to an arrow or activity. As distinguished from an activity, an event is the instant of time at which an activity is just starting or finishing. An activity is preceded by an event and followed by an event; in other words, it has to have both a starting point and a stopping point. The arrows representing activities are *not* vectors, and their lengths and slopes are not significant. Also, unlike precedence diagraming, in arrow diagraming format the activity arrows can be straight, bent, curved, or whatever shapes the user chooses. The real essence of the diagram is the manner in which the activities are joined together into a total operational pattern or network.

Each activity in an arrow network diagram is shown as an arrow along with a pair of circles representing the starting and finishing events of each activity; then, each activity is numbered for reference. In the activity arrow in Figure 14.1, the activity is "place concrete" and its *i–j* designation is (9–10). As an illustration of some of the activities that might be involved in a "project" to drill a hole 660 mm (24 inches) in diameter by 4.6 m (15 ft) deep and fill it with concrete, the activities involved might be as follows:

Approve and sign contract.
Obtain building permit.
Order and deliver drill rig.
Order and deliver concrete.
Locate and lay out the hole.
Drill hole.
Place concrete in hole.
Clean up site.

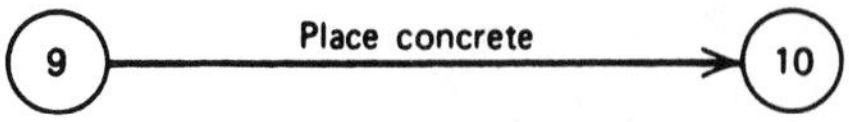

FIGURE 14.1 Activity Arrow.

Job Logic

The job logic or time-sequence relationships among the various activities involved in the previously mentioned project are the next steps to determine. The sequence of operations will be as shown in Figure 14.2. In elementary form, this is the type of information that is generated while a project is being "talked through." For the purposes of CPM, job logic requires that each of the activities in the network have a definite event to mark its starting point. This event may be either the start of the project or the completion of preceding activities. It is not possible in CPM to have the finish of one activity overlap beyond the start of a succeeding activity. When such a condition appears to present itself, it is a sign that work must be further subdivided. It is a fundamental rule that *a given activity cannot start until all those activities immediately preceding it have been completed.* Normally, the job logic is not written down in tabular form. It is only for the purposes of presenting a clear example that a separate tabulation is made here. In practice, an arrow diagram may be drawn along with the development of the job operational plan, as shown in Figures 14.3 and 14.4.

Activity	I-J	Symbol	Predecessor Activity	Successor Activity
Get signed and approved contract	1-2	SC	none	GD, BP, OC
Get building permit	2-3	BP	SC	LH
Get drill rig	2-4	GD	SC	DH
Order and deliver concrete	2-5	OC	SC	PC
Lay out and locate hole	3-4	LH	BP	DH
Drill hole	4-5	DH	GD, LH	PC
Place concrete in hole	5-6	PC	DH, OC	CL
Clean construction site	6-7	CL	PC	none

FIGURE 14.2 Preliminary Activity Tabulation.

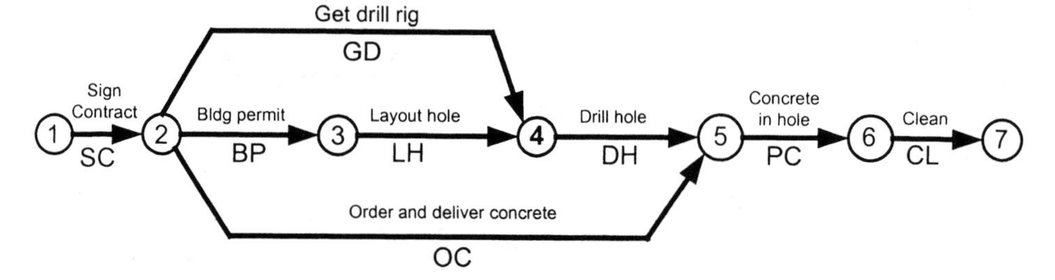

FIGURE 14.3 Activity Diagram Showing Job Logic of a Simple Problem in Arrow Diagraming (*i-j*) Format.

In the course of explaining the structure of a CPM diagram sufficiently to enable the Resident Project Representative to read and understand the symbolism involved, a few special symbols and conventions will be covered in following paragraphs.

Dummy Arrows

Frequently a dotted or dashed arrow will be found in a CPM arrow diagram. These are generally accepted symbols for a *dummy arrow,* which is a way of indicating that the completion of one activity restrains the start of two or more other activities. The dummy is nothing more than a slightly awkward way of representing such a dependency.

In Figure 14.4, the dashed arrow (15–16) is an example of a dummy arrow, because the start of (16–20) cannot begin until both (05–15) and (05–16) have been completed, whereas activity (15–20) can begin immediately after (05–15) has been completed, regardless of whether or not (05–16) has been completed. The direction of the arrow indicates the time flow of construction operations, and the sequential order cannot back up against a dummy arrowhead.

Another common usage of dummy arrows is to give each activity its own numerical *i-j* designation for easy identification as a computer "address." In Figure 14.5 three activities are shown that are parallel (concurrent) to one another and share a common start and finish point. As indicated in the figure, however, each of the three activities would have to be identified by the same *i-j* number of (19–25). Although this might be no particular hardship if the network was being prepared manually and the diagram was always in front of the user, if the CPM was being computer generated, there would be no possible way of separating the three activities in the computer, as they would all appear as the same number, thus canceling one another out.

When a situation is encountered such as that illustrated in Figure 14.5, and it occurs fairly frequently, dummy arrows can be utilized as shown in Figure 14.6 to allow separate *i-j* numbers to be assigned to each of the three activities and still show the dependent relationships.

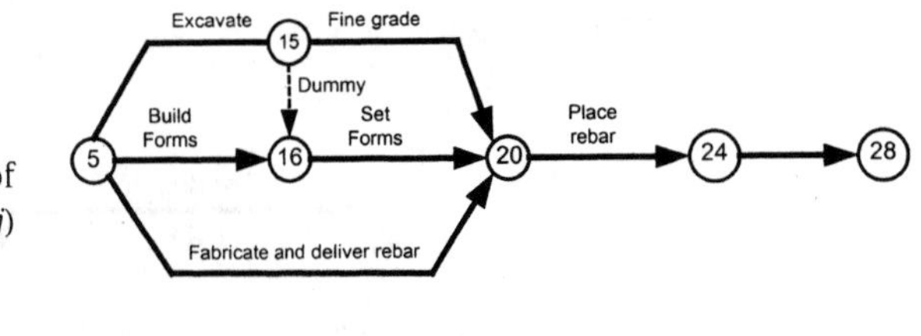

FIGURE 14.4 Activity Diagram Showing Use of a Shared Predecessor in Arrow Diagraming (*i-j*) Format.

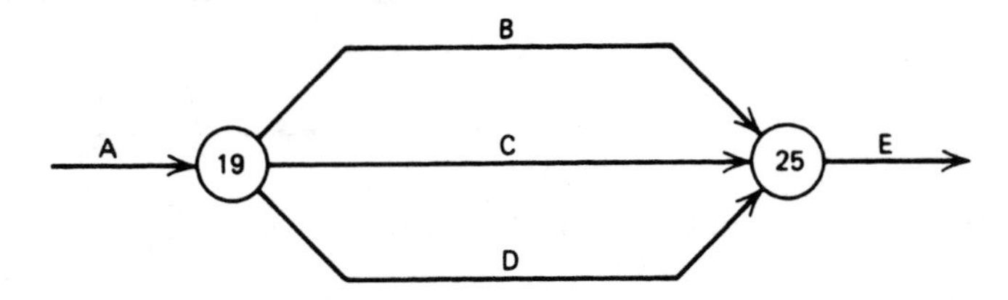

FIGURE 14.5 Three Parallel Activities.

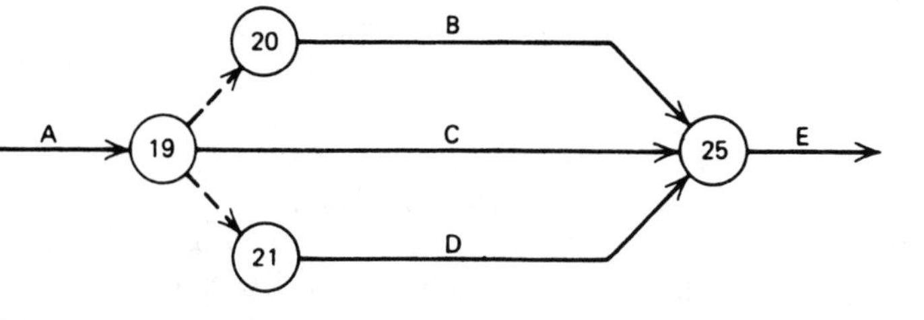

FIGURE 14.6 Use of Dummy Arrows to Preserve Activity Identities.

Thus the three parallel activities shown in Figure 14.5 after the introduction of the dummy arrows shown in Figure 14.6 may now be identified as activity (20–25), (19–25), and (21–25), which will satisfy the computer requirement that all activities must have separate *i-j* number identities, while the dashed dummy arrows retain the original concurrent relationship of each activity to the others.

Events

At the ends of each activity arrow are circles (the most common system) or other geometric figures. These circles are placed at the junction of the arrows and they represent events, or nodes. As mentioned previously, an "event" is the instant of time at which an activity is just starting or finishing. CPM is not basically event oriented; that is, it does not usually emphasize or name the events but merely refers to them as part of the activity ("*start*" or "*finish*"). If some events are particularly important, they may be referred to as "milestone events" and may be specially identified and named on the diagram.

The basic rule applying to an event in arrow diagraming is that all activities leading into a given event must be completed before any other event dependent upon it can occur. This is basic network logic.

Activity Numbering

In CPM diagrams, a number is used to identify each activity. In arrow diagraming format, *i-j* numbers are used, whereas in precedence format a single activity number is used for each activity.

1. Each activity must have its own unique activity designation.
2. When event numbers are assigned, the number of preceding activities should be lower than succeeding activities (preferred rule but not always practical).

Although it is possible for a CPM diagram to be prepared using random numbering, this is especially true where an activity was inserted into a network that was already numbered. If rule 2 were strictly applied, all numbers in the entire network would need to be renumbered to allow the new activity to have consecutive numbers.

Normally, CPM computer scheduling programs will automatically number each activity using precedence numbering format as it is entered through the keyboard. There is actually no significance to the activity numbers themselves except as a means of identifying an activity. Often gaps are left in the numbering system so that spare numbers are available for subsequent work refinements or revisions. One advantage to the use of sequential numbering over that of random numbering of activities is that it is easier to locate activities on the network and other reports. It also helps prevent the inclusion of logical loops, which will be discussed next.

The preferable system of numbering involves the assignment of numbers by skipping gaps so that there is sufficient room to add activities. Numbering in increments of ten, while skipping blocks of 100 from time to time, is the preferred method.

There are two basic practices used in assigning numbers. The most common is by area or phase. The second is by the type of work.

1. In the first method, each separate location or phase of the project would have a unique series. This makes it easy to find the work being performed in a specific area and to know the number of similar activities in other areas. For example, all activities on the first floor of a building could be numbered in the one thousand series, while activities on the second floor would be numbered in the two thousand series. This would make it possible for sheetrock on the first floor to be activity 1120, while activity 3120 would be sheetrock on the third floor.
2. In the second method, numbering may be by type of work. This involves identifying activities found in the same section of the specification with the same prefix. As an example, all activities found in CSI Division 15 sections of the specification would have a prefix of 15. Thus 151222 may designate a certain activity found on the first floor, while activity 157222 would be the same type of work found on the seventh floor.

Logic Loops

The logic loop is a paradox in network planning. It indicates the requirement that an activity be followed by another activity that has already been accomplished. In Figure 14.7, a simple example of a logic loop is shown: where activity B cannot start until activities A and D are finished, and activity C cannot start until activity B is finished. But activity C must finish before activity D can start, and the logic tie D prevents activity B from starting until after activity C is finished; thus activity D is both a predecessor and a successor to activities B and C. The term *logic loop* is a misnomer, since if anything, a logic loop is extremely illogical. Although this is perfectly obvious in the simple example illustrated, logic loops can be inadvertently included in large or complex networks without the scheduler realizing that they are there. Because such loops are representative of impossible conditions, the network planner will take precautions to try to prevent their inclusion. The use of a random numbering system

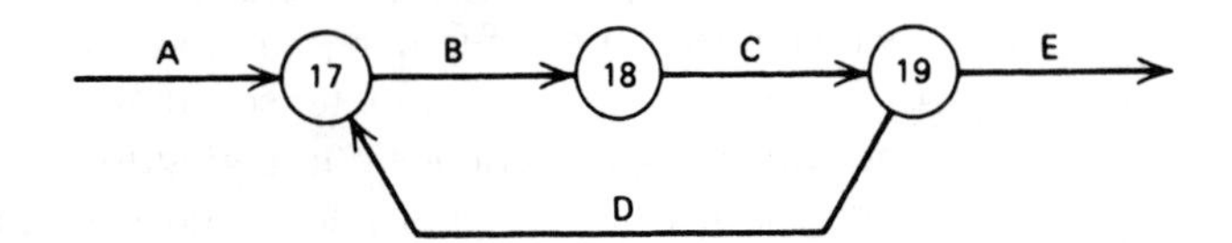

FIGURE 14.7 Logic Loop.

allows a greater likelihood of error by allowing logical loops to remain undiscovered, as a computer printout cannot indicate any clues to their presence under such conditions. When a loop occurs, most Network Analysis (CPM) programs will "freeze." More advanced computer scheduling programs will assist the user in discovering the loop and correcting it. It is wise to study the network diagram carefully at the beginning of the job to confirm the logic of its structure.

Float Time

Float, sometimes called *slack,* can best be described as scheduling leeway. When an activity has float time available, this extra time may be used to serve a variety of scheduling purposes. The contractor uses *float* as a management tool for scheduling and rearranging resources to achieve the most efficient progress on the project. When float is available, the earliest starting time of an activity can be delayed, its duration extended, or a combination of both can occur. To do a proper job of monitoring of the schedules for noncritical items, the Resident Project Representative should understand the working of float times on a project. Briefly, float can best be described (Figure 14.8) as the difference between the time available to complete an activity and the time that is actually required to complete the activity.

Arithmetically, float time is easy to compute, as it is simply the difference between the early and late dates for an activity, as illustrated in Figures 14.9 and 14.10. It represents the available time between the earliest time in which an activity can be accomplished (based upon the status of the project to date) and the latest time by which it must be completed for the project to finish by its deadline. There are three important timing facts that can be determined from a CPM network.

1. The earliest time an activity can start and finish.
2. The latest time an activity can start and finish without delaying the project completion.
3. The amount of leeway available in scheduling an item (the difference between 1 and 2, above).

The two primary types of float are free float and total float. *Free float* is defined as the amount of time that any activity can be delayed without adversely affecting the *early start of the following activity.* In other words, free float is the difference between the early finish of an activity and the early start of all following activities. *Total*

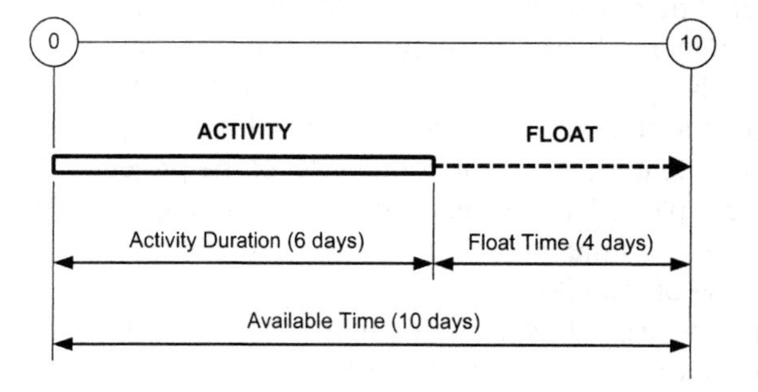

FIGURE 14.8 Graphic Illustration of the Concept of Float.

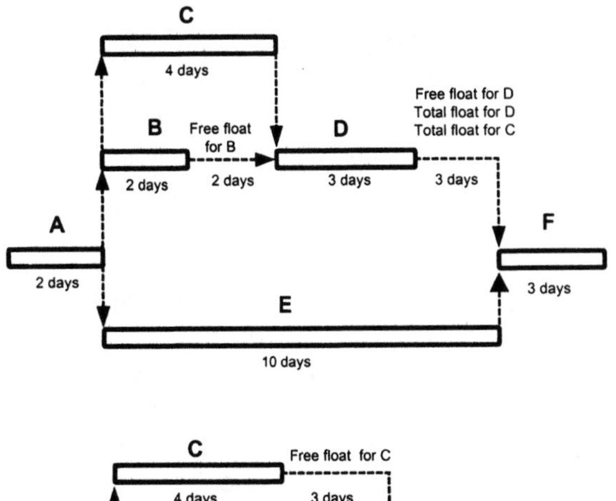

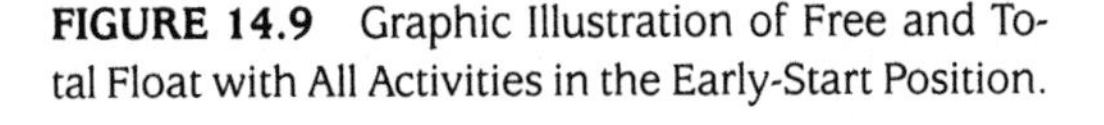
FIGURE 14.9 Graphic Illustration of Free and Total Float with All Activities in the Early-Start Position.

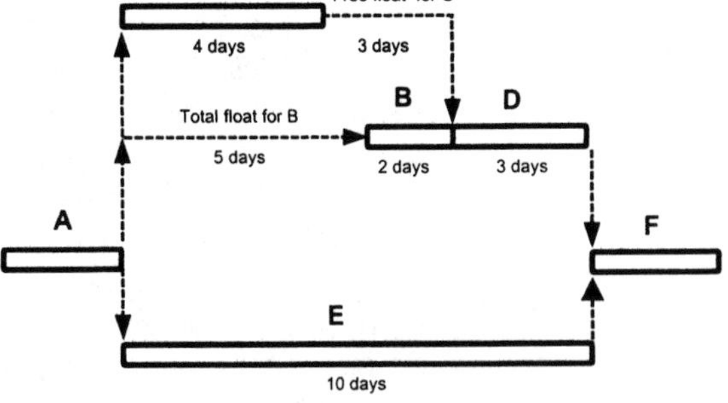

FIGURE 14.10 Graphic Illustration of Free and Total Float with Activities B and D in Late-Start Position.

float is defined as the amount of time an activity can be delayed without adversely affecting the *overall time for the project completion.* It should be understood that just because an activity has a certain amount of *total float,* it does not necessarily mean that activity can use it without creating tighter scheduling restraints on all of the other activities. It must be remembered that *total float* is shared with all other following activities. Free float, however, is not shared with other activities; thus, it provides a true measure of how much an activity can be delayed or extended without adversely affecting any other activity. Why is free float so important anyway? First, it is invaluable in evaluating contractor change orders to determine whether or not to allow additional time on the contract for each change order. Second, it is invaluable to the owner's engineer at the end of the job to assist in evaluating impacts on the contractor's schedule. Neither of these can be accomplished with total float alone. In the accompanying illustration (Figures 14.9 and 14.10), a precedence diagram is used to show a CPM network involving six different activities. The *activity numbers* are shown above the activity bar, and the numbers below the activity bar represent the time durations of each bar and float indicator for completion of the project.

In Figure 14.9 all activities are plotted at their early start position. Figure 14.10 is a graphic illustration of the same example with all noncritical activities plotted in their late start position showing use of total float.

If a series of activities is on the *critical path* of the project, no float exists; the early and late dates are the same. Sometimes, when a project is behind schedule, the earliest time at which an activity can begin is after the latest time it can be done to remain on schedule. Then, not only does no float exist, but also the difference between the early and late dates is less than zero (therefore negative) and is now a measure of how far behind schedule the project is. In this case the late start will actually be shown as earlier than the early start.

Who Owns Float?[1]

Network-based project control systems were developed originally as management tools. They can be abused, however, and some contractors and owners have turned them into weapons, on occasion, for use against one another. The increasing use, or misuse, of network-based management systems often creates as many problems as it solves. We are faced with network schedules prepared more as biased documents to support the originator's right to claims than as management tools to help control the project.

These abuses frequently revolve around float. We define float as scheduling leeway. A more complete definition of float was given earlier in this chapter.

On one hand, a contractor may create an artificial network with multiple critical paths. The intent of the contractor would be to present claims if the owner causes delay on any of the paths. On the other hand, the owner may plan the project duration and then shorten it. The owner's intent would be to obtain a bid on the shortened duration and then to hold the contractor to the time.

Obviously, these practices reduce an otherwise effective management tool to a weapon for justifying or denying claims. The result has been more litigation over more claims than it might even have been possible to resolve by using the management tool correctly.

Differing Viewpoints

Contractors and owners view the problem differently. A contractor may question if an owner has a right to direct starts or delays for specific purposes. Owners, trying to manage project costs and overall scheduling, may ask if they have the right to force the contractor to start an activity before an established late-start date. Too many of these starts could cause problems toward the end of the job.

For the owner, a week's slippage in a power plant's commercial operation date translates into a loss of several hundred thousand dollars. At the same time, improper or premature activity starts can cause serious hardship for a contractor, who may incur additional expense for equipment, material, labor, and other resources.

The owner's engineers will attempt to resolve these and other scheduling problems and will attempt to gain management control. Usually, the engineer will make the request in the contract for one or more of the several network-scheduling techniques currently available. Such specifications would be written with the intent to

[1]Adapted from "Who Owns Float?" an informational brochure by Forward Associates, Ltd., Novato, CA.

cover all bases for the owner's protection. With a combination of appropriate general conditions and supplementary general conditions, such specifications can be quite complex.

The contractor, however, using the techniques just described, will manage to circumvent the specifications of the owner, and the whole process can wind up in claims and litigation. Many court decisions have been rendered on such claims, but none seems to have solved the problem.

Case History

A contractor on a $2-million civil construction project submitted a network that had been prepared with considerable care. The only objective of the plan was to get the job done as efficiently and effectively as possible.

Due to an owner-caused delay, the contractor fell behind schedule. Later, the contractor submitted a claim for additional money and a 35-day extension. The owner's engineers analyzed the network and showed that work on the next major milestone on the critical path was delayed only 2 days. The owner contended that was all the contractor was entitled to. The contractor said that it was only through its diligence in accelerating the work that delay was held to only 2 days.

The contractor could not support its 35-day claim because of improper schedule monitoring. The owner's engineers had been able to do a better job of network analysis than had the contractor, and the latter had to settle for the expenses it could prove plus a small sum for supporting its schedule. The contractor felt, correctly so, that it had been penalized for its diligence in maintaining the schedule, despite owner delay.

Multiple-Critical-Path Case

A contractor on a $12- to $14-million building project submitted a CPM network in compliance with the owner's specifications. There were several zero-float paths—seven critical paths! The contractor was anticipating that the owner would delay one or more of the activities that were a part of these critical paths, thus allowing the contractor to file claims against the owner.

During negotiations, the owner cited the seven critical paths as an unrealistic approach. The contracting firm responded that it was *its* plan of work and the owner had no right to alter it. The negotiation process was soon reduced to a battle of wits.

What had the contractor done to its network, and what might it have achieved? First, it increased the duration of routine concrete pours to three times their normal duration. If requested to shorten this duration, it would request payment for accelerated work. To prevent this, the owner should have the right to refer to standard planning tools such as estimating guidebooks to determine reasonable durations of all activities. Exceptions would be negotiated.

The contractor in the foregoing case used "policy or management constraints" (preferred way of doing work) to consume float. These constraints prevent activities from starting until other preceding activities are sufficiently complete. On detailed analysis, this appeared illogical; however, the contractor was asked to apply work around techniques, and it would counter by saying that the changes to its plan of work would result in extra work and therefore more claims against the owner.

Unless a contractor is able to provide a rational purpose for a logic constraint, the owner *should* have the right to demand that constraints of this type be removed and the work replanned when the constraint threatened the critical path, such as stating how many activities can be within two (or three weeks) of the critical path, for example. Difficulties such as those just described might be overcome with adequate specifications. However, one can hardly hope to anticipate all of the problems that can arise, particularly if the specifications writer has limited knowledge of all of the tricks that can be used.

The specifications should protect both the owner and the contractor. But it should be remembered that the owner pays for the scheduling system and is entitled to get what it pays for. The specifications should set forth restrictions on falsifying networks to eliminate float. They should spell out what rights the owner has to utilize float to its advantage. They should clarify areas where the owner has the right to apply standard estimating techniques to activities where contractors have obviously set overly long durations to eliminate float. Generally speaking, case law supports the position that the owner owns float. A more equitable position, however, would be that whoever gets to the float first should own it. One way to state such a position could be: "time extensions will be granted only if the contractor has used all the float time available for the work involved."

Precedence Diagraming versus *i-j* Diagraming

One method of network diagraming is called *i-j, or activity on arrow.* This method is less commonly used since it is no longer supported by most commercially available software. Most of the rules apply, but instead of numbering the activity, the events are numbered. Also the logic is limited to conventional logic: In arrow diagraming, all activities must be completed before the following activities can begin. The other method of network diagramming is called precedence diagramming, or *activity on node.* In precedence diagramming, the activities are located on the nodes, not on the arrows. Arrow diagraming is easier to learn, but lacks the versatility (and complexity) of precedence diagraming.

In the *i-j* diagram illustrated in Figure 14.11 activity (5–10) is drilled piers, and activity (10–15) is grade beams. The *j* node of the first activity becomes the *i* node of the second activity. This defines the relationship between the two activities, and the software relies on this basic principle to track the relationship between the two activities. Thus it is not necessary to place a logic tie between every two related activities. BASIC RULE: All solid-line arrows in *i-j* diagraming represent both time and

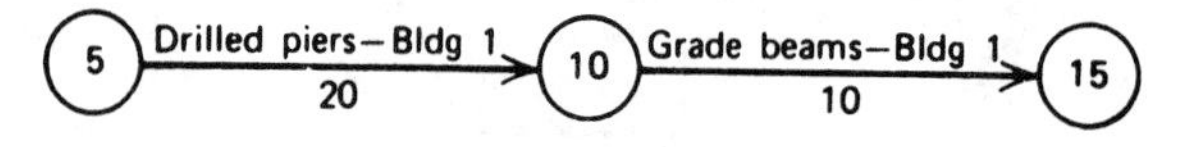

FIGURE 14.11 Arrow Notation of Two Activities in Series.

dependency, and all dotted-line arrows are logic ties only (dummy arrows) and do not represent time.

Figure 14.12 illustrates the same network relationship using a precedence diagram in a box-node format. BASIC RULE: All arrows in precedence networks represent logic ties only. Time is either indicated within the box on box-node format, or by time scaling the box into an activity bar is shown in Figure 14.15.

In arrow diagraming, when four activities are combined in a network where one activity shares a dependency on a predecessor activity with another activity, a dummy arrow is introduced into the network, as illustrated in Figure 14.13, to show the relationship in the illustrated case of activity 5–10 to the following activity 10–15 and 25–30. A dummy arrow differs from an activity in that it does not represent time, only dependency. It is the same as a conventional logic tie in a precedent network in that it constrains the start of a succeeding activity by the finish of its predecessor, shown in Figures 14.14 and 14.15.

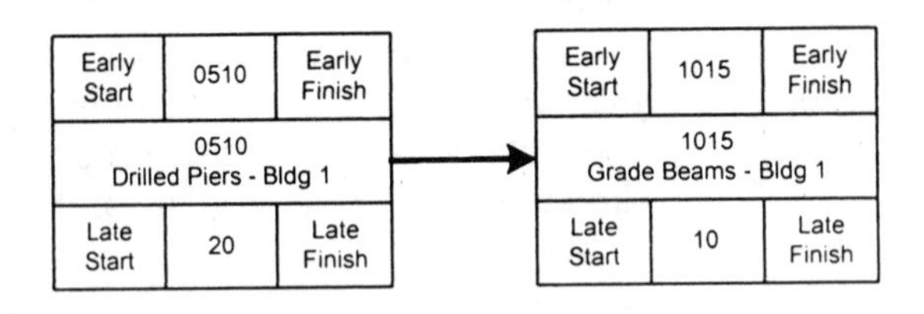

FIGURE 14.12 Precedence Notation of Two Activities in Series.

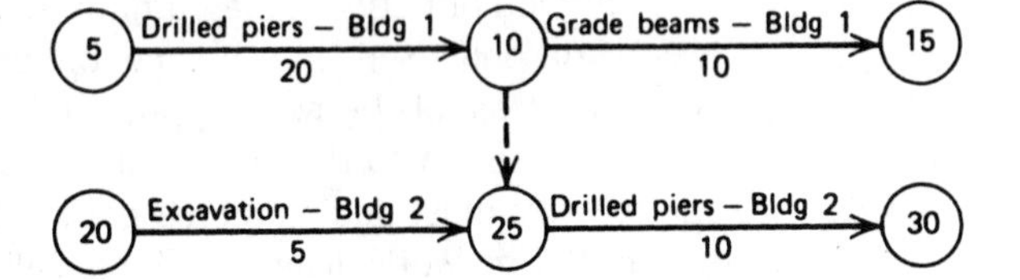

FIGURE 14.13 Arrow Diagram of Four Activities with a Dummy Arrow Showing a Shared Dependency.

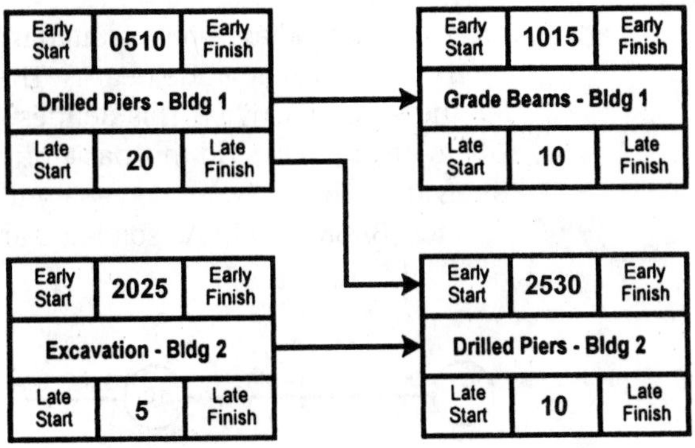

FIGURE 14.14 Box-Node Precedence Diagram of the Same Four Activities Shown in Figure 14.13, with a Restraint.

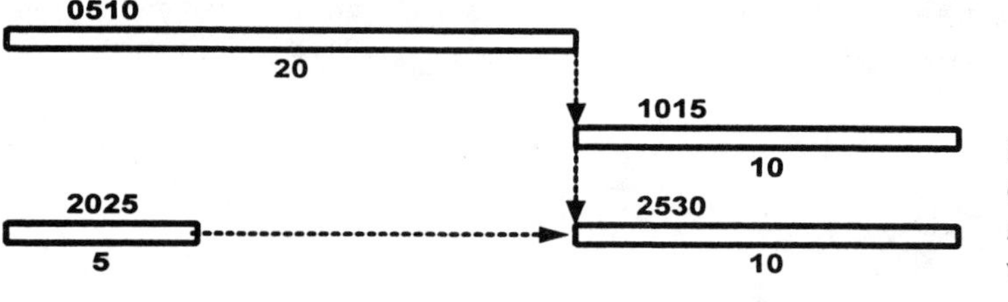

FIGURE 14.15 Time-Scaled Precedence Diagram of the Same Four Activities Shown in Figure 14.13, with a Restraint.

The analogy is apparent. If processed on a CPM program, the output would be identical, except that the *i-j printout* would contain one dummy and the precedence diagram would contain no dummy. However, the printout of the precedence diagram would have to contain a supplementary list of relationships for the user to determine and maintain the logic. With some programs, the supplementary list can be quite awkward to use. But it is important that one be included with every schedule, since it is the only sure way of tracing the logic of the schedule.

Comparison of Precedence and Arrow Diagraming

The primary difference between the two methods can be seen in Figure 14.16. Activities can be overlapped using relationships. This overlapping logic makes it much easier to prepare a precedence diagram. It is not necessary to describe the part of the first activity that must be done before the second activity may start.

Precedence Formats

The emergence of practical computer graphics as a meaningful substitute for hand-drafted network diagrams along with more sophisticated networking programs, and pressure by the software companies, has led to the virtual elimination of the *i-j* method. This has led to different types of formats containing not only the desirable elements of arrow and precedence diagrams, but some additional attributes as well. In this respect this new hybrid format can be superior to either of its predecessors.

Under this later precedence diagram format, which is referred to as *activity-on-node precedence diagraming,* a long, narrow node resembling a bar on a bar chart represents each activity. This format, because it is actually a variation of a precedence diagram, does not use dummy arrows, just dependency arrows. Under this format, float time can be represented as a single line following the activity node, which is represented by a bar.

When this is combined with a programmed time line, the resulting time-scaled network becomes understandable by people of any background. If each bar happened to be located on its own separate line and was properly identified in the left margin, non-CPM-trained users could view the resultant diagram as a simple bar chart by simply ignoring the dependency lines. Some users have actually created what is referred to as "dependency bar charts," which is just another network diagraming format.

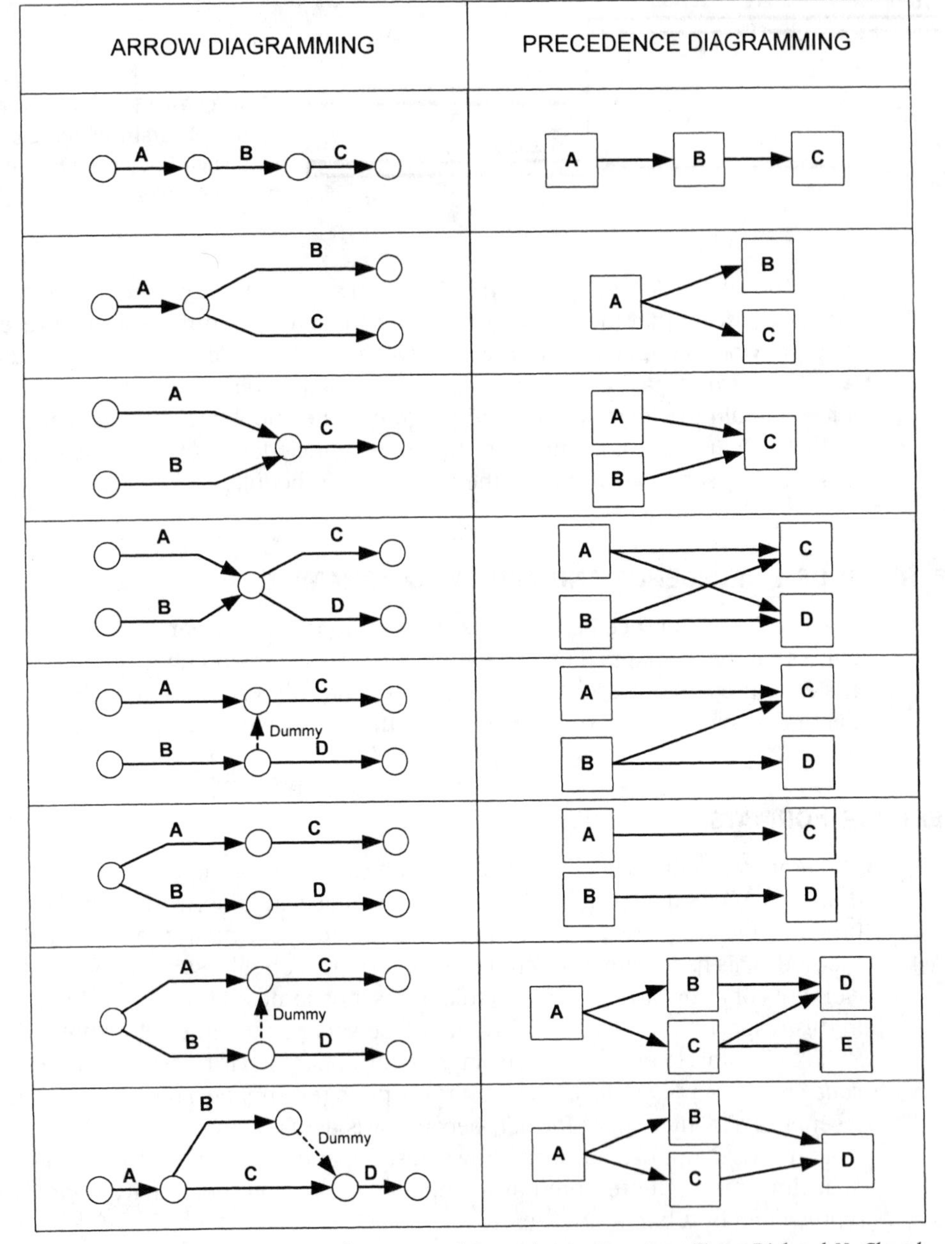

FIGURE 14.16 Comparison of Arrow and Precedence Notation *(From Richard H. Clough, Construction Project Management, © 1972 by John Wiley & Sons, Inc. This material is used by permission of John Wiley & Sons, Inc.)*

Those trained on network diagrams of any format can utilize the new scheduling format just as before, whether trained on arrow or node diagraming formats.

While hand-drawn network diagrams are still preferred by the most experienced schedulers, they are falling out of favor with the do-it-yourself scheduler using PC computers. Although there is no need to draft a finished diagram on most projects, a rough drafted diagram is still the best method of developing proper logic. Logic tends to be lost once a program swallows it. Missing or improper logic is easy to spot as a rough drafted diagram is being prepared. So much rides on the quality of the logic it is best to map it out first. Starting with a keyboard and mouse before the logic is developed and checked is an invitation to the types of errors discussed in earlier parts of this chapter, and the costs can be high.

Reading a Manual CPM Network Schedule

As arrow diagraming is frequently utilized, and the logic of an arrow diagram seems more obvious graphically, the emphasis in this book will be placed upon the reading of CPM arrow diagrams and the associated tabular data. The term *critical path,* of course, refers to that portion or those portions of the work that are the bottlenecks in the construction process. Obviously, the total project cannot be completed earlier than those portions of the work that require the most time to complete. On complex schedules it is impossible to determine by examination of the network diagram which paths represent the critical path—thus computers have come into the picture. In this manner, all possible combinations of tasks and activity times, early and late starts and finish dates, and float times can be analyzed by the computer until all of the key times can be determined from the computer printout. In addition, the computer determines the critical path or paths, as the case may be, and any overruns in time along the critical path will result in a schedule overrun.

If performed manually, CPM does not have the benefit of tabular printouts, which give key time data; thus, the user must rely upon observation of the diagram itself. Although on a very simple arrow diagram it may be possible to determine the critical path, it is normally necessary to construct a time-scaled network from the arrow diagram before any true scheduling can be determined. The arrow diagram of Figure 14.17 simply shows the logical relationships between the various project activities.

Before time scaling the arrow diagram in Figure 14.17, observe the effect of adding activity times from 1-3-6-8-13-16-17-18. Activity B (1–3) has a time of 7 days; activity F (3–6) has an activity time of 4 days; activity L (6–8) has an activity time of 8 days; activity M (8–13) has a time of 10 days; activity U (13–16) has a time of 10 days; activity X (16–17) has a time of 4 days; and activity Y (17–18) has a time of 4 days. Added together they total 47 days. Compare this with the path through 1-3-6-12-16-17-18, which also totals 47 days. From this it might at first appear that both paths are critical and that the minimum time to complete the project was 47 working days.

By plotting each activity on a time-scale chart (Figure 14.18) and adjusting the start and finish times to be compatible with the logic expressed in the arrow diagram,

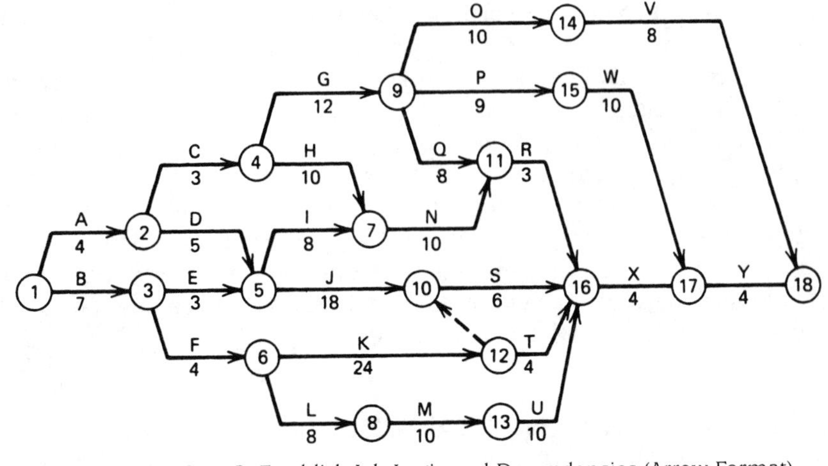

FIGURE 14.17 Step 2: Establish Job Logic and Dependencies (Arrow Format).

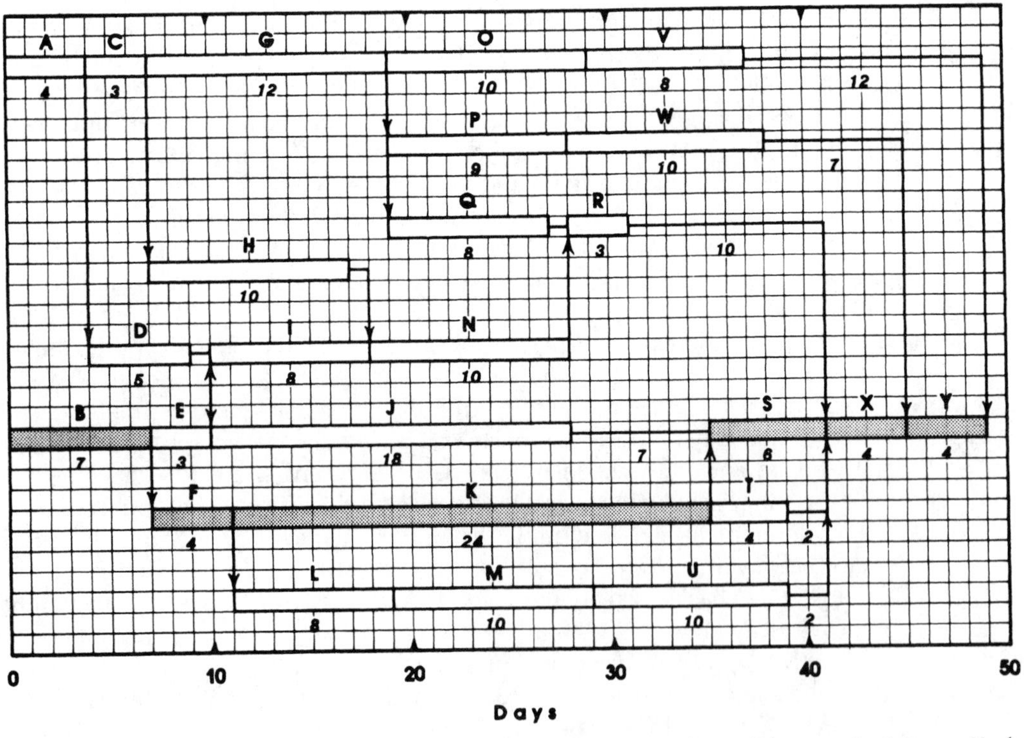

FIGURE 14.18 Step 3: Construct Time-Scaled Diagram (Precedence Diagram Activity-on-Node Format).

it can be seen that some notable limitations come into view. The chart is based upon early start and finish dates with float time indicated by the single lines. In Figure 14.17 the numbers in the circles represent the *i-j* numbers for the activities, based upon early start and late finish. The numbers under the lines represent the amount of time required in days to complete the activity.

First, the path of 1-2-5 would appear from the arrow diagram to be simply the sum of activity times of A and D, which totals 9 days. Yet it should be noted that path 1-3-5 adds up to 7 days plus 3 days, or a total of 10 days to get from event 1 to event 5. Thus, the controlling time to reach event 5 is of course 10 days, not 9 as it might at first appear, and obviously, path 1-2-5 then includes 1 day of float time.

By careful inspection of the time-scaled network, it can be seen that path B, F, L, M, U, X, Y (1-3-6-8-13-16-17-18) contains a total of 2 days of total float time. By continuing the inspection of the arrow diagram and the time-scaled chart, it can be further noted that the path through activities B, E, and J (1-3-5-10) cannot proceed past event 10 until the completion of activity K (6-12), as indicated by the dummy arrow. Therefore, event 10 must be delayed until the completion of event 12. (On the time-scaled chart they are equivalent as far as being completed on the same day.)

It should be noted that the path from event 10 to event 16 is a total of 6 days, in contrast to the 4 days required from event 12 to event 16. Also, since both events are held up until the completion of activity K (event 12), the new path becomes (1-3-6-12-10-16-17-18). The new path then follows through activities B-F-K-dummy-S-X-Y, which totals

$$7 + 4 + 24 + 6 + 4 + 4 = 49 \text{ days}$$

not the 47 days as first supposed!

It now appears that in comparison with the other paths analyzed, the one discussed is the critical path for this sample project, as all of the other paths contain float time.

Many of the questions that arise in construction involve the establishment of early and late start and finish dates for each activity. These dates represent the earliest or latest that any event can be started or finished without changing the critical path of the project. Any change in the starting time of any event on the critical path must result in a change in the completion date of the project. Thus if activity F (3–6) started 3 days late, the time required to complete the project would be 52 days because activity F is on the critical path. However, if event 11 started 10 days late, the completion date of the project would probably not be affected, as it has 10 days of free float time following it on the time-scaled chart. Note, however, that the early start date for activity R cannot change because it is contingent upon the completion of activities I and N, which allow no time variations.

Other questions may be asked, such as: Before activity R can begin, what other activities must be completed?

Working back from event 11 on the arrow diagram of Figure 14.17, it can be seen that activities N and Q must both be completed first, plus all the activities that must necessarily precede activities N and Q, as outlined in what follows:

1. Working back from event 11, note that N and Q must be done.
2. Then, back from event 9, only activity G must be done.
3. Back from event 7, both activities H and I must be completed.
4. From event 4, only activity C must be done.
5. Back from event 5, both activities D and E must be completed.
6. Back from event 3, only activity B must be done.
7. Finally, back from event 2, only activity A must be done.

Do not overlook the significance of the dummy arrow, however. It should be noted that before any work can proceed beyond event 10, it is required that activity K must be completed. To proceed beyond event 10, not only must activities E and J be completed, but also because of the dependency upon completing activity K, all three activities B, F, and K must be completed before activity S can start.

Float Time in the Sample Problem

As stated earlier, the float of an activity represents the potential scheduling leeway. When an activity has float time available, this extra time may be utilized to serve a variety of scheduling purposes. When total float is available, the earliest start of an activity can be delayed, its duration extended, or a combination of both can occur as long as the late finish time is not exceeded. To do a proper job of scheduling of noncritical activities, it is important that the user understand the workings of float or slack time.

As mentioned previously, the total float of an activity is the maximum time that its actual completion can go beyond its earliest finish time and not delay the entire project. It is the time leeway that is available for that activity if the activities preceding it are started as early as possible and the ones following it are started as late as possible. If all of the available total float is used on one activity, a new critical path is created. The free float of an activity is the maximum time by which its actual completion date can exceed its earliest finish time without affecting either the overall project completion or the times of any subsequent activities. If an operation is delayed to the extent of an activity's free float, the activities following it are not affected, and they can still start at their earliest start times.

Thus, in the illustration in Figures 14.17 and 14.18, the total float of the activity T (12–16) would be as follows:

Critical path from event 12 to event 16 is

$$12(10) \text{ to } 16 = 6 \text{ days' time}$$

Completion time for activity T is 4 days

$$\text{Total float time is } (6-4) \text{ days} = 2 \text{ days}$$

Early start time for activity T (from events 12 to 16):

path 1−3−6−12, which totals 35 days

Late start for activity T would be:

path 1−3−6−12 plus float time for activity
T = (35 + 2) = 37 days

Thus, as long as the time required to complete activity T does not exceed (4 + 2) = 6 days, the critical path time will not change. As soon as activity T goes to 7 or more days, the time to complete the work will change. Similarly, if such would occur on a noncritical path and cause that total path to exceed the original critical path, a new critical path is created. The danger is that if one activity uses all of the total float, it is gone for good. From then on the contractor is on a critical path.

A good example can be seen by considering two sequential activities, such as placing conduit for a concrete pour and actually making the pour. Assume that these activities are not on the critical path and that the difference between the early and the late dates is a total of 10 days. Assuming that no other activities will have any effect on the two mentioned, how much total float does the conduit activity have? The answer, of course, is 10 days—the difference between the early and late start or the early and late finish. Now, how much total float does the pour activity have? The answer to that, too, is 10 days, computed in the same manner. Then, how much total float does the entire sequence of activities have—20 days? *No,* because total float cannot be added; the sequence or chain of activities still has only 10 days.

What if the electrician is 3 days late in starting? As of the time started, there are only 7 days of total float. The general contractor has also been reduced to 7 days of total float. What if the electrician finishes 7 days late? Now the general contractor has *no* float—the contractor is on critical path.

The point is that individual activities do not really have total float individually. Chains of activities have total float, and all of the activities in the chain share the same total float time. If one activity uses it up, it is gone for good and is no longer available for the other activities.

Reading a Computerized CPM Network Schedule

By far the majority of cases of CPM scheduling that will be encountered by the Resident Project Representative will involve the use of computer-generated scheduling data required of the contractor by the construction contract. Documents that should be made available to the Resident Project Representative are: the network diagram, the tabular printout of the various types of data, and a narrative explaining any changes to the schedule. It is also wise to require that the contractor provide a licensed copy of the software so that the Resident Project Representative will be able to read the schedule. The contract should also state whether scheduling data should be submitted in digital form, hard copy, or both.

Although some of the tabular reports may include contractor costs and resource data as well as time-scheduling data, most resident engineers or inspectors will probably be interested only in the time-scheduling data and the narrative. The contractor should be required to update the schedule by means of computer on a regular basis so that all information is up to date. Monthly updating is common, but the frequency should be a function of the complexity of the project and its time sensitivity. The level of detail in the updated schedule should be appropriate for project management and can be controlled by stating the length of activity durations, for example, "activities with durations of no more than 10 days" (Associated General Contractors guide specification). Of particular concern would be the redistribution of slack times if portions of total slack have been used by certain individual activities. Such redistribution is important material that should be addressed in the narrative that accompanies the updated schedule. The data are sorted and arranged into user-friendly tabular lists that should all contain both free and total float. Both categories of float must be included in order to fully understand what is happening with the project and what impact change orders have made or may make to the work.

As an example, a CPM schedule prepared for the Aerospace/Mechanical Engineering Building for the University of Arizona provided printouts of the following types of data grouped into separate charts or tabular reports called "sorts," illustrated here.

1. Bar chart (Figure 14.19)
2. Sort by activity or *i-j* numbers (Figure 14.20)
3. Sort by total float/late start dates. (Figure 14.21)
4. Sort by early finish dates (Figure 14.22)

Although the basic principle of CPM is based upon networking of all activities as a means of establishing valid logic relationships between the various work activities, the time-tested bar chart has not gone entirely out of style. As shown in Figure 14.19, it is still the most useful scheduling tool for communicating scheduling requirements to a layperson.

However, a bar chart by itself is not a valid scheduling tool unless it is first preceded by a network schedule to validate the activity logic. In one case (*Mega Construction Co., Inc. v. United States,* 29 Fed Cl. 396 [1993]), the Court of Federal Claims denied a delay claim because the contractor's bar charts failed to establish the logic relationships between disrupted tasks and other activities on the schedule's critical path. The court said that the contractor's bar chart was inadequate because it did not indicate the interdependence of any one or more of the work items. In short, because it was not based upon CPM.

University of Arizona Project

The CPM network in Figure 14.23 was drafted using the *i-j* method on one 762-mm × 1066-mm (30- × 42-inch) summary network and three 762-mm × 1066-mm (30- × 42-inch) sheets of detail net for the Aerospace/Mechanical Engineering Building, supplemented by computer printouts covering each reporting period.

ACTIVITY ID		EARLY START	EARLY FINISH	ACTIVITY (1994: NOV, DEC; 1995: JAN, FEB, MAR, APR, MAY)
				OWNER
1	2	7NOV94	7NOV94	NOTICE TO PROCEED 7NOV94
				GENERAL
978	982	8NOV94	15NOV94	DEMOLITION PERMIT
980	982	8NOV94	22NOV94	MOBILIZATION
1010	1014	9DEC94	22DEC94	ENGINEERING & LAYOUT
				DEMOLITION
982	1010	23NOV94	8DEC94	SITE DEMOLITION
				CAISSONS
1020	1021	23DEC94	10JAN95	DRILL CAISSONS
1022	1026	23DEC94	10JAN95	REINFORCE CAISSONS
1024	1026	23DEC94	10JAN95	ALL CONCRETE CAISSONS
1060	1066	11JAN95	25JAN95	DRILL CAISSONS
1062	1066	11JAN95	25JAN95	REINFORCE CAISSONS
1064	1066	11JAN95	25JAN95	CONCRETE CAISSONS
1200	1206	26JAN95	1FEB95	DRILL CAISSONS 2ND LVL
1202	1206	26JAN95	1FEB95	REINFORCE CAISSONS 2ND LVL
1204	1206	26JAN95	1FEB95	CONCRETE CAISSONS 2ND LVL
				EXCAVATION
1012	1016	9DEC94	15DEC94	EXCAVATION BAS
1030	1033	4JAN95	17JAN95	ALL EXCAVATE FTGS & PITS
1070	1073	26JAN95	31JAN95	ALL EXCAVATE FOOTINGS &
1170	1173	20FEB95	23FEB95	ALL PREP SLAB ON GRADE
1192	1193	18APR95	8MAY95	BACKFILL BASEMENT
				CONCRETE
1036	1040	18JAN95	24JAN95	ALL CONCRETE FOOTINGS & PITS
1152	1154	25JAN95	30JAN95	FORM & PLCE CONCRETE COLUM
1044	1047	23JAN95	10FEB95	ALL FORM & PLCE CONC 1ST LVL
1074	1076	27JAN95	10FEB95	ALL CONCRETE FOOTINGS &
1092	1094	13FEB95	20FEB95	ALL FORM & PLACE CONC. WALLS 2
1082	1094	13FEB95	21FEB95	ALL FORM & PLCE CONC COLUMS 2
1094	1097	22FEB95	24FEB95	FORM & PLACE CONCRETE WALLS 3RD
1102	1104	28FEB95	3MAR95	ALL FORM & PLCE CONC WALLS 4TH LVL
1176	1178	7MAR95	8MAR95	CONCRETE SLAB ON GRADE
1180	1184	9MAR95	29MAR95	ALL FORM SLAB 2ND LVL (SUSPEND
1188	1190	21MAR95	3APR95	ALL CONCRETE SLAB 2ND LVL (SUSPENDED)
1252	1270	4APR95	10APR95	ALL FORM & PLACE CONC. WALLS 2ND LVL SO
1262	1270	4APR95	10APR95	ALL CONCRETE COLUMNS 2ND LVL SO
1254	1255	4APR95	13APR95	2ND LVL SLAB CURE STRIP & BACKSHORE SO
1270	1277	11APR95	8MAY95	ALL FORM 3RD LVL SLABS SOUT
1276	1280	21APR95	12MAY95	ALL CONCRETE SLAB 3RD LVL SOUTH

SLETTEN CONSTRUCTION CO.
AEROSPACE/MECH. ENGINEERING BLDG UofA
BC-01 BAR CHART BY RESPONSIBILITY

(c) Primavera Systems, Inc.

FIGURE 14.19 Computer Printout of Bar Chart. *(CPM by Forward Associates, Ltd. Contractor: Sletten Construction Company. Owner: University of Arizona. Reproduced by permission.)*

SLETTEN CONSTRUCTION CO. U.OF ARIZONA, AEROSPACE & MECH.BLDG

I J SORT

FORWARD ASSOC LTD-CONSTRUCTION CONSULTANTS START DATE 07NOV94 FIN DATE 07SEP96*
CALIFORNIA, FLORIDA, TEXAS, WASHINGTON D.C. REPORT DATE 08MAY95 RUN NO. 66
P.O.BOX 1640,NOVATO,CA,94948 (415) 892-2180 DATA DATE 07NOV94 PAGE NO. 1

I	J	SCHED DUR	REM DUR	PCT	DESCRIPTION		EARLY START	LATE START	EARLY FINISH	LATE FINISH	TOTAL FLOAT	FREE FLOAT
1	2	1	1	0	NOTICE TO PROCEED 7NOV94		07NOV94	07NOV94	07NOV94	07NOV94	0	0
2	978	0	0	0	>		08NOV94	16NOV94	07NOV94	15NOV94	5	0
2	980	0	0	0	>		08NOV94	08NOV94	07NOV94	07NOV94	0	0
978	982	5	5	0	DEMOLITION PERMIT		08NOV94	16NOV94	15NOV94	22NOV94	5	5
980	982	10	10	0	MOBILIZATION		08NOV94	08NOV94	22NOV94	22NOV94	0	0
982	1002	0	0	0	>		23NOV94	30NOV94	22NOV94	29NOV94	3	0
982	1004	0	0	0	>		23NOV94	30NOV94	22NOV94	29NOV94	3	0
982	1010	10	10	0	SITE DEMOLITION		23NOV94	23NOV94	08DEC94	08DEC94	0	0
1002	1010	7	7	0	PLUMBING DEMOLITION		23NOV94	30NOV94	05DEC94	08DEC94	3	3
1004	1010	7	7	0	ELECTRICAL DEMOLITION		23NOV94	30NOV94	05DEC94	08DEC94	3	3
1010	1012	0	0	0	>		09DEC94	16DEC94	08DEC94	15DEC94	5	0
1010	1014	10	10	0	ENGINEERING & LAYOUT		09DEC94	09DEC94	22DEC94	22DEC94	0	0
1012	1016	5	5	0	EXCAVATION	BASEMENT	09DEC94	16DEC94	15DEC94	22DEC94	5	0
1014	1020	0	0	0	>		23DEC94	23DEC94	22DEC94	22DEC94	0	0
1016	1020	0	0	0	>		16DEC94	23DEC94	15DEC94	22DEC94	5	5
1020	1021	10	10	0	DRILL CAISSONS	BASEMENT	23DEC94	23DEC94	10JAN95	10JAN95	0	0
1020	1022	0	0	0	>		23DEC94	23DEC94	22DEC94	22DEC94	0	0
1021	1026	0	0	0	>		11JAN95	11JAN95	10JAN95	10JAN95	0	0
1022	1024	0	0	0	>		23DEC94	23DEC94	22DEC94	22DEC94	0	0
1022	1026	10	10	0	REINFORCE CAISSONS	BASEMENT	23DEC94	23DEC94	10JAN95	10JAN95	0	0
1024	1025	5	5	0	1/2 CONCRETE CAISSONS	BASEMENT	23DEC94	23DEC94	03JAN95	03JAN95	0	0
1025	1026	5	5	0	2/2 CONCRETE CAISSONS	BASEMENT	04JAN95	04JAN95	10JAN95	10JAN95	0	0
1025	1030	0	0	0	>		04JAN95	09JAN95	03JAN95	06JAN95	3	0
1026	1031	0	0	0	>		11JAN95	11JAN95	10JAN95	10JAN95	0	0
1026	1060	0	0	0	>		11JAN95	10FEB95	10JAN95	09FEB95	21	0
1030	1031	2	2	0	1/2 EXCAVATE FTGS & PITS	BASEMENT	04JAN95	09JAN95	05JAN95	10JAN95	3	3
1031	1032	0	0	0	>		11JAN95	13JAN95	10JAN95	12JAN95	2	0
1031	1033	4	4	0	2/2 EXCAVATE FTGS & PITS	BASEMENT	11JAN95	11JAN95	17JAN95	17JAN95	0	0
1032	1033	2	2	0	1/2 REINFORCE FTGS & PITS	BASEMENT	11JAN95	13JAN95	12JAN95	17JAN95	2	2
1033	1036	0	0	0	>		18JAN95	18JAN95	17JAN95	17JAN95	0	0
1033	1038	3	3	0	2/2 REINFORCE FTGS & PITS	BASEMENT	18JAN95	02FEB95	20JAN95	06FEB95	11	0
1036	1037	3	3	0	1/2 CONCRETE FOOTINGS & PITS	BASEMENT	18JAN95	18JAN95	20JAN95	20JAN95	0	0
1037	1038	0	0	0	>		23JAN95	07FEB95	20JAN95	06FEB95	11	0
1037	1042	0	0	0	>		23JAN95	26JAN95	20JAN95	25JAN95	3	0
1037	1044	0	0	0	>		23JAN95	23JAN95	20JAN95	20JAN95	0	0
1038	1040	2	2	0	2/2 CONCRETE FOOTINGS & PITS	BASEMENT	23JAN95	07FEB95	24JAN95	08FEB95	11	0
1040	1043	0	0	0	>		25JAN95	09FEB95	24JAN95	08FEB95	11	8
1040	1046	0	0	0	>		25JAN95	09FEB95	24JAN95	08FEB95	11	11
1040	1150	0	0	0	>		25JAN95	15FEB95	24JAN95	14FEB95	15	0
1040	1152	0	0	0	>		25JAN95	14FEB95	24JAN95	13FEB95	14	0
1042	1043	10	10	0	4/5 REINFORCE 1ST LVL WALLS	BASEMENT	23JAN95	26JAN95	03FEB95	08FEB95	3	0
1043	1047	2	2	0	5/5 REINFORCE 1ST LVL WALLS	BASEMENT	06FEB95	09FEB95	07FEB95	10FEB95	3	3
1044	1046	13	13	0	4/5 FORM & PLCE CONC 1ST LVL WALLS	BASEMENT	23JAN95	23JAN95	08FEB95	08FEB95	0	0
1046	1047	2	2	0	5/5 FORM & PLCE CONC 1ST LVL WALLS	BASEMENT	09FEB95	09FEB95	10FEB95	10FEB95	0	0
1047	1162	0	0	0	>		13FEB95	14FEB95	10FEB95	13FEB95	1	0
1047	1164	0	0	0	>		13FEB95	13FEB95	10FEB95	10FEB95	0	0
1060	1062	0	0	0	>		11JAN95	10FEB95	10JAN95	09FEB95	21	0
1060	1066	10	10	0	DRILL CAISSONS	HI BAY AREA	11JAN95	10FEB95	25JAN95	23FEB95	21	0

FIGURE 14.20 Sort by Activity (*i-j*) Numbers. *(CPM by Forward Associates, Ltd., Novato, California. Contractor: Sletten Construction Co. Owner: University of Arizona. Reproduced by permission.)*

SLETTEN CONSTRUCTION CO. — U.OF ARIZONA, AEROSPACE & MECH.BLDG

TOTAL FLOAT SORT

FORWARD ASSOC LTD-CONSTRUCTION CONSULTANTS — START DATE 07NOV94 FIN DATE 07SEP96*

CALIFORNIA, FLORIDA, TEXAS, WASHINGTON D.C. — REPORT DATE 08MAY95 RUN NO. 64

P.O.BOX 1640,NOVATO,CA,94948 (415) 892-2180 — DATA DATE 07NOV94 PAGE NO. 1

I	J	SCHED DUR	REM DUR	PCT	DESCRIPTION		EARLY START	LATE START	EARLY FINISH	LATE FINISH	TOTAL FLOAT	FREE FLOAT
1	2	1	1	0	NOTICE TO PROCEED 7NOV94		07NOV94	07NOV94	07NOV94	07NOV94	0	0
2	980	0	0	0	>		08NOV94	08NOV94	07NOV94	07NOV94	0	0
980	982	10	10	0	MOBILIZATION		08NOV94	08NOV94	22NOV94	22NOV94	0	0
982	1010	10	10	0	SITE DEMOLITION		23NOV94	23NOV94	08DEC94	08DEC94	0	0
1010	1014	10	10	0	ENGINEERING & LAYOUT		09DEC94	09DEC94	22DEC94	22DEC94	0	0
1014	1020	0	0	0	>		23DEC94	23DEC94	22DEC94	22DEC94	0	0
1020	1022	0	0	0	>		23DEC94	23DEC94	22DEC94	22DEC94	0	0
1022	1024	0	0	0	>		23DEC94	23DEC94	22DEC94	22DEC94	0	0
1024	1025	5	5	0	1/2 CONCRETE CAISSONS	BASEMENT	23DEC94	23DEC94	03JAN95	03JAN95	0	0
1020	1021	10	10	0	DRILL CAISSONS	BASEMENT	23DEC94	23DEC94	10JAN95	10JAN95	0	0
1021	1026	0	0	0	>		11JAN95	11JAN95	10JAN95	10JAN95	0	0
1022	1026	10	10	0	REINFORCE CAISSONS	BASEMENT	23DEC94	23DEC94	10JAN95	10JAN95	0	0
1025	1026	5	5	0	2/2 CONCRETE CAISSONS	BASEMENT	04JAN95	04JAN95	10JAN95	10JAN95	0	0
1026	1031	0	0	0	>		11JAN95	11JAN95	10JAN95	10JAN95	0	0
1031	1033	4	4	0	2/2 EXCAVATE FTGS & PITS	BASEMENT	11JAN95	11JAN95	17JAN95	17JAN95	0	0
1033	1036	0	0	0	>		18JAN95	18JAN95	17JAN95	17JAN95	0	0
1036	1037	3	3	0	1/2 CONCRETE FOOTINGS & PITS	BASEMENT	18JAN95	18JAN95	20JAN95	20JAN95	0	0
1037	1044	0	0	0	>		23JAN95	23JAN95	20JAN95	20JAN95	0	0
1044	1046	13	13	0	4/5 FORM & PLCE CONC 1ST LVL WALLS	BASEMENT	23JAN95	23JAN95	08FEB95	08FEB95	0	0
1046	1047	2	2	0	5/5 FORM & PLCE CONC 1ST LVL WALLS	BASEMENT	09FEB95	09FEB95	10FEB95	10FEB95	0	0
1047	1164	0	0	0	>		13FEB95	13FEB95	10FEB95	10FEB95	0	0
1164	1170	5	5	0	PLUMBING UNDERSLAB	BASEMENT	13FEB95	13FEB95	17FEB95	17FEB95	0	0
1170	1171	2	2	0	1/2 PREP SLAB ON GRADE	BASEMENT	20FEB95	20FEB95	21FEB95	21FEB95	0	0
1171	1172	0	0	0	>		22FEB95	22FEB95	21FEB95	21FEB95	0	0
1172	1173	3	3	0	1/2 REINFORCE SLAB ON GRADE	BASEMENT	22FEB95	22FEB95	24FEB95	24FEB95	0	0
1173	1174	4	4	0	2/2 REINFORCE SLAB ON GRADE	BASEMENT	27FEB95	27FEB95	02MAR95	02MAR95	0	0
1174	1176	2	2	0	TERMITE PROTECTION	BASEMENT	03MAR95	03MAR95	06MAR95	06MAR95	0	0
1176	1178	2	2	0	CONCRETE SLAB ON GRADE	BASEMENT	07MAR95	07MAR95	08MAR95	08MAR95	0	0
1178	1180	0	0	0	>		09MAR95	09MAR95	08MAR95	08MAR95	0	0
1180	1181	5	5	0	1/3 FORM SLAB 2ND LVL (SUSPENDED)	SOUTH	09MAR95	09MAR95	15MAR95	15MAR95	0	0
1181	1184	10	10	0	3/3 FORM SLAB 2ND LVL (SUSPENDED)	SOUTH	16MAR95	16MAR95	29MAR95	29MAR95	0	0
1184	1189	1	1	0	5/5 REINFORCE SLAB 2ND LVL (SUSPENDED)	SOUTH	30MAR95	30MAR95	30MAR95	30MAR95	0	0
1047	1162	0	0	0	>		13FEB95	14FEB95	10FEB95	13FEB95	1	0
1162	1170	4	4	0	ELECTICAL UNDERSLAB	BASEMENT	13FEB95	14FEB95	16FEB95	17FEB95	1	1
1171	1173	2	2	0	2/2 PREP SLAB ON GRADE	BASEMENT	22FEB95	23FEB95	23FEB95	24FEB95	1	1
1031	1032	0	0	0	>		11JAN95	13JAN95	10JAN95	12JAN95	2	0
1032	1033	2	2	0	1/2 REINFORCE FTGS & PITS	BASEMENT	11JAN95	13JAN95	12JAN95	17JAN95	2	2
982	1002	0	0	0	>		23NOV94	30NOV94	22NOV94	29NOV94	3	0
982	1004	0	0	0	>		23NOV94	30NOV94	22NOV94	29NOV94	3	0
1002	1010	7	7	0	PLUMBING DEMOLITION		23NOV94	30NOV94	05DEC94	08DEC94	3	3
1004	1010	7	7	0	ELECTRICAL DEMOLITION		23NOV94	30NOV94	05DEC94	08DEC94	3	3
1025	1030	0	0	0	>		04JAN95	09JAN95	03JAN95	06JAN95	3	0
1030	1031	2	2	0	1/2 EXCAVATE FTGS & PITS	BASEMENT	04JAN95	09JAN95	05JAN95	10JAN95	3	3
1037	1042	0	0	0	>		23JAN95	26JAN95	20JAN95	25JAN95	3	0
1042	1043	10	10	0	4/5 REINFORCE 1ST LVL WALLS	BASEMENT	23JAN95	26JAN95	03FEB95	08FEB95	3	0
1043	1047	2	2	0	5/5 REINFORCE 1ST LVL WALLS	BASEMENT	06FEB95	09FEB95	07FEB95	10FEB95	3	3
2	978	0	0	0	>		08NOV94	16NOV94	07NOV94	15NOV94	5	0
978	982	5	5	0	DEMOLITION PERMIT		08NOV94	16NOV94	15NOV94	22NOV94	5	5

FIGURE 14.21 Sort by Total Float. *(CPM by Forward Associates, Ltd., Novato, California. Contractor: Sletten Construction Co. Owner: University of Arizona. Reproduced by permission.)*

SLETTEN CONSTRUCTION CO. — U.OF ARIZONA, AEROSPACE & MECH.BLDG

EARLY FINISH SORT

FORWARD ASSOC LTD-CONSTRUCTION CONSULTANTS — START DATE 07NOV94 FIN DATE 07SEP96*

CALIFORNIA, FLORIDA, TEXAS, WASHINGTON D.C. — REPORT DATE 08MAY95 RUN NO. 65

P.O.BOX 1640,NOVATO,CA,94948 (415) 892-2180 — DATA DATE 07NOV94 PAGE NO. 1

I	J	SCHED DUR	REM DUR	PCT	DESCRIPTION		EARLY START	LATE START	EARLY FINISH	LATE FINISH	TOTAL FLOAT	FREE FLOAT
1	2	1	1	0	NOTICE TO PROCEED 7NOV94		07NOV94	07NOV94	07NOV94	07NOV94	0	0
978	982	5	5	0	DEMOLITION PERMIT		08NOV94	16NOV94	15NOV94	22NOV94	5	5
980	982	10	10	0	MOBILIZATION		08NOV94	08NOV94	22NOV94	22NOV94	0	0
1002	1010	7	7	0	PLUMBING DEMOLITION		23NOV94	30NOV94	05DEC94	08DEC94	3	3
1004	1010	7	7	0	ELECTRICAL DEMOLITION		23NOV94	30NOV94	05DEC94	08DEC94	3	3
982	1010	10	10	0	SITE DEMOLITION		23NOV94	23NOV94	08DEC94	08DEC94	0	0
1012	1016	5	5	0	EXCAVATION	BASEMENT	09DEC94	16DEC94	15DEC94	22DEC94	5	0
1010	1014	10	10	0	ENGINEERING & LAYOUT		09DEC94	09DEC94	22DEC94	22DEC94	0	0
1024	1025	5	5	0	1/2 CONCRETE CAISSONS	BASEMENT	23DEC94	23DEC94	03JAN95	03JAN95	0	0
1030	1031	2	2	0	1/2 EXCAVATE FTGS & PITS	BASEMENT	04JAN95	09JAN95	05JAN95	10JAN95	3	3
1020	1021	10	10	0	DRILL CAISSONS	BASEMENT	23DEC94	23DEC94	10JAN95	10JAN95	0	0
1022	1026	10	10	0	REINFORCE CAISSONS	BASEMENT	23DEC94	23DEC94	10JAN95	10JAN95	0	0
1025	1026	5	5	0	2/2 CONCRETE CAISSONS	BASEMENT	04JAN95	04JAN95	10JAN95	10JAN95	0	0
1032	1033	2	2	0	1/2 REINFORCE FTGS & PITS	BASEMENT	11JAN95	13JAN95	12JAN95	17JAN95	2	2
1031	1033	4	4	0	2/2 EXCAVATE FTGS & PITS	BASEMENT	11JAN95	11JAN95	17JAN95	17JAN95	0	0
1033	1038	3	3	0	2/2 REINFORCE FTGS & PITS	BASEMENT	18JAN95	02FEB95	20JAN95	06FEB95	11	0
1036	1037	3	3	0	1/2 CONCRETE FOOTINGS & PITS	BASEMENT	18JAN95	18JAN95	20JAN95	20JAN95	0	0
1038	1040	2	2	0	2/2 CONCRETE FOOTINGS & PITS	BASEMENT	23JAN95	07FEB95	24JAN95	08FEB95	11	0
1060	1066	10	10	0	DRILL CAISSONS	HI BAY AREA	11JAN95	10FEB95	25JAN95	23FEB95	21	0
1062	1066	10	10	0	REINFORCE CAISSONS	HI BAY AREA	11JAN95	10FEB95	25JAN95	23FEB95	21	0
1064	1066	10	10	0	CONCRETE CAISSONS	HI BAY AREA	11JAN95	10FEB95	25JAN95	23FEB95	21	0
1070	1071	1	1	0	1/2 EXCAVATE FOOTINGS & GRADE BEANS	HI BAY AREA	26JAN95	24FEB95	26JAN95	24FEB95	21	0
1150	1154	3	3	0	REINFORCE CONCRET COLUMNS	BASEMENT	25JAN95	15FEB95	27JAN95	17FEB95	15	1
1152	1154	4	4	0	FORM & PLCE CONCRETE COLUMNS	BASEMENT	25JAN95	14FEB95	30JAN95	17FEB95	14	0
1074	1075	2	2	0	1/2 CONCRETE FOOTINGS & GRADE BEAMS	HI BAY AREA	27JAN95	08MAR95	30JAN95	09MAR95	28	7
1071	1073	3	3	0	2/2 EXCAVATE FTGS & GRADE BEANS	HI BAY AREA	27JAN95	03MAR95	31JAN95	07MAR95	25	4
1200	1206	5	5	0	DRILL CAISSONS 2ND LVL	SOUTH	26JAN95	09JUN95	01FEB95	15JUN95	96	0
1202	1206	5	5	0	REINFORCE CAISSONS 2ND LVL	SOUTH	26JAN95	09JUN95	01FEB95	15JUN95	96	0
1204	1206	5	5	0	CONCRETE CAISSONS 2ND LVL	SOUTH	26JAN95	09JUN95	01FEB95	15JUN95	96	0
1042	1043	10	10	0	4/5 REINFORCE 1ST LVL WALLS	BASEMENT	23JAN95	26JAN95	03FEB95	08FEB95	3	0
1072	1073	7	7	0	2/3 REINFORCE FTGS & GRADE BEAMS	HI BAY AREA	27JAN95	27FEB95	06FEB95	07MAR95	21	0
1043	1047	2	2	0	5/5 REINFORCE 1ST LVL WALLS	BASEMENT	06FEB95	09FEB95	07FEB95	10FEB95	3	3
1044	1046	13	13	0	4/5 FORM & PLCE CONC 1ST LVL WALLS	BASEMENT	23JAN95	23JAN95	08FEB95	08FEB95	0	0
1073	1075	2	2	0	3/3 REINFORCE FOOTINGS & GRADE BEAMS	HI BAY AREA	07FEB95	08MAR95	08FEB95	09MAR95	21	0
1046	1047	2	2	0	5/5 FORM & PLCE CONC 1ST LVL WALLS	BASEMENT	09FEB95	09FEB95	10FEB95	10FEB95	0	0
1075	1076	2	2	0	2/2 CONCRETE FOOTINGS & GRADE BEAMS	HI BAY AREA	09FEB95	10MAR95	10FEB95	13MAR95	21	0
1090	1093	3	3	0	REINFORCE WALLS 2ND LVL	HI BAY AREA	13FEB95	17MAR95	15FEB95	21MAR95	24	2
1080	1083	4	4	0	REINFORCE CONC COLUMNS 2ND LVL	HI BAY AREA	13FEB95	16MAR95	16FEB95	21MAR95	23	2
1162	1170	4	4	0	ELECTICAL UNDERSLAB	BASEMENT	13FEB95	14FEB95	16FEB95	17FEB95	1	1
1092	1093	5	5	0	2/3 FORM & PLACE CONC.WALLS 2ND LVL	HI BAY AREA	13FEB95	15MAR95	17FEB95	21MAR95	22	0
1164	1170	5	5	0	PLUMBING UNDERSLAB	BASEMENT	13FEB95	13FEB95	17FEB95	17FEB95	0	0
1082	1083	6	6	0	2/3 FORM & PLCE CONC COLUMS 2ND LVL	HI BAY AREA	13FEB95	14MAR95	20FEB95	21MAR95	21	0
1093	1094	1	1	0	3/3 FORM & PLACE CONC. WALLS 2ND LVL	HI BAY AREA	20FEB95	22MAR95	20FEB95	22MAR95	22	1
1170	1171	2	2	0	1/2 PREP SLAB ON GRADE	BASEMENT	20FEB95	20FEB95	21FEB95	21FEB95	0	0
1083	1094	1	1	0	3/3 FORM & PLCE CONC COLUMS 2ND LVL	HI BAY AREA	21FEB95	22MAR95	21FEB95	22MAR95	21	0
1171	1173	2	2	0	2/2 PREP SLAB ON GRADE	BASEMENT	22FEB95	23FEB95	23FEB95	24FEB95	1	1
1094	1097	3	3	0	FORM & PLACE CONCRETE WALLS 3RD LVL	HI BAY AREA	22FEB95	23MAR95	24FEB95	27MAR95	21	0
1096	1097	3	3	0	3/4 REINFORCE WALLS 3RD LVL	HI BAY AREA	22FEB95	23MAR95	24FEB95	27MAR95	21	0

FIGURE 14.22 Sort by Early Finish Dates. *(CPM by Forward Associates, Ltd., Novato, California. Contractor: Sletten Construction Co. Owner: University of Arizona. Reproduced by permission.)*

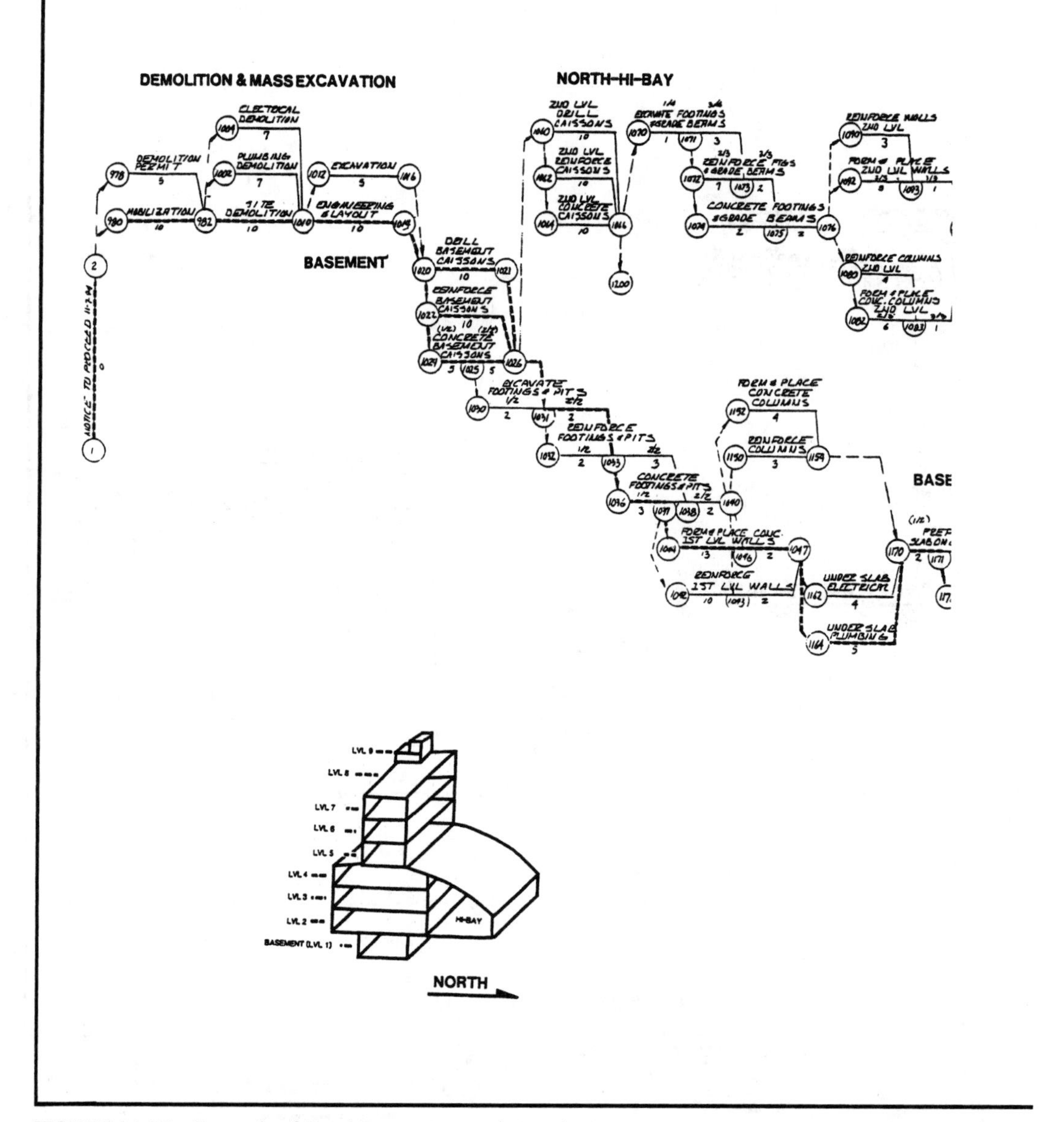

FIGURE 14.23 Example of Hand-Drawn Network Graphics in Arrow Diagraming (*i-j*) Format. *(CPM by Forward Associates, Ltd., Novato, California. Contractor: Sletten Construction Co. Owner: University of Arizona. Reproduced by permission.)*

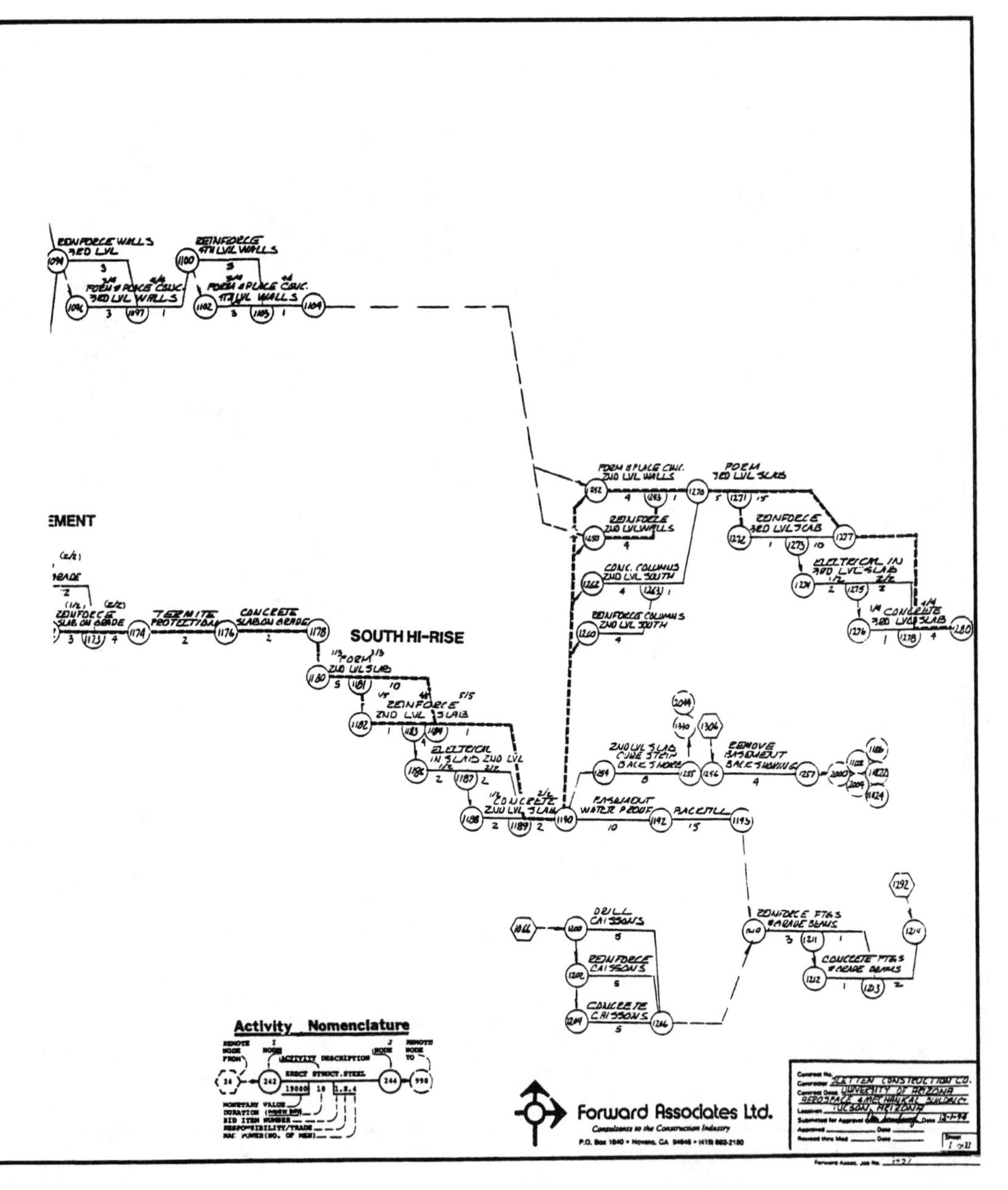

FIGURE 14.23 (continued)

On the detailed network plan shown in Figure 14.23, the critical path is indicated by a heavy, dashed line. Of particular note on this sheet is the fact that at times there is more than one critical path during this phase of the work. Thus, if any change occurs to delay the work along any of the critical paths, that one path will become the one critical path unless similar delays occurred along another path that would exceed those of the one just affected.

It may be noticed that this network was hand-drafted without a time scale, using the *i-j* method. The system of diagraming is an arrow diagram with activity-on-arrow notation. A portion of the same schedule in computer-generated precedence format is shown in Figure 14.24.

Schedule Reports

By observation of each of the computer printouts, it can be seen that the column headings are all identical. The only difference between them is the arrangement of the activities on the sheets. On one, all activities are arranged in numerical order by *i-j* node numbers; in another they are all arranged in the order of dates of early finish of each activity; in another they are all arranged in the order of the amount of float time per activity.

To take just one example, activity 1042–1043 is included in each of the three sample schedule reports shown in Figures 14.20 to 14.22. In each report, the early start date is 23 January 1995 and the early finish date is shown as 3 February 1995. The late start date for the same activity is shown as 26 January and the late finish date as 8 February, the total float time is 3 days, and the free float is shown as zero. Thus, the only difference between the entries on the various report "sorts" is the sequence or arrangement of the data on the sheet.

Note activity 1030–1031 in Figure 14.20. This is a sort by *i-j* numbers. All activities in this report format have been arranged in the numerical order of their activity numbers, as shown in the first two columns headed "I" and "J." Thus, if the user is searching for an activity by its number, it can easily be found by reference to this "sort."

In Figure 14.21, however, the same data used in the sort by *i-j* numbers has now been rearranged by the computer in the order of total float. Thus, if the user reads the two far-right columns headed "TOTAL FLOAT" and "FREE FLOAT," the user can readily locate all activities with the specified amount of float.

Note activity 1042–1043 in Figure 14.21. The total float indicated for this activity is 3 days, yet the amount of free float is zero. This indicates that there is no leeway in the individual schedule for that activity, but any float is the result of slack time somewhere else in the network. By comparison, see activity 1020–1021 in Figure 14.21, which shows that there is no float time of any kind available for this activity. This is an indication that the activity is on the critical path. By locating this activity on the CPM network in Figure 14.21, it can be seen that this is truly the case.

Similarly, in Figure 14.22, the data have again been rearranged, this time in the order of "early finish dates" for each activity. If it is desired to determine all items that will be finished as of any given date, reference to this "sort" will show them.

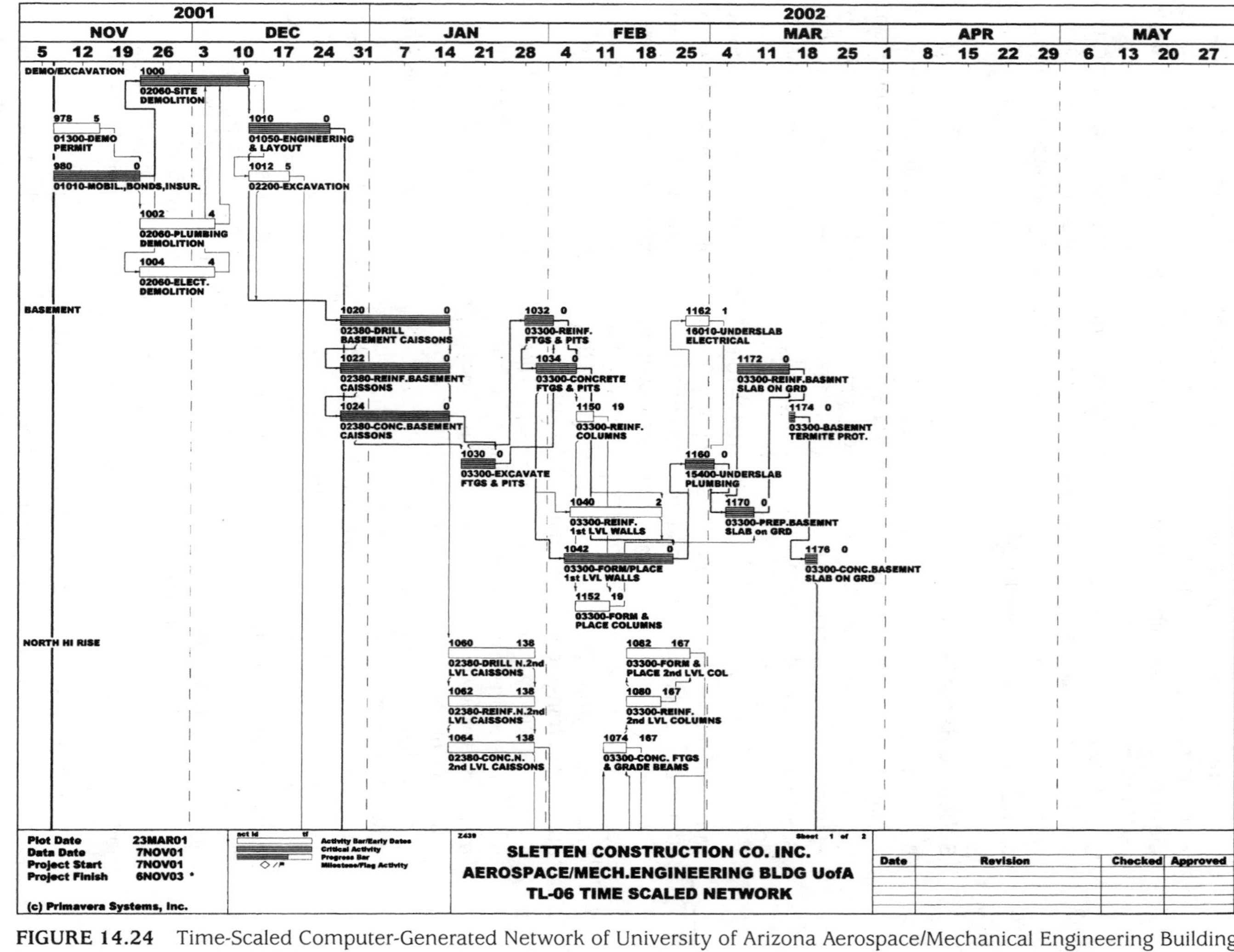

FIGURE 14.24 Time-Scaled Computer-Generated Network of University of Arizona Aerospace/Mechanical Engineering Building in Precedence Format. *(CPM by Forward Associates, Ltd., Novato, California. Contractor: Sletten Construction Co. Owner: University of Arizona. Reproduced by permission.)*

On the schedule reports shown in Figures 14.20 to 14.22 an example of total and free float for a given activity can be shown. Activities 1004–1010 and 1002–1010 are concurrent, followed by successor activity 1010–1014, which is on critical path, a fact that can be observed on both the printouts and the detailed network plan. Activities 1004–1010 and 1002–1010 each show 3 days of total float and 3 days of free float, followed by activity 1010–1014, which has no float because it is on critical path. In this case, delaying either 1004–1010 or 1002–1010 by 3 days will place that activity on critical path. From the network diagram it is evident that 1010–1014 cannot proceed until all three predecessor activities have been completed. The late finish dates of each of the predecessor activities are shown in the schedule report as 8 December 1994. The early/late start date for 1010–1014, which is on critical path, is shown on the schedule reports as 9 December 1994. The early finish dates of predecessor activities 1002–1010 and 1004–1010 are shown as 5 December 1994, a difference of 3 days between the early finish date of 5 December and the late finish date of 8 December, which accounts for the indicated 3 days, float before the early/late start date of critical path activity 1010–1014. Thus any overrun in the completion date of either of the predecessor activities would have an immediate adverse effect on the completion date of the project.

PROBLEMS

1. Name the five basic steps in setting up a CPM schedule.
2. Define *free float* and *total float*.
3. What type of scheduling is defined as an "activity-oriented" system? What type is defined as an "event-oriented" system?
4. In what kind of diagraming does an arrow represent both dependency and duration?
5. Define a *logic loop*.
6. Can a network schedule legitimately contain more than one critical path?
7. Under the guide specifications of the Associated General Contractors (AGC), all work activities should be broken down into work activities not exceeding what maximum number of days?
8. A critical path network is based on which of the following?
   (a) Dates
   (b) Computer programs
   (c) Logical networks
9. Is activity-on-node diagraming more closely related to arrow diagraming or to precedence diagraming?
10. What is a dummy arrow?
11. What is a "critical path"?

15

CONSTRUCTION OPERATIONS

Although most of the material in this book could be classed as part of the contract administrator's and Resident Project Representative's normal construction operations, separate chapters have been presented to cover material that relates to the more complex functions that are necessary, as well as the more lengthy subjects that require special clarification. Thus, the subject of this chapter is of more general nature. The day-to-day considerations experienced by the Resident Project Representative, along with an understanding of who is responsible for what, are the principal items covered. That is not to imply that the material is unimportant or noncontroversial, as some of the day-to-day activities that must be endured by the Resident Project Representative would challenge the patience of a stone statue.

Following are the paraphrased words of Gene Sheley, editor of *Western Construction* magazine,[1] who admittedly was not thinking about inspectors when he wrote:

> The Resident Project Representative must be an expert in engineering, architecture, construction methods, labor relations, barroom brawls, public relations, and should have an extensive vocabulary in graphic and colorful construction terms. He or she must be able to work closely with contractors, subcontractors, engineers, architects, owners, and agree with frequent directives that make no sense from people who know less than he or she does; he or she must be able to understand then ignore environmental impact reports and regulations without the Sierra Club finding out, and he or she must cooperate with construction superintendents and contractor quality control representatives, and avoid backing his or her automobile into the privy while they are inside.
>
> The resident inspector must have skin like an alligator, the stomach of a billie goat, the temperament of a Presbyterian minister, nerves of chrome-molybdenum steel, the fortitude of Job, and the physical strength to take care of a situation when all else fails. He or she should enjoy his or her job, and will probably love the working conditions, except for the dust, numbing cold, searing heat, knee-deep mud, mosquitos, rattlesnakes, scorpions, muck-covered office trailers, questionable toilet facilities, and physical assault by frustrated foremen.
>
> If you still want to be the Resident Project Representative or inspector, the job is yours.

[1] Used with permission of Gene Sheley, Editor, *Western Construction*.

AUTHORITY AND RESPONSIBILITY OF ALL PARTIES

The owner as a contracting party has several especially reserved rights. Depending upon the type of contract and its specific wording, the owner may be authorized to award other contracts in connection with the same work, to require contract bonds from the contractor, to approve the surety proposed, to retain a specific portion of the contractor's monthly progress payments, to make changes in the work, to carry out portions of the work with the owner's own forces in case of contractor default or neglect, to withhold payments from the contractor for adequate cause, and to terminate the contract for cause. The right of the owner to inspect the work as it progresses, to direct the contractor to expedite the work, to use completed portions of the work before contract completion, and to make payment deductions for incomplete or faulty work are also common contractual provisions.

The contract between the owner and the contractor also imposes some responsibilities on the owner. For example, most construction contracts make the owner responsible for furnishing property surveys that describe or locate the project on the site, for making periodic payments to the contractor, and for making land surveys that establish the boundaries of the property upon which the project is to be located. The owner is also obligated to make extra payment in case of eventualities that were not anticipated in the contract, as well as to allow extensions of time to complete the work for such unanticipated conditions.

It is important to note, however, that the owner cannot intrude into the direction and control of the work. By the terms of the usual construction contract, the contractor is classed as an "independent contractor," and even though the owner has certain rights with respect to the conduct of the work, the owner cannot issue direct instructions as to methods or procedures unless specifically provided for under the terms of the contract. The owner does not have the authority to interfere unreasonably with construction operations, or otherwise unduly assume the functions of directing and controlling the work. If the owner were to assume such authority, it would relieve the contractor from the responsibility for the completed work as well as for the negligent acts committed by the contractor in the course of the construction operations—in short, the owner would be acting as a general contractor and would thus have to expect to inherit all of the responsibilities, risks, and liabilities of that position.

The Architect/Engineer as a Separate Design Organization

Except for cases in which both design and construction are performed by the same contracting party (design–build or turnkey construction), or in which the owner has its own in-house design capability, the architect/engineer, as a separate design firm, is not a party to the construction contract, and no contractual relationship exists between the architect/ engineer and the contractor. The design firm is a third party that derives its authority and responsibility from its contract with the owner. When private design professionals are utilized by the owner, the construction contract substitutes the architect/engineer for the owner in many important respects under the contract. However, the jurisdiction of the architect/engineer to make determinations and render decisions binding under the construction contract is limited to the specific terms of the construction contract. The architect/engineer often represents the owner in the administration of the contract and acts for the owner in the day-to-day administration of construction operations. In such contracts, the architect/engineer advises and consults with the owner, and communications between the owner and the contractor are usually made through the architect/engineer. Article 9 of the Standard General Conditions of the Construction Contract of the Engineer's Joint Contract Documents Committee (EJCDC) contains typical provisions regarding the architect/engineer's role in construction contract administration.

Construction contracts of this type impose many duties and bestow considerable authority on the architect/engineer. All construction operations are to be constructed under the surveillance of the architect/engineer, who is generally responsible for overseeing the progress of the work. It is normally the architect/engineer's direct responsibility to see that the quality of work and materials is in conformance with the requirements of the drawings and specifications. To assure fulfillment of these conditions, the architect/engineer firm or its Resident Project Representative (resident engineer or inspector) exercises the right of job inspection and approval of materials. In addition, that firm may exercise the privilege of inspecting the contractor's general program of field procedure and even the equipment that is planned for use, as well as the schedule and sequence of operations to complete the work. Should the work be lagging behind schedule, the design firm or its field representative may reasonably instruct the contractor to speed up the work.

The fact that the architect/engineer retains the privilege of approval of the contractor's methods does not mean that it is assuming responsibility for them. The rights of the architect/engineer are essentially those of assuring that the contractor is proceeding in accordance with the provisions of the contract documents, and that the contractor's methods or equipment are capable of accomplishing this objective.

The contract documents often authorize the architect/engineer to interpret the requirements of the contract. The usual wording is that the "decision of the architect/engineer shall be binding and final, and shall bind both parties." Actually, the jurisdiction of the architect/engineer is limited to the settlement of questions of *fact* only, such as what materials, quantities, or quality are required, or whether the work meets the contract requirements. The answers to questions of fact require the professional knowledge and skill of the architect/engineer, and it is proper that he or she should make such decisions. In the absence of fraud, bad faith, or gross mistake, the decision of the architect/engineer may, in fact, be considered as final unless the terms of the contract contain provisions for appeals or arbitration.

With respect to disputed questions of law, however, the architect/engineer has no jurisdiction. He or she cannot deny the right of a citizen to due process of law, and the contractor has the right to submit a dispute concerning a legal aspect of the contract to arbitration or to the courts. Whether a particular matter is one of fact or one of legal construction can depend upon the language of the contract. Matters concerning time of completion, liquidated damages, and claims for extra work are usually points of law, not fact.

The General Contractor

As you might assume from a document prepared by and especially for the owner, the contractor appears to have fewer rights and more obligations under the contract. The contractor's major responsibility, of course, is to construct the project in accordance with the drawings, specifications, and other contract documents. Despite all the troubles, delays, adversities, accidents, and other misoccurrences that may happen, the contractor is generally expected to deliver a completed project in the allotted time—just as if nothing had happened to slow down the work. Although some casualties are considered to be justifiable reasons to receive more construction time, only severe contingencies such as impossibility of performance can serve to relieve the contractor from contractual obligations.

Regardless of how contracts are awarded for a construction project, there should always be one party placed in the position of responsibility for the management and control of construction operations to maintain established work schedules, promote safe working conditions, and avoid unnecessary disputes and conflicts that result in unjustified claims for extra compensation. Such responsibility is normally placed in the hands of a general contractor, except in those states where local laws require separate prime contracts for mechanical and electrical work. The cost of such management and control of construction operations, including the services of a separate person to act as a mechanical work coordinator for major projects, would become a part of the general contractor's bid price. In general, the project should include the services of a separate mechanical work coordinator, but some projects might not require one because the project is too small to justify one, or the mechanical and electrical systems are uncomplicated. The specifying architects/engineers must make their own determination, taking into consideration the percentage of total work contributed by the mechanical and electrical trades as well as the complexity of the design. In all instances, particularly when separate contracts or assigned contracts occur, the mechanical and electrical specifications should contain provisions requiring full and complete cooperation with the general contractor and its mechanical work coordinator.

Under the provisions of Article 6 of the General Conditions of the EJCDC, and similar documents of other agencies, the contractor is expected to give personal attention to the work, and either the contractor or its authorized representative must be on the site at all times during working hours. The contractor is further required to conform to all laws and ordinances concerning job safety, licensing, employment of labor, sanitation, insurance, zoning, building codes, and other aspects of the work. In many cases, failure of the design firm to research restrictions on its project properly results in designs that in themselves are not in conformance with all applicable regulations. Thus, the contractor really inherits a bucket of worms when the contract calls for conformance to some technical requirement, and at the same time requires the firm to conform to all codes and laws that clearly show the original design to be a violation—sometimes, you just cannot win. Many contracts now include tough new rules designed to decrease air pollution, noise pollution, dust, and similar restrictions as well as rules concerning trash disposal, sanitary wastes, pile driving, blasting, riveting, demolition, fencing, open excavations, traffic control, and housekeeping.

A general contractor is further held responsible for and must guarantee all work and materials on the project, whether constructed by the firm's own forces or by subcontractors, because the subcontractors have no contractual relationship with the owner—only with the general contractor. Every restriction in the construction contract that refers to the "contractor" is binding solely upon the general contractor, as far as the owner is concerned. The general contractor may subcontract portions of the work, but the terms of such subcontracts are not subject to review by the owner and are solely an agreement between the general and the subcontractor. Even though a contractor has no direct responsibility for the adequacy of the plans and specifications, it can incur contingent liability for proceeding with faulty work whose defects should be evident to one in that business. Should an instance occur in which the contracting firm is directed to do something that it feels is not proper and is not within the realm of good construction practice, the contracting firm should protect itself by filing a letter of protest to the design firm or the owner through the resident engineer or inspector, stating its position and the facts as it sees them before proceeding with the matter in dispute. If ordered to proceed by the design firm or the owner, the contractor must continue with the work even if it disagrees, or even in case of dispute for other causes. Settlement of disputes then follows concurrently with the prosecution of the work.

Insurance coverage is an important contractual responsibility of the contractor, both as to type of insurance and the policy limits. The contractor is generally required to provide insurance not only for its own direct and contingent liability, but also frequently for the owner's protection. The contractor firm is expected to exercise every reasonable safeguard for the protection of persons and property in, on, and adjacent to the construction site.

Some of the contractor's most important rights concern progress payments, and the contractor's recourse in case the owner should fail to make such payments, as well as the right to terminate the contract for cause, the right to extra payment and extensions of time as provided in the contract, and appeals from decisions of the owner or the design firm. Subject to contractual requirements and limitations in the contract, the contractor is free to subcontract portions of the work, to purchase materials where it chooses (but not necessarily the right to select such materials), and to proceed with the work in any way or order that it pleases, if permitted under the terms of the contract documents.

The Resident Project Representative; Project Representative

The Resident Project Representative, or Project Representative, normally works as the agent of the owner, the design firm, or a construction management firm, and as such may, under the terms of such agency, exercise such authority as is normally reserved to the owner, design firm, or construction firm, as long as he or she acts within the scope of his or her delegated authority and subject to the restrictions of law that prohibit such persons from performing certain functions reserved by law to the practice of professional engineering or architecture.

The purpose of inspection is to detect, recognize, and report deficiencies in material or workmanship, or noncompliance with applicable plans, specifications, or other contract documents, procedures, standards, codes, or regulations. The resident engineer's or inspector's job is to inspect the workmanship, materials, and manner of construction of all buildings and appurtenant structures, or portions of such structures, to determine whether the requirements described by the plans, specifications, contract documents, codes, ordinances, or other statutory provisions are met by the observed work. This responsibility is basic. Any authority or responsibilities beyond those stated are limited to those delegated by the inspector's employer and should be clearly established before reporting to the job.

The CQC Representative

A contractor quality control (CQC) representative is a position unique to federal contracts, particularly those with NAVFAC, Corps of Engineers, and NASA. Under a CQC provision, a construction contracting firm has the responsibility to inspect its own work on some federal contracts, and to present for federal acceptance only such work that complies with the contract plans and specifications. Under a contract requiring a contractor quality control program (usually construction projects with a budget estimate of over $1,000,000), the contractor is required to assign a responsible and competent individual to the position of CQC representative, and to delegate to that person certain responsibilities and authority. The primary function of a CQC representative at the site is to assure that all inspections and tests are made and to give all approvals unless specifically reserved to the federal agency. This includes the checking of all material and equipment delivered to the site. One of the objectives is to achieve quality construction acceptable to the agency and to contribute to the contractor's profits by preventing defective work rather than discovering deficiencies that may result in costly removal and replacement. In those cases in which unacceptable work is started or completed, the CQC representative must have the authority to take any action necessary to correct the deficiency even though it means stopping the work of the CQC representative's employer on a particular portion of the job. The CQC representative must also coordinate and assure the performance of all tests required by the specifications. As an agent of the contractor, the CQC representative will be held responsible by the federal agency for any fraudulent acts or certifications. In day-to-day contact with the federal agency involved, the CQC representative works through a federal construction representative assigned to the project.

Temporary Facilities Provided by the Contractor

As a part of each construction contract, the contractor is generally obligated to provide certain facilities and services that are not a permanent part of the facilities to be constructed. In some cases the specifications do not mention the specific items involved, thus needlessly complicating the administration of the contract. (See "Provisions for Temporary Facilities" in Chapter 7 for further details.)

All such items, when properly supported by specifications provisions, become a part of the specific contractual obligations of the contractor. Many of the items so listed are also properly considered as a part of the contractor's mobilization requirements, and as such, it is recommended that the payment provisions of the contract specifically itemize all such facilities and services. Then, in case of failure of the contractor to provide any or all of the items listed, the owner may be justified in withholding all or part of the contractor's initial progress payment. (See "Payment for Mobilization Costs" in Chapter 17 for further details.)

Failure of the owner's architect/engineer to provide such controls over the contractor's mobilization activities frequently results in long delays in providing the Resident Project Representative with a field office and field telephone service. Failure to provide timely communications in itself can be a serious obstacle in the path of the architect/engineer for the development of effective project control.

Time of Inspection and Tests

The provisions of the various General Conditions of the Construction Contract generally treat the subject of when and under what conditions the inspections and tests will be performed. Although each standard document now in use seems to treat the subject in different words, they all say essentially the same thing. Generally, the provisions cover the following:

1. Contractor must give the architect/engineer timely notice.
2. Tests and inspections required by public agencies must usually be paid for by the contractor.
3. Tests and inspections, other than those mentioned, that are required by the contract documents will be paid for by the owner.
4. Work covered prior to required inspections must be uncovered for inspection and then recovered, all at the contractor's expense.
5. Failure of an inspector to observe a deficiency does not relieve the contractor of obligation for performance.
6. Extra inspections required as the result of a deficiency noted by the inspector must be paid for by the contractor.
7. The contractor firm must provide all materials for testing at its own cost and expense.

The Resident Project Representative is urged to refer to Articles 13.3 through 13.7 of the EJCDC 1910-8 General Conditions, Article 13.5 of the AIA A201 General Conditions, and Sub-Clauses 36.3 through 37.5 of the FIDIC Conditions of Contract for examples of terms covering such inspections. Similarly, the General Conditions of the Contract for Engineering Construction of the Associated General Contractors in collaboration with the American Society of Civil Engineers places similar restrictions upon the contractor in Section 15 of that document.

Some computerized project management systems have the means of capturing inspection reports, including scanned images. In contracts where the contractor is required to request inspection, such an activity can be done through an *action* item or through a memo with electronic notice to the Resident Project Representative. If the inspection resulted in the need for a Notice of Noncompliance (Chapter 18), one can be generated within the database. A request for inspection can also be made there. A master list of tests, such as those illustrated in Figure 3.13, can be used to set up reminders or action items for each of the listed tests at the varying frequencies tabulated in Figure 3.13. Such a master list could also be used as a computer template.

Contractor Submittals

All submittals from the contractor should be handled in a systematic, consistent, and orderly manner. Changes in systems or procedures during a job lead to confusion, errors, and abuses. There should be no "special cases" or exceptions in the routine established for submittals—these lead to breaks in communication, and occasionally result in gross error. Whenever a Resident Project Representative or Project Representative is provided on a project, the most desirable method is to require that all submittals required under the contract be transmitted directly to that individual, who in turn should forward such items to the project manager of the design firm or owner (see "Handling Job-Related Information" in Chapter 3 and Figure 3.6). This serves a twofold purpose. First, the Resident Project Representative will be fully aware of the status of all phases of the project; second, it will serve to emphasize to the contractor that the Resident Project Representative speaks for the owner and the design firm and that any efforts to bypass that person will be rejected. Furthermore, it eliminates any arguments that certain submittals were "mailed on time" or that the subcontractors submitted their submittals directly to the design firm or owner.

Often, a contractor making a request that has already been denied by the Resident Project Representative may try to approach the design firm or owner directly *without the knowledge* of the Resident Project Representative. If this procedure is followed, the embarrassing condition of a contractor's proposal being accepted by the project manager after it has already been rejected by the Resident Project Representative will occur. There are subsidiary dangers to that situation, also. Once the contractor is successful, the effectiveness of the Resident Project Representative is diminished measurably, and the project manager will find it necessary to conduct most of the field business personally. It would seem that if a job warrants the presence of a Resident Project Representative, he or she should be provided with the authority necessary to conduct the job efficiently and effectively—otherwise, the design firm or owner should save its money and eliminate the position.

Opening a Project

Opening a project requires many details to be completed before the contractor even moves a single piece of equipment onto the project site. Immediately after award of the contract, the contractor is expected to make arrangements for the required policies of insurance, obtain permits, order long-lead purchase items, check the site to determine the availability of storage and work staging areas, make arrangements for off-site disposal of surplus or waste materials, and take care of numerous other tasks.

In addition, the owner and the design firm will usually want to schedule a preconstruction meeting (see Chapters 10 and 12). At this time they will be able to meet with the contractor and other key personnel, identify areas of responsibility, establish job philosophy (set the ground rules), set up requirements for on- or off-site job meetings and set the frequency of such meetings, determine who should be in attendance, point out particular problem areas anticipated in construction and discuss any special methods of treatment of such problems, and if necessary, discuss special sequence of operations or scheduling limitations.

Although the practice is not universally followed, it is desirable for the owner, either directly or through the design firm or construction manager, to issue a written Notice to Proceed to the contractor, which will designate the actual beginning of the contract—very important later when attempting to establish the amount of liquidated damages where the contractor has exceeded the contract time. The *Notice of Award* cannot serve as a valid Notice to Proceed, as no contract will have been executed between the owner and the contractor at that time. The Notice to Proceed sets a precise date that the job began and eliminates any later argument over the time of the contract.

The representative of the design firm or owner should visit the site early after the Notice to Proceed has been issued to assure that all the requirements of the contract documents relating to temporary facilities and utilities are being properly implemented. This includes, particularly for the Resident Project Representative, that the proper field office facilities are being provided, that they are on time, that they are separate from those of the contractor (if specified as such), and that telephone service, temporary power, and sanitary facilities have been arranged for so that they will be installed in time. Where initial temporary fencing is required around construction areas, the Resident Project Representative should closely monitor this requirement.

Each of the construction milestones, such as contractor submittals of key items, materials testing, operational tests, reviews and updates required in schedules, delivery dates of major key pieces of equipment, the beginning of new elements of the work, and similar requirements of the contract should be outlined on paper and a calendar established for the systematic control and monitoring of these functions (see Figure 3.3). The procedures for the handling of all communications should be explained at that time, if not previously covered at the preconstruction meeting, and printed directions issued as a guide to the handling of all construction-related matters in conformance with the owner agency requirements. A simplified diagram showing the routing of all field communications and submittals is shown in Figure 3.6.

It is quite important that once the communications procedures have been agreed upon, neither the Resident Project Representative nor members of the staff should allow informal changes to occur. Such departures from the formally accepted policy can otherwise be justifiably used by the contractor to allege a lack of communication as a defense in case of a dispute.

Job Philosophy

A *firm but fair* policy should be adopted by the Resident Project Representative to control the work and to require proper workmanship and materials as well as compliance with drawings, specifications, and other contract documents. The Resident Project Representative should provide as much assistance to the contractor as possible to alert the firm to special job requirements or portions of the work requiring more critical control, and to provide any known information that may benefit the contractor in the completion of the work within the terms of the contract. Valid claims for extra work beyond the scope of the contract, as well as unforeseen underground conditions, should be fairly reviewed by the design firm and, if valid, presented to the owner with a recommendation for approval. Invalid claims should be rejected.

The requirement that the resident engineer or inspector be fair in all dealings with the contractor does not mean that he or she should be overly lenient or patronizing, as there is no point in having a Resident Project Representative on the job if he or she is not effective. The basic philosophy is to get a good start on the job. The inspector's attitude at the beginning of each job should be one of *firmness*. This will minimize arguments later during the job. If a job is started in a loose fashion, it is almost impossible to regain proper control later in the work, even by replacing the Resident Project Representative. An incorrect method is easier corrected the first time it is practiced than after it has been in use for a while.

Many organizations formalize the process by the use of a policy manual in which instructions to the inspectors and other field personnel are outlined for constant reference by the field personnel, such as the following:

Instructions to Field Personnel

BASIC POLICY

Contractor to be present at meetings and telephone conferences with subcontractors or suppliers.

Job opening philosophy with contractor: Be firm but fair.

RESPONSIBILITY AND AUTHORITY

Be a team member; avoid adversary relationship with the contractor.

Inspections and tests to be made promptly.

Inspect the work as it progresses.

Avoid overly literal specification interpretation.
No field changes without Project Manager approval.
Follow up all required corrective work until completed.
Do NOT supervise any construction nor the contractor's personnel.
No authority to stop the work; notify Project Manager if necessity arises.
No authority to require quality exceeding that covered by the contract.
Instructions to the contractor through Superintendent or Project Manager.
Document all actions taken.

DOCUMENTATION

All field personnel must keep an approved type diary.
Daily and summary reports must be submitted by Resident Project Representative.
Contractor submittals to be documented both coming in and going out.
Substantive content of business telephone calls should be documented in diary.
Keep photographic records of progress and all potential claims issues.
All orders to the contractor must be in writing.

COMMUNICATIONS

Contractor submittals handled only through Resident Project Representative.
Surveys and special inspections requested through Resident Project Representative.
Orders to contractor from ANY source must be submitted through the Resident Project Representative.

CHANGES

Field orders and change orders must be handled through Resident Project Representative.
No changes on oral instructions without written confirmation.
No significant deviations from plans and specifications except by change order—even if no cost or time extension is involved.
Emergency changes by Work Directive Change, followed later by a change order.
Feedback report (Report of Field Problem) to be filed on all correctable problems.

Another important philosophy is that the inspector should not become a creature of habit. Do not get into the swing of a regular routine. No one should be capable of anticipating the inspector's moves from one day to the next. All inspection should be at irregular intervals—and, above all, the inspector should be one of the first ones at the job and one of the last to leave. Many substandard details have been accomplished during the brief time between an inspector's early departure and the contractor's release of the crew for the day. Do not get the idea that the Resident Project Representative should become a police officer, nor that everyone in the contractor's camp is out to defraud. The result will be more tension before the job is half over than either the Resident Project Representative or the contractor can handle. By far the majority of contractors and their employees want to do a good job. Remember that they must usually base their judgments upon previous work that they have personally been involved in, and may not recognize the significance of some of the architect or engineer's design requirements. In addition, in most cases the contractor's project manager is an individual with a practical outlook and may not be easily convinced that the architect or engineer's design theory has approached such a degree of exactness as to justify stringent inspection practices and low-tolerance inspection procedures.

ADMINISTRATIVE ACTIVITIES

The job of the Resident Project Representative involves the handling of numerous administrative responsibilities. All of the items emphasized in the following list are generally necessary to serve the owner's best interests, but care must be exercised to ensure that they are applicable to your situation. Every activity should have a legitimate purpose and objective. There is far too much administrative work needed to be done in the field without being bogged down with "administrivia." Although most of the major tasks have been described in detail, the following list will serve as a summary of the principal administrative activities expected of the Resident Project Representative:

1. Coordinate and provide general direction of work and progress.
2. Review contractor's CPM schedules regularly.
3. Assist in resolution of construction problems.
4. Evaluate contractor claims for the design firm.
5. Maintain log of change orders.
6. Maintain log of contractor submittals.
7. Develop and administer a quality control program.
   (a) Proofs of compliance.
   (b) Qualifications of testing services.
   (c) Define required tests.
   (d) Maintain QC reporting system.
   (e) Maintain QC records of all tests and test results.
   (f) Establish frequency of testing.
8. Physically inspect all construction *every day.*
9. Observe all contractor tests.
10. Maintain daily diary and construction records.
11. Maintain record drawing data.
12. Respond to Requests for Information (RFI).
13. Review contractor progress payment requests.
14. Review contractor's change order requests for design firm.
15. Assure that construction area is safe.
16. Participate in field management meetings.
17. Provide negotiation assistance on contractor claims.
18. Review and recommend contractor value engineering proposals.

19. Supervise inspection forces and field office staff.
20. Report field conditions that prevent original construction.
21. On unit-price projects, obtain accurate field measurements.
22. On all jobs, verify contractor's monthly work quantities.
23. Assist scheduling and ordering required field services.

Although not exhaustive, the foregoing list summarizes the more commonplace activities expected of the Resident Project Representative on a construction project.

Suspension or Termination of the Work

Much has been said about not "stopping the work" but little about the related act of "suspending" it. First, the word *suspension* as used in this context should be very carefully qualified. Work may be suspended in whole or in part, and the nature of a suspension is to cease all or part of the work without actual contract termination.

Suspension of Work by the Owner

The owner may order the contractor in writing to suspend, delay, or interrupt all or part of the work for as long as deemed necessary. However, if the work is delayed or suspended for a longer period than specified, the contractor may claim an adjustment in price for delay damages as well as additional time.

Motives for suspension of the work vary. The owner may have budgetary limits and decide to stop work in certain areas; it may be necessary to update a particular part of some equipment to incorporate state-of-the-art improvements; work may be suspended due to a contractor defaulting or declaring bankruptcy. In the latter case, the bonding company may have to engage another contractor to complete the work—a process that may require additional time and cost.

Under some federal grant-funded programs, the federal-required contract provisions or model subagreement clauses contain a clause that the owner (grantee) has the right to suspend or interrupt portions or all of the work temporarily for an appropriate period of time at the convenience of the owner (grantee). In a few cases, the contractor would be entitled to an equitable adjustment for any additional costs caused by such suspension, particularly if the suspended work is on the critical path of the project schedule.

Temporary suspension is subject to the specific terms of the General Conditions of the Contract for each particular project. In general, however, suspensions can be said to fall into two major categories:

1. The first category relates to the failure of the contractor to carry out orders or to perform any provision of the contract. Any letter ordering such suspension must include reference to applicable sections of the specifications and, if possible, state the conditions under which the work may be resumed. Such action must be taken only after careful consideration of all aspects of the problem, and then only under the direct authority of the design firm,

construction manager firm, or owner. This is a legally risky area, however, and the EJCDC feels that the right to reject defective work was a sufficient weapon for the architect/engineer, and the severe and more risky right to stop the work should be left to the owner. However, as often stated before, in certain circumstances, moral standards or the law may impose a duty on the architect or engineer for the benefit of employees and third parties to stop work that is being carried on in an unsafe manner, or advise the owner to do so. This is so even when the General Conditions state that safety precautions are the sole responsibility of the contractor (see Article 6.20 of the EFCDC General Conditions of the Construction Contract as an example of typical provisions). The resident engineer or inspector as the authorized on-site representative of the design firm or owner is then the only party who will normally observe such hazards and can serve such notices at the site.

2. The second category under which a suspension may be ordered relates to unsuitable weather or conditions unfavorable for the suitable prosecution of the work. Normally, such suspensions are not necessary for periods of 30 days or less, since these are best handled on a day-to-day basis when determining nonworking days.
   (a) *Suspension of an item or an operation.* A suspension may be ordered that affects an item or several items only if desired. This is usually done when either the work or the public will be adversely affected by continuous operation. Such action is recommended for situations where the probable end result is based upon the architect's or engineer's experience and judgment as opposed to factors that are directly specified. Although this type of suspension is an option generally available only to the design firm or owner, the contractor's opinion on such suspension should also be considered.
   (b) *Suspension of the entire project.* In areas subject to severe weather, it is considered permissible to suspend an entire project if this is considered as being in the best interests of the owner. However, the authority of the architect or engineer to suspend is limited to the specific terms of the contract documents. In some cases it might even be necessary to have the contractor's concurrence to suspend an entire project.

The contractor must be advised of the conditions under which maintenance will be performed during any suspension, and who will be responsible for the condition of the unfinished work during any such suspension. It may be possible under many contracts that the contractor may be entitled to receive extra payments to maintain the project during this time.

When the reason for a suspension no longer exists, or when it can be expected that favorable conditions for resumption of the work will soon prevail, the contractor should be notified in writing. The letter should state the date when the count of working days will be expected to be resumed, and the notice should allow a reasonable amount of time for the contractor to regroup its labor forces and equipment. Generally, 10 working days is considered as a reasonable length of time for this.

Termination

Some contract documents contain provision allowing the owner to discontinue all or any part of the work being done by a contractor. This greatly reduces owner risk normally associated with such action in the absence of specific terms and conditions in the contract documents. Such termination may be for reasons such as abandoning the work, bankruptcy or insolvency, unnecessary delay of the work, or displaying other conduct detrimental to the execution of the work. The contractor's conduct must be of a very serious nature, however, to warrant such action.

Terminations, suspensions, and extensions of time will necessitate the issuance of a formal change order. Terminations for convenience or default are usually very complex and may require evaluation of the entire project. Other contract provisions can affect the outcome as well. Extensions of time should be fully evaluated and justified. Under federal grant programs, the owner is cautioned against the granting of unwarranted time extensions, as the costs associated with such extensions may not be interpreted as grant-eligible under the applicable grant program.

Whenever time factors are a part of a change order they should be negotiated together with the other factors, not deferred or separately negotiated. In this way, claims for associated costs such as additional overhead during time extensions can be largely avoided. In some cases, however, a contractor may agree on the price for a given change but will be unwilling to be pinned down to a fixed time impact for a given change order, as it may not be possible to anticipate the impact of the proposed change on other elements of the project. In such cases, it may be wise to settle on the price issue and be willing to negotiate the time after all impacts have been determined.

Construction Services Cost Monitoring

One essential task of the project manager during the construction phase is to maintain control of project field service costs as separate and distinct from architect/engineer design costs. The amount of the field costs is a function of the size and classifications of the on-site field forces assigned to the project, field office overhead costs, materials and supplies, support services from the home office, auto leases and fuel charges, field office share of corporate general and administrative (G&A) costs, outside consultant or contract services, and job profit.

By not only tabulating the monthly accumulations of budget and actual field service costs, but also by graphically plotting each amount on a time versus cost chart, similar in form to the traditional construction S-curve chart, a visual comparison can be made that not only can clearly indicate the status of the contract at any given time, but will also show any change in trend toward either a predicted savings or cost overrun (Figure 15.1). By also plotting a curve representing the amounts invoiced to the owner for such field services, an additional dimension is provided.

To prepare and keep up the chart, regular inputs are required by the project manager from the Resident Project Representative on all field costs and hours of work in each classification at the project site. Arrangements should be made by the project manager to assure that all such data are received from the field on a regular, scheduled basis at the end of each month. Generally, the monthly pay estimates of the contractor's work for partial payment are submitted about the 25th of the month, and all submittals are in to the office for payment before the end of the month. Billings from the architect/engineer to the owner, however, usually are based upon the closing date of the end of the month. Therefore, tabulation of these data should not interfere with the Resident Project Representative's review of the contractor's pay requests.

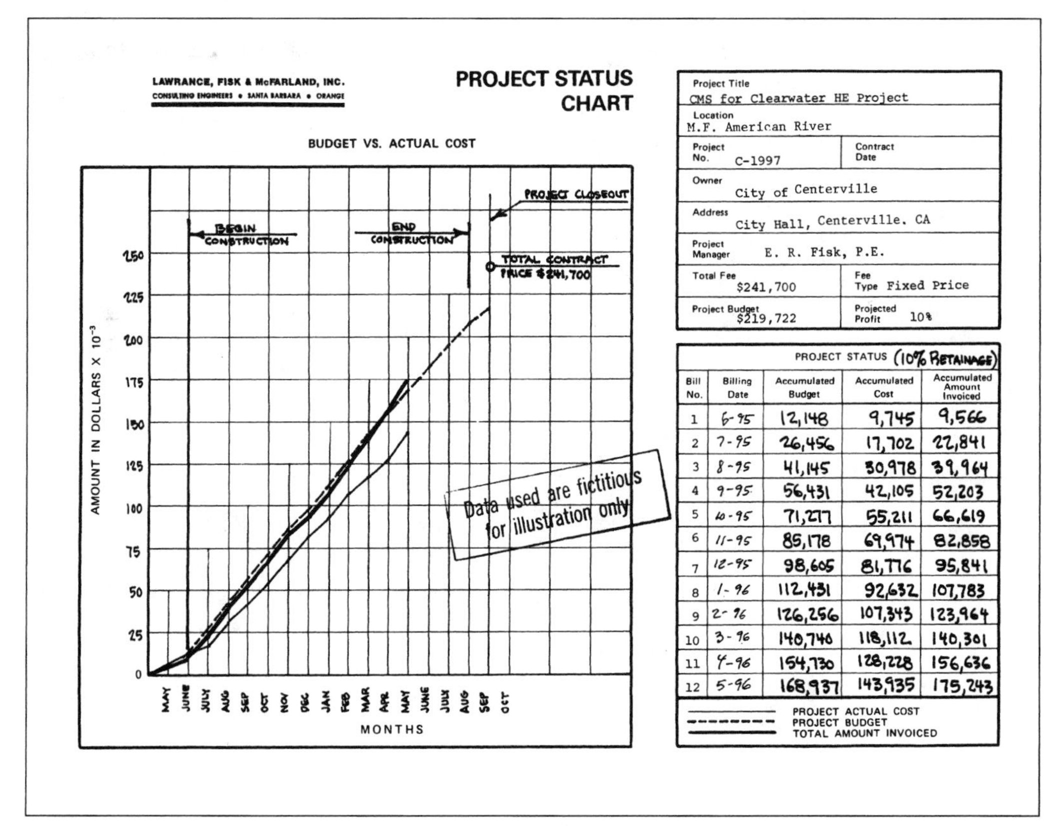

Project Title: CMS for Clearwater HE Project
Location: M.F. American River
Project No.: C-1997 | Contract Date:
Owner: City of Centerville
Address: City Hall, Centerville, CA
Project Manager: E. R. Fisk, P.E.
Total Fee: $241,700 | Fee Type: Fixed Price
Project Budget: $219,722 | Projected Profit: 10%

PROJECT STATUS (10% RETAINAGE)

Bill No.	Billing Date	Accumulated Budget	Accumulated Cost	Accumulated Amount Invoiced
1	6-95	12,148	9,745	9,566
2	7-95	26,456	17,702	22,841
3	8-95	41,145	30,978	39,964
4	9-95	56,431	42,105	52,203
5	10-95	71,277	55,211	66,619
6	11-95	85,178	69,974	82,858
7	12-95	98,605	81,776	95,841
8	1-96	112,431	92,632	107,783
9	2-96	126,256	107,343	123,964
10	3-96	140,740	118,112	140,301
11	4-96	154,730	128,228	156,636
12	5-96	168,937	143,935	175,243

FIGURE 15.1 Example of a Monthly Cost Report of Field Services Performed by a Construction Manager.

PROBLEMS

1. True or false? As a means of minimizing conflicts between the inspector and the contractor, it is better to set absolute dimensions rather than set tolerance ranges, so that the inspector has a specific value to enforce.
2. Describe the kind of risk involved, if any, where the engineer, the inspector, or the Resident Project Representative is contractually given the right to stop the work.
3. If the engineer or the owner disputes the work being performed by a subcontractor, with which party should the owner attempt to resolve the issue?
4. Explain the purpose and function of a CQC representative on a NAVFAC or Corps of Engineers contract.
5. Explain the "firm but fair" policy.
6. Describe the responsibilities of the inspector during construction.
7. Is the Resident Project Representative responsible to see that temporary facilities under the control of the contractor are consistent with the provisions of the specifications?
8. Why is it desirable to have all contractor submittals submitted through the Resident Project Representative or Project Representative?

16

VALUE ENGINEERING

Value engineering—what is it, and what part does the Resident Project Representative play in its application? The latter question will be answered in this chapter. The first question, however, deserves special consideration. It has been said that value engineering is not the proper role for a project administrator, but rather for the decision maker, such as the project manager. As far as the concept of initiating or granting final approval to a value engineering proposal is concerned, this appears to be true. Nevertheless, it may involve construction methods regarding which the inspector may have valuable knowledge to contribute to the decision maker, and, under many federal contracts, it is also a part of the contract requirements of the general contractor to propose value engineering ideas *during the construction phase.* There is no way that a Resident Project Representative can avoid the issue as long as the contractor is involved; thus, it is highly desirable for the inspector to know what value engineering is and how to work with value engineering proposals. In this chapter the resident engineer or inspector will be introduced to the subject in sufficient detail to understand what it is all about and where it fits into his or her responsibility area.

A great deal of lip service has been paid to the concept of value engineering in recent years, and the suggestion is that it is a new, previously undeveloped branch of the construction industry. There are entire engineering organizations devoted solely to the promotion of value engineering, and at least one federal agency that lets contracts for large construction projects now requires its use during the design phase of every project. Value engineering concepts are desirable on any project; however, it must be said that all engineering has always involved *value engineering* concepts, although not under that title. In fact, an architect or engineer who did not consider maximum economy and value in the selection and use of construction materials and methods, within the limits dictated by the design, was simply not doing his or her job. Theoretically, during the *design phase* of a project there should be no need for a separate value engineering effort if the architect or engineer is providing competent

FIGURE 16.1 Typical EPA Clean Water Grant Project Requiring a Separate Value Engineering Contract. *(Photo courtesy of MWH Global, Broomfield, CO.)*

professional services to the client; however, under the old EPA Clean Water Grant Program for projects, illustrated in Figure 16.1, a separate contract with an outside consultant was required to provide value engineering.

The concept of providing the value engineering incentives to the construction contractor during the *construction phase* of a project, however, is a more deserving one. Often the selection of materials and methods by the architect/engineer is dictated by an evaluation of the average market conditions or contractor methods. Upon awarding a contract for construction to a specific contractor, the particular skills, equipment, materials sources, and knowledge possessed by that person of the local trade area and labor market can often be used to reduce beneficially the cost of a portion of a proposed project without necessarily compromising the design concepts involved. The final judgment of the acceptability of such suggestions would be up to the architect or engineer and the owner, of course. An example of a value engineering incentive clause in a construction contract is illustrated in Figure 16.2.

VALUE ENGINEERING INCENTIVE FOR CONSTRUCTION 52.248-3

(a) *General.* The Contractor is encouraged to develop, prepare, and submit value engineering change proposals (VECPs) voluntarily. The Contractor shall share in any instant contract savings realized from accepted VECPs, in accordance with paragraph (g), below.

(b) *Definitions.* Collateral costs, as used in this clause, means agency costs of operation, maintenance, logistic support, or Government-furnished property.

Collateral savings, as used in this clause, means those measurable net reductions resulting from a VECP in the agency's overall projected collateral costs, exclusive of acquisition savings, whether or not the acquisition cost changes.

Contractor's development and implementation costs, as used in this clause, means those costs the Contractor incurs on a VECP specifically in developing, testing, preparing, and submitting the VECP, as well as those costs the Contractor incurs to make the contractual changes required by Government acceptance of a VECP.

Government costs, as used in this clause, means those agency costs that result directly from developing and implementing the VECP, such as any net increases in the cost of testing, operations, maintenance, and logistic support. The term does not include the normal administrative costs of processing the VECP.

Instant contract savings, as used in this clause, means the estimated reduction in Contractor cost of performance resulting from acceptance of the VECP, minus allowable Contractor's development and implementation costs, including subcontractors' development and implementation costs (see paragraph (h) below).

Value Engineering change proposal (VECP) means a proposal that (1) requires a change to this, the instant contract, to implement: and (2) results in reducing the contract price or estimated cost without impairing essential functions or characteristics; provided, that it does not involve a change--

(i) in deliverable end item quantities only; or
(ii) to the contract type only.

(c) *VECP preparation.* As a minimum, the Contractor shall include in each VECP the information described in subparagraphs (1) through (7) below. If the proposed change is affected by contractually required configuration management or similar procedures, the instructions in those procedures relating to format, identification, and priority assignment shall govern VECP preparation. The VECP shall include the following:

(1) A description of the difference between the existing contract requirement and that proposed, the comparative advantages and disadvantages of each, a justification when an item's function or characteristics are being altered, and the effect of the change on the end item's performance,

(2) A list and analysis of the contract requirements that must be changed if the VECP is accepted, including any suggested specification revisions.

FIGURE 16.2 Example of Value Engineering Incentive Provisions in a Federal Construction Contract.

VALUE ENGINEERING INCENTIVE FOR CONSTRUCTION *(Continued)*

(3) A separate, detailed cost estimate for (i) the affected portions of the existing contract requirement and (ii) the VECP. The cost reduction associated with the VECP shall take into account the Contractor's allowable development and implementation costs, including any amount attributable to subcontracts under paragraph (h) below.

(4) A description and estimate of costs the Government may incur in implementing the VECP, such as test and evaluation and operating and support costs.

(5) A prediction of any effects the proposed change would have on collateral costs to the agency.

(6) A statement of the time by which a contract modification accepting the VECP must be issued in order to achieve the maximum cost reduction, noting any effect on the contract completion time or delivery schedule.

(7) Identification of any previous submissions of the VECP, including the dates submitted, the agencies and contract numbers involved, and previous Government actions, if known.

(d) *Submission.* The Contractor shall submit VECP's to the Resident Engineer at the worksite, with a copy to the Contracting Officer.

(e) *Government action.*
(1) The Contracting Officer shall notify the Contractor of the status of the VECP within 45 calendar days after the contracting office receives it. If additional time is required, the Contracting Officer shall notify the Contractor within the 45-day period and, provide the reason for the delay and the expected date of the decision. The Government will process VECP's expeditiously; however, it shall not be liable for any delay in acting upon a VECP.

(2) If the VECP is not accepted, the Contracting Officer shall notify the Contractor in writing, explaining the reasons for rejection. The Contractor may withdraw any VECP, in whole or in part, at any time before it is accepted by the Government. The Contracting Officer may require that the Contractor provide written notification before undertaking significant expenditures for VECP effort.

(3) Any VECP may be accepted, in whole or in part, by the Contracting Officer's award of a modification to this contract citing this clause. The Contracting Officer may accept the VECP, even though an agreement on price reduction has not been reached, by issuing the Contractor a notice to proceed with the change. Until a notice to proceed is issued or a contract modification applies a VECP to this contract, the Contractor shall perform in accordance with the existing contract. The Contracting Officer's decision to accept or reject all or part of any VECP shall be final and not-subject to the Disputes clause or otherwise subject to litigation under the Contract Disputes Act of 1978 (41 U.S.C. 601-613).

(f) *Sharing.*
(1) *Rates.* The Government's share of savings is determined by subtracting Government costs from instant contract savings and multiplying the result by (i), 45 percent for fixed-price contracts or (ii), 75 percent for cost-reimbursement contracts.

FIGURE 16.2 (continued)

VALUE ENGINEERING INCENTIVE FOR CONSTRUCTION *(Continued)*

(2) *Payment.* Payment of any share due the Contractor for use of a VECP on this contract shall be authorized by a modification to this contract to--

(i) Accept the VECP;

(ii) Reduce the contract price or estimated cost by the amount of instant contract savings; and

(iii) Provide the Contractor's share of savings by adding the amount calculated to the contract price or fee.

(g) *Collateral savings.* If a VECP is accepted, the instant contract amount shall be increased by 20 percent of any projected collateral savings determined to be realized in a typical year of use after subtracting any Government costs not previously offset. However, the Contractor's share of collateral savings shall not exceed (1) the contract's firm-fixed-price or estimated cost, at the time the VECP is accepted, or (2) $100,000, whichever is greater. The Contracting Officer shall be the sole determiner of the amount of collateral savings, and that amount shall not be subject to the Disputes clause or otherwise subject to litigation under 41 U.S.C. 601-613.

(h) *Subcontracts.* The Contractor shall include an appropriate value engineering clause in any subcontract of $50,000 or more and may include one in subcontracts of lesser value. In computing any adjustment In this contract's price under paragraph (f) above. the Contractor's allowable development and implementation costs shall Include any subcontractor's allowable development and Implementation costs clearly resulting from a VECP accepted by the Government under this contract, but shall exclude any value engineering incentive payments to a subcontractor. The Contractor may choose any arrangement for subcontractor value engineering incentive payments; provided, that these payments shall not reduce the Government's share of the savings resulting from the VECP.

(i) *Data.* The Contractor may restrict the Government's right to use any part of a VECP or the supporting data by marking the following legend on the affected parts:

> "These data, furnished under the Value Engineering-Construction clause of contract shall not be disclosed outside the Government or duplicated, used, or disclosed, in whole or in part, for any purpose other than to evaluate a value engineering change proposal submitted under the clause. This restriction does not limit the Government's right to use information contained in these data if it has been obtained or is otherwise available from the Contractor or from another source without limitations."

If a VECP is accepted, the Contractor hereby grants the Government unlimited rights in the VECP and supporting data, except that, with respect to data qualifying and submitted as limited rights technical data, the Government shall have the rights specified in the contract modification implementing the VECP and shall appropriately mark the data. (The terms unlimited rights and limited rights are defined in part 27 of the Federal Acquisition Regulation.)

(End of clause)

FIGURE 16.2 (continued)

Definition

The first question to arise is usually: What is value engineering, anyway? *Value engineering* is a systematic evaluation of a project design to obtain the most value for every dollar of cost. By carefully investigating costs, availability of materials, construction methods, shipping costs or physical limitations, planning and organizing, cost/benefit values, and similar cost-influencing items, an improvement in the overall cost of a project can be realized.

The entire value engineering effort is aimed at a careful analysis of each function and the elimination or modification of anything that adds to the project cost without adding to its functional capabilities. Not only are first costs to be considered, but also even the later in-place costs of operation, maintenance, life, replacement, and similar characteristics must be considered. Thus, although the name is new, value engineering is simply a *systematic* application of engineering economy as taught in every engineering course long before anyone ever thought up a catchy name for it.

The principal difficulty of applying value engineering principles to construction is the problem of having a third party, who may not possess the same degree of expertise in the subject area that the architect or engineer does, cause changes in design that simply substitute the value engineer's judgment for that of the designer—a risky process at best. Theoretically, the value engineer does not actually "cause" a design change, but by placing the architect or engineer in the position of having to defend the original design, the entire production schedule is threatened.

In the construction phase, however, it is a completely new ball game. Now the value engineer is the contractor, whose experience is in construction methods, techniques, and costs. The contractor's input can often offer the benefit of construction experience and a knowledge of the marketplace and labor force that the designer did not possess. It is here that the greatest cost benefit can result with a minimum of conflict with the designer. Often on federally funded construction contracts, a value engineering incentive clause may be provided, in which the government will allow the contractor to retain a portion of the cost savings realized in any value engineering proposal submitted by the contractor that is accepted and implemented.

Where a voluntary value engineering incentive clause is used in federal contracts, the government is required to share the cost savings with the contractor on a 50/50 percent basis. Where the program is mandatory, the government/contractor share is on a 75/25 percent basis [48 CFR 48.104-1].

In public works projects at lower levels of government, application of the principles of sharing value engineering savings varies from state to state. In California, for example, contractor participation in value engineering is optional; however, if it is implemented, the public agency must share the savings with the contractor in the amount of 50 percent of the net savings [CA Public Contracts Code § 7101].

The Role of the Resident Project Representative

The Resident Project Representative, although generally having no part in the actual submittal of value engineering proposals, may frequently be called upon by the architect or engineer to evaluate a value engineering (VE) proposal that has been submitted by the contractor. It is not being suggested here that the resident engineer or inspector should render judgment in the matter of whether a cost savings proposal is valid, as this would be no better than leaving the matter up to a single individual who did not take part in the actual design, and quite probably did not possess sufficient technical and professional background in the subject area to justify the acceptance of his or her judgment over that of the design engineer. In addition, it should always be remembered that the design engineer is the *engineer of record,* with a license and professional reputation at stake, not that of a value engineer consultant. The inspector's part in this process is to relate to the real world surrounding the assigned project, and evaluate the probable effect of the proposal on the project if it is accepted. All such observations should be written into a memorandum form and submitted to the architect or engineer so that when the final evaluation is made, it can be based upon a full knowledge of both the design conditions and the field conditions as communicated by the Resident Project Representative.

Fundamentals of Value Engineering

Function

In value engineering, the *function* is defined as the specific purpose or use intended for something. It describes what must be achieved. For value engineering studies, this function is usually described in the simplest form possible, usually in only two words; a verb and a noun. "Support weight," "prevent corrosion," and "conduct current" are typical expressions of function.

Worth

Worth refers to the least cost required to provide the *functions* that are required by the user of the finished project. Worth is established by comparison, such as comparing it with the cost of its functional equivalent. The worth of an item is not affected by the possibility of failure under the value engineering concept embraced by the federal government. Thus, if a bolt supporting a key joint in a large roof truss fails, the entire roof of the structure may be caused to fail. Nevertheless, the worth of the bolt is the lowest cost necessary to provide a reliable fastening.

Cost

Cost is the total amount of money required to obtain and use the functions that have been specified. For the seller, this is the total cost in connection with the product. For the owner, the total cost of ownership includes not only the purchase price of the product, but also the cost of the paperwork of including it in the inventory, operating it, and providing support in the form of maintenance and utility services for its total usable life. The cost of ownership may also include a proportional share of expenditures for development, engineering, testing, spare parts, and various items of overhead expense.

Value

Value is the relationship of *worth to cost* as realized by the owner, based upon his or her needs and resources in any given situation. The ratio of *worth to cost* is the principal measure of *value.* Thus, a "value equation" may be used to arrive at a *value index* as follows:

$$\text{value index} = \frac{\text{worth}}{\text{cost}} = \frac{\text{utility}}{\text{cost}}$$

The value may be increased by doing any of the following:

1. Improve the utility of something with no change in cost.
2. Retain the same utility for less cost.
3. Combine improved utility with less cost.

Optimum value is obtained when all utility criteria are met at the lowest overall cost. Although *worth* and *cost* can be expressed in dollars, *value* is a dimensionless expression showing the relationship of the other two.

The Philosophy of Value

If something does not do what it is intended to do, no amount of cost reduction will improve its value. Any "cost reduction" action that sacrifices the needed utility of something actually reduces its value to the owner. However, costs incurred to increase the functional capacity of something beyond that which is needed amounts to "gilding the lily" and provides little actual value to the owner. Therefore, anything less than the necessary functional capacity is unacceptable; anything more is unnecessary and wasteful.

Types of Value Engineering Recommendations

Within the Department of Defense and some other federal agencies, there are two types of recommendations that are the result of a value engineering effort:

1. *Value Engineering Proposal (VEP).* A value engineering recommendation that originates from within the government agency itself, or one that was originated by the contractor and may be implemented by unilateral action. A VEP can only relate to changes that are within the terms of the contract and specifications and thus would not require a change order to implement.
2. *Value Engineering Change Proposal (VECP).* A value engineering recommendation by a contractor that requires the owner's approval, and that, if

accepted, requires the execution of a change order. This would apply to any proposed change that would require a change in the contract, the specifications, the scope of the work, or similar limits previously established by contract.

Areas of Opportunity for Value Engineering

Value Engineering by the Architect or Engineer

Value engineering is a basic approach that takes nothing for granted and challenges everything on a project, including the necessity for the existence of a product or project, for that matter. The cost of a project is influenced by the requirements of the design and the specifications. Prior to completing the final design the architect or engineer should carefully consider the methods and equipment that may be used to construct the project. Requirements that increase the cost without producing equivalent benefits should be eliminated. The final decisions of the architect or engineer should be based upon a reasonable knowledge of construction methods and costs.

The cost of a project may be divided into five or more items:

1. Materials
2. Labor
3. Equipment
4. Overhead and supervision
5. Profit

Although the last item is beyond the control of the architect or engineer, there is some control possible over the cost of the first four items.

If the architect or engineer specifies materials that must be transported great distances, the costs may be unnecessarily high. Requirements for tests and inspections of materials may be too rigid for the purpose for which the materials will be used. Frequently, substitute materials are available nearby that are essentially as satisfactory as other materials whose costs are considerably higher. The suggestions of the contractor can be of value here.

The specified quality of workmanship and methods of construction have considerable influence on the amount and class of labor required and upon the cost of labor. Complicated concrete structures are relatively easy to design and draw but may be exceedingly difficult to build. A high-grade concrete finish may be justified for exposed surfaces in a fine building, but the same quality of workmanship is not justified for a warehouse. The quality of workmanship should be in keeping with the type of project involved.

Architects and engineers should keep informed on the developments of new construction equipment, as such information will enable them to modify the design or construction methods to permit the use of economical equipment. The resident engineer or inspector is a vital link in the chain of information that supplies the architect and engineer with the latest up-to-date data on construction methods and equipment. The normal *daily construction report* (Figure 4.1) contains sufficient information to keep the architect and engineer adequately informed, and upon noting new methods or equipment in the inspector's report, these leads can be followed to determine the specific capabilities and advantages of each case. For example, the use of a dual-drum concrete paving mixer instead of a single-drum mixer can increase the production of concrete materially, and for most projects will reduce the cost of pavement construction. The use of a high-capacity earth loader and large trucks may necessitate a change in the location, size, and shape of the borrow pit, but the resulting economies may easily justify the change.

The utilization of higher-capacity delivery equipment does not always result in a cost saving, however. On some recent projects in a hot, dry climate, as an example, the use of 5.4 m^3 (7 cubic yard) transit mixers for delivering concrete to a project proved far more costly than making twice as many trips using 2.3 m^3 (3 cubic yard) mixers. The placing requirements involved the construction of thin concrete columns and wall sections with a high concentration of reinforcing steel, with the resultant reduction in the rate of placement. Before a 5.4 m^3 (7 cubic yard) truckload of concrete could be completely discharged under the existing conditions, the concrete mix started to set in the delivery vehicles because of the high temperature, low humidity, and three-hour time span from the addition of water to the mix to the final placement in the forms. Long delivery routes are often the cause of this, combined with slow pour conditions. The resultant frequency of rejection of portions of the 5.4 m^3 (7 cubic yard) load (retempering was prohibited by specifications) was more costly than the increase in the number of deliveries and the smaller batch size that allowed the use of a fresh load after every 2.3 m^3 (3 cubic yards) placed.

The following are some of the methods that the architect or engineer may use to reduce the cost of construction:

1. Design concrete structures with as many duplicate members as is practical to allow the reuse of forms without rebuilding.
2. Confine design elements to modular material sizes where possible.
3. Simplify the design of the structure wherever possible.
4. Design for the use of cost-saving equipment and methods.
5. Eliminate unnecessary special construction requirements.
6. Design to reduce the required labor to a minimum.
7. Specify a quality of workmanship that is consistent with the quality of the project.
8. Furnish adequate foundation information wherever possible.

9. Refrain from requiring the contractor to assume the responsibility for information that should have been furnished by the architect or engineer, or for the adequacy of the design.
10. Use local materials when they are satisfactory.
11. Write simple, straightforward specifications that state clearly what is expected of the contractor. Define the results expected, but within reason permit the contractor to select the methods of accomplishing the results.
12. Use standardized specifications that are familiar to most contractors whenever possible.
13. Hold preconstruction conferences with contractors to eliminate any uncertainties and to reduce change orders resulting from misunderstandings to a minimum.
14. Use inspectors who have sufficient judgment and experience to understand the project and have authority to make decisions.

Value Engineering by the Contractor

One desirable characteristic of a successful contractor from the standpoint of value engineering is a degree of dissatisfaction over the plans and methods under consideration for constructing a project (a characteristic not always appreciated by the architect or engineer). However, complacency by members of the construction industry will not develop new equipment, new methods, or new construction planning, all of which are desirable for providing continuing improvements in the construction industry at lower costs. A contractor who does not keep informed on new equipment and methods will soon discover that his competitors are underbidding him.

Suggestions for possible reductions in construction costs by the contractor include, but are by no means limited to, the following items:

1. Study the project before bidding, and determine the effect of:
   (a) Topography
   (b) Geology
   (c) Climate
   (d) Sources of materials
   (e) Access to the project
   (f) Housing facilities, if required
   (g) Storage facilities for materials and equipment
   (h) Labor supply
   (i) Local services
2. The use of substitute construction equipment that has greater capacities, higher efficiencies, higher speeds, more maneuverability, and lower operating costs.
3. Payment of a bonus to key personnel for production in excess of a specified rate.

4. The use of radios as a means of communications between headquarters office and key personnel on projects covering large areas.
5. The practice of holding periodic conferences with key personnel to discuss plans, procedures, and results. Such conferences should produce better morale among the staff members and should result in better coordination among the various operations.
6. The adoption of realistic safety practices on a project as a means of reducing accidents and lost time.
7. Consideration of the desirability of subcontracting specialized operations to other contractors who can do the work more economically than the general contractor.
8. Consideration of the desirability of improving shop and servicing facilities for better maintenance of construction equipment.

Improvements in the *methods* of construction—long the sole domain of the contractor—can result in significant savings in the cost of the project. This type of cost saving, if implemented after the award of a contract, is seldom if ever shared with the owner. However, such cost-reducing considerations are an integral part of the competitive bidding system. Thus, the owner benefits in lower bid costs. As an example, an estimator for a contracting firm prepared a bid for a project. When the bids were opened, it was discovered that his firm's bid was so low that the other members of the firm feared that a serious error had been made in preparing the bid. The estimator was called in and asked if he thought that he could actually construct the project for the estimated cost. He replied that he could if he were permitted to adopt the construction methods that he used in estimating the cost. The firm agreed; he was placed in charge of the construction of the project and he completed the work with a satisfactory profit to the contractor. At the same time, the owner benefitted by receiving its project at a low cost.

Field Responsibility in Value Engineering

As mentioned previously, the resident engineer or inspector's part in the value engineering process is an indirect one. However, participation may be requested in either of two areas: first to assist the architect or engineer in providing valuable, up-to-date construction information on materials, methods, availability, the capabilities of a particular contractor or contractors, and similar field data. In addition, a contractor's judgment may be solicited during both the *information* and *speculation* phases to assist the architect or engineer in his or her own VE proposals. Second, the contractor's comments may be requested to evaluate the contractor's VE proposals and submit commentary to the architect or engineer. In this manner, the Resident Project Representative may perform the function of one of the "specialists" that the architect or engineer may wish to consult during the analysis and evaluation of the contractor's VE proposals.

Problems

1. What is value engineering?
2. What is value?
3. How may value be increased?
4. Is the practice of value engineering limited to the design professionals, for example engineers or architects?
5. What are the four fundamental concerns that must be addressed by the value engineer?
6. Explain the philosophy of value.

17

MEASUREMENT AND PAYMENT

CONTRACTS FOR CONSTRUCTION

What makes a contract? How is it different from a promise or an agreement? Many definitions have been given to the term *contract* over the years. A contract is a promise, or a set of promises, that the law will enforce.

A contract is said to come into being when an offer is made followed by an acceptance. In construction, a contract consists of promises that are made by the owner to the contractor in exchange for return promises that are made by the contractor to the owner. A contract may consist of a single promise by one person to another, or there may be any number of persons or any number of promises.

Legally, there is a difference between Contract and Agreement. Agreement is a much broader term, since it encompasses promises that the law will not enforce, as well as those that the law will enforce. The difference between legal concept and the kinds of promises that the law enforces are those that it deems of enough social or economic importance to warrant it. What makes a contract?

Types of Construction Contracts

There are many different types of construction contracts, distinguished primarily by the method of determining the final contract price. Regardless of the method used, the goal is the same quality construction completed on time and to all specifications for the lowest possible price while allowing the contractor an opportunity to make a fair profit. To encourage the parties to meet this goal, several different types of contracts have evolved; the most commonly used pricing methods are discussed in the following paragraphs. The type of contract chosen may depend on several factors, including the identity and relationship, if any, of the owner and contractor; the completeness of the design and its complexity; the type of work being done; and the need or desire for competitive pricing.

Fixed-Price Contracts

Lump-sum Contracts. One form of fixed-price contract is a lump-sum contract; it is one in which the contractor agrees to do specified construction for a fixed price set forth in the contract. The only changes allowed to the fixed price are for extras or change orders. Lump-sum contracts are commonly used on both private and public works contracts. Bids are requested based on a complete set of plans and specifications, thus allowing for easy comparison of bid prices and fostering competition.

Unit-price Contracts. Another form of fixed price contract is a unit-price contract; it sets forth the price for each unit of work constructed. The unit may be, for example, the number of meters of pipe furnished and installed, a manhole structure, the number, each, of pumps, a modular building, or a cubic meter of excavation. The contract may specify a particular number of units or may state that the supplier will furnish all units needed or a specified percentage of needed units. For example, if the contract is for excavation, it might state that the excavator will be paid $14.70 per cubic meter of material excavated from the site. Anything that can be measured in units can be the basis of a unit-price contract.

Unit-price contracts are most often used in heavy construction and public works contracts, such as pipelines, highways, earthworks, bridges, tunneling, and transit facilities—situations where it is difficult to calculate quantities in advance.

Fixed-price Incentive Contracts. A fixed-price incentive contract is a fixed-price contract that provides for adjusting profit and establishing a final contract price by application of a formula based on the relationship of total final negotiated cost to total target cost. The final price may be subject to a price ceiling, negotiated at the time of entering into contract. The concept can be structured in either of two forms, for example, firm target price or successive target prices. A fixed-price incentive contract is appropriate when a firm-price contract is not suitable (c.f. Federal Acquisition Regulations 48 CFR 16.403).

Cost-Reimbursable Contracts

Cost-plus contracts are those in which the contractor is paid its actual costs of the construction plus a specified markup to cover overhead and profit. Typically, the contract defines costs as including all expenses incurred in the construction, including expenses for materials, labor (including payroll taxes), and subcontractors and suppliers. The contract should specify, in detail, what are and what are not eligible costs. One contract specifies that costs include wages for labor; salaries of field office personnel and support personnel to the extent attributable to the job; payroll taxes and contributions; related travel and subsistence expenses of the contractor and its employees; all materials, supplies, and equipment incorporated into the work and cost of their transportation; payments to subcontractors; cost of materials, supplies, equipment, temporary facilities, and hand tools consumed by the job and cost less salvage value of such items used on, but not consumed by, the job; rental charges or value if owned by the contractor; bond and insurance premiums; sales and use tax; permit fees and royalties; losses not compensated by insurance or otherwise that are related to the job and are not due to the contractor's actions; telephone service at the site, long distance charges, mail charges, and similar petty cash items; debris removal costs; emergency costs; and other costs approved in writing in advance by the owner. The contract also specifies what are not eligible costs for reimbursement purposes. The following discussion sets out several variations that are commonly used.

Time-and-materials or Cost-plus-percentage-of-cost. In this type of contract, the contractor is paid its actual costs plus a specified percentage of those costs for overhead. Thus, the contract would specifically exclude actual overhead expenses from the definition of eligible costs. To the total of costs and the overhead is then added a specified percentage for profit. A typical cost-plus-percentage-of-cost contract might provide for 15 percent overhead value and a 10 percent profit, resulting in a total of 26.5 percent markup to the contractor. Assume that a contractor expended $100,000 in materials and labor to build a home for an owner. The owner would pay the contractor a total of $126,500, computed as follows:

$100,000	labor and materials
+ 15,000	15 percent overhead
$115,000	subtotal
+ 11,500	10 percent profit
$126,500	total

Cost-plus-fixed-fee. In this type of contract, the contractor is paid its actual costs plus a fixed fee that is set in advance. The contract may or may not specify that costs include a set daily rate for overhead.

Cost-plus-incentive-fee. In this type of payment structure, the contract specifies time and quality criteria. If the contractor meets those criteria, it is paid its costs plus a set fee. If the contractor exceeds those criteria, perhaps by completing the job early, the contractor is paid an additional fee based on a scale set forth in the contract. If the contractor does not meet those criteria, the fee is less. This type of fee arrangement encourages early, quality work (c.f. Federal Acquisition Regulations 48 CFR 16.404).

Cost-plus contracts are appropriate where, due to an incomplete or very complex design, a contractor would be unable to give a lump-sum price without including a large contingency for unknown factors.

Guaranteed Maximum Price Contracts

A guaranteed maximum price contract is a variation of the cost-plus contract. In this type of contract, the owner and contractor agree that the project will not cost the owner more than a set price, the guaranteed maximum. The contractor is paid on a cost-plus fixed fee or percentage of cost basis, but in no event more than the set maximum price.

In some guaranteed maximum price contracts, a savings clause provides that if the project costs the owner less than the guaranteed maximum price, the owner and contractor are to split the difference between the costs and the guaranteed maximum price; typical splits are 50/50 percent and 60/40 percent (owner/contractor). For instance, assume that a contract specifies a guaranteed maximum price of $500,000 and the cost-plus basis turns out to be $400,000. Under a 50/50 percent split, the contractor would be entitled to $450,000 for its work. Guaranteed maximum price contracts give contractors great incentive to keep costs as reasonable as possible to ensure themselves as much profit as possible. They also encourage contractors to value engineer the project.

Construction Progress Payments

Among the most important items to the contractor are those provisions of the construction contract governing the making of monthly progress payments to the contractor throughout the job (Figure 17.1). The contractor will be expected to submit a request for payment, stating amounts of work completed and the estimated value of such work, to the Resident Project Representative approximately 10 days before the payment due date. At this point, the work of the Resident Project Representative is very similar for both a lump-sum and a unit-price contract. Before transmitting the payment request to the owner's or designer's project manager, the Resident Project Representative is generally delegated the task of verifying and evaluating the contractor's payment request. Because of nearness to the work, the Resident Project Representative is often the only representative of the owner or the design firm who can truly verify the actual quantities of the various items of work completed and determine the probable value of such work.

Although a unit-price contract requires extreme care in the verification of the contractor's payment claim, a lump-sum job can usually be handled more informally, as the overage claims will be compensated for ultimately by the fixed-price nature of the contract. Nevertheless, it is undesirable to be so careless as to allow payments in excess of the amounts of work actually done, as first, it cancels the effect of the retention, if any, and second, if the contractor were to default on the contract, the inspector might be in the position of having approved for payment more work than was actually built, thus allowing the defaulting contractor a tidy, unearned profit.

Approval of Payment Requests

The responsibilities of the Resident Project Representative in the approval stage of contractor partial payment requests are limited to checking quantities and costs prior to submittal to the owner's or designer's project manager with a recommendation for payment if warranted.

The following is a list of possible tasks that might be delegated to the Resident Project Representative during the validation phase of the contractor's payment request:

1. Quantity takeoff of work actually completed as of date of request.
2. Inventory of equipment and materials delivered but not yet used in the work.

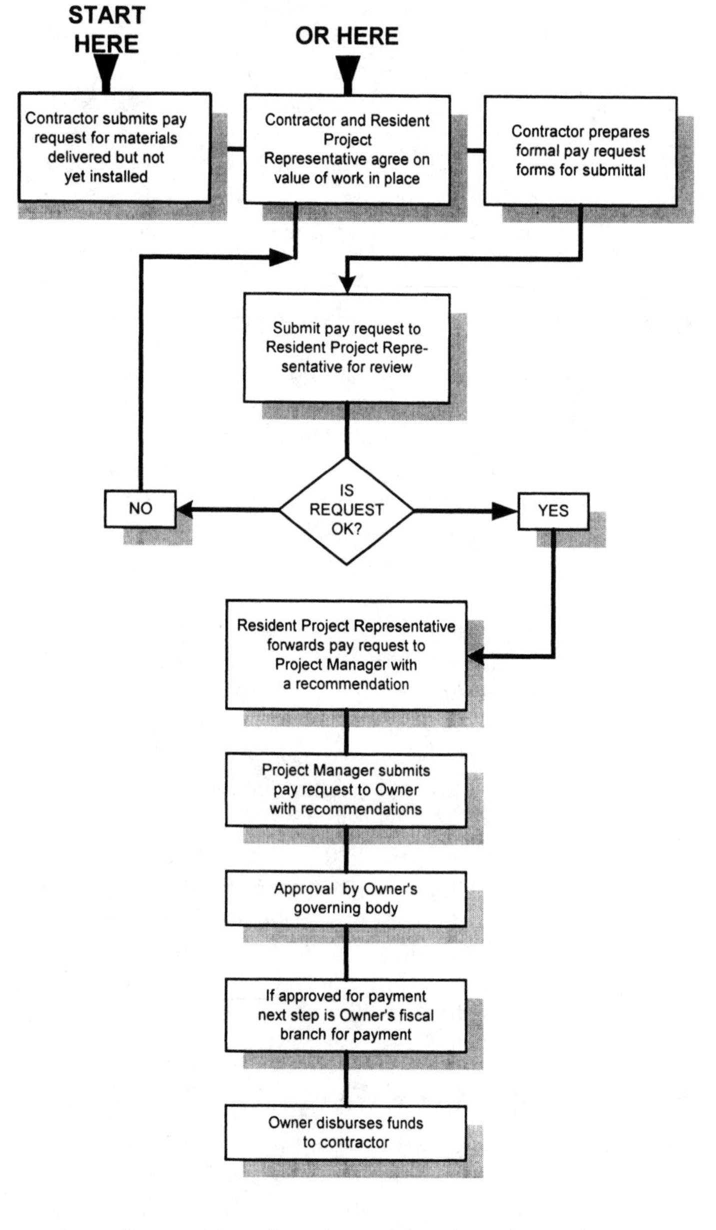

FIGURE 17.1 Progress Payment Flow Diagram.

3. Field measurements of quantities of work completed or claimed.
4. Construction cost estimate of all completed work, using unit prices in bid or in cost breakdown submitted at beginning of job.
5. Audit of invoices and costs (cost-plus jobs only).
6. Review of claims for extra work and completed change orders.

7. Check of retention amount.
8. On extra work and change orders, check of the method used to determine profit and overhead, material costs, and proper application of each in accordance with the conditions of the contract.
9. Recommendation to the project manager, submitted together with the contractor's payment request (Figures 17.2 and 17.3 or Figure 17.4).

LAWRANCE, FISK & McFARLAND, INC.
CONSULTING CIVIL & ENVIRONMENTAL ENGINEERS
Santa Barbara, California

MONTHLY PAYMENT ESTIMATE SUMMARY

Project Title: Oakbrook Valve Structure		Project No.: 94-132
Contractor: ABC Constructors, Inc.	Address: Thousand Oaks, CA	
Estimate No. Period 5-26-95 to 6-26-95	Contract or Spec. No.	Date: 26 June 1995

Description	Amounts		
	Previous	This Month	To Date
Total Earnings for Work and Materials Installed	2 328 796.17	160 643.90	2 489 440.07
Plus 80 % of Invoice Amt for Mat'ls Stored at Site	1 416.07	155.79	1 571.86
Less Adjustment for Previous Mat'ls invoices	<-0.00>	<-1 416.07>	<-1 416.07>
Sub Total	2 330 212.24	159,383.62	2 489 595.59
Less 10 % Retainage	<-233 021.22>	<15 938.36>	<-248 959.59>
Net Earning	2 097 191.02	143 445.26	2 240 636.27
Less Previous Payments			<-2 097.191.02>
Plus: Reduction of Retainage			0.00
Net Payment this Estimate			143 445.25

Data used are fictitious for illustration only

Percent of Time Elapsed: 40%	Percent of Work Completed: 47%

Contract Completion Data		Contract Cost Data	
Notice to Proceed Date	15 Dec 1994	Amount of Original Contract	4 671 360.00
Orig. Contract Completion Time	527 days	Approved Change Orders	138 785.00
Approved Time Extensions	42 days	Total Contract Amount	4 810 145.00
Contract Completion Date	5 Jul 1996		

Submitted by (Signature & Title): M. O. Parker, Resident Engineer	Administrative Review - Final Payments
Approved by (Signature & Title): R. E. Barnes, Project Manager	Approved by - Final Payments

Fisk Form 14-1

FIGURE 17.2 Contractor's Payment Request/Approval Summary Sheet.

LAWRANCE, FISK & McFARLAND, INC.
CONSULTING CIVIL & ENVIRONMENTAL ENGINEERS
Santa Barbara, California

MONTHLY PAYMENT ESTIMATE

Project No. 95-130 Contract No. 95-9345 Sheet 1 of 1

Project Title Oakbrook Dam and Appurtenant Works Contractor ABC Constructors, Inc. Date 30 June 1995 Estimate No. 15

PAY ITEM NO.	PAY ITEM	CONTRACT				THIS MONTH		TO DATE	
		QUANTITY	UNIT	UNIT PRICE	ITEM TOTAL	QUANTITY	EARNINGS	QUANTITY	EARNINGS
1	Construction Facilities	---	LS	---	700 000 00	0.7%	4 900 00	84.9%	594 300 00
2	Dewatering	---	LS	---	400 000 00	1.25%	5 000 00	77.5%	310 000 00
3	General excavation	707 800	M3	2.00	1 415 600 00	1 191	2 382 00	498 892	497 784 00
4	Structure excavation	76 312	M3	8.00	610 496 00				
5	Ditch excavation	8 400	M3	8.00	67 200 00				
6	Overhaul	966 690	M3/KM	0.15	145 003 00	—		510 507	76 576 05
7	Compacted embankment	236 400	M3	1.00	236 400 00	—		140 808	140 808 00
8	450 mm I.D. CMP	120	M	10.00	1 200 00	26	260 00	26	260 00
9	Soil cement	776	M3	10.00	7 760 00				
10	Structural concrete	1 200	M3	90.00	108 000 00				
11	Backfill concrete	3 097	M3	65.00	201 305 00	1 565	101 725 00	1 565	14 725 00
12	Access road	---	LS	---	500 000 00				
13	Misc. Concrete	1 543	M3	65.00	100 295 00	—		1 013.5	65 877 50
14	Steel Rebar	289 475	KG	0.38	110 000 00	—		135 854	51 624 52
15	Cement	930	TONN	70.00	65 100 00	184.67	12 926 90	435	30 450 00
16	Sump pump	1	EA	---	3 000 00	1	3 000 00	✓	3 000 00
	ORIGINAL CONTRACT TOTAL				4 671 360 00				
	CONTRACT CHANGE ORDERS								
CO1	Modify field ofc size	---	LS	---	<-10 210 00				<10 210 00>
CO2	Extend 150 mm SN piping	---	LS	---	3 995.00 00			3 995	3 995 00
CO3	Drainage modifications	---	LS	---	145 000 00	21%	30 450 00	85%	123 250 00
	TOTAL OF APPROVED C.O.'s TO DATE				138 785 00				
	TOTAL CONTRACT AMOUNT				4 810 145 00				
	TOTAL FOR WORK & MATLS INSTALLED						160 643 90		2 489 440 07
	MATLS DELIVERED & STORED ON SITE								
16	SUMP PUMP		EA	1575.35		—	—	1	1 575 35
8	150mm CMP		M	7.44		26	194 74	52	389 48
	TOTAL INVOICE AMT FOR MATLS STORED AT SITE						194 74		1 964 83

Data used are fictitious for illustration only

FIGURE 17.3 Contractor's Pay Estimate.

Basis for Payment Amounts

The basis of all progress payments is a determination, in the field, of the actual quantities of work that have been accomplished as of the date that the payment request is submitted. The accuracy required of these field measurements is determined by the type of contract involved. As described earlier in this chapter, there are numerous types of construction contracts, distinguished primarily by the method of determining the final contract price; they include:

1. Fixed-price contracts
   - Lump-sum contract
   - Unit-price contract
   - Fixed-price incentive contract
2. Cost-reimbursable contracts
   - Time-and-materials (Cost-plus-percentage-of-cost) contract
   - Cost-plus-fixed-fee contract (CPFF)
   - Cost-plus-incentive-fee contract (CPI)
3. Guaranteed maximum price contracts (GMP)

APPLICATION FOR PAYMENT NO. ____________

To ____(OWNER)
Contract for ____ .
OWNER's Project No. ____________________ . ENGINEER's Project No. ____________________ .
For Work accomplished through the date of ______________________________________ .

ITEM	CONTRACTOR's Schedule of Values			Work Completed	
	Unit Price	Quantity	Amount	Quantity	Amount
	$		$		$
Total (Orig. Contract)			$		$
C.O. No. 1					
C.O. No. 2					

Accompanying Documentation:

GROSS AMOUNT DUE .... $________
LESS ____% RETAINAGE .... $________
AMOUNT DUE TO DATE .... $________
LESS PREVIOUS PAYMENTS .... $________
AMOUNT DUE THIS APPLICATION .... $________

CONTRACTOR'S Certification:

The undersigned CONTRACTOR certifies that (1) all previous progress payments received from OWNER on account of Work done under the Contract referred to above have been applied to discharge in full all obligations of CONTRACTOR incurred in connection with Work covered by prior Applications for Payment numbered 1 through ________inclusive; and (2) title to all materials and equipment incorporated in said Work or otherwise listed in or covered by this Application for Payment will pass to OWNER at time of payment free and clear of all liens, claims, security interests and encumbrances (except such as covered by Bond acceptable to OWNER).

Dated ______________________, 19____ ______________________________
CONTRACTOR

By ______________________________

Payment of the above AMOUNT DUE THIS APPLICATION is recommended.

Dated ______________________, 19____ ______________________________
ENGINEER

By ______________________________

EJCDC No. 1910-8-E (1983 Edition)
Prepared by the Engineers' Joint Contract Documents Committee and endorsed by The Associated General Contractors of America.

FIGURE 17.4 EJCDC Form of Application for Payment, to Be Used with Payment Request/Approval Form Shown in Figure 17.2 or Equivalent Document. *(Copyright © 1990 by National Society of Professional Engineers.)*

APPLICATION FOR PAYMENT

INSTRUCTIONS

A. GENERAL INFORMATION

This standard form is intended as a guide only. Many projects require a much more extensive form with space for numerous items, descriptions of Change Orders, identification of variable quantity adjustments, summary of materials and equipment stored at the site and other information. It is expected that a separate form will be developed by Engineer or Contractor at the time Contractor's Schedule of Values is finalized. Note also that the format for retainage must be changed if the Agreement permits (or the Law provides), and Contractor elects, the deposit of securities in lieu of retainage. Refer to Article 14 of the General Conditions for provisions concerning payments to Contractor.

B. COMPLETING THE FORM

The Schedule of Values, submitted and approved as provided in paragraphs 2.6.3 and 2.9 of the General Conditions, should be copied in the space indicated on the Application For Payment form. Note that the cost of materials and equipment is often listed separately from the cost of their installation. All Change Orders affecting the Contract Price should be identified and include such supplemental Schedules of Values as required for progress payments.

The form is suitable for use in the Final Application for Payment as well as for Progress Payments; however, the required accompanying documentation is usually more extensive for final payment. All accompanying documentation should be identified in the space provided on the form.

C. ENGINEER'S REVIEW

Engineer *must* review all Applications for Payment with care to avoid recommending any payments not yet earned by Contractor. All accompanying documentation of legal nature, such as lien waivers, should be reviewed by an attorney, and Engineer should so advise Owner.

FIGURE 17.4 (continued)

Although each type of contract may be let on the basis of competitive bids, with the lowest bidder getting the job, several significant differences exist. On a lump-sum contract, the contractor must complete the work for the fixed price shown as long as the scope of the contract has not been altered by change orders. Any cost overruns must come out of the contractor's pocket; similarly, any money that is saved on the job (as long as it conforms to the plans and specifications) belongs to the contractor.

On a unit-price contract, the design firm provides a list of all individual bid items, along with an "engineer's estimate" of the quantities involved. Blanks are provided in the proposal document for the bidder to insert a fixed price per unit for which the contractor agrees to build the work, which when multiplied by the quantity shown in the engineer's estimate indicates the total amount of its bid for each item. In the example shown in the accompanying illustration (Figure 17.5), the bid sheets that the contractor must fill out to submit the bid already contain a typed-in quantity. The contractor then fills in the price that will be charged for each bid item.

The amount that the contractor shows in the "unit price" column is actually the firm bid. The amount under the "total" column is merely the product of the unit price

BID SCHEDULE

Schedule of Prices for Construction of

PORTER STREET IMPROVEMENTS

Item No.	Description	Quantity and Unit	Unit Price	Amount
1.	Excavation and removal of existing asphalt concrete and aggregate base	1050 m^3	$ 3.92	$ 4 116.00
2.	Asphalt concrete paving	500 t	$ 31.30	$ 15 650.00
3.	6-in Class II Aggregate base	1500 m^2	$ 4.31	$ 6 465,00
4.	Junction structure	1 each	$ 2 000.00	$ 2 000.00
5.	Reinforced concrete pipe 600 mm diameter	120 m	$ 146.25	$ 17 550.00
6.	Reinforced concrete headwall		Lump Sum	$ 2 000.00

Data used are fictitious for illustration only

TOTAL $ 47 781.00

NOTE:

All amounts and totals in the Bid Schedule will be subject to verification by the Owner. In case of variation between the unit price and the total shown by the bidder, the unit price will be considered as the bid.

The quantities listed in the Bid Schedule are supplied to give an indication of the general scope of the Work, but the accuracy of these figure is not guaranteed and the bidder shall make his or her own estimates from the drawings.

* * * * *

PROPOSAL FORMS
PAGE B-3

FIGURE 17.5 Unit-Price Bid.

bid multiplied by the *engineer's quantity estimate*. The actual amount of money that the contractor will finally receive to do the work will be based on the actual *field-measured quantities*, *not* the quantities shown in the engineer's estimate. However, for the purposes of determining the lowest bid, the quantities shown in the engineer's estimate are used during the bidding so that all bids can be compared on the same basis.

On a cost-plus-fixed-fee project, the contractor agrees to a fixed profit level and is reimbursed for all costs of labor and material at their actual cost, plus the addition of a fixed fee rate for profit. Under this arrangement, the contractor's books must be open to the owner or its representative, and all of the contractor's costs must be regularly audited to establish the amount of progress payments.

Schedule of Values versus Cost-Loaded CPM Schedule

Often, due to the unfamiliarity of the specification writer with the principles of contract administration, the contract documents ask for information from the contractor that is wholly inconsistent with the type of contract involved. For example, on unit-price contracts, all prices are based upon an itemized list of all of the work items or tasks that must be performed under the contract, each with a unit price. In this manner the final price of any line item will simply be the product of the unit price times the actual quantity of that item that was installed or completed. Under this concept, the controlling price is the unit price of each item. Many get mixed up with lump-sum procedures and ask for a schedule of values as well. This is redundant, as the unit-price bid sheet *is* a schedule of values that is tendered with the contractor's original bid.

On a lump-sum project, however, the contractor agrees to do all the work that is defined in the plans and specifications for a fixed price, regardless of quantity variations that are not the result of owner or architect/engineer changes in the work. As a convenience, a schedule of values is invariably requested. In appearance, a schedule of values looks exactly like a unit-price bid, but with several significant differences; prices shown are for convenience in making monthly progress payments only and have no other contractual significance. They cannot legitimately be used to price change orders, for example.

Sometimes confusion results from requests for cost-loaded CPM networks under the assumption that they always result in easier payment administration. This may be true on a lump-sum project, but it has exactly the opposite effect on a unit-price contract, as the cost-loaded network activity items never coincide with the pay line items on unit-price contracts. Some guidelines are as follows:

Unit-Price Contracts

1. Do *not* call for a cost-loaded CPM schedule; it will conflict with the line items on the unit-price bid sheet.
2. Do *not* call for a Schedule of Values; the original bid sheet serves this purpose.
3. A breakdown of all line items is submitted with the contractor's original bid and is binding for pay purposes throughout the job.

Lump-Sum Contracts with Cost-Loaded CPM

1. Do *not* ask for a Schedule of Values; the cost-loaded CPM takes its place.
2. Request that a cost-loaded CPM be submitted after execution of the Agreement but before the end of the first pay period.

Lump-Sum Contracts without Cost-Loaded CPM

1. Do *not* ask for a cost-loaded CPM schedule.
2. Do *not* ask for a bid breakdown with the bid.
3. Request a Schedule of Values after execution of the Agreement but before starting work on the project.

EVALUATION OF CONTRACTOR'S PAYMENT REQUESTS

Submittal Requirements

At a prearranged date each month, the contractor is expected to submit a request for payment for all of the work performed during the preceding month (Figure 17.2 or 17.4). Prior to forwarding these payment requests to the owner through the design firm or construction management firm for payment, the Resident Project Representative must check to assure that all items for which the contractor has claimed payment have actually been completed. If a project was contracted for on a lump-sum basis, a reasonable estimate of the fair value of the work that was accomplished might be very difficult, so under the terms of the General Conditions of the Contract for both the EJCDC and the AIA, the contractor is required to submit a schedule of values of the various portions of the work, including quantities where required, so that these values can be used as the basis for determining the amount of progress payments to be made to the contractor each month (see again Figure 12.8). The contractor is normally required to show evidence at the time of submittal of the price breakdown that the pricing is correct, and that the total aggregate amount will equal the sum of the total contract amount. Each item should contain its own share of profit at overhead.

Unit-Price Contracts

If a project is based upon a unit-price bid, as many engineering contracts are, the actual unit prices stated in the original bid will be held as the basis for computation of all progress payment amounts. It is essential to note, however, that because a unit-price contract does not have a fixed price ceiling, but the final cost to the owner will be determined by the quantities actually completed, the determination of field quantities must be very precise. This is in direct contrast to the administration of a lump-sum contract, where even if a small error was made one month, it would be compensated for in later payments. The contractor cannot receive more money than the stated fixed price, even if some of the final quantities varied somewhat from the anticipated amounts. Under the concept of a lump-sum contract, the contractor agrees to build a complete project that will perform the intended function indicated in the plans and specifications. Anything that is necessary to accomplish this must be considered as part of the contract, even if not specifically stated. On a unit-price contract, however, the contractor may still be required to construct a complete functional project, but if any variation occurs in the quantities of any of the separate bid items listed, the contractor is entitled to an amount of money equal to the unit price of the bid item multiplied by the actual quantity of that item that was constructed or furnished. In this manner, if the quantity of a bid item exceeds what the original quantity estimate indicated, the contractor will receive more money than that shown in the bid; and if it is less than in the original bid, the amount received will be less than the amount of the bid. All such quantity differences must be based on the unit price stated by the contractor in the bid, as that alone is the controlling amount, not the extension obtained by multiplying the bid price by the anticipated quantity. The one notable exception to this rule is if the quantity overrun or underrun is in excess of 25 percent of the bid item. Then the contractor is entitled to renegotiate its unit price for the item. Although a 15 to 25 percent figure is frequently mentioned in contract documents, in situations where it was omitted, the principle of allowing a price renegotiation in cases in which the actual quantities differed from the anticipated quantities by over 25 percent has been upheld by some state courts. The Associated General Contractors (AGC), on the other hand, recommends a value of 15 percent.

Customary pay line items and units of measure under SI metric are illustrated on the following pages.

Equipment and Materials Delivered but Not Yet Used in the Work

The resident engineer or inspector should check the specifications carefully on this point before starting to total the amount of payment due the contractor. Often, certain fabricated items of equipment or certain products are delivered to the site a considerable amount of time before they are actually needed or installed in the work. In some contracts no payment will be made for any such material until it is actually used in the work (such as large pipe mains, which may not be paid for until they have been laid, jointed, backfilled, tested, and paved over; or large pieces of equipment such as motors, generators, valves, or steel plate fabrications that had to be made up especially for the project as opposed to an off-the-shelf item). A second approach to this problem is to allow partial payment for materials or equipment as soon as they have been delivered to the site. In any case, the decision is not for the Resident Project Representative to make, but must be determined by the provisions of the specifications.

Unit prices are often quoted for items as "installed-in-place." Wherever this is the case, it is not reasonable to allow full payment for any material solely on the basis of its having been delivered to the site. If the contract allows any payment at all, and the material is priced as in-place, the contractor should provide a copy of the invoice for the material to support its claim for payment of the delivery and storage of the material at the site. There is considerable justification for allowing some payment for certain types of materials, products, or equipment because the contractor is expected to pay the suppliers within 30 days after delivery in most cases. Thus, if a fabricated item such as mortar-lined-and-coated steel pipe, or precast prestressed roof members, or special pumps or large valves are ordered, there is often a long lead time required in placing the order to be certain of delivery in time. Often, when the item is actually delivered, the work may not yet be far enough along to allow its immediate installation; yet the contractor has already been obligated to make payment. Nevertheless, there is nothing that the inspector can do to alleviate the problem except to read the provisions of the specifications carefully to make certain that such payments can be made under the terms of the contract. It can save a lot of arguments if the inspector can show the contractor the provisions of the specifications that control such payments.

A word of caution at this point. As long as a contract is based upon a lump-sum price, or as long as a unit-price contract shows separate line items for a material and its installation, no particular problem exists other than a need to exercise care in evaluating the amount of work and materials claimed for payment. However, on a unit-price contract where a single line item covers both the cost of the material and its installation, a serious risk of overpayment exists.

Take the example of an underground pipeline under a roadway. Often, a single price may be quoted to cover the cost of the pipe and its installation, including all earthwork and even the pavement over the trench, as would be the case for item 5 in the bid sheet shown in Figure 17.5. In such cases, the price per linear foot of completed pipeline already includes the price of the pipe material. If early payment was allowed for pipe that was delivered to the site but not yet installed, the cost of the pipe would have to be subtracted from the price claimed later for construction of the pipeline in place; otherwise, the contractor would be receiving payment twice for the same pipe.

Force Account

Occasionally, there appears to be some misunderstanding about the meaning of the term *force account* in the construction industry. According to the VNR Dictionary of Civil Engineering, it is simply the U.S. term for *daywork* or for any type of *cost-plus* payment. Daywork is the customary term for such work on international (FIDIC) contracts. In most construction contracts in the United States, this method of payment is associated with extra work and change orders. Because of this, most Department of Transportation (DOT) Standard Specifications classify force account under Section 9, Measurement and Payment, while occasionally it may be specified under the heading of Changes and Extra Work instead. Under some contracts, markup rates are not established in advance. Instead a contract may state that ". . . a reasonable allowance for overhead and profit shall be added to the contractor's cost. . . ." It would be interesting to note how long it might take to arrive at mutually agreeable terms under those conditions . . . although the U.S. Army Corps of Engineers seems to be able to make it work.

Force Account as a Payment Method

When it is difficult to provide adequate measurement or to estimate the cost of certain items of work, force account may be used to pay the contractor for performing the work. Some contract items may even be set up in advance to be paid by force account. Also, some change orders may require payment by force account. However, force account should be the last choice when setting up items for payment on the original contract or when determining an equitable adjustment for a change. Generally, most DOT Standard Specifications will describe an allowable payment procedure and limits for force account work (Figure 17.6).

When added work is to be paid by force account, a change order should be prepared detailing the added work to be performed and the estimated cost.

Generally, force account payment is not authorized for superintendents or other employees engaged in general supervisory work. Allowance for their pay is included in the contractor's percentage for overhead and profit. However, a foreman devoting full time to the force account work would be eligible for payment under force account.

The project manager has the authority to direct every aspect of force account work. The specifications should provide options for the prices to be paid on force account. Therefore, before any work is performed on a force account basis by the contractor, the project manager and the contractor must review and agree upon the following:

1. **Labor**—The classification and approximate number of workers to be used; the wage rate to be paid those workers; whether or not travel allowance and subsistence is applicable to those workers; and what foremen, if any, will be paid for by force account.
2. **Equipment**—The equipment to be used including the size, rating, capacity, or any other information to indicate the equipment is proper for the work to be performed; whether the equipment to be used is owned by the contractor or is to be rented; the cost per hour for the equipment to be used.
3. **Materials**—The material to be used including the cost and any freight charges; whether the material is purchased specifically for the project or comes from the contractor's own supply.

The project manager should prepare lists of the equipment and labor classifications actually used and the rates for each after the work has started. These lists should be filed and become a part of the documents in support of the force account item. Such lists should include the following information:

EQUIPMENT LIST

The equipment list must include the complete nomenclature of the equipment, to establish the proper rental rate. Equipment rates should be those that are included in the Rental Rate Book referred to in the contract. There are several published books of equipment rental rates on the market, as well as published DOT equipment rental rates established by the state DOT for highway and bridge work.[1] Special equipment rates that are not set forth in the Rental Rate Book should require the approval of the project manager before they are used.

---

[1]Examples include: the *AED Green Book*, published by Machinery Information Division of K-III Directory Corporation, San Jose, Ca.; *Contractors Equipment Cost Guide* and *Rental Rate Blue Book*, both published by Dataquest (a Dunn and Bradstreet Company); *Labor Surcharge and Equipment Rental Rates*, published by the California Department of Transportation; *AGC/WSDOT Equipment Rental Agreement*, published by the Washington Department of Transportation; and similar publications.

Washington State Department of Transportation

Page 1 of 2

DAILY REPORT OF FORCE ACCOUNT WORKED

CONTRACT NO	DATE	ITEM NUMBER	ITEM NAME
3711	1-12-90	156	Additional Erosion Control

PRIME CONTRACTOR	SUB CONTRACTOR / AGENT
Ace Construction Co	Coast Brothers Inc

LINE / STATION	GROUP	BASIS OF MATERIAL ACCEPTANCE	RAMS NO
L5 Line Sta 137+00	3	Test, Manufacturer's Cert	3650, 3666

DESCRIPTION OF WORK PERFORMED: Placing rock for slope protection, laying 6" drain pipe

TIME WORKED RECORD

	CODE OR CRAFT GROUP	WORKMAN AND/OR EQUIPMENT WORKING	OCCUPATION OF WORKMAN OR EQUIPMENT SIZE	HOURS WORKED REG	HOURS WORKED OT	RATE	AMOUNT
1		Labor:					
2	350-0821	Leo Hellinan	General Laborer	8.0		18.84	151 —
3	350-1215	Tim Craig	Pipelayer	8.0		19.49	156 —
4	570-1201	Wayne Hagerty	Dozer Operator	8.0		22.80	182 —
5	570-6610	Loren Olsen	Dragline Operator	8.0		23.10	185 —
6	730 0310	Bob Morris	Trucker	8.0		20.98	168 —
7		Add Med Aid, Indus. Ins, Supp. Pension		40.0		0.6152	25 —
8		Add travel pay, 2 @ 4.80/day					20 —
9			Labor Sub Total				887 —
10							
11		Equipment:					
12		Tractor Crawler, P.S. Case 450 w/dozer		8.0		19.70	158 —
13		Excavator, Crawler, Northwest Model 6		8.0		67.50	540 —
14		Add for 3 C.Y. Dragline Bucket		8.0		4.60	37 —
15		Truck, Belly Dump, Mack R-600 #25		8.0		22.64	181 —
16		Truck, Pickup, Ford F-250 3/4 Ton		8.0		4.90	39 —
17			Equipment Sub-Total				955 —
18							
19		Material:					
20		Rock, 122 tons @ 3.60/ton					439 —
21		Gravel Backfill for drains, 12 C.Y. @ 5.20/yd.					62 —
22		6" drain pipe, 40 L.F. @ 3.22/ft. invoice attached					129 —
23			Material Sub-Total				630 —
24							
25			Sub-Total				2,472 —
26							
27		Add 20% of Labor for OH & Profit					177 —
28		Add 15% of Equipment for OH & Profit					143 —
29		Add 15% of Material for OH & Profit					95 —
30							
31			Sub Total				2,887 —
32		UTC Equipment:					
33		Dumptruck, International, 10 yd. #32 (Jim Smith)		2.0		60.29	121 —
34			Sub Total				3,008 —
35							
36			Continued next page				
37							
38							
39				From page 2 →			

Calculated By	Date	Checked By	Date	TOTAL
ER	1-16-90	DS	1-16-90	3,482 —

CAPS ENTRY NO	ENTERED BY	DATE	ENTRY VERIFIED	DATE
1134	FR	1-17-90	DS	1-18-90

INSPECTOR	CONTRACTOR'S REPRESENTATIVE	TITLE
John Doe	John Goodfellow	Supt.

DOT 422-008 Revised 1/90

SUPERSEDES PREVIOUS EDITIONS

FIGURE 17.6 Example of Documentation of Force Account Work.

Washington State
Department of Transportation

Page 2 of 2

DAILY REPORT OF FORCE ACCOUNT WORKED

CONTRACT NO	DATE	ITEM NUMBER	ITEM NAME
3711	1-12-90	156	

PRIME CONTRACTOR	SUB-CONTRACTOR / AGENT

LINE / STATION	GROUP	BASIS OF MATERIAL ACCEPTANCE	RAMS NO

DESCRIPTION OF WORK PERFORMED

TIME WORKED RECORD

	CODE OR CRAFT GROUP	WORKMAN AND/OR EQUIPMENT WORKING	OCCUPATION OF WORKMAN OR EQUIPMENT SIZE	HOURS WORKED REG	HOURS WORKED O.T.	RATE	AMOUNT
1		Sub Total brought forward from page 1					3,008 —
2							
3		Add 5% for Sub-Contractor Ins, Tax & Bonding					150 —
4		Sub-Contractor Sub Total					3,158 —
5							
6		Add 5% for Prime Contractor Admin cost					158 —
7		Sub Total					3,316 —
8							
9		Add 5% for Prime Contractor Ins, Tax & Bonding					166 —
10		Total					3,482 —
11							
12							
13							
14							
15							
16							
17							
18							
19							
20							
21							
22							
23							
24							
25							
26							
27							
28							
29							
30							
31							
32							
33							
34							
35							
36							
37							
38							
39							

Calculated By	Date	Checked By	Date	TOTAL	

CAPS ENTRY NO	ENTERED BY	DATE	ENTRY VERIFIED	DATE

INSPECTOR	CONTRACTOR'S REPRESENTATIVE	TITLE

DOT 422-008 Revised 1/90

SUPERSEDES PREVIOUS EDITIONS

FIGURE 17.6 (continued)

LABOR LIST

The list for labor should include the labor classification and the composite hourly rate. The composite hourly rate is the current basic wage the contractor is obligated to pay for each classification and all added costs for labor listed in the DOT Standard Specifications.

Force Account Payment

When extra work is to be paid for on a force account basis, the labor, materials, and equipment used should be determined as follows:

1. To the direct cost of labor, materials, and equipment used in the work, a markup is allowed that shall include full compensation for all overhead costs that are deemed to include all items of expense not specifically designated as cost or equipment rental. The total payment is deemed to be the actual cost of the work and is considered under the contract to constitute full compensation for all such work. Markup amounts vary widely from agency to agency, but generally fall within the following ranges:

   Labor: From 15 to 33 percent, with 20 percent most often used.

   Materials: From 15 to 20 percent, with 15 percent most often used.

   Equipment rental: From 15 to 20 percent, with 15 percent most often used.

2. When extra work to be paid for on a force account basis is performed by a subcontractor, an additional markup is usually allowed. The additional markup is to reimburse the contractor for additional administrative costs, and no other payment should be made due to the performance of extra work by a subcontractor.

   Subcontractor markup: From 0 to 15 percent, with 5 percent most often used.

   A word of caution here; specifications should state that the subcontractor markup applies to first-tier subcontractors only, otherwise the owner may be required to pay the specified markup for every level of subcontract involved. With a little creative management by the general contractor, this could become a lucrative source of extra money.

Payment for Extra Work and Change Orders

Although in a unit-price contract a prearranged value has been established for each item of work to facilitate the determination of the amount owed to the contractor, and on a lump-sum job, a bid breakdown is often requested to establish a fair price for any work completed, as a means of arriving at the amount to be paid for each monthly progress payment, the cost of extra work that was not included in the original contract must be negotiated separately or a fair method agreed upon for determining its value.

Obviously, if no groundwork were laid to establish a procedure for evaluating and determining the cost of such extra work, the contractor would have the owner "over a barrel," so to speak. Thus, it is common in all standard forms of the General Conditions of a construction contract to specify the means of determining such costs. It should be kept in mind, however, that not all General Conditions treat this subject area in the same way. Many such documents, because they must be universally acceptable, have been so watered down in their provisions that extensive Supplementary General Conditions must be prepared to adapt these provisions to a particular job.

General Conditions documents such as the AIA General Conditions advocate the principle of completing such work on a cost-plus basis with "a reasonable allowance for overhead and profit." Unfortunately, what is reasonable to the contractor may not be considered as reasonable by the owner. Most of the General Conditions used by public agencies for their own use have established fixed policies, and their General Conditions reflect this. An example can be seen in the provisions of the EJCDC *Standard General Conditions of the Construction Contract* for changes in contract price and time, shown in Figures 17.7 and 17.8, where a maximum of 15 percent of the actual cost of the work is allowed to cover the cost of general overhead and profit.

Similar, but sometimes considerably more comprehensive, are the provisions for overhead and profit allocation of the various state department of transportation standard specifications for change orders or force account work. The markups involve a percentage added to the cost of labor, which generally varies from agency to agency but is usually 15, 20, 24, or 33 percent, with 20 percent as the most frequently allowed markup. An additional markup of 15 percent is generally applied to the cost of materials and another 15 percent to the equipment rental. Most agencies also allow the general contractor to mark up first-tier subcontractor bids by 5 percent, with no markup allowed for second-tier or lower subcontractors. There occasionally is some argument concerning the issue of whether or not foremen and superintendents are included in the markup or are part of home office overhead.

As it can be determined from the provisions shown in the accompanying examples, some questions may still be unanswered and are thus subject to negotiation at the time of the owner's incurring the cost of extra work. If all the possible contingencies and conditions could be anticipated, then and only then could a document be produced that would provide a definite answer to any question relating to the cost of such extra work. Obviously, this is an impossible task.

As a part of the contractor's payment request, if extra work authorized by the issuance of a valid change order was completed during the payment period, the resident engineer or inspector not only must evaluate the regular bid items and quantities, but also, under the terms of the specific contract documents controlling the project, must determine the validity of the claims of the contractor for the amount of money claimed for completing such extra work. In each case, the inspector should

ARTICLE 10 - CHANGES IN THE WORK; CLAIMS

10.01 Authorized Changes in the Work

A. Without invalidating the Agreement and without notice to any surety, OWNER may, at any time or from time to time, order additions, deletions or revisions in the Work by Written Amendment, a Change Order, or a Work Change Directive. Upon receipt of any such document, CONTRACTOR shall promptly proceed with the Work involved which will be performed under the applicable conditions of the Contract Documents (except as otherwise specifically provided).

B. If OWNER and CONTRACTOR are unable to agree on entitlement to, or on the amount or extent, if any, of an adjustment in the Contract Price or Contract Times, or both, that should be allowed as a result of a Work Change Directive, a Claim may be made therefor as provided in paragraph 10.05.

10.02. Unauthorized Changes in the Work

A. CONTRACTOR shall not be entitled to an increase in the Contract Price or an extension of the Contract Times with respect to any work performed that is not required by the Contract Documents as amended, modified or supplemented as provided in paragraph 3.04, except in the case of an emergency as provided in paragraph 6.16 or in the case of uncovering Work as provided in paragraph 13.04.B.

10.03. Execution of Change Orders

A. OWNER and CONTRACTOR shall execute appropriate Change Orders recommended by ENGINEER (or Written Amendments) covering:

1. changes in the Work which are: (i) ordered by OWNER pursuant to paragraph 10.01.A, (ii) required because of acceptance of defective Work under paragraph 13.08.A or OWNER's correction of defective Work under paragraph 13.09, or (iii) agreed to by the parties;
2. changes in the Contract Price or Contract Times which are agreed to by the parties; and
3. changes in the Contract Price or Contract Times which embody the substance of any written decision rendered by ENGINEER pursuant to paragraph 10.05; provided that, in lieu of executing any such Change Order, an appeal may be taken from any such decision in accordance with the provisions of the Contract Documents and applicable Laws and Regulations, but during any such appeal, CONTRACTOR shall carry on the Work and adhere to the progress schedule as provided in paragraph 6.18.A.

10.04. Notification to Surety or Insurer

A. If notice of any change affecting the general scope of the Work or the provisions of the Contract Documents (including, but not limited to, Contract Price or Contract Times) is required by the provisions of any Bond to be given to a surety or by the provisions of any insurance policy to be given to any insurer, the giving of any such notice will be CONTRACTOR's responsibility. The amount of each applicable Bond will be adjusted to reflect the effect of any such change.

FIGURE 17.7 Example of Provisions for Changes in the Work from EJCDC Standard General Conditions of the Construction Contract 1910-8 (1995). *(Copyright © 1995 by the National Society of Professional Engineers.)*

10.05. Claims and Disputes

A. Notice: Written notice stating the general nature of each Claim, dispute or other matter shall be delivered by the claimant to ENGINEER and the other party to the Contract promptly (but in no event later than thirty days) after the start of the event giving rise thereto. Notice of the amount or extent of the Claim, dispute or other matter with supporting data shall be delivered to the ENGINEER and the other party to the Contract within sixty days after the start of such event (unless allows additional time for claimant to submit additional or more accurate data in support of such Claim, dispute, or other matter). A Claim for an adjustment in Contract Price shall be prepared in accordance with the provisions of paragraph 12.01.B. A Claim for an adjustment in Contract Time shall be prepared in accordance with the provisions of paragraph 12.02.B. Each Claim shall be accompanied by claimant's written statement that the adjustment claimed is the entire adjustment to which the claimant believes it is entitled as a result of said event. The opposing party shall submit any response to ENGINEER and the claimant within thirty days after receipt of the claimant's last submittal (unless ENGINEER allows additional time).

B. *ENGINEER's Decision:* ENGINEER will render a formal decision in writing within thirty days after receipt of the last submittal of the claimant or the last submittal of the opposing party, if any ENGINEER's written decision on such Claim, dispute, or other matter will be final and binding upon OWNER and CONTRACTOR unless:

   1. an appeal from ENGINEER's decision is taken within the time limits and in accordance with the dispute resolution procedures set forth in Article 16, or
   2. if no such dispute resolution procedures have been set forth in Article 16, a written notice of intention to appeal from ENGINEER's written decision is delivered by OWNER or CONTRACTOR to the other and to ENGINEER within thirty days after the date of such decision and a formal proceeding is instituted by the appealing party in a forum of competent jurisdiction, within sixty days after the date of such decision or within sixty days after Substantial Completion, whichever is later (unless otherwise agreed in writing by OWNER and CONTRACTOR), to exercise such rights or remedies as the appealing party may have with respect to such Claim, dispute, or other matter in accordance with applicable Laws and Regulations.

C. No claim for an adjustment in Contract Price or Contract Times (or Milestones) will be valid if not submitted in accordance with this paragraph 10.05.

FIGURE 17.7 (continued)

consult the project manager of the design firm or the owner to determine the exact terms of the agreement that cover the construction of such extra work, and apply these terms to the evaluation. In any case, the terms of the contract General Conditions should be followed carefully, as they form the basis for any specific agreement with the owner for establishing the costs of extra work.

Whenever extra work is being performed under the terms of the contract, the Resident Project Representative should keep a daily record of all such work performed and materials furnished for use in checking the contractor's payment requests. Such a record may also be of considerable value in the settlement of claims. A simplified way of documenting such costs is to utilize a daily extra work report designed for this purpose, as illustrated in Figure 17.9, which was derived from a form used by California DOT for documentation of extra work. Figure 17.9 illustrates a form used by the State of Washington for documentation of WSDOT extra work costs under force account.

ARTICLE 12 - CHANGE OF CONTRACT PRICE; CHANGE OF CONTRACT TIME

12.01. Change of Contract Price

A. The Contract Price may only be changed by a Change Order or by a Written Amendment. Any Claim for an adjustment in the Contract Price shall be based on written notice submitted by the party making the claim to the ENGINEER and the other party to the Contract in accordance with the provisions of paragraph 10.05.

B. The value of any Work covered by a Change Order or of any Claim for an adjustment in the Contract Price will be determined as follows:

   1. where the Work involved is covered by unit prices contained in the Contract Documents, by application of such unit prices to the quantities of the items involved (subject to the provisions of paragraph 11.03); or

   2. where the Work involved is not covered by unit prices contained in the Contract Documents, by a mutually agreed lump sum (which may include an allowance for overhead and profit not necessarily in accordance with paragraph 12.01.C.2); or

   3. where the Work involved is not covered by unit prices contained in the Contract Documents and agreement to a lump sum is not reached under paragraph 12.01.B.2, on the basis of the Cost of the Work (determined as provided in paragraph 11.01) plus a CONTRACTOR's fee for overhead and profit (determined as provided in paragraph 12.01.C).

C. CONTRACTOR's Fee: The CONTRACTOR's fee for overhead and profit shall be determined as follows:

   1. a mutually acceptable fixed fee; or

   2. if a fixed fee is not agreed upon, then a fee based on the following percentages of the various portions of the Cost of the Work:

      a. for costs incurred under paragraphs 11.01.A.1 and 11.01.A.2, the CONTRACTOR's fee shall be fifteen percent;[1]

      b. for costs incurred under paragraph 11.01.A.3, the CONTRACTOR's fee shall be five percent;

      c. where one or more tiers of subcontracts are on the basis of Cost of the Work plus a fee and no fixed fee is agreed upon, the intent of paragraph 12.01.C.2.a is that the Subcontractor who actually performs the Work, at whatever tier, will be paid a fee of fifteen percent of the costs incurred by such Subcontractor under paragraphs 11.01.A.l and 11.01.A.2 and that any higher tier Subcontractor and CONTRACTOR will each be paid a fee of five percent of the amount paid to the next lower tier Subcontractor;

      d. no fee shall be payable on the basis of costs itemized under paragraphs 11.01.A.4, 11.01.A.5 and 11.01.B;

      e. the amount of credit to be allowed by CONTRACTOR to OWNER for any change which results in a net decrease in cost will be the amount of the actual net decrease in cost plus a deduction in CONTRACTOR's fee by an amount equal to five percent of such net decrease; and

      f. when both additions and credits are involved in any one change, the adjustment in CONTRACTOR's fee shall be computed on the basis of the net change in accordance with paragraphs 12.01.C.2.a through 12.01.C.2.e, inclusive.

FIGURE 17.8 Provisions for Payments for Extra Work from the EJCDC Standard General Conditions of the Construction Contract 1910-8 (1995). *(Copyright © 1995 by the National Society of Professional Engineers.)*

E. R. FISK & ASSOCIATES
P.O. Box 6448 • Orange, CA 92613-6448

DAILY WORK REPORT

PROJECT TITLE PARK ANNEX LIFT STATION
PROJECT NO. F 293 LOCATION CITY OF MERCED, CA
WORK PERFORMED BY ABC CONSTRUCTORS, INC.
DESCRIPTION OF WORK HAUL & ERECT CONSTRUCTION SIGNS

DATE PERFORMED 7 AUG 2000
DATE OF REPORT 9 AUG 2000

REPORT NO. 49
AMOUNT AUTHORIZED $
PREVIOUS EXPENDITURE $ 0
TODAY $
TO DATE $
CONTRACTOR JOB NO. 87-0157
CONTRACTOR REPORT NO.

Equip. No.	Equipment	Hours	Hourly Rate	Extended Amounts
222	FLATRACK 19,500 GVW	8 (10)	22.80	228.00
111	PICKUP UNDER 12,000 GVW	6 (10)	13.05	130.50
R-1	HAND-HELD AUGER UNDER 5-HP	1-DAY	PER DAY	6.65
				365.15

Data used are fictitious for illustration only

P.R. No.	Labor		Hours	Hourly Rate	Extended Amounts
	TRUCK DRIVER	O.T.			
	PAUL MILLIGAN	Reg.	8	17.25	138.00
	LABOR FOREMAN	O.T.			
	JOHN RUSSELL	Reg.	6	16.44	98.64
	LABORER	O.T.			
	KAHNM TRAN	Reg.	8	15.94	127.52
	LABORER	O.T.			
	GUADALUPE RODRIGUEZ	Reg.	6	15.94	95.64
	LABORER	O.T.			
	GERALD MADLEY	Reg.	6	15.94	95.64
		O.T.			
		Reg.			
				SUB TOTAL	555.44

MATERIAL and/or WORK done by specialists

Description	No. Unit	Unit Cost	
LUMBER & HARDWARE SEE ATTACHED INVOICE #21704		225.55	225.55
TOTAL COST OF EQUIPMENT, MATERIALS AND WORK			

Added Percentage—(See Special Conditions)		
Subsistence ___ No. ___ @$ ___		—
Travel Expense ___ No. ___ @$ ___		—
Other		—
TOTAL COST OF LABOR	A	644.31
→	B	590.70
+ 20 % ON LABOR COST	(A)	128.86
+ 15 % ON EQUIP., MAT'L AND WORK COST	(B)	88.61
TOTAL THIS REPORT		1,452.48

G. Osborne
Submitted—Contractor

R. L. Barnes
Recommended—Res. Proj. Representative

DISTRIBUTION: 1. Proj. Mgr. 2. Field Office 3. File

For Office Use Only
Pd. on est. no. ___
Checked by ___

Wiley-Fisk Form 15-5

FIGURE 17.9 Record of Extra Work or Disputed Work Costs. *(Fisk, Edward R.,* Construction Engineer's Complete Handbook of Forms, *1st edition, © 1993. Reprinted by permission of Pearson Education, Inc., Upper Saddle River, NJ.)*

Payment for Mobilization Costs

There seems to be an unjustifiable amount of controversy surrounding the concept of allowing a "mobilization" payment to a contractor. By definition, mobilization costs are those initial expenditures that a contractor is obligated to make even before qualifying for a penny of progress payments. Mobilization costs include the setting up of the field offices; delivering equipment to the job site; obtaining bonds, insurance, and permits; and similar costs. On a heavy construction job the cost can be very large.

Those who naively suggest that it is solely the contractor's problem fail to realize that nothing that the contractor is obligated to do on your project is going to be provided to the owner in the form of a donation. Failure to provide a line item for payment of mobilization simply forces the contractor to prorate its cost over all of the earliest items of construction to get as early a return on its investment as possible. Unfortunately, this also means that the bids will be unbalanced to show a disproportionately high unit cost on many early construction items. Then, in case of a quantity overrun on those items, the owner will be paying a higher cost for the project than is necessary.

It is therefore sensible to provide for payment of mobilization to a contractor as a separate line item. But as we all know, any contracting firm worth its salt will take every opportunity to inflate the unit prices of all first items of construction to obtain front money. Then, if the mobilization item is left open as a competitively bid item, a serious imbalance could result. The solution is simple; include a line item for *mobilization* but type in a fixed *allowance* on the bid sheet for this item. Then every bidder will be able to include the fixed amount of this allowance as a part of its bid, but will be unable to inflate it.

The author once experienced a case involving two earth dams, where the owner's representatives were not very sophisticated in construction contract administration. Initially, they opposed the concept of any mobilization payment but were later convinced that failure to do so on a project involving extensive amounts of heavy earthmoving equipment could have a serious effect on the unit prices of the remaining bid items. The author had careful cost estimates made to show the probable real cost to the contractor of mobilization for each of the dams and included $50,000 for the larger structure and $25,000 for the smaller dam. The amounts involved so shocked the owner's representatives that they insisted upon leaving the mobilization amount blank and calling for competitive bids on the mobilization item. Unfortunately, the local contractors who bid the project took advantage of the situation, as one might expect, and the low bidder on the project had listed $100,000 for mobilization for the larger dam and $75,000 for the smaller one. Thus, the retainage normally collected by the owner became a farce. The contractor was well ahead of the game financially as soon as the first payment was received at the end of the first month of operations.

Some other difficulties have been experienced in the payment of mobilization, too, and perhaps that is one of the reasons for the frequent negative reactions to such payments by architect/engineers. This is the fact that by failure to specify what constitutes "mobilization," the architect/engineer is placed in the position of making a full payment for this line item at the end of the month without even being fully aware of whether or not it had been fully earned.

The solution is simple. Itemize everything in the specification that will be defined as mobilization and then either prorate the mobilization payment at the end of the month for failure to complete all of the listed items, or as the author prefers, provide a lump-sum payment for completion of all of the mobilization items, complete, to qualify for payment of the lump-sum amount for the item.

Although the list of legitimate mobilization items will vary widely from job to job, there are certain basic items of interest to the architect/engineer firm that they may well find it to their advantage to list as required mobilization items. For example, the author has often experienced delays in setting up the engineer's field office and telephone service. If these items were included as mobilization items included in a lump-sum quantity of work that must be completed before the contractor can qualify for any of its mobilization payment amount, there would be more of a tendency on the part of the contractor to meet the required schedule.

Figure 17.10 is a sample specification provision for mobilization that is provided as a guide for the development of similar provisions in the specific job specifications of applicable lump-sum projects. On unit-price projects, a separate specification section entitled "Mobilization" is not used; instead, the provisions for mobilization are included in the section entitled "Measurement and Payment." The user is cautioned that the example shown in Figure 17.10 may not be suitable for all types of projects in its present form, but may require editing to fit the requirements of each project. The following list sums up two important differences between lump-sum and unit-price projects.

Lump-Sum Projects

1. Mobilization terms must be in a separate technical section. Mobilization may be computed as a percentage of project cost.
2. Do *not* include a Section for Measurement and Payment; this is inconsistent with the lump-sum payment principle and does not apply to the Schedule of Values.

Unit-Price Projects

1. Mobilization terms are appropriately covered as a part of Measurement and Payment. Mobilization is included for payment as a separate pay line in the Bid Sheet, with an engineer-determined lump-sum allowance shown in the extension column.
2. Do *not* include a section entitled "Mobilization," as all mobilization terms must be covered in the Measurement and Payment section.

On unit-price contracts where mobilization is included as a bid item, instead of including a section entitled "Mobilization," the measurement and payment section is used to specify the terms and conditions under which the initial payment will be

SECTION 01500

MOBILIZATION

1.01 GENERAL

A. Mobilization shall include the obtaining of all bonds, insurance, and permits; moving onto the site of all plant and equipment; and the furnishing and erecting of plants, temporary buildings, and other construction facilities; all as required for the proper performance and completion of the Work. Mobilization shall include but not be limited to the following principal items:

1. Moving on to the site of all Contractor's plant and equipment required for first month's operations.
2. Installing temporary construction power, wiring, and lighting facilities per Section 01510.
3. Establishing fire protection system per Section 01510.
4. Developing and installing construction water supply per Section 01510.
5. Providing field office trailers for the contractor and the architect/engineer, complete with all specified furnishings and utility services including telephones, telephone appurtenances, and copying machines per Section 01550.
6. Providing all on-site communication facilities and potable water facilities as specified per Section 01510
7. Providing on-site sanitary facilities and potable water facilities as specified per Section 01510.
8. Furnishing, installing, and maintaining all storage buildings or sheds required for temporary storage of products, equipment, or materials that have not yet been installed in the Work. All such storage shall meet manufacturer's specified storage requirements, and the specific provisions of the specifications, including temperature and humidity control, if recommended by the manufacturer, and for all security per Section 01600.
9. Arranging for and erection of Contractor's work and storage yard per Section 01550.
10. Obtaining and paying for all required permits.
11. Posting all OSHA required notices and establishment of safety programs.
12. Have the Contractor's superintendent at the job site full time.(See Article 6 of the General Conditions)

FIGURE 17.10 Sample Provisions in CSI Format for Specifying Mobilization in Lump-Sum Contracts.

13. Submittal of detailed Preliminary Construction Schedule for Architect/Engineer's approval within 7 days after execution of the Agreement per Section 01310.

14. Submittal of Initial Construction Schedule, embodying all corrections required by the Engineer, within 25 days of date of Notice to Proceed. No payment for Mobilization can be made until this has been approved and submitted per Section 01310.

15. Planning and submittal of proposed design for a Project Sign in accordance with Section 01300, and fabrication and erection of the approved Project Sign in accordance with the requirements of Section 01580.

16. Submittal of a Preliminary Schedule of Values in accordance with the requirements of Article 14.1 of the General Conditions and Section 01300 of the Technical Specifications within 7 days following the date of Notice to Proceed.

B. In addition to the requirements specified above, all submittals shall conform to the applicable requirements of Section 01300 "Contractor Submittals."

1.02 PAYMENT FOR MOBILIZATION

A. The Contractor's attention is directed to the condition that 10 percent of the total Contract Price will be deducted from any money due the Contractor as initial progress payments until all mobilization items listed above have been completed as specified. The aforementioned amount will be retained by the Owner as the agreed, estimated value of completing all of the mobilization items listed. Any such retention of money for failure to complete all such mobilization items as a lump-sum item shall be in addition to the retention of any payments due to the Contractor as specified in Article 14 of the General Conditions.

B. Payment for mobilization will be made in the form of a single, lump-sum, non-proratable allowance determined by the Architect/Engineer, and no part thereof will be approved for payment under the Contract until all mobilization items listed above have been completed as specified.

- END OF SECTION -

FIGURE 17.10 (continued)

made to the contractor. Figure 17.11 provides an example from a measurement and payment section (the list of 16 required items has been omitted for convenience). In a final specification, the provisions would include the same list as contained in the lump-sum version, except that item 16 is deleted, as there is no requirement for a schedule of values in a unit-price project.

Figure 17.12 illustrates how mobilization is shown on a unit-price bid schedule.

SECTION 01028

MEASUREMENT AND PAYMENT

PART 1 - GENERAL

1.01 SCOPE

A. Payment for the various items of the Bid Sheets, as further specified herein, shall include all compensation to be received by the Contractor for furnishing all tools, equipment, supplies, and manufactured articles, and for all labor, operations, and incidentals appurtenant to the items of work being described, as necessary to complete the various items of work as specified and shown on the drawings, including all appurtenances thereto, and including all costs of compliance with the regulations of public agencies having jurisdiction and the Occupational Safety and Health Administration of the U.S. Department of Labor (OSHA). No separate payment will be made for any item that is not specifically set forth in the Bid Sheet(s), and all costs therefor shall be included in the prices named in the Bid Sheet(s) for the various appurtenant items of work.

PART 2 - PAYMENT SCHEDULE

2.01 BID SCHEDULE

A. All pay line items will be paid for at the unit prices named in the Bid Sheets for the respective items of work. The quantities of work or material stated as unit price items on the Bid Sheets are supplied only to give an indication of the general scope of the Work; the owner does not expressly nor by implication agree that the actual amount of work or material will correspond therewith, and reserves the right after award to increase or decrease the quantity of any unit price item of work by an amount up to and including 25 percent of any bid item, without a change in the unit price, and shall have the right to delete any bid item in its entirety, or to add additional bid items up to and including an aggregate total amount not to exceed 25 percent of the contract price.

B. Quantity variations in excess of the allowable quantity changes specified herein shall be subject to the provisions of Article 10 of the General Conditions.

2.02 MOBILIZATION (Bid Item No. 1)

A. Measurement for payment for mobilization will be based upon completion of such work as a lump sum, non-proratable pay item, and shall require completion of all of the listed items during the first 25 days following Notice to Proceed.

B. Payment for Mobilization will be made at the lump sum allowance named in the Bid Sheets under Item No. 1, which price shall constitute full compensation for all such work. Payment for mobilization will be made in the form of a single, lump-sum,

FIGURE 17.11 Sample Provisions in CSI Format for Specifying Mobilization in Unit-Price Contracts.

non-proratable payment, no part of which will be approved for payment under the Contract until all mobilization items listed herein have been completed as specified. The scope of the work included under Pay Item No. 1 shall include the obtaining of all bonds, insurance, and permits; moving onto the site of all plant and equipment; and the furnishing and erecting of plants, temporary buildings, and other construction facilities; all as required for the proper performance and completion of the Work. Mobilization shall include but not be limited to the following principal items:

C. In addition to the requirements specified above, all submittals shall conform to the applicable requirements of Section 01300 Contractor Submittals.

D. No payment for any of the listed mobilization work items will be made until all of the listed items have been completed to the satisfaction of the Engineer.

E. The aforementioned amount will be retained by the Owner as the agreed, estimated value of completing all of the mobilization items listed. Any such retention of money for failure to complete all such mobilization items as a lump-sum item shall be in addition to the retention of any payments due to the Contractor as specified in Article 14 of the General Conditions. The following is an example of a Mobilization section of the Specifications in a lump sum public works project:

END OF SECTION

FIGURE 17.11 (continued)

BID SCHEDULE

Item No.	Description	Estimated Quantity	Unit	Unit Price	Amount
1.	Initial Mobilization	---	LS	Allowance	$15 000
2.	Sheeting, shoring, and bracing or equivalent method	---	LS	---	$________
3.	Clearing, grubbing, and removals	---	LS	---	$________
4.	Remove existing PCC and AC pavement	1 162	M2	$______	$________

FIGURE 17.12 Example Showing How Initial Mobilization Allowance Would Be Shown on a Unit-Price Bid Schedule.

Partial Payments to the Contractor

Waiver of Lien Procedure

Liens exist throughout the United States, but they are not uniform from state to state. As a means of providing some measure of owner protection in the enforcement of contractor and supplier lien rights, many states have enacted preliminary notice requirements (see "Preliminary Notice of Potential Claim" in Chapter 20).

Although it is common to require only a monthly pay estimate from the contractor as a prerequisite to payment, without any specific assurance that the subcontractors and suppliers have been paid for their work, the practice of requiring partial waivers of lien from the prime contractor and all of its subcontractors and suppliers prior to release of partial payments seems to be gaining ground (Figures 17.13 and 17.14). An attorney should be consulted to ascertain how the procedure may be implemented in each state.

The practice of requiring partial waivers of lien from the prime contractor and each of the subcontractors and suppliers based upon current payment requests, however, imposes harsh penalties upon each of those parties by requiring the prime contractor to pay its subcontractors and suppliers prior to having been paid by the owner, or by requiring the subcontractors and suppliers to waive their lien rights before being paid by the prime contractor, for the prime contractor to receive payment from the owner. The capital required by the prime contractor to pay its subcontractors and suppliers prior to obtaining waivers is quite substantial, and such a practice is unfair to the prime contractor. As a result, the subcontractors and suppliers are frequently pressured into releasing their waivers of lien to the prime contractor without having received payment for the work covered by that specific release and without further assurance that they will be paid when the prime contractor receives payment from the owner.

The Construction Industry Affairs Committee of Chicago studied the problem and offered the following recommendations. The first payment request should be accompanied by the prime contractor's partial waiver of lien only. Each subsequent partial payment should then be accompanied by the prime contractor's partial waiver and the partial waivers of all subcontractors and suppliers who were included in the *immediately preceding payment request,* to the extent of that payment. Thus, the prime contractor must submit waivers on a current basis, but the subcontractors and suppliers may not be more than one payment late with their partial waivers. Request for final payment would then be accompanied by the final waivers from the prime contractor, the subcontractors, and the suppliers who had not previously furnished such waivers.

Under this procedure, the owner is afforded considerable protection. The owner will not be asked to make final payment until all necessary waivers have been received; the prime contractor will not be asked to make advance payments to secure waivers; and the subcontractors and suppliers will not be placed in a position of surrendering their legal rights without proper consideration.

CONDITIONAL WAIVER
AND RELEASE UPON PROGRESS PAYMENT
[California Civil Code §3262(d)1 amended 1993]

Upon receipt by the undersigned of a check from ______________________________
(Maker of Check)
in the sum of $________________ payable to ______________________________
(Amount of Check) (Payee or Payees of Check)
and when the check has been properly endorsed and has been paid by the bank upon which it is drawn , this document shall become effective to release any mechanic's lien, stop notice, or bond right the undersigned has on the job of

__________________________ located at ______________________________
(Owner) (Job Description)
to the following extent. This release covers a progress payment for labor, services, equipment, or material furnished to

__________________________ through ______________________________
(Your Customer) (Date)

only and does not cover any retentions retained before or after the release date; extras furnished before the release date for which payment has not been received; extras or items furnished after the release date. Rights based upon work performed or items furnished under written change order which has been fully executed by the parties prior to the release date are covered by this release unless specifically reserved by the claimant in this release. This release of any mechanic's lien, stop notice, or bond right shall not otherwise affect the contract rights, including rights between parties to the contract based upon a rescission, abandonment, or breach of the contract, or the right of the undersigned to recover compensation for furnished labor, services, equipment, or material covered by this release if that furnished labor, services, equipment, or material was not compensated by the progress payment. Before any recipient of the document relies on it, said party should verify evidence of payment to the undersigned.

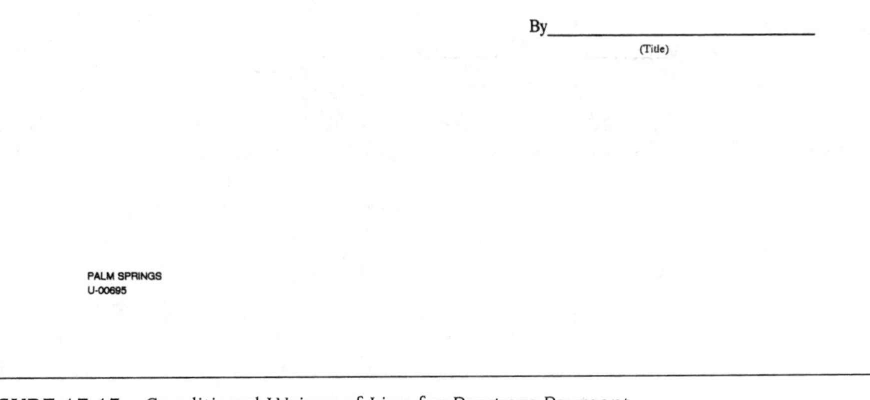

Dated__________________________ ______________________________
(Company Name)

By______________________________
(Title)

PALM SPRINGS
U-00695

FIGURE 17.13 Conditional Waiver of Lien for Progress Payment.

UNCONDITIONAL WAIVER AND RELEASE UPON PROGRESS PAYMENT

[California Civil Code §3262(d)2 amended 1993]

The undersigned has been paid and has received a progress payment in the sum of:

$ 1,275.25 for labor, services, equipment, and material Furnished to:

XYZ CONSTRUCTORS, INC.
(Your customer)

on the job of CITY OF PALM SPRINGS
(Owner)

located at RAMON ROAD
(Job Description)

and does hereby waive and release any mechanic's lien, stop notice, or bond right that the undersigned has on the above referenced job to the following extent. This release covers a progress payment for labor, services, equipment, or material furnished to:

XYZ CONSTRUCTORS, INC. through 25 JUNE 1999

only and does not cover any retentions retained before or after the release date; extras furnished before the release date for which payment has not been received; extras or items furnished after the release date. Rights based upon work performed or items furnished under a written change order which has been fully executed by the parties prior to the release date are covered by this release unless specifically reserved by the claimant in this release. This release of any mechanic's lien, stop notice, or bond right shall not otherwise affect the contract rights, including rights between parties to the contract based upon a rescission, abandonment, or breach of the contract, or the right of the undersigned to recover compensation for furnished labor, services, equipment, or material covered by this release if that furnished labor, services, equipment, or material was not compensated by the progress payment.

Dated 2 JULY 1999 ABC PAVING CONTRACTORS, INC.
(Company Name)

By [signature]
(Title)

Data used are fictitious for illustration only

NOTICE TO PERSONS SIGNING THIS WAIVER: THIS DOCUMENT WAIVES RIGHTS UNCONDITIONALLY AND STATES THAT YOU HAVE BEEN PAID FOR GIVING UP THOSE RIGHTS. THIS DOCUMENT IN ENFORCEABLE AGAINST YOU IF YOU SIGN IT, EVEN IF YOU HAVE NOT BEEN PAID. IF YOU HAVE NOT BEEN PAID, USE A CONDITIONAL RELEASE FORM.

FIGURE 17.14 Unconditional Waiver of Lien for Progress Payment.

Copies of sample forms for partial waiver of lien are shown in Figures 17.13 and 17.14. An attorney should be consulted before use, as lien laws vary from state to state.

Total Cost Pricing of Change Orders

Contractors are sometimes allowed to use the so-called "total cost method" for pricing change orders or delayed work. This approach compares the bid price of the work as bid with the actual cost of doing the work, and charges the increase to the project owner. It is permitted in situations where (1) the nature of the claim makes it very difficult to establish a price within a reasonable degree of accuracy, (2) the contractor's bid price was realistic, (3) the contractor's actual costs were realistic, (4) the contractor's actual costs were reasonable, and (5) the contractor was not responsible for any of the increased costs.

The use of the total cost method reduces the contractor's obligation to link specific costs to specific causes, but it in no way eliminates the need for accurate record keeping. Without good job records, it is impossible to judge how reasonable the contractor's alleged actual costs are or to compare those costs with the original bid. If the cost records are inadequate or nonexistent, a court will not allow the contractor to use the total cost method for pricing its claim [*Appeal of J.D. Abrams, Inc.,* ENG BCA No. 4332 (November 28, 1988)].

Forward Pricing of Change Orders

Project owners sometimes ask contractors to "forward-price" change order work, that is, establish a mutually agreed lump-sum price increase for doing the changed work. The advantage for contractors is that they are relieved of the need to maintain detailed, segregated cost records. The disadvantage is that a misjudgment may prove costly. If unanticipated factors, such as delay and schedule impact, drive the costs up, the contractor will still be limited to the agreed-upon fixed-price increase. In such a case, the contractor may wish it had gone through the trouble of tracking and documenting actual costs.

It should be emphasized that a contractor cannot be forced to forward-price change order work. A contractor's refusal to submit to a nonbinding change order estimate or cost proposal may be considered as a breach of contract [*Palmer & Sicard, Inc., v. United States,* 6Cl.Ct.232(1934)]. A contractor cannot be forced to sign off on a forward-priced change order [*Appeal of Centex Construction Co., Inc.,* ASBCA No. 26830 (April 29, 1983)].

Retainage

The majority of construction projects, particularly public works projects, involve some retention of a portion of earned funds of the contractor. Typically, this has been 10 percent, although 5 percent is not uncommon. In some cases, the retained funds are held throughout the life of the job; in other cases, the amount may be reduced or actually eliminated at some point after the midpoint of the project.

The purpose of the retention of funds is to provide an amount of money that can be available to the owner for the satisfaction of lien claims because of the failure of the prime contractor to pay its subcontractors or suppliers, and as a means of holding the profit that the contractor makes on a given project. In case of default by the contractor, these funds may be utilized by the surety to complete the work.

There is some dispute over the validity of the concept involved, as a project that is protected by payment and performance bonds is already secured against such claims. Because of this, there has been some consideration in recent years to reduce or even eliminate retainage after a project is half over (Figure 17.15). One contractor told me, however, that although a reduction in retainage was welcome, its elimination was not. Otherwise, the argument was that there would be no leverage for withholding a similar amount from the subcontractors.

The amount withheld may, after over a year on a large project, accumulate to a rather large sum. Thus, there may well be some validity in the reduction concept. It should be kept in mind, however, if a project is covered by performance and payment bonds, the permission of the surety company is essential prior to the release or even the reduction of the retention amount. Convenient forms for this purpose are provided, as shown in Figure 17.16.

There are two methods of computing retainage. By one method, the amount held at the end of the project is equal to, for example, 10 percent of the entire project cost. On a recent project that the author was involved with, that amounted to over $250,000. Another method allows the retainage to be calculated as a percentage of the current month's billing. In this manner, the owner never holds more than perhaps 10 percent of any month's earnings. In most construction contracts it is the former method that is used.

Both the U.S. government and the American Institute of Architects have approved a procedure recognizing the validity of the concept that the holding of 10 percent during the first half of the work is adequate to assure completion and protection against claims, and that no further retainage is needed. Thus, the owner will be holding a retainage amounting to 5 percent of the entire project cost at the end of the project. Another version of this concept involves the retention of 10 percent retainage until the halfway point, then a refund to the contractor of all but 5 percent. From that point on, the owner continues to withhold 5 percent until the end of the job. This policy applies only if in the judgment of the owner's agent (architect or engineer), performance has been proper and satisfactory progress is being made in the work.

A similar provision should be made for payments to subcontractors by a prime contractor. It is a good idea to furnish to a subcontractor, upon request, information on all payments approved and paid to the prime contractor by the owner as it relates to the work of that subcontractor.

Finally, approximately 35 days after the completion of the work, if all lien releases are in and all work completed, and all subcontractors, suppliers, and employees have been paid, obtain the approval of the surety and then release the retention money except for any amounts representing the value of uncompleted or substandard work or materials.

REQUEST FOR ADJUSTMENT OF RETAINAGE

PROJECT TITLE Emerson Pumping Plant PROJECT NO. 89/90-00

OWNER City of Thousand Oaks CONTRACT NO. ______

PROJECT MANAGER R. E. Barnes

CONTRACTOR XYZ Constructors, Inc., Thousand Oaks, CA

The Contractor, XYZ Constructors, Inc., hereby requests that the percentage of partial payment estimates retained by the Owner under the provision of the contract be REDUCED to 50 % / ~~DISCONTINUED~~.

by [signature] Date 8-10-90
Contractor Representative

The Surety on the Performance Bond for said project ABC Indemnity Company, Inc., Los Angeles, CA hereby approves the foregoing request.

Power of Attorney must be attached to original copy

by [signature] Date 8-14-90
Attorney-in-Fact

Approval ~~IS~~ / IS NOT recommended. The percentage of completion as of 18 July 1990 is 40.4 % and the present percentage of elapsed time as of 18 July 1990 is 56.3 %

by [signature] Date 8-15-90
Resident Project Representative

Approval ~~IS~~ / IS NOT recommended;

by [signature] Date 20 Aug 1990
Project Manager

~~APPROVED~~ / DISAPPROVED

by C. D. Collins Date 8-24-90
Owner's Representative

Distribution: 1. Proj Mgr.
2. Field Ofc
3. File
4. Owner

Wiley-Fisk Form 14-4

FIGURE 17.15 Example of a Request for Adjustment of Retainage. *(Fisk, Edward R.,* Construction Engineer's Complete Handbook of Forms, *1st edition, © 1993. Reprinted by permission of Pearson Education, Inc., Upper Saddle River, NJ.)*

City of Thousand Oaks

CONSENT OF SURETY
For Reduction of or
Partial Release of Retainage

Project Name Moorpark Road Improvements

Location Between Thousand Oaks Blvd to Janss Rd

Project No. 89/90-00 Contract No. ____

Type of Contract Lump Sum

Amount of Contract $12,345,570.00

In accordance with the provisions of the above-named contract between the Owner and the Contractor, the following named surety:

Guaranteed Indemnity Corp., Los Angeles, CA

on the Bond of the following named Contractor:

XYZ Constructors, Inc., Thousand Oaks, CA

hereby approves a reduction of or partial release of retainage to the Contractor as set forth below:

Continued retention of all monies withheld through 24 Aug 1990. Reduction of future retention funds to 5 percent monthly for all monies earned subsequent to 24 Aug 1990.

The Surety Company hereby agrees that such reduction of or partial release of retainage to the Contractor shall not relieve said surety company of any of its obligations to the following named Owner as set forth in said Surety Company's bond:

City of Thousand Oaks, a legal entity organized and existing in the County of Ventura, State of California.

IN WITNESS WHEREOF, the Surety Company has hereunto set its hand and seal this 12 day of July 1990

Guaranteed Indemnity Corp.
(Name of Surety Company)

(Signature of Authorized Representative)

TITLE Vice President

(Affix corporate seal here)

Wiley-Fisk Form 17-4

FIGURE 17.16 Consent of Surety for Reduction or Partial Release of Retainage Funds by the Owner. *(Fisk, Edward R.,* Construction Engineer's Complete Handbook of Forms, *1st edition, © 1993. Reprinted by permission of Pearson Education, Inc., Upper Saddle River, NJ.)*

Liquidated Damages During Construction

Traditionally, the assessment of liquidated damages against a contractor for monetary losses suffered by the owner as the result of a contractor's failure to meet the contractual completion date has been computed only for the excess construction time beyond the final specified completion date. Recently, there have been successful actions to collect liquidated damages for failure of a contractor to meet various specified key dates for completion of certain specified portions of the work affecting the ability of other prime contractors to deliver their portion of the work on time. The effect of delaying other prime contractors can result in added charges for owner-caused delays and in rendering the specified completion dates of the other affected contractors as unenforceable.

Some owners have successfully specified and collected liquidated damages from some of their prime contractors for failure to meet such key dates, and the amount of the liquidated damages was simply deducted from the amount of the progress payment owed the contractor for work done that month. Where the amount of the liquidated damages exceeded the amount of the earned progress payment, it was simply charged as a debit against the contractor's account.

Standard Contract Provisions for Measurement and Payment

Although each set of contract documents treats this subject slightly differently, the provisions of the project documents must be the final determining factor in establishing the method and procedure for handling contractor payment requests. The following references are provided as an aid to comparing the provisions of several General Conditions documents in current use:

1. *EJCDC:* Articles 14.1 through 14.13 of the General Conditions (1996 ed.)
2. *AIA:* Articles 9.1 through 9.10.4 of the General Conditions (1987 ed.)
3. *GSA:* Article 7 of the General Provisions (federal contracts)
4. *FIDIC:* Articles 60.1 through 60.10 of the Conditions of the Contract for Works of Civil Engineering Construction (1987 ed.)

The foregoing are typical of most stock forms of General Conditions, or General Provisions as they are sometimes called.

Similar provisions are made in the various standard specifications, as these documents generally include all general provisions as well as technical ones. The principal difference is that the majority of the standard specifications also contain provisions for measurement methods to be used in determining the payment quantities—an excellent way of minimizing field disputes resulting from disagreements over the method of measurement, which can result in significant differences. The following references are provided as an aid in comparing the measurement and payment provisions of some typical standard specifications:

1. American Public Works Association/Associated General Contractors Standard Specifications for Public Works Construction ("Greenbook")
   Measurement of Quantities for Unit-Price Work: Section 9-1
   Measurement of Quantities for Lump Contracts: Section 9-2
   Payment: Section 9-3
   Payment for Extra Work (Force Account): Section 3-3.2
2. California Department of Transportation Standard Specifications
   Measurement of Quantities: Section 9-1.01
   Payment: Sections 9-1.02 through 9-1.09
   Force Account Payment: Section 9-1.03
3. Florida Department of Transportation Standard Specifications for Road and Bridge Construction
   Measurement of Quantities: Section 9-1
   Payment: Sections 9-2 through 9-10
   Force Account Work: Section 9-4
4. Washington State Department of Transportation, Standard Specifications for Road, Bridge, and Municipal Construction
   Measurement of Quantities: Section 1-09.1
   Scope of Payment: Section 1-09.3
   Force Account: Section 1-09.6
5. U.S. Department of Transportation, Federal Highway Administration, Standard Specifications for Construction of Roads and Bridges on Federal Highway Projects
   Measurement of Quantities: Section 109.01
   Scope of Payment: Section 109.08
   Force Account: Section 109.06

INTERPRETING THE CONTRACTOR'S BID

Interpretation of Bidding Errors

If the final-measured amount of earth excavation shown in Figure 17.17 was determined to be 40,000 m^3 (52,318 cubic yards) instead of the 39,757 m^3 (52,000 cubic yards) estimated by the engineer, the actual payment to the contractor for completing that item would be 40,000 $m^3 \times 3.60$ (52,318 cy $\times$ 2.75) and he or she would be paid $144,000 instead of the $143,125.20 shown in the totals column of its bid sheet. If the contractor knew in advance that the engineer's estimate was incorrect on this item, it might be worthwhile to place a higher unit price on the item, thus assuring higher profits without placing the firm in a noncompetitive position against the other bidders (see the following section, "Unbalanced Bids").

Often, errors are made in bids submitted and a determination must be made as to the true value of a bid. One common error involves a multiplication of the unit

Item No.	Description	Quantity and Unit	Unit Price	Amount
BALANCED BID BASED UPON COMPETITIVE UNIT PRICES				
1.	Site Preparation	L.S.		$ 17 500.00
2.	Temporary Sheeting	1 133 m^2	$ 0.11	$ 124.63
3.	Permanent Sheeting	6 661 m^2	$ 19.50	$ 129 889.50
4.	Earth Excavation	39 757 m^3	$ 3.60	$ 143 125.20
5.	Rock Excavation	6 881 m^3	$ 21.25	$ 146 221.25
6.	Reinforcing Steel	393 Tonne	$ 495.80	$ 194 849.40
7.	Conduit Bedding	1 009 m^3	$ 17.66	$ 17 818.94
8.	Concrete Conduit	3 884 m^3	$ 170.03	$ 660 396.52
			TOTAL $	1 309 925.44

FIGURE 17.17 Balanced Bid Based upon Competitive Unit Prices.

Item No.	Description	Quantity and Unit	Unit Price	Amount
8.	Concrete Conduit	5 080 m^3	$ 113.20	$ 175 056.00

FIGURE 17.18 Example of an Error in Bid Computation.

price and the estimated quantity. In the example in Figure 17.18 it can be seen that the proper product of the unit price times the quantity should be $575,056 instead of the $175,056 shown.

In some cases, contractors have been allowed to withdraw a bid after opening where a clerical error existed; however, in an apparent equal number of such cases, such withdrawal has been successfully refused. Recent legislative action in some states has made it possible for any contractor to withdraw such an erroneous bid.

Similarly, where a bid sheet asks for the amount in "words and figures," the *words* will govern over the figures, as there is less chance of error this way. This type of error is likely in either a lump-sum or a unit-price bid amount. If the amount of the bid read

> Four thousand two hundred and forty-seven dollars ($40,247.00)

the final amount of the bid would be held as $4,247.00, not $40,247.00.

Unbalanced Bids

For unit-price contracts, a balanced bid is one in which each bid item is priced to carry its share of the cost of the work and also its share of the contractor's profit. Occasionally, contractors will raise prices on certain items and make corresponding reductions of the prices on other items without changing the total amount of the bid for the project. The result is an unbalanced bid. In general, extremely unbalanced bids are considered as undesirable and should not be permitted when detected, although the practice sometimes seems justified from the contractor's viewpoint. Some of the purposes of unbalancing bidding are as follows:

1. To discourage certain types of construction and encourage others that may be more favorable to the contractor.
2. When the contractor believes that the engineer's estimate for certain items is low, by unbalancing the bid in favor of such items, an increased (unearned) profit in the actual payment of the work can be obtained without increasing the apparent total amount of the bid.
3. Unreliable contractors may increase their bid prices for the first items of work to be completed, with corresponding reductions elsewhere in the bid, with the intention of receiving excessive early payments, then defaulting on the contract. This could leave the surety to complete the contract with insufficient funds remaining in the contract.
4. By unbalancing the bid in favor of items that will be completed early in the progress of the work, a contractor can build up its working capital (front money) for the remainder of the work. This can also serve to eliminate the financial squeeze caused by the usual 10 percent retention money. This is a fairly common practice.

Of all of the reasons mentioned for unbalancing a bid, only the last item seems to have some justification when dealing with reliable contractors. The expenses of mobilizing the construction plant, bringing the equipment and materials to the site, and the general costs of getting the work started are significant. These items often do not appear on the bid as separate bid items, and therefore are paid for only by adding their cost to the items actually listed. This usually means, however, that they would be paid for only as the work progresses, even though the actual cost to the contractor was all incurred at the beginning of the job. This can cause a hardship on the contractor in that the working capital would be tied up in the work to the sole advantage of the owner.

The prevention of unbalanced bids requires a knowledge of construction costs in the project area so that unreasonable bids on individual items may be detected. An obvious case of unbalanced bidding should be considered as grounds for rejecting the entire bid.[2]

[2]R. W. Abbett, *Engineering Contracts and Specifications* (New York: Wiley, 1963).

Example of an Unbalanced Bid

Initially, a contractor must figure its bid normally without unbalancing it to produce a competitive price (Figure 17.17). The price thus determined will become the bottom-line price for the future unbalanced bid. The following example is based upon a partial list of bid items from the low bid for a reinforced box culvert bid in Connecticut in 1976. All original bid items and the quantities are furnished in the bidding documents by the engineer, and the unit prices and extensions are all filled in by the bidder.

As an example, assume that a bidder, upon carefully studying the plans, discovers that the engineer has made an error in the quantities shown for temporary and permanent trench sheeting. Instead of 1,133 m^2 (12,200 sq ft) of temporary sheeting and 6,661 m^2 (71,700 sq ft) of permanent sheeting as the engineer's estimate indicates, totaling 7,794 m^2 (83,900 sq ft) of trench sheeting all together, let us assume that the bidder has discovered that although the total of 7,794 m^2 (83,900 sq ft) of sheeting is correct, the engineer has the individual amounts wrong. According to the bidder's estimate, there are actually 6,661 m^2 (71,700 sq ft) of temporary sheeting and only 1,133 m^2 (12,196 sq ft) of permanent sheeting. Now, although the price of the temporary sheeting is low because of the contractor's ability to reuse the material, the opportunities presented by the knowledge that the quantities are in error might suggest to the bidder that if its price on this item is high enough, the added quantity over that indicated in the engineer's estimate could be a financial windfall. The bidder, of course, also knows that if that bid item alone is raised, the total bid price will be too high and it may not get the job. Therefore, the prices of certain other items must be reduced to compensate for the raise in the unit price for temporary sheeting. When completed, the new bid must have the same bottom-line total as the balanced bid.

In the unbalanced bid in Figure 17.19 the bidder has raised the unit prices of bid items 2 and 3, and compensated by lowering the unit prices for bid items 6 and 8. At the same time the bidder has increased the amount of money bid for site preparation because it will be the first item completed. In this way it could provide the bidder with additional working capital (front money). Note that the total bid price (bottom line) has remained the same as it was in the original balanced bid in Figure 17.17.

By holding the original competitive bid price, the bidder assures itself of a fair chance of being awarded the job. Then, if this bidder gets the job, the payments to that firm as contractor will be based upon the *actual* quantities of each item of the bid sheet completed (Figure 17.20). Thus, the high bid price on the quantity that the engineer showed as low will yield high unearned profits, which are not reduced significantly by the redistribution of the other bid prices in the unbalanced bid.

A quick comparison of the amount of the contract price that the contractor would have received if the quantities estimated by the engineer were correct with the amount that would actually be claimed by the contractor is shown in the following:

\$1,412,470.93	Amount claimed by the contractor for actual quantities
\$1,309,925.44	Original bid price for the project (based upon engineer's quantities)
102,545.49	Additional unearned profit due to false unit prices

UNBALANCED BID BASED UPON KNOWN QUANTITY ERRORS IN ENGINEER'S ESTIMATE

Item No.	Description	Quantity and Unit	Unit Price	Amount
1.	Site Preparation	L.S.		$ 23 925.63*
2.	Temporary Sheeting*	1 133 m^2	$ 45.12	$ 51 120.95*
3.	Permanent Sheeting*	6 661 m^2	$ 27.23	$ 181 379.03*
4.	Earth Excavation	39 757 m^3	$ 3.60	$ 143 125.20
5.	Rock Excavation	6 881 m^3	$ 21.25	$ 146 221.25
6.	Reinforcing Steel	393 Tonne	$ 435.80	$ 171 269.40
7.	Conduit Bedding	1 009 m^3	$ 17.66	$ 17 818.94
8.	Concrete Conduit	3 884 m^3	$ 148.06	$ 575 065.04
			TOTAL	$ 1 309 925.44

(* Denotes unbalanced bid items)

FIGURE 17.19 Unbalanced Bid Based upon Known Quantity Errors in the Engineer's Estimate.

ACTUAL PAYMENTS TO THE CONTRACTOR

Item No.	Description	Quantity and Unit	Unit Price	Amount
1.	Site Preparation	L.S.		$ 23 925.63
2.	Temporary Sheeting*	6 865 m^2	$ 45.12	$ 309 748.80
3.	Permanent Sheeting*	929 m^2	$ 27.23	$ 25 296.67
4.	Earth Excavation	39 757 m^3	$ 3.60	$ 143 125.20
5.	Rock Excavation	6 881 m^3	$ 21.25	$ 146 221.25
6.	Reinforcing Steel	393 Tonne	$ 435.80	$ 171 269.40
7.	Conduit Bedding	1 009 m^3	$ 17.66	$ 17 818.94
8.	Concrete Conduit	3 884 m^3	$ 148.06	$ 575 065.04
			TOTAL	$ 1 412 470.93

FIGURE 17.20 Actual Payments to the Contractor.

Note that the amount of increase in the price bid for site preparation did not alter the final payment amount; it only provided early money for the contractor to use in its operations.

Detection of an Unbalanced Bid

There are several ways of detecting an unbalanced bid, depending upon the type of imbalance involved. Using the illustrated example in Figures 17.17, 17.19, and 17.20, however, it is relatively easy to detect a flaw. Generally, an unbalanced bid such as that in the illustration is an indication that the architect/engineer made an error in the preparation of the plans and specifications.

Typically, a bid such as that illustrated stands out when compared with the bids of all other bidders on the same line items. For example, for line items 2 and 3 for Temporary Sheeting and Permanent Sheeting, respectively, the bidder quoted prices of $45.12 and $27.23 per square meter, respectively. While that may not alarm the architect/engineer in that context, if compared with the same items as quoted by all other bidders, the flaw really stands out. For example, on temporary sheeting, if there were five bidders, the bids might be $0.50, $0.10, $1.10, $45.12, and $0.70 per square meter for the same line item. If the architect/engineer tabulates all bids on a Summary of Proposals Received form (Figure 12.7), it will immediately become evident that one contractor's bid appears to be unusually high on this item simply by reading horizontally across the page. Similarly, by comparing the bids for permanent sheeting a similar discrepancy may be noted. Those lines in the form might look like the following:

No.	Item	Bids for Line Items 2 and 3				
		Bidder A	Bidder B	Bidder C	Bidder D	Bidder E
2	Temporary sheeting	$0.50	$ 0.10	$ 1.10	$ 45.12	$ 6.70
3	Permanent sheeting	$13.50	$18.10	$14.50	$27.23	$19.90

By comparing the prices bid for each individual line item on the bid summary (spreadsheet), it quickly becomes evident that one particular bidder appears to have discovered a discrepancy in the documents. This discovery can be put to good use by the architect/engineer, as the bidder has not only made the architect/engineer aware of a probable error in the plans and specifications, but has even told the architect/engineer exactly where the error lies.

The solution depends upon when the discovery was made by the architect/engineer. If discovered prior to award of the contract, the wisest move is to reject all bids, search out and correct the error in the documents, and readvertise the job. If discovered after a contract was let, the only solution is the execution of a change order.

On private work, it is simple merely to reject the bid tendered by the contractor who submitted the unbalanced bid. On public works, however, the task is a little more difficult. As proving the existence of an imbalance is difficult, and because of the necessity of accepting the lowest dollar bidder, the better way is to reject all bids, correct the mistake, and readvertise.

Resolving the Problem of an Unbalanced Bid

It is not enough to simply detect the problem of an unbalanced bid. You must decide what to do next. In the private sector it may not pose a serious problem, as you may simply reject the unbalanced bid, whether or not it is the low bidder. On public works, however, it is not that simple. Unless you can show probable evidence of attempted fraud, you may lack authority to reject a low bidder.

The author has long advocated a procedure for use in the public sector to handle a problem such as this. Upon observation of an obviously unbalanced bid that is sufficiently extreme as to suggest the existence of a problem such as the one described in this chapter, you should simply exercise the owner's right to reject all bids. Then, take the matter to the engineer, point out the area of suspicion, and tell the engineer that you suspect the possibility of attempted fraud in the pricing of a particular bid item. Tell the engineer that the bidder has shown exactly where the problem lies, and that you want the engineer to restudy the affected plans and specifications, find the area in question, and make corrections in the documents that will eliminate the loophole that the bidder has apparently discovered. Upon correcting the documents, readvertise the project and you are back to normal . . . except, of course, for the delay involved in finding and making the correction and readvertising the project.

The author was involved in a situation closely paralleling this, involving AC paving removal on a city project. Examination of the bids showed an exceptionally high unit price for AC paving removal. It was suspected that the bidder had discovered a discrepancy between the documents and the actual conditions at the site. In anticipation of a high quantity overrun for the paving removal, the bidder overpriced the pay line item for AC pavement removal. Upon completing a comparative bid review, I advised the city to reject all bids, correct the problem, and readvertise.

Unfortunately, the city failed to heed my advice. They held the original bids while they searched for a discrepancy. Upon receiving the engineer's denial that an error or omission existed in the plans and specifications, award was made to the low bidder. The result was predictable. An exceptionally high overrun did exist, and even by falling back on the traditional 25 percent guarantee provision allowing renegotiation of a unit price when the engineer's estimate for a quantity was exceeded by 25 percent or more, the city was unable to negotiate a reasonable settlement, resulting in a considerable financial loss to the city. The city next considered legal action against the engineering firm that designed the project for denying that an error existed and refusal to correct their documents even after having the location of the error pointed out to them. All of this could have been avoided by following some simple procedural steps.

The following guidelines are recommended in case you suspect the existence of a serious imbalance in the low bidder's proposal. This is especially important on public works projects:

1. Examine the bid spread for evidence of a comparatively high unit price.
2. Note that the bidder may not only have attempted to capitalize on an apparent engineer's error, but also he or she has shown you exactly where the problem lies.

3. Point out the location of the apparent discrepancy to the design engineer and request a review and, if justified, correction of the problem.
4. Have the problem corrected in the documents.
5. Readvertise the project.

Measurement for Payment

At first a person may be tempted to think, "What is so difficult about field measurements?" The actual problem is neither with the techniques nor with the accuracy of the measurements as taken, but rather the fact that certain types of measurements may not be representative of the true pay quantities. An example might be the determination of the amount of pipe to be paid for under the construction contract. Often, the contract documents may specify that payment will be made based on the lengths *indicated on the drawings.* It should be noted that the lengths on the drawing are generally shown in *stations,* 100-meter (328.08-ft) increments. Instead of the familiar 100-ft stations still widely used in the United States, conversion to SI metric will utilize 100-meter (328.08-ft) stations. Under SI a station value will look something like 2180.374. Because 100 meters is such a large distance, station tick marks will be at 20-m intervals with annotation at 20-m intervals. The first number in the Station value is the Station number (in 100 m), the next two digits are in meters, and the last three digits are in millimeters. The reason that measurements are to the millimeter is to retain the degree of accuracy indicated by measurements to the nearest 100th of a foot.

Now, lengths shown in stations are always *horizontal dimensions.* Thus, they are not representative of the actual lengths of pipe furnished by the contractor! If the pipe is laid in trenches that have a steep profile grade, the contractor must furnish a longer pipe than will show in the plan view, and its bid price must reflect this difference. An inspector who determines the amount of pipe to be paid for on this type of contract by measuring the actual lengths of pipe laid in trenches may be approving an overpayment.

Determination of Pay Quantities for Pipelines

The example shown in Figure 17.21 is an actual case involving an overflow pipeline from the surge chamber of a hydroelectric powerhouse project. In plan it can be seen that the pipeline is comprised of three reaches of pipe, each at a different slope in a trench on a hillside. The length of the pipe in plan, as determined from the indicated stations, is 175.87 m (577.00 ft) less the length of the upper structure. However, all lengths in stations are horizontal measurements only; thus, the true length is a calculated one. In the example shown, the lower reach of 1200-mm (48-in) diameter pipe has a slope of S 5 0.384 (same as the tangent of the angle measured off the horizontal). The true length of this pipe *in place* is computed as follows:

End station	5 + 77
Upper station	3 + 20
	2 + 57 stations = 257.00 ft (78.3 m) horizontal length

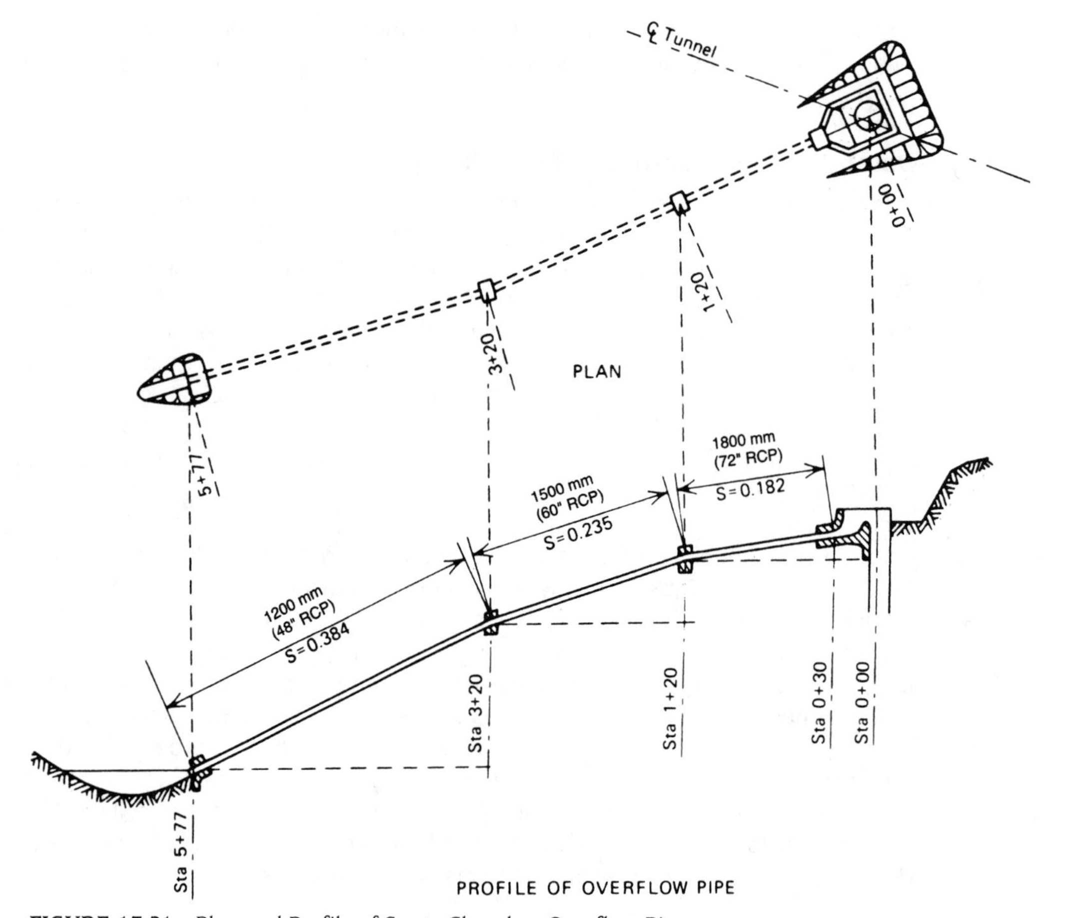

FIGURE 17.21 Plan and Profile of Surge Chamber Overflow Pipe.

For $S = 0.384$: vertical slope angle = 21.00678943°

$$\text{True length} = \frac{257}{\text{Cos } 21.00678943°} = 275.30 \text{ ft } (83.9 \text{ m})$$

18.3 ft (5.58 m) longer than indicated in plan view

By computing each reach of the entire length of the pipeline in the same manner, it is found that the actual length of the pipeline *in place* is 602.72 ft (183.71 m) minus the length of the upper structure, or 572.22 ft (174.41 m). This is a total of 25.22 ft (7.69 m) more pipe than indicated from the horizontal dimensions as determined from the plan.

The emphasis on the "in-place" length of a pipe is based upon the fact that the actual delivered lengths of pipe will add up to even more than the indicated 572.22 ft (174.41 m), to allow for fitting the spigot ends into the bell ends of each length of pipe. Thus, the inspector who measures delivered lengths of pipe for payment runs a serious risk of overpayment unless the specifications specifically call for payment to be made on the basis of lengths of pipe as delivered instead of laying length (in-place dimension).

Determination of Earth and Rock Pay Quantities

It is appropriate to warn of the pitfalls involved in the determination of the quantities of earth or rock excavation or backfill. This is one of the most likely areas for miscalculation. In determining the amounts of excavation and embankment material in construction, an allowance must be made for the difference in space (volume) occupied by the material before excavation and the same material after it has been compacted in embankment. The various earth materials will be more compact in embankment (will occupy less space than they did in their original state), and rock will be less compact (will occupy more space than it did in its original state). This difference in volume between the same material as excavated and as replaced in fill is called *shrinkage* in the case of earth materials, and *increase* or *swell* in the case of rock materials.

The amount of shrinkage depends upon the kind of material and the method of placing it in the fill. Thus, a borrow area needed to provide 19 000 m^3 (24 850 cu. yd.) of fill material may have to be capable of yielding 20 600 m^3 (26 944 cu. yd.) of borrow material. As can be seen, the Resident Project Representative who is on a unit-price job must be very careful to measure the material in strict accordance with the measurement for payment instructions of the job specifications. Obviously, a contractor who is excavating borrow material to be used as fill on its project would much rather be paid for the volume of earth embankment material hauled.

The actual percentages can be determined only by a qualified soils engineer; however, the following percentages are from the average of general experience. They express shrinkage in volume of several classes of materials:

Fine sand	6%
Sand and gravel	8%
Ordinary clay	10%
Loam	12%
Surface soil	15%

If an understanding is reached in the contract to begin with, it does not matter which types of measurements are agreed upon. Thus, the contractor would be able to structure its bid accordingly. But watch for the contractor who wants to change methods of measurement later. Above all, the resident inspector should not yield to any pressure to count scraper or truck loads based upon loose volume!

Another method occasionally suggested by the contractor is that of paying for truckloads of loose haul material by weight. This sounds acceptable at first, but with earth materials this is an extremely unreliable method. The principal problem lies in the variations of moisture content of the material being weighed. If the contractor uses its water truck extensively during excavation and loading into truck from the borrow area, which might be necessary to loosen some materials, the added weight of the damp or wet materials will be paid for as earth material, when in fact the added weight was actually due to the moisture content of the material. Although the contractor must water the material to achieve proper compaction when he or she is constructing earth embankments, this is usually done at the point of deposit. If, instead, water is added before hauling, on a job where haul materials are paid for by weight, the contractor can realize a tidy profit on the deal. The other way around, money might be lost if unit weight was based upon optimum moisture content and the material was removed and weighed at a lower moisture content.

One common way of combatting the problem is to pay for materials in place *as calculated from the dimensions shown on the plans.* In this manner, the drawings represent "pay lines" and any material in excess of that shown on the plans would not be paid for. This has a disadvantage to the owner: on a heavy earthworks project it would be possible to unbalance the bid to provide a high unit-price on backfill material; then if the contractor overexcavated by a few inches of depth where unit-price payment is also made for excavation, the owner might end up paying twice for the same material.

Determination of Paving Quantities

On asphalt concrete pavement jobs where the asphalt concrete material is to be paid for by the ton, the inspector should watch closely for overexcavation by as little as 13 mm (1/2 in.) depth. On a large roadway project the cost difference could be great. Similarly, if the price of asphalt concrete is priced by the square yard of material placed to a specific depth, any reduction in pavement thickness would simply assure the contractor the price for a full-depth pavement while allowing the difference between the estimated amount of material specified and the amount actually placed to be pocketed by the contractor.

Many materials are subject to special conditions for the determination of pay quantities. Asphaltic prime coat, for example, is usually measured in gallons. However, the inspector should be certain that the volumes being paid for are based upon the volume that the material occupied at 60°F. If the material is placed hot and the volumetric measurements are made at that time, overpayment could result.

Measurement Guidelines for Determination of Unit-Price Pay Quantities

Bid Items Based upon Linear Measure

Pipelines. These are often paid for in terms of length in stations (horizontal measure), which is determined from the plans. Could possibly involve field measurements by a survey crew (also horizontal measurements).

Also paid for by measurement along the top of the pipe in place. This method will yield actual laying length. Do not allow measurement along the side of the pipe on the outside of curves. Do not accept lengths of pipe for measurement prior to laying.

Sometimes paid for by measurement in the field, *horizontally* along the centerline alignment of the pipe in place. This should result in the same quantity that would result from the stations indicated on the plans.

Curbs. Curbs are generally measured in the field by measurement along the top edge facing the street. Watch for measurements made at the flow line, as the slope of the curb face may yield a slightly greater quantity this way under certain conditions. Watch for measurements made at the sidewalk side of the top edge of the curbing. In some cases this can cause erroneous lengths also. The important thing is not so much *where* the measurements have been taken, but that all measurements are taken at the *same place* throughout the life of the job.

Channels. Where flood control channels or similar structures are to be lined, be sure to measure at the same location each time. If the point of reference is the toe of a slope, do not permit the measurements to be made on opposite sides, as this will alter the indicated length.

Sewer Lines. Measurement of VCP sewer lines is generally made by measurement of the pipe in place. However, do not allow measurements to be made through sewer manholes, as the separate price paid for manhole construction already covers this cost.

Fencing. Measurement of chain-link fencing can be accomplished by horizontal measurements or by measuring along the top rail of the fence in place, depending upon the method specified. As in the case of pipelines, the length indicated will be less when measured horizontally.

Bid Items Based upon Area Measurements

Pavements. As mentioned before, the measurement of areas for payment usually presents no problem, but particular care must be exercised to assure that the proper pavement thicknesses have been attained. Watch, too, for underruns where the contractor is being paid both for excavation of the roadbed on a volume basis and pavement surfacing on an area basis.

A particular risk is involved where there are separate payments for pavements over small areas, such as trench resurfacing, and for larger paved areas that can be done with a paving machine. On one actual pipeline job in a city street, the unit price for repavement over trenches was quoted somewhat higher than roadway pavement because it would have to be done using hand methods. The contractor for this project, being an enterprising person, carefully studied the specifications and noted that the earthwork provisions permitted a slope of that portion of the trench walls that were above a plane lying 300 mm (1 ft.) above the pipe, provided that any excess excavation resulting from such methods was to be at the contractor's own expense. The contractor sloped all of the trench walls in the city street area, opening up the entire width of one half of the city street, which he then repaved using a regular paving machine. The specifications for repavement of the area over the pipe trench were based upon the area to be paved, rather than upon the length of pavement to be constructed over trenches. The result was that the contractor not only eliminated the added cost of trench shoring in deep trenches, but also "bought" a street repaving job to repave an entire half of the city street using a paving machine, but performing the work *at unit prices intended for hand work!* The only excess cost to the contractor was the extra labor of the added excavation involved in the sloping of the sides of the pipe trench and the removal of the additional existing paving. The cost to the city was an additional $15,000, as there was nothing in the contract that would provide legal relief. Thus, the method of measurement for payment purposes can mean significant differences in a project cost. In this case, it was the specifier's fault for not coordinating the paving specification with the earthworks specification. An alert Resident Project Representative noticed this early, but unfortunately, under the terms of the contract specifications, the contractor was fully within legal rights to do this (Figure 17.22).

Retaining Walls. If a retaining wall is to be constructed around the periphery of a property, the inspector should be exceptionally careful of the method of making wall area measurements. If all measurements are around the outside of the wall, the cost to the owner will be excessive. However, if all measurements were on the inside face, the cost to the contractor would be unfair. Each wall surface should be taken as a prism, and any space occupied by the previously measured prism should not be included in any other measurement of adjoining surfaces.

Volume Measurements. Be certain of the method of measurement, particularly in the case of earth and rock materials and of materials that must be placed at high temperatures. In many cases, volume measurements will have to be made using a survey crew to take cross sections of the affected area.

If a volume of material is to be placed in accordance with the lines shown on the drawings, be sure to get a survey crew on the site before the work begins, to establish the exact profile of the ground as it existed before the work began; otherwise, disputes may arise concerning the quantities because of a difference of opinion as to the condition that existed prior to the beginning of the work.

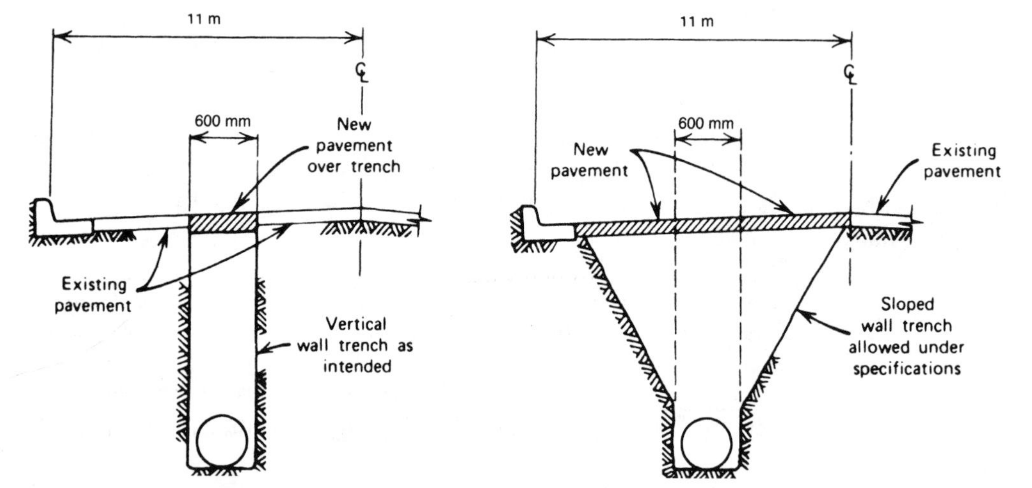

FIGURE 17.22 Effect on Contractor's Earned Payment of Sloping Trench Walls Where Payment for AC Paving Was Based upon Surface Area of Tonnage Instead of Linear Feet.

Weight. Establish at the beginning of the job what the basis of weight measurement is to be. Under the old English system, which has been converted to metric, a "ton" could have been either the *short ton* of 2000 pounds (907.1847 kg) or the *long ton* of 2240 pounds (1017.0469 kg). Similarly, smaller units of weight needed to be clearly defined under the English system.

The foregoing is not intended to be an exhaustive list of all the special areas of concern, but rather a sampling of some of the more common measurement problems encountered under a unit-price contract project.

Final Payment to the Contractor

After the Certificate of Completion or Substantial Completion has been filed (Chapter 21), the contractor will apply for final payment. Although the form used is the same as that used previously for monthly progress payments, there are several additional requirements that must be complied with before the architect/engineer should issue a final certificate for payment.

1. The contractor must pay all bills or other indebtedness.
2. Under certain contracts he or she must submit receipts, releases, and waivers of liens.
3. Consent of surety must be obtained where a surety is involved.

Under the provisions of the AIA and the EJCDC General Conditions, the final payment is withheld only until the contractor has provided evidence that each of the items in the foregoing list has been complied with. Under the AIA and EJCDC provisions, the making and acceptance of the final payment constitutes a waiver of all claims by the owner against the contractor other than those arising from unsettled liens, from defective work appearing after final payment, or from failure to comply with the requirements of the contract documents or the terms of any special guarantees that are a part of the contract. It is also a waiver of all claims by the contractor against the owner other than those previously made in writing that are still unsettled (Figure 17.23).

Although the holding time for the release of the contractor's final payment under the AIA and the EJCDC contract provisions is not specific but is subject to the time it takes the contractor to submit its documentation, most public agency contracts note a specific period of time before release of the final payment to allow time for any lien holders to file before the owner releases the final payment to the contractor. In this manner, if liens have been filed, or valid claims are presented to the owner for unpaid bills, the owner can pay such indebtedness and deduct all the sums from the money due to the contractor.

Under the provisions for final payment of many other public agency contracts, the owner retains the right to withhold funds to satisfy liens or outstanding bills. In some states, however, this creates a potential problem; the period for the filing of liens against a construction project in many states where a Certificate of Completion or Substantial Completion has been executed and recorded has been established by

14.07 Final Payment

A. Application for Payment

1. After CONTRACTOR has completed all such corrections identified during the final inspection to the satisfaction of ENGINEER and delivered in accordance with the Contract Documents all maintenance and operating instructions, schedules, guarantees, Bonds, certificates or other evidence of insurance certificates of inspection, marked-up record documents (as provided in paragraph 6.12) and other documents, CONTRACTOR may make application for final payment following the procedure for progress payments. The final Application for Payment shall be accompanied (except as previously delivered) by: (i) all documentation called for in the Contract Documents, including but not limited to the evidence of insurance required by subparagraph 5.04.B.7; (ii) consent of the surety, if any, to final payment; and (iii) complete and legally effective releases or waivers (satisfactory to OWNER) of all Liens arising out of or filed in connection with the Work.
2. In lieu of the releases or waivers of Liens specified in paragraph 14.07.A.I and as approved by OWNER, CONTRACTOR may furnish receipts or releases in full and an affidavit of CONTRACTOR that: (i) the releases and receipts include all labor, services, material and equipment for which a Lien could be filed; and (ii) all payrolls, material and equipment bills and other indebtedness connected with the Work for which OWNER or OWNER's property might in any way be responsible have been paid or otherwise satisfied. If any Subcontractor or Supplier fails to furnish such a release or receipt in full, CONTRACTOR may furnish a Bond or other collateral satisfactory to OWNER to indemnify OWNER against any Lien.

B. Review of Application and Acceptance

1. If, on the basis of ENGINEER's observation of the Work during construction and final inspection, and ENGINEER's review of the final Application for Payment and accompanying documentation as required by the Contract Documents, ENGINEER is satisfied that the Work has been completed and CONTRACTOR's other obligations under the Contract Documents have been fulfilled, ENGINEER will, within ten days after receipt of the final Application for Payment, indicate in writing ENGINEER's recommendation *of* payment and present the Application for Payment to OWNER for payment. At the same time ENGINEER will also give written notice to OWNER and CONTRACTOR that the Work is acceptable subject to the provisions of paragraph 14.09. Otherwise, ENGINEER will return the Application for Payment to CONTRACTOR, indicating in writing the reasons for refusing to recommend final payment, in which case CONTRACTOR shall make the necessary corrections and resubmit the Application for Payment.

FIGURE 17.23 Provisions for Final Payment to the Contractor from the EJCDC Standard General Conditions of the Construction Contract. *(Copyright © 1990 by the National Society of Professional Engineers.)*

statute. Because of this fact, many agencies set a holding period for all retention money equal to the statutory period plus 5 or 10 days to allow the lien filing period to close before releasing the contractor's money. In this way, the owner can be assured that all potential lien holders have been satisfied before making final payment to the contractor. Another alternative would be a waiver of claims (see Chapter 21).

As recommended in Chapter 21, it is desirable that a formal Certificate of Completion or Substantial Completion be executed *and recorded* in the county recorder's office. If this is not done, no reasonable retention period of the contractor's final payment may sometimes suffice. It would normally appear to be in the

C. Payment Becomes Due

1. Thirty days after the presentation to OWNER of the Application for Payment and accompanying documentation, the amount recommended by ENGINEER will become due and, when due, will be paid by OWNER to CONTRACTOR.

14.08. Final Completion Delayed

A. If, through no fault of CONTRACTOR completion of the Work is significantly delayed and if ENGINEER so confirms, OWNER shall, upon receipt of CONTRACTOR's final Application for Payment and recommendation of Engineer and without terminating the Agreement, make payment of the balance due for that portion of the Work fully completed and accepted. If the remaining balance to be held by OWNER for Work not fully completed or corrected is less than the retainage stipulated in the Agreement, and if Bonds have been furnished as required in paragraph 5.01, the written consent of the surety to the payment of the balance due for that portion of the Work fully completed and accepted shall be submitted by CONTRACTOR to ENGINEER with the Application for such payment. Such payment shall be made under the terms and conditions governing final payment, except that it shall not constitute a waiver of Claims.

14.09. Waiver of Claims

A. The making and acceptance of final payment will constitute:

1. a waiver of all Claims by OWNER against CONTRACTOR, except Claims arising from unsettled Liens, from detective Work appearing after final inspection pursuant to paragraph 14.06, from failure to comply with the Contract Documents or the terms of any special guarantees specified therein, or from CONTRACTOR's continuing obligations under the Contract Documents; and
2. a waiver of all Claims by CONTRACTOR against OWNER other than those previously made in writing which are still unsettled.

FIGURE 17.23 (continued)

owner's best interest to see that a Certificate of Completion or Substantial Completion is filed and recorded within the statutory time allowed after substantial completion of the work, or to be sure of receiving a release or waiver of claims from the general contractor, all subcontractors, and material suppliers prior to releasing retention money.

Final Progress Payment

Final payment as referred to in many General Conditions of the Contract is the last progress payment made to the contractor, less retainage, and should not be made until after execution of the Notice of Completion (Substantial Completion). This notice constitutes formal "acceptance" of the Work by the owner.

When computing the contractor's eligibility for the last payment, several items should be considered for deduction *in addition to normal retainage:*

1. Any liquidated damages due to the date of Notice of Completion (Substantial Completion).
2. Twice the value of all outstanding punch-list items (some states limit this to 1.5 times). (In case the punch-list items are not completed within the agreed time, the contractor may forfeit this amount.)

3. The value of any lien claims already on file. (In some states, final payment terminates liability of the owner *only* for those contractor claims filed prior to acceptance of final payment by the contractor.)

Acceptance of final payment by the contractor may terminate the liability of the owner in some jurisdictions. The final payment should not be approved for payment until after formal "acceptance" of the Work by the owner as evidenced by the execution of a formal Notice of Completion (Substantial Completion). From that date forward, the lien law clock starts running, which limits the time for filing of liens or Stop Notices against the project.

PROBLEMS

1. True or false? Front loading of a bid is always an illegal or unethical form of unbalancing a contractor's bid.
2. Under a unit-price contract with a 25 percent guarantee requirement of all unit prices quoted, the engineer's estimate of earthwork quantities totals 9,270 cubic meters. If the actual quantities measured in the field for pay purposes total 10,650 cubic meters, is a change order justified for the quantity change?
3. Are liquidated damages intended only for missing the completion date of a project, or may they also be established for failure to meet midproject interface dates?
4. On a unit-price contract, what is the "contract price"?
5. Is measurement for payment of underground pipelines constructed in place in hilly terrain generally based upon true length of the pipe laid or horizontal projected length (station lengths)?
6. On a unit-price contract, if a contractor bids $145.00 per meter for 765 meters of pipe to be constructed in place and incorrectly shows the total amount as $110,952.00, how much money is the contractor entitled to receive?
7. What is a waiver of lien?
8. What means are available to a contractor bidding a unit-price project for recovery of "General Conditions Work" (mobilization, bonds, insurance, cost of permits, and other preconstruction expenses)?
9. Is the final amount paid to a contractor under a unit-price contract limited by the quantities of materials or work shown in the original Bid Schedule?
10. What type of items are recommended for payment to the contractor upon delivery, even before assembly into the constructed work?
11. Is it advisable to use cost-loaded CPM for progress payments to the contractor on unit-price contracts?

18

CONSTRUCTION MATERIALS AND WORKMANSHIP

Construction materials control is fully one-half of what construction inspection is all about. The other half is control over workmanship. Many inspectors understand quality construction when they see it, or proper materials when they see them, but fail to understand fully what their authority and responsibility requires them to do. The implication often too firmly implanted in some inspectors' minds is that they are on the job to assure that the project will be constructed with only the "best" quality materials and the "highest" quality of workmanship. On numerous occasions, inspectors have been heard to remark, or even interrupt their instructions to say, supposedly reassuringly, "Don't worry about the specs; I'll see that you get a first-class job!"

Reassuring as it may sound at first, it is *not* the proper approach for an inspector to take. The inspector's real authority is *limited* to requiring the contractor to provide all that has been agreed to in writing in the contract documents. If an owner, through lack of funds or otherwise, chooses to purchase less than top-quality goods or to accept adequate but somewhat less than first-class workmanship, it is its prerogative to do so. It would be an overstep of the authority of the inspector to attempt to require the contractor to provide anything in excess of the terms of the approved contract—and certainly would be unfair to the contractor. Remember, one of the basic principles of the law of contracts is that it must involve a meeting of the minds. Thus, anything that is not part of the written terms of the contract is not within the authority of the inspector to attempt to require without a supplemental agreement, such as a change order. If, however, an inspector observed a condition that, if performed in strict accordance with the plans and specifications, would result in an unsafe condition or that might be considered as being of questionable judgment, the obligation of the inspector is to bring the matter to the direct attention of the architect or engineer or the owner. Then, if in the judgment of the architect or engineer or the owner a change should be made, it will be executed as a formal change order, with a possible appropriate adjustment in the contract price.

It all comes down to a simple axiom. If a person agrees to buy a Ford, the dealer should not be forced to deliver a Continental for the same price, just because it might represent a higher-quality product in the eyes of the buyer's agent. Similarly, if it is found that the Ford will not do the job but the Continental is required, the buyer must make a new agreement to purchase and pay for the higher-priced product.

In short, the inspector is not on the job to enforce what he or she believes to be proper construction, but rather to obtain the type and quality of construction that has been called for in the plans and specifications. This cannot be emphasized too strongly, as the wise contractor will generally provide exactly what the inspector demands, then file claims for extras for the cost differences between what was provided at the inspector's direction and what was called for in the plans and specifications—and will be entitled to get it.

Materials and Methods of Construction

Most specifications, in their General Conditions, provide that unless specified otherwise, all workmanship, equipment, materials, and articles incorporated in the work are to be of the best available grade in the local trade area. Materials called for on the drawings that are not called for in the specifications, but that are known to be required for a complete project, are similarly required to be of comparable quality. These phrases are usually found in specifications prepared by someone who does not really know what it takes to make a "complete" project, or how properly to specify quality in a product. The result is one of the all-encompassing generalities such as that noted. It may get the job done, but not without many arguments in the field. It also saves the specifier the embarrassment of having to tell someone how to build something that he or she may not know the first thing about. Apparently, the concept is legally sound, but it puts a considerable added burden upon the Resident Project Representative, who must be the one to interpret the terms "best" and "quality." Generally, an inspector can plan on some arguments over either of these terms, because that which one person considers high quality another may consider substandard and because the construction contract was supposed to represent a "meeting of the minds." Unless both parties agree as to what the terms mean, there may be some doubt concerning that portion of the contract.

Often the specifications provide that all materials furnished must be free of defects or imperfections, and must normally be all new materials. Workmanship, similarly, is often stated in unenforceable terms. Phrases to the effect that "workmanship shall be of highest quality" or that something should be built using the "best workmanship" are almost useless, as the terms cannot be precisely defined, and therefore are largely unenforceable conditions. About all that can be rejected by the inspector under such provisions is craftsmanship that is so obviously defective that even a layperson can recognize it.

With the proper specification terms, the inspector has a useful tool that can be used to great advantage. Under the specific terms of a specification that clearly defines the quality of workmanship, an inspector has every right to reject all that does not meet the specified standards—in fact, he or she has an obligation to do so. Specifications such as these are the products of professionals. A good specifications writer is one who has a good understanding of field construction, contract law, and design principles, and a writing ability—a rare breed.

Interpretation of the Specifications

It should be kept in mind that the designers are the people most familiar with the intent of the contract documents and their provisions. To them, none of the terms seems vague or ambiguous because they understood what it was they were trying to say when they wrote the terms into their specifications. The contractor, however, must attempt to interpret the strict wording of the specifications to prepare the bid, and thus must rely on the ability of the specifications writer to communicate accurately the designer's intent through the wording of the specifications. As a contractor once stated, after being told what was intended by the terms of a specification: "I don't care what was intended—this is *what it says!*" The contractor was right, of course. The contractor was to provide what it *said, not* what it should have said. If the designer or owner chooses to interpret the terms differently, he or she must expect to pay for the privilege.

The courts also respect this concept. They have generally held that, all other things being equal, in case of a dispute over the meaning of the wording of the specifications provisions, the binding interpretation will be in favor of the party who did not write the contract. This means that the judgment stands a good chance of being in the contractor's favor, as the contractor is generally not the party who wrote the contract provisions. In this manner, it is possible that in an imperfectly written set of documents, the contractor may actually be entitled to extra to build the work according to the meaning that the designers intended when they wrote the specifications.

Requests for Substitutions of Materials

By far one of the most frequent requests received on the job will be requests by the contractor to use substitute materials for those actually specified by the architect or engineer. The conditions controlling the use of such alternative choices of materials differ somewhat between public and private construction contracts, and must be considered separately.

Whenever a substitute is offered, the contractor is obligated to give adequate notice of an intention to offer a substitute—not wait until it is already too late to get delivery in time for the product actually called for. Then, after submittal, sufficient time must be allowed for the architect or engineer to review the technical data submitted by the contractor in support of his or her claim that the product is the equal of the specified one.

On a private project, the design firm may specify a single proprietary item for every item on the project if it chooses to do so, and there is no obligation to anyone except the owner to consider substitutes unless the architect or engineer wishes to do so. Such instructions must be communicated to the inspectors so that they can properly respond to the contractor in case of attempted submittals for consideration as substitutions. Furthermore, if in the judgment of the architect or engineer no substitutes may be considered, then only the specific brand or model of the specified product may be used in the work. All products delivered to the site for use on the work must be rejected if they fail to conform to the specific terms of the plans and specifications.

On public works projects, certain limitations exist all over the United States and in many foreign countries that limit the power of an architect or engineer to specify a single name brand of a product *if equivalent products are on the market.* All specifications for proprietary products in many states, as well as those for use either on federal contracts or contracts by other agencies in which part of the funding is from a federal agency, are required by law to name one or two brand names and the words "or equal" of a product called for by brand name. Also, a prescriptive specification that upon analysis can be seen as applying only to a single brand name is considered the same as calling out only one brand name, and is thus considered illegal. The majority of states have similar laws governing public projects within their jurisdiction, so even without federal funds, these conditions usually prevail on public projects. In certain cases, the law controlling the specifying of brand names allows an exception to the rule (1) if the product specified is required to be compatible with existing facilities, and (2) if the product specified is unique and no other brand is made.

Furthermore, it is the opinion of many competent legal authorities that the specification of several brand names, where one brand is called for by manufacturer and that manufacturer's catalog number, and the remaining brands are called for only by manufacturer's name, is in violation of the intent of the law and will be judged as if no alternatives had been offered. Each product named must include the description in comparable terms. It sounds unreasonable at first, but consider the frequent case of a product called for by catalog number, such as

> Well pump for the emergency water system shall be Jacuzzi 75S6AV15, Byron-Jackson, or approved equal . . .

In the first place, only one *product* has actually been specified. There is no catalog number for the Byron-Jackson unit; thus it is not a product but only the name of a potential *manufacturer* of an "equal" product. Often, a little research turns up the fact that some of the manufacturers named in specifications in this manner do not even manufacture a similar product. The other possibility is that the alternative "product" opens up the specifications so broadly that hardly any offered substitute can be excluded. One of the inherent dangers of the "or equal" concept is that the products named may not be equal at all. Often the designer selected exactly what was wanted in one catalog, then hastily selected what appeared to be an equal from another manufacturer. In doing this, the specifications may have unintentionally broadened to allow any other brand whose characteristics were anywhere within the extremes possessed by either brand. A product that contained all the worst features of both would still have to be considered as acceptable, as the simple naming of a second product that omitted a feature that was in the designer's primary choice of products automatically eliminated that particular feature as a minimum prerequisite for acceptance of a contractor-offered substitute. In some cases, there are also provisions for a public agency to specify a single proprietary product as part of a research or experimentation program in which the single product specified is the article being researched.

Traditionally, the architect or engineer is considered to be the final judge of the quality of a product, and the courts have often upheld this concept. If in their determination the product offered is not equal, they have the power to reject it summarily. Furthermore, the contractor may be required to carry the burden of the cost that may be necessary to prove equivalency where a laboratory analysis or similar certification is required. A product may be judged as not being equal on the basis of physical or chemical properties, performance, selection of materials, or even due to dimensional incompatibility with the design of the finished structure where the use of the alternative product may require redesign of portions of the structure to accommodate the substitute product. One case in Los Angeles, however, failed to uphold the owner's engineering staff in its refusal of a substitute product when its rejection was on wholly aesthetic grounds, because the specifications in this case had not cited aesthetics as being a proper criterion for the determination of product equivalency [*Argo Construction Co., Inc.*, v. *County of Los Angeles,* Court of Appeals, 2d Civil No. 32568 (1969)].

In any case, the product offered as a substitute to a specified article must be submitted to the architect or engineer or to the owner, through the Resident Project Representative, for consideration and approval before such a substitute product may be used in the work. The inspector *must* reject any article that fails to satisfy one of the following two requirements:

1. It is the specific product called for in the plans and specifications.
2. It is a substitute product that the architect or engineer or the owner has approved in writing to the contractor.

In the absence of either of these conditions, the substitute product must be rejected by the inspector and required to be removed from the construction site. Failure to observe this requirement may be grounds for withholding partial payment to the contractor, and the inspector should make a careful survey at the time of payment requests to assure that all materials that have been installed or delivered to the site are in fact the same materials that have been specified or allowed by the design firm or the owner in writing as an approved substitute. If the contractor fails to meet these requirements, the inspector should submit a recommendation to the design firm or the owner along with the contractor's payment request that payment be disallowed for the nonconforming portion of the work.

Time to Consider Substitutions

The construction industry needs generally to be made aware that substitutions proposed by bidders and contractors unduly disrupt the normal bidding and construction processes. Too often, valuable time and efforts of key personnel are wasted in the consideration of such requests that are originated by the proposer solely for his or her own financial benefit. Two factors should be kept in mind before a contractor proposes a substitution. First, there may be several perfectly valid but undisclosed reasons why the selection or specification was established as it was in the first place. Second, and in any case, time and effort will be required for the architect/engineer's and owner's investigation of the proposed substitute product, for which the personnel must be paid, and during which the work and other necessary activities may be delayed.

The claim is often made that all attempts to limit the consideration of substitutions result in a stifling of competition and loss of economy to the owner. This argument might be valid if the efforts of everyone concerned with the problem were without cost and if economics were the only interest of each owner. Everyone knows that the situation is not that simple, yet reasonable competition leading to economy is an important consideration for almost every owner.

There are several distinct periods during the life of a project when unjustified requests for substitutions can be expected to be submitted.

1. Design phase
2. Bidding phase
3. Time between bid opening and award of contract
4. Construction phase

The only time that consideration should be given to evaluation and acceptance of a substitute "or equal" product is during either the design phase or the construction phase just listed. No consideration should be made of any submittal or request for consideration of a substitute product during *any* of the other listed times.

Consideration of an "or equal" product during the bidding phase not only is unfair to the architect/engineer, because it allows too little time for a fair appraisal, but actually places the specifier in a high-risk situation for the benefit of a third party, and it is also quite unfair to the vendors who were willing to spend their own time and money during the design phase to provide data and consultations with the designers for proper application of the products specified (many of which are nonproprietary by nature). Vendors who request consideration of their product as an "equal" during the bidding phase are little more than freeloaders capitalizing on the efforts of the vendors who provided the preliminary design assistance.

Furthermore, these freeloaders are often able to underbid their competitors, as they can deduct an amount from their bid that is equivalent to the money spent by the others in providing design services. The author was personally involved in a case that later ended in litigation, in which consideration of another product was made during the bidding period, and an addendum was issued to the specifications to cover it. The latecomer's product was the lowest priced, of course, but unfortunately, it later failed (even before completion of the project), causing extensive financial loss to both the owner and the architect/engineer, as well as to the contractor and the supplier. Had adequate time been available for study, or had the vendor worked with the architect/engineer firm's specifications engineers during the design phase, the difficulty might never have developed.

The author recalls a recent court case in which the judge ruled against an engineer when a product that had been specified during the bidding phase failed, by stating that it was the engineer's professional obligation to specify only those products that he or she was sure of, and that a specification of one was in a sense a guarantee that the product was suitable for the purpose for which it was specified—despite the fact that the engineer had argued that there was insufficient time to make a proper evaluation of the proposed substitution during the bid period when it had been offered. In case of a vendor's or contractor's inquiry during the bidding phase, the only safe course is to advise all bidders to bid the products specified or proceed at their own risks, and that any requests for substitutions would have to wait until after award before consideration.

On public works projects, there should be no consideration of any substitutions between the bid-opening time and the date of the actual award of the contract. It is, however, both possible and practical in public works construction to limit consideration of proposed substitutes to a 35- to 45-day period just following the signing of the contract, if desired. This would effectively eliminate one of the "squeeze" plays used by some contractors to effect a substitution of a product for their sole personal gain. By this method, the contractor may submit a proposed substitute product at the last minute, claiming that the originally specified product is not available or has a long-lead purchase time that would result in delays to project completion. However, the contractor may offer the substitute product by stating "I can get this right away!" Nevertheless, if it had wanted to do so, the contractor could generally have ordered the specified product months before and met the schedule. As an added squeeze to the owner, this type of contractor might also threaten that if the owner insists on the product originally specified, the long-lead purchase item required will delay the completion of the work, and the owner will be held liable for the contractor's delays.

In such cases, it is the author's viewpoint that if insufficient time is allowed for proper consideration of an "or equal" product, the contractor should be required to submit the originally specified product and be held liable for any delays that might have been caused by its failure to order in time. Furthermore, the contractor should be informed that it will be subject to payment of liquidated damages for any overrun of the project schedule resulting from inability to obtain the product on time due to its failure to place the order early. In practice, it has often been found that this position materially changes the predicted long-lead purchase time, and often the item arrives right on schedule.

Access to the Work by Quality Assurance Personnel

The contractor is obligated at all times to provide access to the work to the architect/engineer and the owner or their authorized representatives, and the contractor is responsible for their safety while they are at the site. Of course, the contractor may require that all such persons coming onto the site wear appropriate safety devices and conduct their operations in a safe manner. Similarly, the inspector is entitled, normally, to have access to the place of manufacture of materials or equipment that is

FIGURE 18.1 Offshore Construction Presents Special Problems for Access to the Work by the Inspectors, as well as Increased Insurance Risks.

to be used in the work. Work on offshore facilities may present a problem, however, if the designer or specifications writer neglected to require the contractor to furnish transportation to such facilities (Figure 18.1). In that case, the design firm's or owner's representatives may just have to provide their own boats to get to the project site. The contractor's obligation *to allow* inspector access to the site is not an obligation to provide transportation for the inspector as well, although few contractors will want to risk the adversary relationship that would be the inevitable result in case of refusal. Nevertheless, this is an added risk and cost to the contractor, and where it is a significant amount, the contractor will be justified in claiming extra payment for the service.

Inspection of Materials Delivered to the Site

It is the responsibility of the inspector to inspect promptly all materials delivered to the site *prior* to their being used in the work. The practice of withholding inspection until the job is done, then announcing to the contractor that the work fails to conform to the specifications, is totally unacceptable conduct for a Resident Project Representative. Certain types of intermittent inspection as performed by government agencies, such as building departments, permit this type of inspection, but it is only because the responsibility for on-site quality control is that of the contractor and the owner's representative. A building department's responsibility is limited to assurance that all requirements of the code and the approved plans and specifications have been followed.

In certain cases it may be desirable to perform inspections of materials or fabricated products prior to their delivery at the site. A case in point would be an inspection of the precasting operation at a concrete precasting plant (Figure 18.2). Usually, the product remains in the casting yard for an extended period before delivery to the site, and failure to make early discovery of patent defects may hold up a project for several months while the precasting yard clears the casting beds to work in a new casting schedule and set up the new forms on the beds between other scheduled operations just to recast defective work. In addition, the placement of stirrups and similar conventional reinforcement in pretensioned, precast, prestressed concrete structural members must be carefully checked at the precasting yard before placing concrete, just as it must for cast-in-place conventional concrete. All too often the work of placing such reinforcing steel is not accurately done and can result in major structural failures.

The author was personally involved in one project in which over 80 percent of all roof members on a 61-m × 61-m (200-ft × 200-ft) roof developed progressive failures that were traced to improper placement of stirrups at the precasting yard.

FIGURE 18.2 Precast, Prestressed Concrete Members at Casting Yard.

In addition to checking of stirrups, the plant inspection will provide an opportunity to measure the net length of all prestressed beams and girders stored at the plant in time to compare their dimensions with the design spacing of supports to see that adequate bearing will be provided at the ends of all beams and girders. Failure to do this has also resulted in failures at several sites in the past.

Rejection of Faulty Material

As described before, the inspector not only has a right to reject faulty materials, but also is obligated to do so. Upon the rejection of nonconforming items, they should be clearly and indelibly marked in such manner that the article cannot be used in the work without the mark being clearly visible to the inspector (Figure 18.3). Such marks can be made with an indelible felt-tip pen, paint, or impression-type markers. The inspector should require that all rejected articles be removed from the construction area immediately and placed in a separate pile to be transported off the site the same day. The inspector should assure himself that the rejection marks cannot easily be erased and the nonconforming articles returned to the site as "new material."

Acceptability of any material, article, or equipment should be based upon accepted standards of the industry, such as ASTM or trade association standards for the products involved (see "Special Material and Product Standards" in Chapter 7). If additional restrictions are imposed as acceptance criteria, they should clearly be spelled out in the specifications unless the requirements are so common in the industry that they are considered as unquestioned standards of trade.

The contractor should be provided with written notice in each case of noncompliance (Figure 18.4) and each such incident documented and retained in the project files (Figures 18.5a and 18.5b) as the data could later be necessary in defense of claims.

NON-CONFORMING

DO NOT USE

The accompanying article fails to conform to the contract requirements and shall be removed from the site unless approved deviation or corrective action has been accomplished (KSC Form 8-69)

ITEM ____________________

DEFECT ____________________

INSP. ____________________ **DATED** __________

WAIVER REQUEST KSC FORM 869 NO. __________

FIGURE 18.3 Example of an Inspector's Rejection Tag.

E. R. FISK & ASSOCIATES

NOTICE OF NON-COMPLIANCE

To L&M Construction Company, Inc.
1234 Fontana Street
Marysville, CA 90000
Attn: Paul Marikian

Date 15 May 1995
Time 09:45 hrs
Inspector Jackson
ABC Labs, Inc.

Project Title Interceptor & Collector Sewers Project No. F-345

Contractor Jorgensen & Sons Contract No. 95-0005

You are hereby notified that ☒ tests ☒ inspection that the trench backfill and compaction between Stations 67+45 and 85+20

does not conform to the requirements of the Contract Documents. The provision violated is:

Section 02450 Article/Paragraph 3.4(A) under the provisions of the Contract Documents, the requirements are sand maust be placed in pipe zone, and after consolidation backfill must be placed in lifts not exceeding 300 mm in depth and compacted to 90 percent per ASTM D 1557-78.

Noncomplying work may be required to be removed and replaced at no cost to the Owner.

It shall be your responsibility to determine corrective action necessary, and to determine whether you wish to discontinue operations until additional investigations by the Owner or Engineer confirm or refute the initial findings.

R. E. Barnes
Resident Project Representative

Data used are fictitious for illustration only

Non-compliance notice was received by the Contractor on 5-15-95
Date

By Harold Martinson

Title Proj. Manager

Organization L&M Const., Inc

FISK FORM 7-6

FIGURE 18.4 Notice to the Contractor That Work Does Not Comply with Specification Requirements. *(Fisk, Edward R.,* Construction Engineer's Complete Handbook of Forms, *1st edition, © 1993. Reprinted by permission of Pearson Education, Inc., Upper Saddle River, NJ.)*

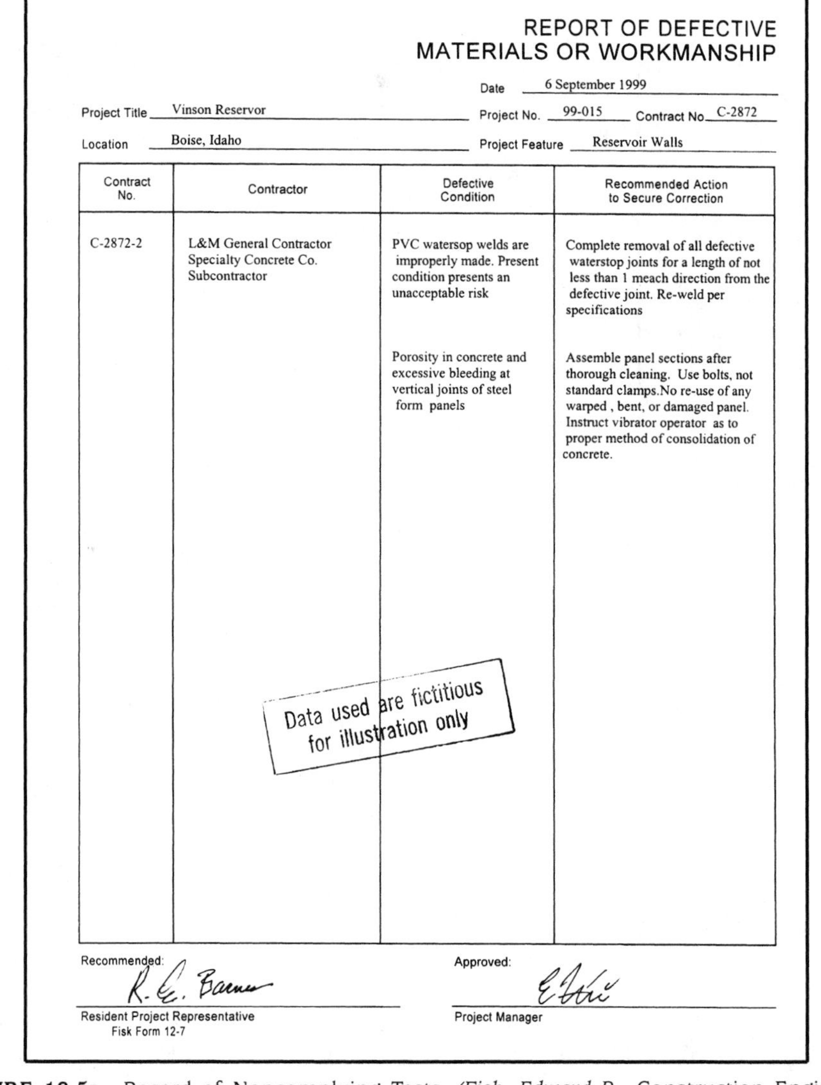

REPORT OF DEFECTIVE
MATERIALS OR WORKMANSHIP

Date 6 September 1999

Project Title Vinson Reservor — Project No. 99-015 — Contract No. C-2872

Location Boise, Idaho — Project Feature Reservoir Walls

Contract No.	Contractor	Defective Condition	Recommended Action to Secure Correction
C-2872-2	L&M General Contractor Specialty Concrete Co. Subcontractor	PVC watersop welds are improperly made. Present condition presents an unacceptable risk	Complete removal of all defective waterstop joints for a length of not less than 1 meach direction from the defective joint. Re-weld per specifications
		Porosity in concrete and excessive bleeding at vertical joints of steel form panels	Assemble panel sections after thorough cleaning. Use bolts, not standard clamps.No re-use of any warped , bent, or damaged panel. Instruct vibrator operator as to proper method of consolidation of concrete.

Data used are fictitious for illustration only

Recommended: R. E. Barnes
Resident Project Representative
Fisk Form 12-7

Approved: [signature]
Project Manager

FIGURE 18.5a Record of Noncomplying Tests. *(Fisk, Edward R.,* Construction Engineer's Complete Handbook of Forms, *1st edition, © 1993. Reprinted by permission of Pearson Education, Inc., Upper Saddle River, NJ.)*

Construction Equipment and Methods

Generally, selection of the type of equipment required to do a job is the responsibility of the contractor. However, if it can be clearly established that the use of a certain piece of equipment to do a specified job will in all probability result in inferior construction quality, it is within the authority of the architect/engineer or owner to require that appropriate changes be made. These instructions would normally be issued through the Resident Project Representative. Also, if the inspector notes that

RECORD OF
NONCOMPLYING TESTS

Project Title Taylor Avenue Reservoir Project No. 00-004 Contract No. W35476

Contractor ABC Constructors, Inc. Type of Work Concrete & Earthworks

GENERAL INFORMATION AS TO TYPE OF TEST, RESULTS, AND OTHER AVAILABLE PERTINENT DATA (Cite ASTIVI, ACI, ANSI, AWS, etc., as applicable)	QUANTITY INVOLVED	ACTION TAKEN
High slump (140 mm) Maximum allowable 65 mm ASTM C 143-90a	7 M3	Rejected and returned load No adjustment allowed
Core taken from panel lining Permeable concrete in reservoir lining	2 panels	Field order issued to remove existing non-complying panels and recast. No adjustment allowed
Backfill in intermediate pipe zone tests at 84 percent compaction per ASTM D 1557-91 mod. to 3 layers	Sta 5 + 36 to Sta 5 + 74	Required conditions to be corrected to specified 90 percent. No adjustment allowed
Excavation of reservoir basin shows unforseen large deposit of expansive claay	2785 M2	Change Order #4 issued to presoak soil, then trim w/Gradall; seal with Visqueen and cast lining concrete while subgrade is wet. Contarct price increased by $23 450.00 plus additional time of
		21 work days. Justification per PCC 7104 for Type 2 DSC

Data used are fictitious for illustration only

Fisk Form 12-8

FIGURE 18.5b Record of Defective Materials or Workmanship. *(Fisk, Edward R., Construction Engineer's Complete Handbook of Forms, 1st edition, © 1993. Reprinted by permission of Pearson Education, Inc., Upper Saddle River, NJ.)*

certain equipment, such as a crane, is overloaded and can possibly lead to a serious accident, there is some obligation to interrupt the use of the particular piece of equipment until the hazard has been eliminated.

One example of the right of a design firm or owner to limit the type of equipment to be used on a project concerns a concrete-lined reservoir where a 127-mm (5-inch) thick concrete lining was being placed in 7.3-m (24-ft) square panels joined with PVC waterstops sealed with polysulfide joint sealant. The side walls were sloped to 1-1/2 to 1, and the basin was 4.9 m (16 ft) deep. The concrete lining was to be placed over a layer of polyethylene sheet, and the design requirements prohibited any penetration through the plastic sheet into the earth below. Thus, when the contractor planned to use 3.7-m (12-ft) span mechanical vibratory screeds to span each panel, it was obvious that each panel would have to be screeded using two passes. This would require the use of steel stakes to support a wood screed upon which one end of the vibratory screed was expected to ride. The result would have required penetration through the polyethylene sheet, as well as some risk of substandard slab thickness in the area of the screed pins due to incomplete filling and consolidation of the screed pin holes. The contractor was informed that the equipment planned on was unsatisfactory and that an alternative method would have to be provided. It should be noted here that the Resident Project Representative did not tell the contractor which method to use—only which method was unsatisfactory. It was still up to the contractor to select equipment and methods as long as they were capable of doing a satisfactory job within the terms of the specifications.

Similarly, any Resident Project Representative can influence the methods of construction to be used by the contractor if it can be shown that the proposed method of the contractor will not provide a satisfactory product. As mentioned, however, the resident engineer or inspector must be extremely careful as to how these statements are worded to the contractor. The resident engineer or inspector may indicate that a particular method of construction or piece of equipment is unsatisfactory, but must not go on to the next step and tell the contractor how the work should be done or which particular piece of equipment should be used.

Sometimes an engineer or inspector may be required under the contract to inspect the contractor's equipment. This is a potentially risky position to get into as the average engineer or inspector would not seem to be qualified to make such inspections. It would seem that the type of person most qualified to perform equipment inspections might be either an equipment mechanic or a safety engineer. Picture the D-11 CAT illustrated in Figure 18.6. If you were assigned to perform such an inspection, where would you start? The author was placed in exactly such a situation on one project, but was fortunate in being able to negotiate terms that later relieved him of that responsibility.

Quality Level and Quality Assurance

Quality Level

Quality level is the specific degree of excellence, basic nature, character, and kind of performance possessed by a particular item or group of items required by the designer. The minimum quality levels are those called for in the specifications for the project. The items that control quality include the following:

1. Location of the project
2. Magnitude of each phase of construction
3. Availability of local materials
4. Contemplated life of the construction
5. Climatic and operating conditions
6. Cost limitations
7. The desires of the architect or engineer

FIGURE 18.6 D-11 CAT Working on a Major Highway Job. Is an Inspector or a Construction Engineer Qualified to Inspect such Equipment?

Quality Assurance

Each quality level requires sufficient quality control to assure that the established quality levels are met. Quality assurance may be by visual inspection, tests, certifications, reports, shop drawings, and similar procedures. Like the quality level, the quality assurance will vary from project to project. The quality standards that are the means to assure that construction is in conformance with the contract specifications and drawings are the responsibility of the Resident Project Representative. Other quality assurance is provided by the results of tests, samples, shop drawings, and similar procedures.

Quality Assurance Provisions

The following quality control provisions are considered separately from the visual inspection requirements of the Resident Project Representative. They are included in the specifications to assure that the quality of the items specified is actually provided by the contractor. When quality cannot be verified by routine field inspection, the quality assurance for material or equipment must be established by other means to assure satisfactory performance.

Testing

Testing is limited to those laboratory or field tests actually called for in the specifications, or allowed by them. Such tests may be performed by the contractor, the architect/engineer, the owner, or commercial testing laboratories. The commercial laboratories may be hired by the contractor, the owner, or the design firm, but in each case the project manager and the Resident Project Representative should receive a copy of the test results.

Testing is required for items of work that are critical and are particularly susceptible to unsatisfactory levels of quality, and that cannot generally be detected by observation (Figure 18.7). Examples are concrete, soil compaction, and similar materials. The specifications determine which items are to be tested, which tests and procedures apply, and the required levels of performance. The specification also determines who should perform the test. Testing is necessary for any work that has a history of poor performance and involves an assembly of products furnished by more than one contractor, where the end result is critical.

Installation in Accordance with the Product Manufacturer's Instructions

In many cases, the specification requires that the product manufacturer provide instructions for the method of installation of products that are installed by subcontractors who are not directly affiliated with the product manufacturer. This provision

FIGURE 18.7 Ultrasonic Testing of Structural Field Welds.

is supposed to be used only where rigid adherence to the manufacturer's instructions is critical, where product composition and construction create limitations not likely to be understood by the installer, and where installation procedures are complex or subject to significant variations between different manufacturers. Unfortunately, it is also used by lazy specifications writers who either do not know enough about a product to specify it, or who will not take the time to research it.

Experience Qualification

An experience qualification is a requirement that a firm performing a certain type of work have an established reputation for the successful completion of similar work elsewhere for a specified amount of time. The use of an experience requirement is limited to those fields of work in which the ability to do a certain amount of work in the time normally allowed, as well as competence in performing installations and services, requires a considerable amount of previous experience. Examples of such fields of work include:

1. Metal curtain walls
2. Foundation piles
3. Dewatering
4. Precast architectural concrete
5. Calking and sealing
6. Spray-on fire protection
7. Laboratory equipment
8. Mechanical and electrical equipment

This type of provision is devised to prevent "fly-by-night" firms from performing work for which they are not qualified or which they may not be able to complete. The disadvantage is that there is still no assurance that the firm selected will remain in business or will stand behind the work it performs. The provision is generally used for work that requires special qualifications for which there are a number of firms of long standing who are generally recognized in the industry as having this special capability, and where there is a history of fly-by-night operators.

Factory Inspection

Occasionally, a construction contract calls for inspection of production and fabrication facilities at a manufacturer's plant as a part of the quality control requirements of the project. The specification must tell the specific type of inspection to perform. This provision is used for assuring the quality control of custom products of such nature that on-site inspection or testing is either impossible or impractical. This provision is used mostly for large prefabricated products that are fabricated especially for each project, where it is impractical to perform tests and inspection at the job site or at a testing laboratory. Examples of such products are:

1. Precast concrete piles
2. Architectural precast concrete

FIGURE 18.8 Precast, Prestressed Concrete Members at Casting Yard.

3. Precast, pretensioned concrete members (Figure 18.8)
4. Fabricated steel plate specials
5. Pump station manifolds
6. Concrete or asphalt concrete batch plant facilities
7. Welded steel tanks
8. Hydroelectric project equipment (Figure 18.9)

Matching Samples on Display during Bidding

Sometimes a contractor is asked to base the quality of a product upon a sample that is placed on display during bidding. This provision is used only where important aesthetic considerations cannot be adequately specified in words, and for which no known or local appearance standard can be furnished for prospective bidders to examine. To maintain effective quality control during construction, each such sample must later be stored at the construction site for ready reference as a basis of acceptance. Examples of such materials or items are:

1. Natural stone
2. Precast concrete panels with exposed aggregate
3. Concrete finishes
4. Special wood finishes and cabinetry standards

This method allows an effective means to assure that the desired visual characteristics of highly textured or grained materials, which are often difficult to describe verbally, can be provided. Additional advantages when used in public works construction are that trade names can be avoided and that disputes between the Resident Project

FIGURE 18.9 Factory Inspection of a Hydroelectric Turbine Shaft and Thrust Bearing Runner Prior to Shipment.

Representative and the contractor over the appearance of a material surface that may otherwise meet specification requirements can be avoided. The disadvantage is that facilities must be provided in one central location for displays of all samples during bidding, and again at the site during construction. Care must be taken to assure that sample panels represent the full range of colors and textures permitted, and that matching material is obtainable from more than one source in the case of public agency contracts.

Mock-up

Another method of quality control is the requirement for the contractor to construct a mock-up or a prototype construction assembly that, after approval, will serve as the standard for the same type of construction throughout the project. This method is best used where it is impossible or impractical to specify critical aesthetic characteristics that may be desired, or where an assembly is too large or complex to be fabricated prior to award of the contract. Examples are:

1. Certain masonry or natural stone assemblies
2. Architectural precast concrete

Mock-ups by the contractor prior to construction can be advantageous to control the quality of a complex construction system where the aesthetic and other requirements cannot be described accurately in words or in drawings. The disadvantages are that:

1. The standard of production is established *after award* of the contract.
2. Numerous arguments over fine points of aesthetics may arise.
3. The finally approved mock-up may not conform to other specification requirements.
4. Delays may result while awaiting approvals.
5. Additional costs are normally reflected in the bids.

The use of this system is usually confined to complex exterior wall assemblies on monumental buildings where aesthetics are a major consideration, and where there is no other adequate way to describe the desired appearance.

Proven Successful Use

Under this provision, the contractor is asked to provide proof that the same type of product or similar products or equipment have been used successfully in similar construction for a specific period of time. The use of this provision is generally limited to mechanical and electrical equipment that requires proof of safe, dependable, continuous operation for a number of years. This provision is not used for those items that would benefit by innovation. The primary advantage of this type of provision is that the risks usually associated with the experimental use of new products are avoided. It has the disadvantage of discouraging innovation, however, and it offers no insurance against faulty installation. It is generally used to specify items such as elevators, electrical equipment, large pumps, water and sewage treatment plant equipment, and similar items that must be in continuous and reliable service for many years. It is not generally used for architectural items or for products for which it is not possible to determine "successful use" clearly.

Qualified Products List

A *qualified products list* is a provision of the specifications that requires that the procurement of certain contract items be restricted to those items that have previously been tested and approved, and have been included in a list of approved items in the specifications. A qualified products list is not necessarily limited to products that are named in the specifications, but may also be lists of products compiled by independent authorities, such as the Underwriters' Laboratories, which assumes the responsibility for testing and updating their lists. The advantage of using a qualified products list is that the sometimes long, complex, or expensive tests that may otherwise be required have already been performed and do not need to be done for each project. The disadvantages of a qualified products list are:

1. Possibility of less competition
2. Possibility of disclaimer by the contractor for defects in materials furnished under this system
3. Administrative difficulties in maintaining up-to-date lists

Certified Laboratory Test Reports

This is a requirement by which the contractor is asked to provide a certificate that indicates that a product meets specified quality requirements for performance or physical or chemical standards when the submitted sample is tested in accordance with certain specified laboratory standard tests. Submittal of test reports is required for those standard items for which there is a need for quality control testing but for which the testing of the actual item to be installed cannot be justified. Requirements for this provision normally include a statement calling for exact test methods, minimum level of performance, and identification of the product to be tested to be sure that it is the same as the one to be used in construction. In addition, the tests are required to be performed by a recognized independent testing laboratory acceptable to the design firm or the owner. Examples of materials that may require such certified reports are:

1. Concrete reinforcing steel
2. Structural steel
3. Sound control ratings of materials
4. Fire-spread ratings of materials
5. Polyvinyl chloride materials for waterstops
6. Masonry units

Where the contractor is responsible for obtaining the test reports, the disadvantage is the possible danger of apparent conflict of interest due to the contractor furnishing both the material and the testing. Additionally, the reliability of the testing laboratory could be open to question.

The manner in which this provision appears in specifications varies widely. Some of the more common requirements for certified laboratory test reports that are included in specifications are:

1. Test reports shall be based upon results of tests that have been made within a certain time on materials representative of those proposed for use; or
2. Test reports shall be based upon results of tests made on samples taken from materials proposed for delivery to the job site; or
3. Test reports shall be based upon results of tests made on items installed at the job.

Test reports included in the manufacturer's literature are often worthless, as the tests from which the data were derived are often nonstandard or are designed to dramatize certain properties and to conceal undesirable properties.

Certificate of Compliance or Conformance

Under these provisions the contractor is required to provide a certificate that says that the product complies with a specified reference standard. It is necessarily limited to products of standard manufacture for which quality can be clearly assured by the manufacturer, installation is not critical, and job testing is neither necessary nor justified. Examples of such products would include:

1. Glass
2. Paint
3. Aluminum windows
4. Wood

The primary advantage is that certificates can usually be obtained with very little if any increase in the price of the product. The disadvantage is that their validity and reliability depend entirely upon the integrity and knowledge of the certifier.

The requirements of the specifications that call for certificates of compliance are normally included only for those times for which the extra costs of certified laboratory costs cannot be justified. Usually, such certificates are reliable when they come from a member of an industry or a trade that has a strong association that exerts some policing effort to maintain quality. When they do not come from such a source, their reliability may be subject to question. Generally, a resident engineer or inspector is protected by clauses in the specifications that reserve the right for them to inspect and to test any article over and above the test requirements specified, and that the results of such inspections and tests are also binding upon the contractor. In such cases, the extra testing is usually at the expense of the owner unless it turns up a defect; then, often the contract provisions require payment by the contractor of both the corrective measures and the tests that disclosed the defect.

Warranties; Guarantees

The terms *warranty* and *guarantee* are often erroneously used interchangeably in construction contracts and often are used to refer to the maintenance and repair obligations of the contractor for a specific period of time after the completion of construction. The General Conditions of the contract on most projects include specific requirements governing contractor warranties or guarantees. For the purposes of this definition, the term *warranty* will be used to describe this provision.

Two types of warranties are recognized under the law:

1. Implied warranties
2. Express warranties

The term *implied warranties* means that the goods must be capable of passing in trade under the contract description and are fit for the purpose intended. *Express warranties* are those that are specifically set forth in the contract itself; they are in common use for many construction contracts. Warranties are generally for packaged items such as water heaters and compressors. Where the industry practice is to furnish a warranty for an item, the requirement for such a warranty may be included in the specifications. An express warranty is a means of achieving good procurement results by making the contractor responsible for its work and for failures of its work during some part of its useful life. The primary disadvantage of warranties is that they are often unenforceable. Moreover, a warranty clause costs money in the form of higher bid prices, and it cannot be demonstrated that the owner recovers the cost of warranty.

There are several obstacles to the strict enforcement of warranties; some of these obstacles follow:

1. Even in what appears to be a clear-cut contractor responsibility under a warranty, certain action on the part of the owner or design firm may cloud the issue and result in litigation.
2. After acceptance of an item by the owner, the operation and maintenance of the item is performed by other than the contractor's or manufacturer's personnel. Thus, many defects that occur during the warranty period can be argued to result from faulty operation and maintenance by others rather than from a defect in the item itself.
3. Industry is becoming reluctant to accept several of the warranty clauses now in use. In particular, it will not accept the provision relating to third-party damages and the responsibility for an entire building and its contents, including damages to personnel, resulting from the failure of a single item.
4. Warranties of sole-source items are generally unenforceable.
5. In practical application, warranties are generally enforceable only as to defects existing at the time of delivery and acceptance.

Although the warranty clauses usually require the contractor to obtain and enforce warranties normally furnished by manufacturers and suppliers, the exact nature of the warranty is usually not stipulated in the specifications. Therefore, their effectiveness is subject to the wording used by the guarantor and can be expected to be something less than specific. Such a warranty depends almost entirely upon the integrity of the guarantor (the manufacturer or supplier). There are substantial differences in the warranty requirements of public and private agencies in their construction contracts. They range from requiring the contractor to warrant all work as to materials, workmanship, and contractor's design, to requiring the contractor to warrant only mechanical and electrical work. One agency may require the contractor to remedy all damage to equipment, site, buildings, and their contents resulting from a defect, whereas another agency may require the contractor to act as the owner's agent and obtain and enforce the subcontractors' and suppliers' warranties.

Ownership of Materials

Generally, it may be said that the ownership of all materials used in the work or stored at the site is vested in the contractor until final acceptance of the work by the owner at the end of the job. Thus, any risk associated with the protection of the work and the repair of damaged work, or the delivery of damaged materials, is usually the responsibility of the contractor, not the owner. Similarly, the contractor's insurance carrier will normally be called upon to pay the costs of any such claims.

Delivery and Storage of Materials

There is no firm formula for the determination of which facilities will be accorded the contractor as a working or staging area or for its storage of construction materials. Generally, it is the contractor's responsibility to make its own arrangements for such facilities if provisions have not been made by the owner. Although it is true that most owners do provide space, in some areas no such space is under the owner's direct control; in such cases the contractor must make arrangements for space. Occasionally, this is done by the contractor actually entering into a rental agreement for space, which must accordingly be taken into consideration at the time of figuring the bid.

In no case should any contractor assume that it has the right to block public thoroughfares or to use public property of any kind, including parking lots, for its construction purposes, even when the work is being performed for the owner of the property under consideration, unless specific written authority has been granted. As a means of protecting the owner from such claims, the Resident Project Representative should be assured that the space being used by the contractor for its work area has been properly granted by the owner of the affected property.

Handling of Materials

The resident engineer or inspector must be concerned not only with the quality of the materials as delivered and their methods of installation, but also should require that all such materials be properly handled during delivery, unloading, transporting, storage, and installation so that undue stresses will not result in latent defects that will not be detected until after the project has been signed-off. When in doubt, the resident engineer or inspector should have the design firm or the owner contact the manufacturer of the affected material, who is generally just as interested in its proper handling in the field as is the owner. This is because the manufacturer is often the victim of unjust claims for defective materials when, in fact, the problem may have been due to improper handling during construction.

Problems

1. True or false? As an accepted standard of quality, a supplier's certificate is an acceptable substitute for a manufacturer's mill test certificate.
2. List two times during a project (conception to completion) when it is not justified for the engineer or architect to require the contractor to submit its requests for substitutions of products under the provisions for "or equals."
3. Name at least six types of products or fabrications that would be suitable for the engineer or architect to call for inspection at the manufacturing or fabrication plant.
4. True or false? The inspector has no interest in the manner or location of storage of construction materials delivered and stored at the site.

5. Generally, is ownership of materials to be used in the work, and stored at the site, vested in the owner or the contractor?
6. Name the 11 means of establishing quality assurance or compliance described in the text.
7. True or false? An inspector is responsible for securing that the highest possible quality of work be done.
8. Who is in control of construction means and methods in a construction contract?
9. Name one set of circumstances where it is permissible in a public contract to specify a sole-source product without the words *or equal.*

19

CHANGES AND EXTRA WORK

CONTRACT MODIFICATIONS

A change order is a written agreement to modify, add to, or otherwise alter the work from that set forth in the contract documents at the time of opening bids, provided that such alteration can be considered to be within the scope of the original project; otherwise, a contract modification may be required. It is the only legal means available to change the contract provisions after the award of the contract. Functionally, a change order accomplishes after execution of the Agreement what the specifications *addenda* do prior to bid opening (Figure 19.1), except that an accompanying price change may be involved in a change order. A price change is not necessarily always in the contractor's favor, however, as it could also be in the form of a cash credit to the owner, or it may involve no price change at all.

It is standard practice in construction contracts to allow the owner the right to make changes in the work after the contract has been signed and during the construction period. Depending upon the contract and its specific terms, such changes might involve additions to or deletions from the work, changes in the methods of

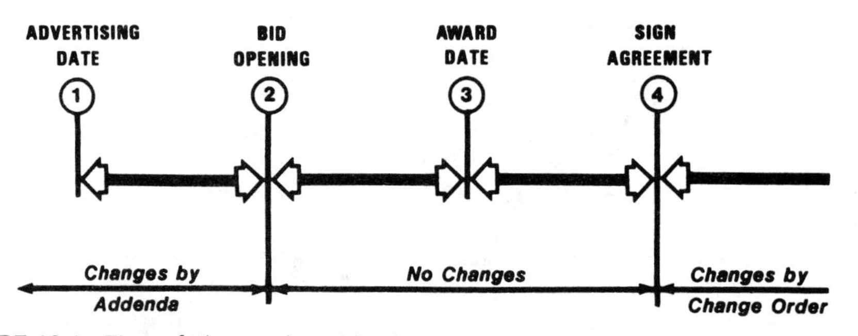

FIGURE 19.1 Time of Changes by Addenda versus by Change Order.

construction or manner of work performance, changes in owner-furnished materials or facilities, or even changes in the contract time or order of the work. Changes may also be executed to correct errors in the plans or specifications, or they may be the direct result of contractor suggestions that are approved by the owner and the architect/engineer.

Changes in the Work

Impact Costs

Often, an owner or architect/engineer fails to realize that the cost of changes in the work may well exceed the cost of the immediate change itself. Many change order forms contain an exculpatory clause that precludes a contractor from recovery of impact costs. In some cases, the two parties already agree on the price of a change in both time and dollars, but the contractor wants to reserve the right to file for impact costs. If this is all that the contractor wants, and all of the other terms of the document are acceptable, it may be advisable to accept the change order with that reservation. If the owner does not, the contractor can file a claim anyway (and will, too), but the amount of the new claim will not only be the cost of any potential impact costs, but also will include the now-inflated value of the time and dollars that had originally been agreed to. A contractor would be ill-advised to sign off on a change order if it expects to claim any future impact costs.

A California court in *Vanlar Construction, Inc. v. County of Los Angeles,* 217 Cal. Rptr.53 (Cal. App. 1985) ruled that a contractor who signed off 81 change orders and 7 supplemental agreements had waived its right to any impact costs arising out of the cumulative effect of the numerous changes.

The County of Los Angeles awarded a contract to Vanlar Construction, Inc., to construct a public building. During the course of construction, the county issued 81 change orders and 7 supplemental agreements. Each of these documents stated that it covered "all charges direct or indirect arising out of this additional work." Vanlar officials signed off on each document.

After completion of the work, Vanlar brought a claim for the impact costs that were not included in the individual change orders. Vanlar argued that it was impossible to measure the cumulative impact of the changes as each change order was issued, making it necessary to wait until completion of the construction. Vanlar also stated that it was a well-recognized custom of the industry to wait until project completion to submit a claim for impact costs.

The California court ruled that Vanlar's execution of the change orders and supplemental agreements precluded subsequent claims for increased costs arising out of those changes. The documents clearly stated that they covered indirect costs arising out of the extra work. The court also said that because the change orders were clear and unambiguous, Vanlar could not resort to industry custom to explain the intended meaning of those documents.

In the court's words: "If Vanlar contemplated any future claim for impact costs, whether at the time of the first, fiftieth, or the eighty-first change order, it was legally obligated to request that a suitable reservation clause be inserted in each change order and supplemental agreement prior to accepting and signing the change order or supplemental agreement."

If a project owner refuses to agree to appropriate reservation language, the contractor may perform the changed work under protest without signing the change order or agreeing to a price.

Oral Change Orders

This is risky turf. Despite the fact that the documents require all change orders to be in writing, the actions of both the owner's representative and the contractor can constructively waive that requirement.

In the case of *Udevco, Inc. v. Wagner* 678 P.2d 679 (Nev. 1984), the Nevada Supreme Court held that a project owner orally waived a requirement that no extra work be performed without a written change order. In that case, Wosser-Laster Enterprises was awarded a contract by Udevco, Inc., to perform framing work on a condominium project being developed by Udevco. Due to an error in the owner's plans and specifications, the prefabricated roof trusses did not fit. Additionally, the framing had to be modified to accommodate cabinets and doors that did not match the original framing design.

Wosser performed this extra work without first obtaining a written change order as required by the contract. It did so at the oral direction of Udevco's project superintendent, who assured Wosser that it would be paid for the work. When Wosser invoiced the owner for the extra work, however, Udevco denied responsibility for work performed without a change order. Wosser argued that the requirement had been waived, but a trial court disagreed. The trial court ruled that because Udevco never paid for any extra work performed without a written change order, the requirement had not been waived.

Upon appeal, the Supreme Court of Nevada reversed this ruling. One way to waive a change order requirement is to pay for extra work performed without written authorization. The requirement may also be waived by oral agreement, however. The court said: "Whether Udevco made payments or not for extra work is not solely controlling of whether it waived the written change order requirement. After Udevco's express oral waiver and Wosser's reliance thereon, Udevco at that time was bound to pay for extra work, regardless of whether it later failed to pay. We conclude under these facts, as a matter of law, that the parties mutually intended to waive the written change order condition."

Change Orders

A change order is the formal document that alters some condition of the contract documents. The change order may alter the contract price, schedule of payments, completion date, or the plans and specifications.

Bilateral Change Orders. The term *change order,* as normally referred to in all except federal contracts, refers to a *bilateral* agreement between the owner and the contractor to effect a change in the terms of the contract. In federal contracts, this document is called a *contract amendment,* as the term *change order* in federal contracts refers to a *unilateral* order to effect a change (Figure 19.2).

	TYPE OF CHANGE	PUBLIC (NONFEDERAL) CONTRACTS AND PRIVATE CONTRACTS	FEDERAL CONTRACTS ONLY (FAR 43.101 and 43.103)
CONTRACT MODIFICATIONS	Bilateral	Change Order	Contract Amendment (or Supplemental Agreement)
	Unilateral	Work Change Directive (sometimes called other names)	Change Order

FIGURE 19.2 Chart Showing the Terminology Used for the Various Change Documents.

Unilateral Change Orders. A unilateral change in nonfederal contracts is referred to in the EJCDC contract documents as a *Work Change Directive* (Figure 19.3), and in AIA contract documents as a *Construction Change Directive* (Figure 19.4). In federal contracts, a unilateral change is referred to as a *Change Order* (Figure 19.2). A unilateral contract modification is intended to expedite issuance of a change order to perform emergency work or protested work, and must be replaced by a regular bilateral change order that addresses the effect of the change on contract cost and time before payment can be made to the contractor.

When approved by the owner, the provisions of a change order become a part of the contract. If an approved change order is executed by the contractor, all of its provisions and terms are as binding upon the parties as are those of the original contract.

Each change order must be evaluated individually; however, there are certain basic principles that apply to the handling of all types of change orders.

1. No work should be included beyond the scope of the base contract, particularly on public works contracts.
2. The identity of the individuals authorized to request and approve change orders should be established early. Such information should be provided to the contractor's superintendent or foreman and the owner's Resident Project Representative.
3. During the preconstruction conference, a meeting should be held to establish the change order handling procedures.
4. All changes in the work must be authorized in writing prior to the execution of any change.
5. The scope of a change order must be clear, and a request for a change order proposal should contain enough information to enable the contractor to make a realistic estimate.
6. The contractor should submit its proposal to execute a change order as soon as possible after receiving the request, and the owner's approval or rejection should follow as soon as possible.
7. The proposal should be fair. It should recognize the contractor's right to include overhead and profit percentages in change order estimates and in time and material change order billings. It should also be recognized that the contractor is entitled to compensation for legitimate time-delay claims, processing of deduct change orders, costs of disposing of removed material, and all other legitimate costs incurred in the execution of the change.

LAWRANCE, FISK & McFARLAND, INC.
CONSULTING ENGINEERS • SANTA BARBARA • ORANGE

WORK CHANGE DIRECTIVE

(Instructions on reverse side) No. 7

PROJECT ABC Reservoir & Pumping Plant

DATE OF ISSUANCE 8 June 1990 EFFECTIVE DATE 8 June 1990

OWNER City of Thousand Oaks, Dept. of Utilities

OWNER's Contract No. 89/90-00

CONTRACTOR XYZ Constructors, Inc. ENGINEER Lawrance, Fisk & McFarland, Inc.

You are directed to proceed promptly with the following change(s):

Description: Install pipe closure on water line shown as a bandoned on Dwg. No. C-12 and remove and dispose of damaged materials; Repair damage from water to facilities & access road per Attachment A.

Purpose of Work Change Directive: Unforseen active water line listed as abandoned in City records. Line was actually under pressure & was damanged due to removal attempt.

Attachments: (List documents supporting change)
Attachment A (Sketch SK-25) showing method & location of closure

If a claim is made that the above change(s) have affected Contract Price or Contract Times any claim for a Change Order based thereon will involve one or more of the following methods of determining the effect of the change(s).

Method of determining change in Contract Price:

- ☐ Unit Prices
- ☐ Lump Sum
- ☒ Other Time & materials

Method of determining change in Contract Times:

- ☐ Contractor's records
- ☒ Engineer's records
- ☐ Other

Data used are fictitious for illustration only

Estimated increase (decrease) in Contract Price:
$ 2300.00
If the change involves an increase, the estimated amount is not to be exceeded without further authorization.

Estimated increase (decrease) in Contract Times:
Substantial Completion: unknown days;
Ready for final payment: unknown days.
If the change involves an increase, the estimated times are not to be exceeded without further authorization.

RECOMMENDED:
Lawrance, Fisk & McFarland, Inc.
ENGINEER
By: (Authorized Signature)
C. H. Lawrance, Project Manager

AUTHORIZED:
City of Thousand Oaks
OWNER
By: (Authorized Signature)
Senior Civil Engineer
Dept. of Utilities

EJCDC No. 1910–8-F (1990 Edition)
Prepared by the Engineers Joint Contract Documents Committee and endorsed by The Associated General Contractors of America.

FIGURE 19.3 Example of EJCDC Work Change Directive. *(Copyright © 1990 by National Society of Professional Engineers; from Fisk, Edward R.,* Construction Engineer's Complete Handbook of Forms, *1st edition, © 1993. Reprinted by permission of Pearson Education, Inc., Upper Saddle River, NJ.)*

WORK CHANGE DIRECTIVE

INSTRUCTIONS

A. GENERAL INFORMATION

This document was developed for use in situations involving changes in the Work which, if not processed expeditiously, might delay the Project. These changes are often initiated in the field and may affect the Contract Price or the Contract Times. This is not a Change Order, but only a directive to proceed with Work that may be included in a subsequent Change Order.

For supplemental instructions and minor changes not involving a possible change in the Contract Price or the Contract Times a Field Order may be used.

B. COMPLETING THE WORK CHANGE DIRECTIVE FORM

Engineer initiates the form, including a description of the items involved and attachments.

Based on conversations between Engineer and Contractor, Engineer completes the following:

METHOD OF DETERMINING CHANGE, IF ANY, IN CONTRACT PRICE: Mark the method to be used in determining the final cost of Work involved and the estimated net effect on the Contract Price. If the change involves an increase in the Contract Price and the estimated amount is approached before the additional or changed Work is completed, another Work Change Directive must be issued to change the estimated price or Contractor may stop the changed Work when the estimated price is reached. If the Work Change Directive is not likely to change the Contract Price, the space for estimated increase (decrease) should be marked "Not Applicable."

METHOD OF DETERMINING CHANGE, IF ANY, IN CONTRACT TIMES: Mark the method to be used in determining the change in Contract Times and the estimated increase or decrease in Contract Times. If the change involves an increase in the Contract Times and the estimated times are approached before the additional or changed Work is completed, another Work Change Directive must be issued to change the times or Contractor may stop the changed Work when the estimated times are reached. If the Work Change Directive is not likely to change the Contract Times, the space for estimated increase (decrease) should be marked "Not Applicable."

Once Engineer has completed and signed the form, all copies should be sent to Owner for authorization because Engineer alone does not have authority to authorize changes in Price or Times. Once authorized by Owner, a copy should be sent by Engineer to Contractor. Price and Times may only be changed by Change Order signed by Owner and Contractor with Engineer's recommendation.

Once the Work covered by this directive is completed or final cost and times are determined, Contractor should submit documentation for inclusion in a Change Order.

THIS IS A DIRECTIVE TO PROCEED WITH A CHANGE THAT MAY AFFECT THE CONTRACT PRICE OR THE CONTRACT TIMES. A CHANGE ORDER, IF ANY, SHOULD BE CONSIDERED PROMPTLY.

FIGURE 19.3 (continued)

CHANGE ORDER REQUEST

Project Title: Tri-City RTD Bus Maintenance Facility

Project No. 94-2417 Contract No. DAA-0-689-94 Contract Date 2-18-94

Contractor: ABC Constructors and D&F Construction (Jt Venture)

Proposed By: M. O. Parker, Resident Engineer (Name) Date 7-28-95

Submitted By: E. R. Fisk, Project Manager (Name) Date 7-31-95

☐ Owner ☒ Arch. & Engr. ☐ Construction

Actual job conditions in area of proposed change: Electrical service not yet completed.
All masonry work completed. Rolling steel door installed & operable.

Change order justification:
Area served by rolling door has no other access or opening. Design of roll-up door installation calls for operating controls to be installed inside the room. Door cannot be operated from outside, as constructed.

Contractor authorized to proceed with this change ☒ YES ☐ NO on issuance of change order

Other contracts involved are as follows (List Contracts by No.): none Is Dwg. Req.? ☐ NO ☒ YES E-14 (Sheet No.)

Description of Work to be Performed:
Drill masonry wall to allow conduit installation from maintenance area into the secure room and install electrical wiring to allow electrical motor operation or the door to the secure room from the adjacent maintenance room. Install key-operated switch for actuating rolling steel door to the secure area.
Power from circuit J-3.

Data used are fictitious for illustration only

Estimated effect on costs of: A&E $1,080.00 Inspection: $ $125.00

Approved [signature] for Tri-City RTD 8-11-95 (Date)

Fisk Form 15-1

FIGURE 19.4 Initiator Change Order Request. *(Fisk, Edward R.,* Construction Engineer's Complete Handbook of Forms, *1st edition, © 1993. Reprinted by permission of Pearson Education, Inc., Upper Saddle River, NJ.)*

Change orders create additional work for everyone in the industry. One of the most aggravating conditions is the length of time that elapses between the time that a proposed contract modification is announced until the matter is finally rejected or approved as a change order. If the procedures discussed in this chapter are followed, it is possible to minimize the inequities otherwise possible in the administration of a contract change order.

In some cases, a situation may arise at the construction site where immediate instruction must be given, and revised drawings and specifications may be unnecessary. In such instances, the Resident Project Representative, upon receiving proper authorization, should prepare an *Initiator Change Order Request* (Figure 19.4) with instructions as to whether the work should proceed immediately or wait until the cost proposals have been approved and a formal contract change order (Figures 19.5 and 19.6) issued.

Whereas it is possible, under emergency conditions, to start such extra work under a *Work Change Directive* (Figure 19.3), it must be emphasized that any work performed beyond the scope of the original contract provisions must be followed up with a confirming change order clearly citing the effect of the change upon both project cost and time of completion. *In no case* should any work beyond the original contract scope or requirements be executed solely upon the authority of a Field Order or Work Change Directive. All such work *must* be supported by a change order. This is for the protection of both the owner and the contractor. The owner has the obligation to review promptly any change order proposals and recommendations and direct the architect/engineer to reject or authorize such change.

Changes frequently have an impact upon the performance of other work that is not in itself changed. The term *impact* refers to the indirect delay or interference that a change on one phase of the work may create on another phase, and the costs of such delay or interference should be recognized as a consequential cost to be considered as a part of direct cost expenses of the contractor, and consideration must be given to allowing for payment of these costs to the contractor.

TYPES OF CHANGES

Directed versus Constructive Change

Before addressing the specific elements of a changes clause in a contract, it is useful to define two basic types of change: directed and constructive.

Directed Changes. A directed change is easy to identify. The owner directs the contractor to perform work that differs from that specified in the contract or is an addition to the work specified. A directed change may also be deductive in nature; that is, it may reduce the scope of work called for in the contract.

In the case of a directed change, there is no question that a change occurred. Disagreements tend to center on questions of financial compensation and the effect of the change on the construction schedule.

Constructive Changes. A constructive change is an informal act authorizing or directing a modification to the contract caused by an act or failure to act. In contrast to the mutually recognized need for a change, certain acts or failure to act by the owner that increase the contractor's cost and/or time of performance may also be considered grounds for a change order. This is termed a *constructive change* and must be claimed in writing by the contractor within the time specified in the contract

E. R. FISK & ASSOCIATES
PO Box 6448 • Orange, CA 92613-6448

CHANGE ORDER

PROJECT TITLE Dalles Hydroelectric Project

PROJECT NO. F-409 CONTRACT NO. W34-6759 CONTRACT DATE 29 Oct 1990

CONTRACTOR International Constructors, Inc.

The following changes are hereby made to the Contract Documents:

Construction of access bridge abutment No. 1 drainage system; and
Reset two penstock bearing plates. All in accordance with revised DWG S-17209
Revision 3, dated 28 August 1991.

Data used are fictitious for illustration only

Justification:

Unforseen soil conditions

CHANGE TO CONTRACT PRICE

Original Contract Price: $ 13,231,053.00

Current contract price, as adjusted by previous change orders: $ 13,257,760.00

The Contract Price due to this Change Order will be [increased] [decreased] by $ 14,342.00

The new Contract Price due to this Change Order will be: $ 13,272,102.00

CHANGE TO CONTRACT TIME

The Contract Time will be [increased] [decreased] by 21 calendar days.

The date for completion of all work under the contract will be 24 June 1992

Approvals Required:

To be effective, this order must be approved by the Owner if it changes the scope or objective of the project, or as may otherwise be required under the terms of the Supplementary General Conditions of the Contract.

Requested by Proj Mgr - E.R. Fisk & Associates date 28 Aug 1991

Recommended by E. R. Fisk & Associates date 28 Aug 1991

Ordered by Dalles Power Company date 2 Sept 1991

Accepted by International Constructors, Inc. date 9 Sept 1991

Wiley-Fisk Form 15-3

FIGURE 19.5 Change Order. *(Fisk, Edward R.,* Construction Engineer's Complete Handbook of Forms, *1st edition, © 1993. Reprinted by permission of Pearson Education, Inc., Upper Saddle River, NJ.)*

LAWRANCE, FISK & McFARLAND, INC.
CONSULTING ENGINEERS • SANTA BARBARA • ORANGE

CHANGE ORDER

(Instructions on reverse side) No. 13

PROJECT Hydroelectric Project

DATE OF ISSUANCE 14 May 1990 EFFECTIVE DATE 15 May 1990

OWNER Oakville Power Company

OWNER's Contract No. 89/90-00

CONTRACTOR XYZ Constructors, Inc. ENGINEER Lawrance, Fisk & McFarland, Inc.

You are directed to make the following changes in the Contract Documents.

Description: Item 1 - Exploratory excavation
Item 2 - Pipe relocation
Item 3 - Delete lateral 4A

Reason for Change Order:
Corrective work

Attachments: (List documents supporting change)
Attachment B (Sketch SK-30)

Data used are fictitious for illustration only

CHANGE IN CONTRACT PRICE:	CHANGE IN CONTRACT TIMES:
Original Contract Price $ 65,123,405.00	Original Contract Times Substantial Completion: 1400 Ready for final payment: ____ days or dates
Net changes from previous Change Orders No. 1 to No. 12 $ 62,037.00	Net change from previous Change Orders No. 1 to No. 12 ____ days
Contract Price prior to this Change Order $ 65,185.442.00	Contract Times prior to this Change Order Substantial Completion: 1410 Ready for final payment: 1440 days or dates
Net Increase (decrease) of this Change Order $ 1726.00	Net Increase (~~decrease~~) of this Change Order 4 days days
Contract Price with all approved Change Orders $ 65,187,168.00	Contract Times with all approved Change Orders Substantial Completion: 1414 Ready for final payment: 1434 days or dates

RECOMMENDED: By: Engineer (Authorized Signature) Date:

APPROVED: By: Owner (Authorized Signature) Date:

ACCEPTED: By: Contractor (Authorized Signature) Date:

EJCDC No. 1910–8-B (1990 Edition)
Prepared by the Engineers Joint Contract Documents Committee and endorsed by The Associated General Contractors of America.

FIGURE 19.6 Example of the EJCDC Change Order Form. *(Copyright © 1990 by National Society of Professional Engineers; from Fisk, Edward R.,* Construction Engineer's Complete Handbook of Forms, *1st edition, © 1993. Reprinted by permission of Pearson Education, Inc., Upper Saddle River, NJ.)*

CHANGE ORDER

INSTRUCTIONS

A. GENERAL INFORMATION

This document was developed to provide a uniform format for handling contract changes that affect Contract Price or Contract Times. Changes that have been initiated by a Work Change Directive must be incorporated into a subsequent Change Order if they affect Contract Price or Times.

Changes that affect Contract Price or Contract Times should be promptly covered by a Change Order. The practice of accumulating change order items to reduce the administrative burden may lead to unnecessary disputes.

If Milestones have been listed any effect of a Change Order thereon should be addressed.

For supplemental instructions and minor changes not involving a change in the Contract Price or Contract Times, a Field Order may be used.

B. COMPLETING THE CHANGE ORDER FORM

Engineer initiates the form, including a description of the changes involved and attachments based upon documents and proposals submitted by Contractor, or requests from Owner, or both.

Once Engineer has completed and signed the form, all copies should be sent to Contractor for approval. After approval by Contractor, all copies should be sent to Owner for approval. Engineer should make distribution of executed copies after approval by Owner.

If a change only applies to Contract Price or to Contract Times, cross out the part of the tabulation that does not apply.

FIGURE 19.6 (continued)

documents in order to be considered. The owner should evaluate a change order proposal based on such a claim and can use the same reasoning process as with any other proposal. The types of constructive changes may include:

Defective plans and specifications
Engineer's interpretation
Higher standard of performance than specified
Improper inspection and rejection
Change in method of performance
Change in the construction sequence
Owner nondisclosure
Impossibility/impracticability of performance

Constructive changes are a major source of construction disputes. A constructive change arises when the contractor alleges that something that the owner has done, or failed to do, has resulted in a de facto change in the contract requirements. The argument, of course, is that the contractor is entitled to extra compensation for performing the work. The owner frequently disagrees that a change in the contract requirements has occurred.

Most constructive changes disputes center around the interpretation of the plans and specifications. It is not surprising that owners and architect/engineers tend to interpret the contract in the manner that will be most beneficial to the project. Contractors, on the other hand, are inclined to read the plans and specifications in a manner that will minimize performance costs. Obviously, the most effective way to prevent constructive change disputes is to have a detailed, carefully prepared scope of work. This does not always happen, of course. Even when it does, disputes may still arise. It is for this reason that a comprehensive changes clause should be included in the General Conditions of any construction contract.

Elements of a Change Order

A change order specifies the agreed-upon change to the contract and should include the following information:

Identification of change order
Description of change
Reason for change
Change in contract price
Change in unit prices (if applicable)
Change to contract time
Statement that secondary impacts are included
Approvals by owner's and contractor's representatives

The owner may require the engineer's signature for internal control, but there is no substitute for the owner's signature.

A change order is a written agreement between the owner and the contractor authorizing an addition, a deletion, or a revision in the work and/or time of completion within the limits of the terms of the construction contract after it has been executed. It is a specific type of contract modification that does not go beyond the general scope of the existing contract.

The change order generally originates as a claim or recommendation for a change from the contractor or as a request-for-proposal from the owner, seeking a change to the existing contract documents. The change order is necessary to increase and decrease the contract cost or work, interrupt or terminate the project, revise the completion date, alter the design, or in general to implement any deviation from the original contract terms and conditions.

Because change orders are responsible for more disputes than any other single aspect of a construction project, they should be carefully discussed in complete detail so that there is a complete understanding by both the owner and the contractor. Some of the items that are recommended by the Associated General Contractors for discussion include:

1. Percentages for overhead and profit to be applied to change orders; what costs will or will not be included in the change order price.
2. Length of time that a change order proposal price is to be considered as firm.
3. Determination of the individual representative of the owner who is authorized to approve change orders.
4. Procedures to be followed in the submittal of change order proposals.
5. Change order forms to be used (i.e., AIA, NSPE, Federal, Fisk,[1] contractors, or others).
6. Time extensions required, if any—requests made by contractors due to changes in the plans or specifications.
7. The detail required of contractors when submitting change order proposals—will a complete breakdown of all costs be required? Brief description—descriptive drawings.
8. Overtime necessary due to change orders. Consideration of decreased productivity.
9. When materials or equipment is to be removed due to a change, which party owns the removed items, and who removes them from the site of the job?
10. Responsibility for record drawings brought about due to the change orders.

[1] E. R. Fisk, *Construction Engineer's Form Book,* New York: Prentice Hall, 1981, 1992; Englewood Cliffs, NJ: Prentice Hall, 1992.

Evaluating the Need

A contract change order is always used to effect a change within a contract. Such changes should always be in writing. Standard forms such as those shown in Figures 19.5 and 19.6 are readily available for this purpose; however, standard forms are not necessary, as a change order can be in the form of a letter, if desired.

The following are some of the purposes served by change orders:

1. To change contract plans or to specify the method and amount of payment and changes in contract time therefrom.
2. To change contract specifications, including changes in payment and contract time that may result from such changes.
3. To effect agreements concerning the order of the work, including any payment or changes in the contract that may result.
4. For administrative purposes, to establish the method of extra work payment and funds for work already stipulated in the contract.
5. For administrative purposes, to authorize an increase in extra work funds necessary to complete a previously authorized change.
6. To cover adjustments to contract unit prices for overruns and underruns, when required by the specifications.
7. To effect cost-reduction incentive proposals (value engineering proposals).
8. To effect payment after settlement of claims.

A contract change order is used in most instances when a written agreement by both parties to the contract is either necessary or desirable. Such use further serves the purpose of notifying a contractor of its right to file a protest if it fails to execute a change order.

Considerations for Evaluation

The following are common categories or conditions that generally give rise to the need for a contract change order:

Differing site conditions
Errors and omissions in plans and specs
Changes instituted by regulatory agencies
Design changes
Overruns/underruns in quantities beyond limits
Factors affecting time of completion

It is possible that some change order proposals may fall outside these categories; however, others commonly have characteristics similar to the categories mentioned. Therefore, they can be related to the reasoning process developed in one or more of the six categories mentioned.

Evaluation of Delays in the Work

Before it can be determined that a delay in the work was compensable, thereby justifying the issuance of a change order granting both additional time and money to the contractor, the following questions need to be answered:

1. Was the cause of the delay beyond the contractor's control? Did the contractor fail to take normal precautions?
2. Was the contractor ready and able to work?
3. Did the contractor submit a detailed schedule projecting project completion within the allotted time? Was the schedule updated regularly? Did the updated schedule justify a time extension? Bar charts and dates "scratched out" on napkins by the superintendent while at the local diner are insufficient to justify or support change orders granting additional time because they don't show *why* more time is necessary!
4. Did this schedule contain a critical path analysis or equivalent?
5. Has the contractor maintained sufficient forces in those operations along the critical path where needed to meet target dates?
6. How have causes, other than normal weather, beyond the control and without the fault or negligence of the contractor affected the target dates along the critical path?
7. Has the contractor proven "unusually severe weather" with such information as climatological data, return probability of severe storms, or flood depth data?
8. Did the weather phenomenon actually delay operations along the critical path or in secondary operations?
9. Was the contractor shut down for other reasons?

Change Orders for Differing Site Conditions

Justification for the issuance of a change order to compensate the contractor for a differing site condition will be influenced by the specific terms of the Contract Documents. Some guidelines follow that suggest special areas of concern to be evaluated prior to issuance of a change order to cover extra work resulting from a differing site condition:

1. What is the differing site condition?
2. What is the site condition shown on the plans?
3. What sections of the specifications are applicable?
4. Do the parties have the latest revised issues of the plans and specifications?
5. Have the parties checked all of the plan references? Plans are usually grouped by specialty: architectural, site, plumbing, mechanical, structural, electrical, special and standard details. A critical note or dimension on one of the specialty drawings may solve the whole puzzle of need and responsibility.

6. Has the inspection staff provided adequate data? If not, correct that situation immediately. The next claim could be much larger.
7. Why was the condition shown differently by the design?
8. Was provision made for this situation in the contract documents?
9. Did the construction contractor encounter unstable soils, rock excavation, or subsurface structures where no careful prebid site inspection and contract documents could have predicted their existence?
10. Was the construction contractor forced to employ unusual construction techniques and equipment to overcome the obstacles encountered?
11. Can the construction contractor's performance, selection of construction procedures, and responses to site conditions be evaluated by the architect/engineer (or possibly a third party) experienced in modern construction techniques?

Subsurface Investigation

Prior to the determination of the validity of a contractor's request for a change order to compensate the contractor for a differing site condition, the architect/engineer and the owner should consider the following questions before assuming that the responsibility is all on the shoulders of the contractor:

1. Were borings made or test holes dug, displayed correctly, available to all bidders, of proper kind and depth, spaced at reasonable intervals? Contract documents often have disclaimers regarding differing subsurface conditions, but these are not always as enforceable as they may appear.
2. Was the overall subsurface investigation in proportion to the type and magnitude of the project?
3. Was the geologic history of the site incorporated into the subsurface investigation data displayed in the contract documents?
4. Although not part of the evaluation, the owner should consider whether additional subsurface investigation should be initiated immediately to minimize future claims and delays.

Inclusion of Soil Reports in Contract Documents

Whenever the owner or the architect/engineer knows of or is in the possession of soils reports, logs of borings, asbestos reports, hazardous waste reports, or similar knowledge about a construction site, it is under obligation to make full disclosure of all such documents to the bidders on a project. This is normally done through the medium of identifying them by title, origin, and date in the Notice Inviting Bids, and at the same time cautioning the bidder that they are *not a part of the contract documents,* and are identified for the convenience of the bidders. Usually, exculpatory language is included to the effect that the owner or its architect/engineer makes no representations regarding such data, and that the bidder must interpret and draw his or her own conclusions therefrom. Recent issues of EJCDC General Conditions, however, do allow the contractor's reliance on certain limited technical information, but not upon any interpretations contained in the report (Figure 19.7). It is again emphasized in the contract documents that the soils reports, for example, should *not* be a part of the contract documents. Inclusion of such documents as a part of the contract documents invariably leads to claims for extra work and change orders.

ARTICLE 4 - AVAILABILITY OF LANDS; SUBSURFACE AND PHYSICAL CONDITIONS; REFERENCE POINTS

4.01 *Availability of Lands*

A. OWNER shall furnish the Site. OWNER shall notify CONTRACTOR of any encumbrances or restrictions not of general application but specifically related to use of the Site with which CONTRACTOR must comply in performing the Work. OWNER will obtain in a timely manner and pay for easements for permanent structures or permanent changes in existing facilities. If CONTRACTOR and OWNER are unable to agree on entitlement to or on the amount or extent, if any, of any adjustment in the Contract Price or Contract Times, or both, as a result of any delay in OWNER's furnishing the Site, CONTRACTOR may make a Claim therefor as provided in paragraph 10.05.

B. Upon reasonable written request, OWNER shall furnish CONTRACTOR with a current statement of record legal title and legal description of the lands upon which the Work is to be performed and OWNER's interest therein as necessary for giving notice of or filing a mechanic's or construction lien against such lands in accordance with applicable Laws and Regulations.

C. CONTRACTOR shall provide for all additional lands and access thereto that may be required for temporary construction facilities or storage of materials and equipment.

4.02 *Subsurface and Physical Conditions*

A. *Reports and Drawings*: The Supplementary Conditions identify:

I those reports of explorations and tests of subsurface conditions at or contiguous to the Site that ENGINEER has used in preparing the Contract Documents; and

2. those drawings of physical conditions in or relating to existing surface or subsurface structures at or contiguous to the Site (except Underground Facilities) that ENGINEER has used in preparing the Contract Documents.

B. *Limited Reliance by CONTRACTOR on Technical Data Authorized:* CONTRACTOR may rely upon the general accuracy of the "technical data" contained in such reports and drawings, but such reports and drawings are not Contract Documents. Such "technical data" is identified in the Supplementary Conditions. Except for such reliance on such "technical data," CONTRACTOR may not rely upon or make any Claim against OWNER, ENGINEER, or any of ENGINEER's Consultants with respect to:

1. the completeness of such reports and drawings for CONTRACTOR's purposes, including, but not limited to, any aspects of the means, methods, techniques, sequences, and procedures of construction to be employed by CONTRACTOR, and safety precautions and programs incident thereto; or

2. other data, interpretations, opinions, and information contained in such reports or shown or indicated in such drawings; or

3. any CONTRACTOR interpretation of or conclusion drawn from any "technical data" or any such other data, interpretations, opinions, or information.

FIGURE 19.7 Provisions for Contractor Reliance on Soils Report Data in the 1996 Edition of the EJCDC Standard General Conditions of the Construction Contract. *(Copyright © 1990 by National Society of Professional Engineers.)*

Evaluation of a Claim of Differing Site Conditions

Once the information on differing site conditions has been assembled, the owner should be in a position to determine if the site conditions differed materially from those indicated in the contract or those ordinarily encountered and whether this will justify an increase in the construction contractor's cost or the time to complete the work.

The owner or its architect/engineer should be able to determine whether a change order is justified by a careful review of the contractor's supporting documentation, including reports, pictures, plans, and whatever is appropriate to demonstrate where and how actual conditions deviated from the plan or from conditions that should have been anticipated (see "Differing Site Conditions," Chapter 7).

Other Influences on Change Order Justification

Prior to filing a request for a change order due to alleged differing site conditions, the architect/engineer should determine whether or not the contractor complied with the following:

1. Did the construction contractor promptly and before such conditions were disturbed comply with the owner's modification requirements?
2. Did the construction contractor make a site inspection and evaluation of conditions prior to bidding?
3. Is the construction layout correct? Who performed the construction staking; who checked the staking?
4. Did utility companies respond and locate their respective facilities? Did the construction contractor protect or obliterate utility reference stakes and use suitable construction techniques to protect structures and utilities?

Starting the Change Order Process

Initiation of Change Orders

Change orders are usually initiated by construction personnel at the project site (Figure 19.8). However, changes are also requested from various other sources, such as the contractor, the design firm, outside public agencies, or private individuals. In short, any of the parties can initiate (propose) a change order; however, only the owner can authorize a change order. A proposed change order is written only after the designers have given consideration as to the necessity, propriety, other methods of accomplishing the work, method of compensation, effect on contract time, estimate of cost, the contractor's reaction to the proposed change, and the probability of final approval.

Any change in the work that involves a change in the original contract price must be approved in writing by the owner before a change order can be executed. If it is not the owner who signs, then the party who does sign for him or her must have written authorization from the owner to sign on his or her behalf. The architects or engineers of record, by virtue of their positions alone, have no authority to order

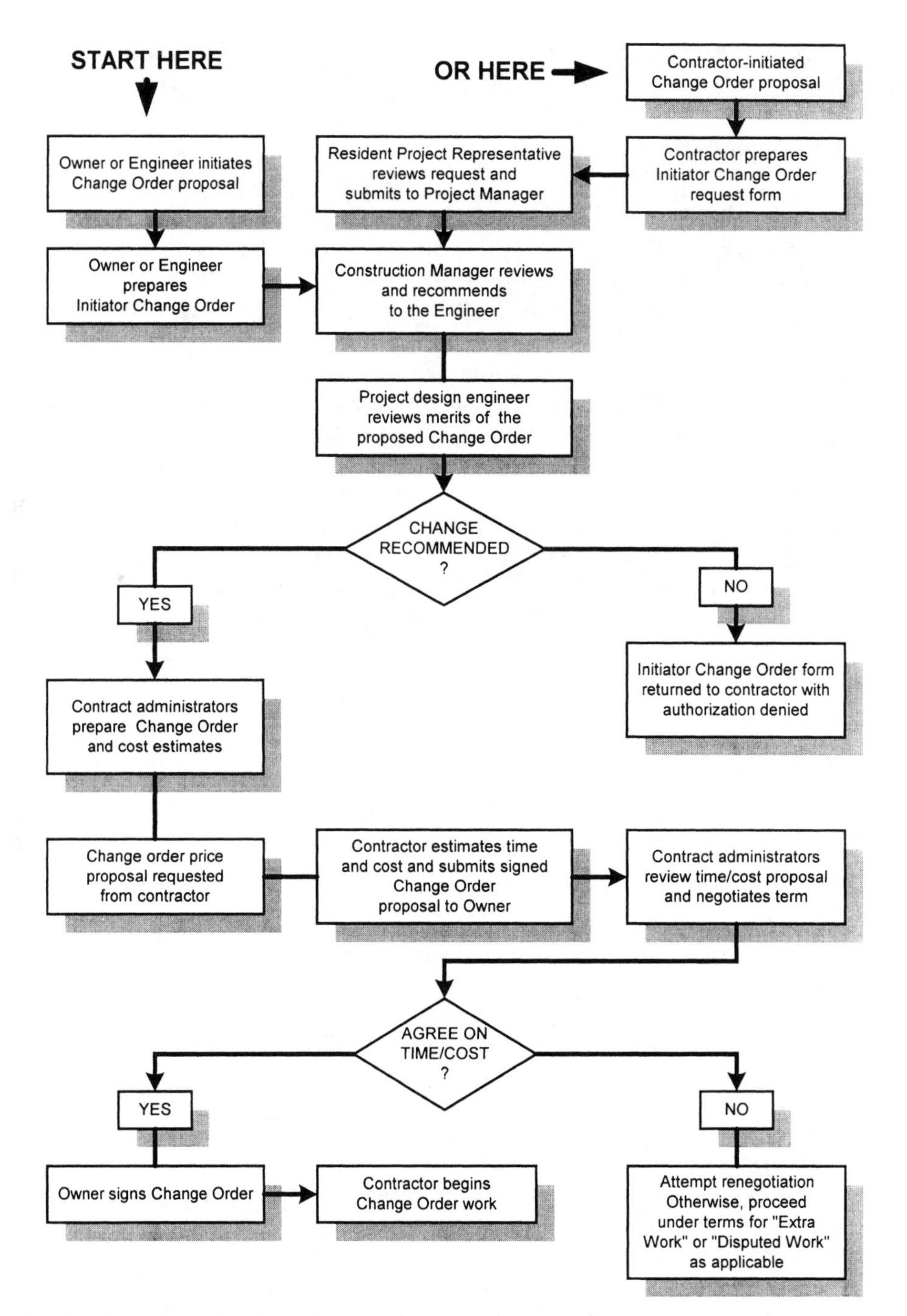

FIGURE 19.8 Flow Diagram Showing Change Order Procedure.

changes to the contract. If they do act in this manner, they must be authorized in some way to act in the owner's behalf.

Change Order Preparation

A change order must be clear, concise, and explicit. It must tell the contractor what is to be done, where or within what limits, when the work is to be performed, if the order of the work is affected, how the contractor will be paid, and what consideration will be given to contract time (extensions, etc.).

For users of the current EJCDC documents, the Work Change Directive (Figure 19.3) was developed for use in situations involving changes in the Work that, if not processed expeditiously, might delay the project. The AIA counterpart is called a Construction Change Directive. Such changes are often initiated in the field and may affect contract price or time. This form is not a change order, however, but only a directive to proceed with the Work that may be included in a subsequent change order. Unfortunately, the AIA form confuses the issue somewhat by having a block where the contractor can sign the document. If both parties are to sign the change document, you may as well issue a regular change order; after all, the purpose of that document is to allow the issuance of a unilateral order to perform the Work. The EJCDC, AIA and federal forms for implementing changes in the contract are:

Bilateral Contract Modification

Change Order EJCDC Document 1910-8-B

Change Order AIA Document G701

Federal Contract Modification

Change Order—Fisk Form 15-2

Unilateral Contract Modification

Work Change Directive EJCDC Document 1910-8-F

Construction Change Directive AIA Document G714

Federal Change Order

Work Directive Change—Fisk Form 15-3

Although the use of the change order and the Work Change Directive is quite clear and it satisfies all the requirements for documenting and executing an amendment to the contract provisions, the same cannot necessarily be said of a Field Order. The Field Order is supposed to be an order from the architect/engineer to the contractor that either interprets the contract documents or orders minor changes in the work without changing either the contract sum or the contract time. As a document to interpret the contract provisions, it serves its purpose admirably well. However, any document that changes any of the provisions of the plans, specifications, or other contract documents *is* a change order no matter what you may choose to call it. As such, in the absence of a written authorization from the owner that allows the architect/engineer unilaterally to make such changes, only the owner can execute this document.

On a recent project where a Field Order was used, it was desired to make a minor change in the work, for which there would be no additional charge by the contractor. When confronted with this document, the contractor asked that a formal change order be issued instead, even though the terms had been agreed to previously and no charge was being made for the changes.

The contractor was not only legally right in requesting a formal change order, but also wise to do so. A formal change order provides the written authorization of the owner to make a change in the contract provisions; it states the effect on the contract sum, if any; it specifies the effect on contract time, if any; and it could serve to release the contractor from liability that might result from making the change.

Any document that alters the terms of the original contract *in any way* must be considered as a change order. As such, the document must be signed either by the owner or by someone with the power of attorney to act for him or her. The architect/engineer's Field Order only provides for the signature of the architect/engineer and thus does not satisfy this requirement unless the architect/engineer also possesses written authority of the owner to execute such changes on his or her behalf. Even then, if a change in the work is intended, as opposed to an interpretation of the contract documents, it is still far more desirable to utilize the change order form for this purpose as a protection to all parties to the contract, even though there will be no change in contract price or time.

The AIA Construction Change Directive differs, principally, from the form of the EJCDC Work Change Directive (Figure 19.3) in that it provides for the written acceptance by the contractor.

There are only four basic steps involved in the execution of a nonemergency change order, although the in-house procedures of an organization may expand on this to satisfy its own procedural requirements. The four basic steps are:

1. Change order request by initiator (may be initiated by contractor or architect/engineer) for architect/engineer's and owner's approval.
2. Upon approval of initiator request, discuss with contractor and draft a change order proposal document citing the effect of the change on contract time and cost.
3. Contractor submits a signed change order proposal to the owner, showing all costs and additional time required.
4. Owner accepts proposal by signing the proposal and orders execution of the work described.

Cost of Delays Caused by Change Orders

Often, the execution of a change order involves slowdowns or delays of the contractor's operations. It should be kept in mind that all such delays involve extra cost to the contractor and, as such, the costs are recoverable by the contractor. In a case involving construction delays resulting from defective specifications for an FAA Air Force Traffic Control Center, the U.S. Court of Claims ruled that a contractor working on a government contract is entitled to the extra costs incurred as a result of the government's defective specifications, even if only part of the job is delayed [*Chaney and James Construction Co. Inc.,* v. *United States,* U.S. Court of Claims No. 150-67 (Feb. 20, 1970)].

This lends even more importance to the inspector's obligation to record all equipment and workers at the site, and to note whether such equipment is being used on that day or not (Chapter 4). Such information can be the basis for defending claims of excessive charges for delays caused by extra work.

PROBLEMS

1. A Work Change Directive under the EJCDC documents must be signed by which party or parties?
2. Who can initiate (propose) a change order—the owner, engineer, Resident Project Representative, or contractor?
3. What is a change order?
4. True or false? A change order can be issued at any time after award of the contract.
5. What is a Work Change Directive?
6. Under what conditions may an oral change order be valid and binding, even if the contract states that all changes must be in writing?
7. What is a constructive change?
8. Name at least three conditions that may justify the issuance of a change order.
9. May change orders be issued between the time of opening bids and the execution of the Agreement?
10. What are the principal issues that must be addressed in a change order?

20

CLAIMS AND DISPUTES

Five Principles of Contract Administration

Construction contracts are subject to broader principles of interpretation than most other contracts. There are five basic rules of contract interpretation that are quite important in construction contracts. The courts are frequently unfamiliar with the specialized rules that have evolved in the construction industry and often rely upon the testimony of experts in the subject area to guide them in forming a decision.

A written contract is merely a form of documentation of the conditions agreed upon by the parties. However, it reflects the understanding that each party had with regard to the wording of the documents. If every written contract was clear, unambiguous, and complete, the rules of contract interpretation would be unnecessary, as the intent and understanding of the parties would be self-evident. Unfortunately, this is not the case.

In any construction contract, it is almost inevitable that the written documents will not adequately address every single matter. There may be gaps, conflicts, or subtle ambiguities. The five principles mentioned here are used to help resolve such problems. They are little more than applications of common sense in an effort to determine the intentions of the parties from a written document that does not adequately express those intentions.

1. The Document Must Be Read as a Whole

Often in a construction dispute, each party concentrates on some narrow provision of the specifications, general conditions, or drawings that seems to support his or her position. The attitude is: "There it is in writing."

The law recognizes that the intentions of the parties cannot be determined by examining one small part of a document by itself. Every provision of the document is presumed to have meaning; therefore, no provision should be presumed to be meaningless or superfluous. The document must be considered in its entirety.

2. The Document Will Be Construed against the Drafter

This is a well-known principle but must be recognized in any dispute as to the meaning of the wording used in the document. There may be latent ambiguity; the contract may be subject to more than one reasonable interpretation, and a court is often faced with the burden of choosing between the two.

In such a situation, the document is interpreted in favor of the party that did not draft the document. The reasoning behind this principle is that the party that drafted the document had ample opportunity to avoid ambiguity and clearly express the intended meaning, but failed to do so. The nondrafting party did not have such an opportunity; it reasonably assumed that the contract language meant a certain thing, and in court that interpretation must prevail.

This means that in construction contracts the contractor's reasonable interpretation will usually prevail over that of the owner or the architect/engineer, and in the case of prime contractor/subcontractor relations, the subcontractor's reasonable interpretation will usually prevail over that of the prime contractor. Although not necessarily so in all cases, contracts are usually prepared by owners or architect/engineers, and subcontracts are usually prepared by prime contractors.

3. The Document Supersedes All Previous Discussions

The bottom line here is that any oral commitments made prior to the execution of an agreement are nonbinding, as the written document is presumed to constitute the entire agreement.

Basically, the rule is that the document speaks for itself. However, when it does not speak for itself, or when it is so inarticulate that no one knows what it means, oral testimony is permitted to determine the intended meaning of the parties. This is a well-known exception to the *parol evidence* rule. After hearing the expert testimony, the final interpretation of the contract will probably be decided based upon the respective credibility of each party's expert testimony. Unfortunately, this does not assure that justice will be served, as in the author's experience a case was reviewed in preparation for an appeal of an earlier decision in which the court heard conflicting opinions of two experts. The court said that it had heard two experts testify as to the opposite opinion on a crucial issue, and the court was being called upon to determine which expert was correct. Unfortunately, said the court, lacking specific technical knowledge of the subject matter, the court's only recourse was to accept the opinion of the expert "who sounded most credible." In our review of the case, it was found that the most credible, or most articulate, was the one whose testimony was in error. Unfortunately, it was too late to change that, as the decision had already been rendered, so the case was decided on the strength of erroneous testimony.

4. Specific Terms Govern over General Terms

Sometimes there will be a conflict between two different provisions of a written agreement. The courts rule that the more specific term should govern over the general term. The reasoning behind this is that a narrowly drafted provision was customized to fit a specific situation and thus reflects more accurately the parties' intentions than a "general" or "boilerplate" provision of the contract.

On construction contracts, this is often applied by making handwritten or typewritten provisions prevail over conflicting provisions found in the preprinted sections of the contract. Similarly, job-specific Supplementary General Conditions will prevail over the General Conditions.

5. The Document Must Be Read in the Context of the Trade

This is another commonsense rule. If a word or phrase has a commonly accepted meaning in the construction industry, that meaning must be applied when reading and interpreting construction contracts.

Construction Problems

Every construction project seems to include its share of problems. To one way of thinking, perhaps if it were not for these problems, the job of Resident Project Representative might never have been conceived. A good Resident Project Representative is not simply a person who has the unique ability to solve problems, for generally the contractor can do this. It is the inspector's ability to apply personal experience and knowledge to the project, to look ahead, anticipate the occurrence of various events resulting from the various approaches that the contractor might take, evaluate these conditions, and offer constructive assistance in seeing that the potential problems either never occur or are minimized. An inspector who lacks this quality contributes nothing to the project—but is simply an added cost.

To help you understand fully their application to the terms of the construction contract, several terms frequently used in this book will be defined. This chapter is divided into two major classifications. The first deals with the definition and the administrative procedures involved in the handling of protests, claims, and disputes under the terms of the construction contract; the second addresses itself to the methods available for the settlement of such differences, as well as an explanation of the principle of arbitration under the rules of the American Arbitration Association, which is receiving growing respect as a fast, economical, and fair method for the resolution of construction contract disputes.

Protests

The term *protests* as used here refers to disputes arising out of the issuance of a contract change order by the architect or engineer against the objections of the contractor. The terms of the contract documents for the orderly filing of written protests by the contractor as a means of establishing its claims for additional compensation or time to complete the work or claims of unfair treatment under the terms of the contract, short of turning the problem over to arbitration, are normally covered in the General Conditions (boilerplate). In Articles 9.11 through 9.12 and Article 16 (including Exhibit GC-A) of the EJCDC Standard General Conditions of the Construction Contract (1990 edition), and in Sub-Clauses 67.1 through 67.4 of the FIDIC Conditions of Contract (1987 edition), the procedures for the handling of protests and disagreements are specified in detail, up to and including referring the matter to binding arbitration. An example of the current provisions of the 1990 edition of the EJCDC General Conditions for dispute resolution is illustrated in Figure 20.1.

The normal procedure is for the Resident Project Representative to discuss with the contractor any objections to a particular change order or other directions of the architect/engineer. If the contractor's objections can be satisfied by minor changes in the provisions of a proposed change order or direction and such changes do not violate the contract provisions (this includes plans, specifications, addenda, previous change orders, codes, permit provisions, or similar constraints), the change should normally be discussed with the architect/engineer's or owner's project manager, and with their concurrence the change should normally be made. This will often avoid a formal protest, which is often costly even to the winner.

In the event that the Resident Project Representative cannot satisfy the contractor's objections, the project manager should be consulted to explore other possible means of settlement. When neither the Resident Project Representative nor the project manager can resolve the problem through normal procedures, the pending change order or other direction of the architect/engineer or the owner should be issued in its original form and the contractor reminded of the provisions of the contract documents that require the contractor's conformance with any such order of the architect/engineer or owner even if the contractor disagrees with it, under penalty of being considered in default on the contract. It should also be carefully noted in the Resident Project Representative's or inspector's diary that such a reminder was given to the contractor. This does not take away the contractor's right to file a letter of protest that could be used as a later basis of claims against the owner, provided that such letter is submitted within the time schedules that are specified in the General Conditions of the construction contract.

When a protest letter is received from the contractor, it should be examined carefully prior to acknowledgment of its receipt to assure that the basic requirements of the specifications are included in the contractor's letter. The architect/engineer or owner should then review the merits of the contractor's protest. If the architect/engineer or owner decides that the protest is without merit, it should issue a letter to the contractor advising of its rights under the contract to file legal claims. Normally, if the contractor fails to file claims in accordance with the provisions of the contract, especially as to the time of filing, it may waive its rights under the contract.

EXHIBIT GC-A to General Conditions of the Agreement Between OWNER and CONTRACTOR Dated________
For use with EJCDC No. 1910-8 (1990 ed.)

DISPUTE RESOLUTION AGREEMENT

OWNER and CONTRACTOR hereby agree that Article 16 of the General Conditions to the Agreement between OWNER and CONTRACTOR is amended to include the following agreement of the parties:

16.1. All claims, disputes and other matters in question between OWNER and CONTRACTOR arising out of or relating to the Contract Documents or the breach thereof (except for claims which have been waived by the making or acceptance of final payment as provided by paragraph 14.15) will be decided by arbitration in accordance with the Construction Industry Arbitration Rules of the American Arbitration Association then obtaining, subject to the limitations of this Article 16. This agreement so to arbitrate and any other agreement or consent to arbitrate entered into in accordance herewith as provided in this Article 16 will be specifically enforceable under the prevailing law of any court having jurisdiction.

16.2. No demand for arbitration of any claim, dispute or other matter that is required to be referred to ENGINEER initially for decision in accordance with paragraph 9.11 will be made until the earlier of (a) the date on which ENGINEER has rendered a written decision or (b) the thirty-first day after the parties have presented their evidence to ENGINEER if a written decision has not been rendered by ENGINEER before that date. No demand for arbitration of any such claim, dispute or other matter will be made later than thirty days after the date on which ENGINEER has rendered a written decision in respect thereof in accordance with paragraph 9.11; and the failure to demand arbitration within said thirty days' period will result in ENGINEER's decision being final and binding upon OWNER and CONTRACTOR. If ENGINEER renders a decision after arbitration proceedings have been initiated, such decision may be entered as evidence but will not supersede the arbitration proceedings, except where the decision is acceptable to the parties concerned. No demand for arbitration of any written decision of ENGINEER rendered in accordance with paragraph 9.10 will be made later than ten days after the party making such demand has delivered written notice of intention to appeal as provided in paragraph 9.10.

16.3. Notice of the demand for arbitration will be filed in writing with the other party to the Agreement and with the American Arbitration Association, and a copy will be sent to ENGINEER for information. The demand for arbitration will be made within the thirty-day or ten- day period specified in paragraph 16.2 as applicable, and in all other cases within a reasonable time after the claim, dispute or other matter in question has arisen, and in no event shall any such demand be made after the date when institution of legal or equitable proceedings based on such claim, dispute or other matter in question would be barred by the applicable statute of limitations.

16.4. Except as provided in paragraph 16.5 below, no arbitration arising out of or relating to the Contract Documents shall include by consolidation, joinder or in any other manner any other person or entity (including ENGINEER, ENGINEER's Consultant and the officers, directors, agents, employees or consultants of any of them) who is not a party to this contract unless:

16.4.1. the inclusion of such other person or entity is necessary if complete relief is to be afforded among those who are already parties to the arbitration, and

16.4.2. such other person or entity is substantially involved in a question of law or fact which is common to those who are already parties to the arbitration and which will arise in such proceedings, and

16.4.3. the written consent of the other person or entity sought to be included and of OWNER and CONTRACTOR has been obtained for such inclusion, which consent shall make specific reference to this paragraph; but no such consent shall constitute consent to arbitration of any dispute not specifically described in such consent or to arbitration with any party not specifically identified in such consent.

16.5. Notwithstanding paragraph 16.4 if a claim, dispute or other matter in question between OWNER and CONTRACTOR involves the Work of a Subcontractor, either OWNER or CONTRACTOR may join such Subcontractor as a party to the arbitration between OWNER and CONTRACTOR hereunder. CONTRACTOR shall include in all subcontracts required by paragraph 6.11 a specific provision whereby the Subcontractor consents to being joined in an arbitration between OWNER and CONTRACTOR involving the Work of such Subcontractor. Nothing in this paragraph 16.5 nor in the provision of such subcontract consenting to joinder shall create any claim, right or cause of action in favor of Subcontractor and against OWNER, ENGINEER or ENGINEER's Consultants that does not otherwise exist.

16.6. The award rendered by the arbitrators will be final, judgment may be entered upon it in any court having jurisdiction thereof, and it will not be subject to modification or appeal.

FIGURE 20.1 Provisions for Dispute Resolution in the 1990 Edition of the EJCDC General Conditions of the Construction Contract. *(Copyright © 1990 by National Society of Professional Engineers.)*

16.7. OWNER and CONTRACTOR agree that they shall first submit any and all unsettled claims, counterclaims, disputes and other matters in question between them arising out of or relating to the Contract Documents or the breach thereof ("disputes"), to mediation by The American Arbitration Association under the Construction Industry Mediation Rules of the American Arbitration Association prior to either of them initiating against the other a demand for arbitration pursuant to paragraphs 16.1 through 16.6, unless delay in initiating arbitration would irrevocably prejudice one of the parties. The respective thirty and ten day time limits within which to file a demand for arbitration as provided in paragraphs 16.2 and 16.3 above shall be suspended with respect to a dispute submitted to mediation within those same applicable time limits and shall remain suspended until ten days after the termination of the mediation. The mediator of any dispute submitted to mediation under this Agreement shall not serve as arbitrator of such dispute unless otherwise agreed.

ARTICLE 16—DISPUTE RESOLUTION

If and to the extent that OWNER and CONTRACTOR have agreed on the method and procedure for resolving disputes between them that may arise under this Agreement, such dispute resolution method and procedure, if any, shall be as set forth in Exhibit GC-A, "Dispute Resolution Agreement," to be attached hereto and made a part hereof. If no such agreement on the method and procedure for resolving such disputes has been reached, and subject to the provisions of paragraphs 9.10, 9.11, and 9.12, OWNER and CONTRACTOR may exercise such rights or remedies as either may otherwise have under the Contract Documents or by Laws or Regulations in respect of any dispute.

Decisions on Disputes:

9.11. ENGINEER will be the initial interpreter of the requirements of the Contract Documents and judge of the acceptability of the Work thereunder. Claims, disputes and other matters relating to the acceptability of the Work or the interpretation of the requirements of the Contract Documents pertaining to the performance and furnishing of the Work and Claims under Articles 11 and 12 in respect of changes in the Contract Price or Contract Times will be referred initially to ENGINEER in writing with a request for a formal decision in accordance with this paragraph. Written notice of each such claim, dispute or other matter will be delivered by the claimant to ENGINEER and the other party to the Agreement promptly (but in no event later than thirty days) after the start of the occurrence or event giving rise thereto, and written supporting data will be submitted to ENGINEER and the other party within sixty days after the start of such occurrence or event unless ENGINEER allows an additional period of time for the submission of additional or more accurate data in support of such claim, dispute or other matter. The opposing party shall submit any response to ENGINEER and the claimant within thirty days after receipt of the claimant's last submittal (unless ENGINEER allows additional time). ENGINEER will render a formal decision in writing within thirty days after receipt of the opposing party's submittal, if any, in accordance with this paragraph. ENGINEER's written decision on such claim, dispute or other matter will be final and binding upon OWNER and CONTRACTOR unless: (i) an appeal from ENGINEER's decision is taken within the time limits and in accordance with the procedures set forth in EXHIBIT GC-A, "Dispute Resolution Agreement," entered into between OWNER and CONTRACTOR pursuant to Article 16, or (ii) if no such Dispute Resolution Agreement has been entered into, a written notice of intention to appeal from ENGINEER's written decision is delivered by OWNER or CONTRACTOR to the other and to ENGINEER within thirty days after the date of such decision and a formal proceeding is instituted by the appealing party in a forum of competent jurisdiction to exercise such rights or remedies as the appealing party may have with respect to such claim, dispute or other matter in accordance with applicable Laws and Regulations within sixty days of the date of such decision, unless otherwise agreed in writing by OWNER and CONTRACTOR.

9.12. When functioning as interpreter and judge under paragraphs 9.10 and 9.11, ENGINEER will not show partiality to OWNER or CONTRACTOR and will not be liable in connection with any interpretation or decision rendered in good faith in such capacity. The rendering of a decision by ENGINEER pursuant to paragraphs 9.10 or 9.11 with respect to any such claim, dispute or other matter (except any which have been waived by the making or acceptance of final payment as provided in paragraph 14.16) will be a condition precedent to any exercise by OWNER or CONTRACTOR of such rights or remedies as either may otherwise have under the Contract Documents or by Laws or Regulations in respect of any such claim, dispute or other matter pursuant to Article 16.

FIGURE 20.1 (continued)

CLAIMS

Potential Claims

The term *potential claim* applies to any differences arising out of the performance of the work that *might* reasonably lead to the later filing of a formal claim by the contractor if the difference cannot be resolved in the field.

It should be the policy of the architect/engineer or owner to consider the merits of a potential claim at the earliest possible time. As soon as the Resident Project Representative has knowledge of the existence of a dispute that may lead to the filing of a potential claim, the situation should be discussed with the contractor. If the Resident Project Representative determines that the contractor's preliminary arguments are valid, such corrective measures should be taken as are within the scope of the inspector's authority under the contract, including the possibility of making recommendation to the architect/engineer or owner to submit change orders to alleviate the problem.

In the event that the Resident Project Representative cannot resolve the differences in the field, the problem should be discussed with the architect/engineer or the owner. If the differences still cannot be resolved within the terms of the contract, the contractor should be reminded of the provisions of the contract documents relating to the time and methods for it to file claims, and such reminder should be recorded in the Resident Project Representative's diary.

Upon receipt of a potential claim in writing from the contractor, and prior to submitting it to the architect/engineer or owner for review, it should be reviewed by the Resident Project Representative to see that it contains the basic information necessary, such as the reasons that the contractor believes additional consideration is due. It should also be reviewed to determine that the timeliness of the submittal is in accordance with the terms of the contract documents. If the contractor's letter is not sufficiently complete with respect to the nature of the claim, the architect/engineer or owner should request in writing that the contractor submit additional data before further processing of the contractor's claim.

Each potential claim should be reviewed by the architect/engineer or owner. If after this review the potential claim is considered as being without merit, an answering letter should be sent to the contractor. The letter should include the statements referred to under "protests" in previous paragraphs. If the contractor's notice of potential claim does not meet the requirements for timeliness as set forth in the specifications, the architect/engineer or owner should advise the contractor in writing of this deficiency.

Early Claims Reporting

The professional liability underwriters would prefer to have claims overreported rather than underreported, as it is better to report a dispute situation before it becomes an actual claim than to wait until it is a very hot issue and the battle lines have been drawn. The claims department of a professional liability insurance company can usually give counsel on methods of reducing exposure, or at least on keeping things fair. In addition, their claims department can advise on methods for mitigating damages (if any) so that whoever is finally held responsible, the financial burden will be less. By spotting a dispute before it becomes a claim and reporting it to the architect/engineer or owner, a monumental amount of later grief may be avoided.

Insurance Claims

Sometimes, through oversight or lack of understanding of contract practice, issues can arise that may spell latent financial disaster for the owner. On one occasion, the author was engaged by an insurance company in southern California to investigate a project that had just suffered major fire damage resulting from an arsonist who set the blaze about 4:00 P.M. on a Thanksgiving day when no one was at the site, resulting in a near total loss of a project that was at that time about 70 percent completed (Figure 20.2). The initial thrust of the investigation was to determine whether the project was a total loss, hence a higher payoff by the insurance carrier, or whether any remaining portions of the structure were salvageable. Copies of the policies and project drawings were made available and a detailed field investigation was started.

FIGURE 20.2 Fire Damage to a Project under Construction. Owner Suffered Total Loss Due to Failure to Include Policy Provisions in the Construction Contract.

Assisted by CTL laboratories (a subsidiary of the Portland Cement Association), the author made a thorough examination of the structural properties of the remaining portions of the concrete and concrete block portions of the structures, with a determination that there was a salvage value of over $300,000. All wood and steel members were a total loss, and the only surviving remnants of the structure were the reinforced concrete parking garage structure and some peripheral concrete block buildings.

The surprise came, however, when the insurance policy and the construction contract documents were examined. The policy involved appeared to be a standard builder's-risk fire policy. Although it was purchased by the owner, not the contractor, this is not unusual. The insurance industry informed me that often, the owner buys the policy; otherwise, the contractor buys the policy. The problem occurred because the policy terms and conditions were not included in the construction contract documents by the architect, and the endorsements called for a 1.8-m (6-ft) chain-link fence to be erected around the project and *24-hour security guard service.* Investigation showed that the project was fenced but no security guard service was provided. However, because the terms of the policy were not specified as a part of the contractor's construction documents, it was not the contractor's responsibility to provide either guard service or fencing. The contractor provided fencing on its own to protect its equipment, but was possibly completely unaware of the insurance endorsement requirements. Thus, the obligation to assure that the job was properly policed rested completely upon the shoulders of the owner, who unknowingly thought the contractor was obligated to provide security guards. The author's recommendation here was that the insurance company deny the entire claim because of the breach. A court of law later reached the same conclusion.

The lesson here is that whenever the owner elects to purchase its own policy of insurance on a construction project, it should make certain that its architect/engineer includes the terms and conditions imposed in the policy as a part of the contractor's obligation under the construction contract.

Claims and Disputes

The term *claim* applies to differences that *are* developed during the life of the contract under protests and potential claims, and that are not yet resolved at the time the contractor returns the proposed final estimate of the amount of additional money or time asked for. In other words, a protest or potential claim does not become a claim until the contractor repeats its objections by notifying the architect/engineer or owner at the time the proposed final estimate for such claim is returned to the architect/engineer.

Whether it is the competitive bidding process, increased competition, or just part of the growing trend toward more litigation, more and more projects are being affected by claims and disputes. Very often, a majority of contractor claims and disputes arise out of poorly drafted or ambiguous contract documents. Disagreements between the contractor and the owner, and increasingly including the architect/engineer, regarding interpretation of the plans and specifications, what should be considered as extra work to the contract, payment for contract work and change orders, extensions of time, damages for delay caused by either the owner or the contractor, changed or unforeseen conditions, performance of subcontractors, compliance with contractual requirements, partial acceptance of a project by the owner, and similar problems seem to be increasing. When such claims or disputes arise, the owner and architect/engineer must pay careful attention to the procedures set forth in the contract documents for the handling of claims and disputes.

Contractor Must Alert Owner

Typically, contract documents require a contractor to alert the owner and/or the architect/engineer immediately, or within a certain number of days, of any potential claim or dispute when it arises. Contracts normally require that any claim or dispute must be submitted to the architect/engineer or owner's representative as the first step in the claims-resolution process. Contracts often set forth an appeals procedure so that the contractor can proceed with its claim or dispute if the architect/engineer's decision is unfavorable. When all administrative steps set forth in the contract documents have been exhausted, the contractor may then, and only then, resort to the courts or arbitration, if provided for under the contract. The procedural steps set forth in the contract must be followed by the contractor. Failure to follow these steps could in itself be considered as a breach of the contract by the contractor.

Administrative Procedures

A contractor must exhaust all of its administrative remedies before resorting to the courts and must strictly adhere to any claim procedures mandated by law. For example, in addition to the claims and disputes provisions outlined in the typical owner contract, some states require that a claim first be filed by the contractor against the appropriate owner, with that owner subsequently rejecting the claim within a prescribed time, before the contractor has the right to take the matter to court. A contractor's failure to follow mandated government claims procedures can result in the contractor being barred from recovery on its claim.

Work Performed under Protest

Except for total nonpayment by the owner, or due to particular circumstances under a specific contract, the contractor cannot refuse to proceed with the work on the project when a claim or dispute arises. Normally, the contractor must continue its operations despite any claims or disputes. Generally, the contractor must perform any disputed work under protest and must still complete the project on schedule, leaving any unresolved claims or disputes until after the project has been completed.

When a claim or dispute arises, or when the contractor foresees that it will arise, the contractor must take two immediate steps. First, the contractor must give the owner notice of any claim or dispute, or any potential claim or dispute, as soon as possible. Most contracts contain very specific requirements for "Notice of Claims" or "Notice of Potential Claims" that set forth specific time frames within which a contractor must notify the owner. Although the courts have varied in their interpretation and enforcement of such time requirements, a contractor runs a risk when its claims are not filed in accordance with contractual requirements.

Owner Must Have Opportunity to Correct

Regardless of any contractual language regarding the time for making claims, common sense and equitable principles dictate that once a contractor becomes aware of a claim or a potential claim, it must be communicated to the owner. For example, if the owner's actions in some way delay the contractor and the project, the owner must be advised of the situation in order for the owner to have an opportunity to correct or alleviate the problem. If a contractor does not give notice to the owner that it is being delayed at the time that the delay is occurring, the contractor will have very little chance of recovering damages incurred as a result of such delay after the project is complete. Without proper notice, the owner will argue, and rightfully so, that it cannot be held liable to the contractor for a delay that it was never made aware of and never had an opportunity to do anything about.

Contractor's Right to File Claims

Contractors are entitled to file any real and provable claim that arises during the course of a project. Claims should not be made or taken "personally," as they are simply the means available to the parties to the contract to be able to adjust the contractual and economic relationship between them to meet changing conditions. However, the filing of claims is not a proper way for the contractor to make up for a bad bid, or to recoup losses incurred on the project. Ultimately, it is the integrity of the contractor that dictates what claims are and are not made against an owner. In general, if a contractor presents only claims to an owner that are legitimate and verifiable, it will stand a better chance of having those claims, and all other claims it files, settled in an equitable manner. If, on the other hand, the contractor makes it a practice to file claims for any and all reasons, some less legitimate than others, the contractor runs the risk of prejudicing its legitimate claims and reducing the possibility of resolving them without litigation—a very costly and delayed process for both parties.

A contractor is entitled to pursue every legitimate claim that it has, but must recognize that the owner is entitled to proper notice and adequate proof of the claim before the owner has any obligation to pay on the claim. Contrary to a bidding situation where the contractor assumes only the best-case circumstances, a contractor's claim is generally estimated by assuming only the worst-case circumstances in the determination of estimated impact costs yet to be incurred.

Particular attention should be paid to situations where claims are directly related to the actions or inactions of the owner's representatives or agents. Claims that sometimes arise as a result of the obligations of the owner's architect/engineer to design a portion of a project properly are assertable against the owner, and under some circumstances, directly against the architect/engineer.

Although claims often provide the contractor with an opportunity to attempt to make up for mistakes made in its bid or in the construction of the project, it is not a good practice for a contractor to make more claims simply to make up for mistakes and errors. No purpose will be served by a contractor who turns a project into a battleground to cover losses incurred as a result of its own mistakes and errors.

Differences between the Parties

In general, most claims issues will fall into approximately 10 classifications, with numerous variations possible within each category.

1. Owner-caused delays in the work.
2. Owner-ordered scheduling changes.
3. Constructive changes.
4. Differing site conditions.
5. Unusually severe weather conditions.
6. Acceleration of the work; loss of productivity.
7. Suspension of the work; termination.
8. Failure to agree on change order pricing.
9. Conflicts in plans and specifications.
10. Miscellaneous problems.

Owner-Caused Delays

The majority of all claims involve at least some elements of delay, even if the primary issue is one of the other categories. Most contractors fail to realize the potential for recovery of losses in this sensitive claims area until they have consulted a claims specialist.

Delay claims fall into three categories: nonexcusable, excusable, or compensable. Nonexcusable delay is one that is caused by factors within the contractor's reasonable control. Essentially, this means that the delay is the contractor's fault, so the contractor will be unable to recover additional time or additional compensation.

A delay that is caused by factors beyond the contractor's reasonable control, but is not the result of the owner's actions or failure to act, is considered excusable. An excusable delay entitles the contractor to an extension of time but to no additional compensation for the cost of the delay (Figure 20.3).

A compensable delay occurs when the owner fails to meet an obligation stated or implied in the contract. The owner must grant additional time and money for the resulting delay.

Contractor Entitled to Complete Early

An unusual twist on the delay issue occurs when a contractor plans to finish a project ahead of schedule. In some cases, an owner has been known to object to early completion of a project for varied reasons. However, it should be kept in mind that an owner may be held liable for hindering early completion of a project [*Appeal of CWC, Inc.*, ASBCA No. 264342 (June 29, 1982)]. CWC, Inc., was prime contractor for renovation of certain military buildings. The contract required the government to

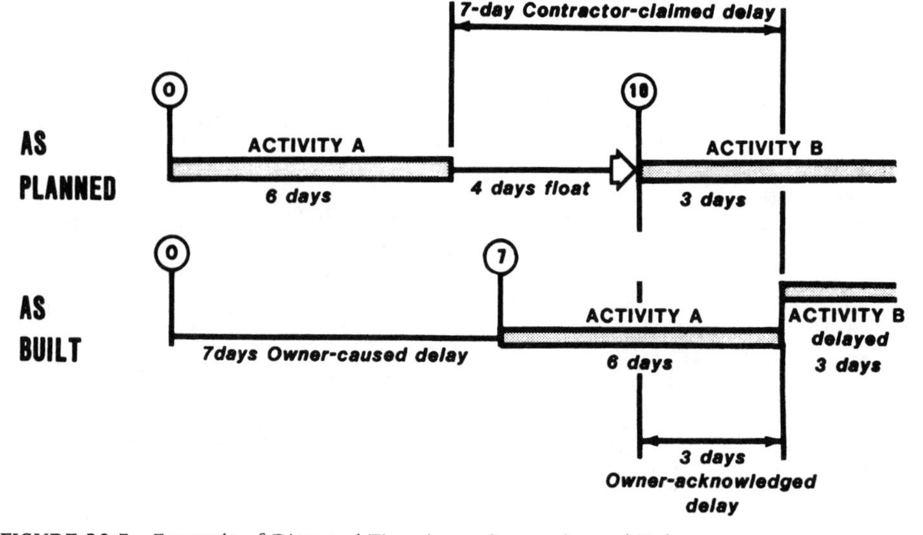

FIGURE 20.3 Example of Disputed Time in an Owner-Caused Delay.

arrange access to the individual buildings within two weeks after receiving a request from the contractor. In several instances, however, the government failed to provide access in a timely manner.

The contractor finished the entire project in less than the allowed time and prior to the completion date indicated in the contractor's own schedule. Nevertheless, the contractor submitted a claim for the increased costs caused by the government's delay. The Armed Services Board of Contract Appeals recognized the claim and issued the following statement: "Barring express restrictions in the contract to the contrary, a construction contractor has the right to proceed according to his [or her] own job capabilities at a better rate of progress than represented by his [or her] own schedule. The government may not hinder or prevent earlier completion without incurring liability." The bottom line here is that if the owner has reason to require the contractor to conform to a scheduled completion date, allowing neither delays nor early completion, this requirement must be clearly spelled out in the contract documents.

Sources and Causes of Time-Related Disputes

All too often, owners and architect/engineers automatically blame the contractor for virtually all construction delays. Although that is often a well-founded accusation, there are many sources and causes of time-related disputes, as indicated in the following list:

Source	Cause
Owner	1. Lack of expertise
	2. Long line of authority in project organization

	3. Delayed approvals of schedules and change orders
	4. Slow change order processing
	5. Failure to obtain permits
	6. Irrelevant milestone dates in documents
Contract Documents	1. Inadequate scheduling clauses
	2. Directing work sequence by owner or architect/engineer
	3. Drawings not indicating interfaces
	4. Permitting responsibilities vague
	5. Milestone dates and interface clauses
	6. Leverage for enforcement of schedule specification
	7. Coordination inadequately specified
Construction Manager	1. Lack of expertise in schedule management by designer
	2. Implementation of specifications
	3. Inadequate record keeping
	4. Inadequate schedule updates and progress monitoring
	5. On-site coordination
	6. Job meetings
Contractor	1. Noncompliance with specifications
	2. Schedule updates not done
	3. Reluctance to cooperate and coordinate
	4. Failure to meet milestone dates
	5. Not following permit requirements

Compensable versus Noncompensable Delays

A delay that is considered compensable is one where the owner has failed to meet an obligation stated or implied in the construction contract. If a delay is considered compensable, the owner must grant a time extension and reimburse the contractor for the increased cost caused by the delay. If a contractor experiences concurrent delays where one is compensable and the other is merely excusable, no compensation is allowed. Similarly, on concurrent delays where one is caused by the contractor and the other by the owner, the delay is considered neither excusable nor compensable to the extent of the overlap or concurrency [*Appeal of Rivera General Contracting*, ASBCA No. 25888 (April 30, 1985)]. It should be stressed that these definitions are general in nature. The parties have great latitude to contractually determine whether or not the contractor will be entitled to additional compensation, an extension of time, or nothing at all.

If an event delays an activity on the critical path, the entire project completion date will be delayed. To avoid delay, it is necessary to accelerate the performance of a subsequent critical activity or adjust the sequence of work activities to compensate for the delay. If a delay occurs to an activity not on the critical path, however, the impact is cushioned by the available free-float time. Delays to a noncritical activity, which do not exceed the amount of free-float for that activity, will not result in a delay to the completion date of a project. If, however, a delay to a noncritical activity directly caused an increase in the contractor's costs, such costs may be recovered by the contractor if the delay was the owner's responsibility. This, of course, is a matter requiring very clear proof to overcome the presumption that delays to noncritical activities do not result in delay-related costs. Although an owner-caused delay to a noncritical activity that used only free-float time may not be justification for an extension of time, such delays may still be compensable when they cause the contractor to accelerate the work, cause inefficiency or disruption, or delay some specific activity and thus cause extended costs, such as cost of equipment rental. In addition, such delay is not necessarily measured from the time that the contract documents required the project to be completed, but rather from the time that the contractor could have completed the project, from a completion date earlier than the one indicated in the contract documents [*Metropolitan Paving Co. v. United States,* 325 F.2d 241 (1963)]. The most valuable tools for the defense of a claim are properly prepared CPM schedules.

Delays are the most prevalent problem on construction projects. A variety of factors in an ongoing project can lead to a delay. Typical types of delays that might be caused by an owner include the following:

1. Late approval of shop drawings and samples
2. Late approval of laboratory tests
3. Delays in answers to field inquiries by the contractor
4. Changes in the contractor's method of doing the work
5. Variation in estimated quantities
6. Interference with the contractor during construction
7. Owner-caused schedule changes
8. Design changes
9. Changes in inspection level
10. Failure to provide for site access
11. Lack of required rights-of-way
12. Interference by other contractor's or owner's forces

A contractor may jeopardize its claim, however, by being the cause of nonexcusable delays of its own. Such delays are frequently the result of any of the following causes:

1. Late submittal of shop drawings
2. Late procurement of materials or equipment

3. Insufficient personnel
4. Unqualified personnel
5. Inadequate coordination of subs or other contractors
6. Subcontractor delays
7. Late response to owner and architect/engineer inquiries
8. Construction not conforming to contract requirements, making repeated reworking necessary

Attendance at Site Visit

From time to time, owners and architect/engineers feel that they will better protect the owner's interests by requiring all bidders to attend a prebid site visit. This is obviously based upon the presumption that the contractor might have a stronger claim for differing site conditions if he or she failed to attend a scheduled prebid site tour. However, it appears that the owner may not be as vulnerable as at first thought. In a 1986 case, the Comptroller General ruled that a contractor's failure to attend a prebid site visit is not a valid reason to reject an otherwise acceptable low bid [*Matter of Arrowhead Construction, Inc.*, Comp. Gen. No. B-220386 (January 8, 1986)].

In the subject case, the government requested bids for construction of a paint booth. The Notice Inviting Bids required all bidders to attend a site visit on a designated date. When the government announced its intention to award the contract to a bidder that did not attend the site visit, the second-lowest bidder protested. The Comptroller General denied the protest.

The decision was that a bidder that fails to attend a prebid site visit does so at its own risk and will be held responsible for any information that it could reasonably determine from that inspection. According to the Comptroller General, the site inspection does not relate to the responsiveness of the bid. Even when the Notice Inviting Bids expressly requires attendance at a prebid inspection, failure to attend is not a valid basis for rejection of an otherwise responsive bid.

Home Office Overhead

The costs of delays involve many elements. Naturally, the direct costs are affected, but a contractor's work efficiency, construction schedule, available favorable weather, impact costs, and even home office overhead may be affected. Normally, most contract documents specify the limits of overhead charges that may be levied for force account work, but it should be kept in mind that such limitations apply only to a contract that is completed within the specified time. If the owner causes a compensable delay that forces the contract into a longer time frame, the contractor may be entitled to unabsorbed or extended home office overhead.

The Eichleay Formula for Home Office Overhead

The most common way of computing the value of extended home office overhead is based upon the use of the *Eichleay formula*. The validity of this formula was challenged successfully in February 1983 but was reborn a year later when the U.S. Court of Appeals for the Federal Circuit reinstated it [*Capital Electric Co. v. United States*, Appeal No. 83-965 (February 7, 1984)]. The Eichleay formula is a method of calculating home office overhead damages in delays, suspensions, or extensions of work. Such damages have usually been called "extended" home office overhead.

The Eichleay formula is commonly expressed as follows:

$$\text{(a)}\quad \frac{\text{Contract billings}}{\text{total billings for the contract period}} \times \frac{\text{total overhead for}}{\text{the contract period}} = \begin{matrix}\text{overhead allocable}\\ \text{to the contract}\end{matrix}$$

$$\text{(b)}\quad \frac{\text{Allocable overhead}}{\text{Days of performance}} = \text{daily contract overhead}$$

$$\text{(c)}\quad \begin{matrix}\text{Daily contract}\\ \text{overhead}\end{matrix} \times \text{number of days of delay} = \begin{matrix}\text{amount claimed}\\ \text{(total added overhead)}\end{matrix}$$

The Eichleay formula requires proof of actual damage, however. The Armed Services Board of Contract Appeals ruled that although the use of the Eichleay formula is well recognized in a suspension-of-work situation, the contractor must still show that it was unable to fill in with other work during the suspension period.

In one instance, a contract awarded to Ricway, Inc., at an Army ordnance facility, the government issued a stop work order during construction, which remained in effect for 95 days. Ricway later submitted a claim for its unabsorbed home office overhead, computed according to the Eichleay formula.

The board acknowledged that Eichleay is an appropriate measure of unabsorbed home office overhead but said that it cannot be applied automatically. First, the contractor must show that it incurred actual damage, that is, that the contractor was unable to obtain other contracts that could have absorbed the contractor's overhead during the suspension period [*Appeal of Ricway, Inc.*, ASBCA No. 29983 (February 1986)].

Unabsorbed Home Office Overhead

Unabsorbed home office overhead is computed in a similar manner but relates only when the direct cost base is not large enough to absorb the fixed overhead (an indirect cost) at the contractor's "normal" absorption rate. Fixed overhead is an indirect cost in that it is a cost incurred for the benefit of more than one cost objective or project. Direct costs, of course, are always identified with separate projects. However, indirect costs cannot be identified in this manner, and as a result must be allocated or distributed in a logical manner among all of the contractor's projects.

The allocation of indirect costs is usually done by using a certain "base," for example, labor cost, contract billings, machine hours, or a similar base. The Eichleay formula uses *contract billings* as a base. A ratio of the base for a specific project to the total base for all projects is used to allocate the indirect cost. This is called *absorption costing;* the distributed cost is absorbed by the projects.

Underabsorption of Overhead

One of the problems with the Eichleay formula as a means of calculating underabsorbed home office overhead is that it does not truly reflect underabsorption of overhead. Assuming that a contractor can offer proof of entitlement, the Eichleay formula results in recovery regardless of whether there was an actual underabsorption of home office overhead. For example, if the delay is caused by the influx of a large amount of extra direct costs as a result of numerous change orders, the contractor's base for absorbing home office overhead may actually increase to the extent that overhead is overabsorbed despite the extended duration of the project. In other words, the contractor was not prevented from obtaining additional work and forced to allocate an unfair amount of overhead to the project. Instead, the contractor may have obtained plenty of additional work through additional change orders to which the contractor may have been able to allocate home office overhead.

As an example, consider a fictitious project that we shall call Project A. If work on Project A is suspended, the contractor cannot earn any revenue, and its billings decrease. However, during the suspension period, the contractor continues to incur the fixed overhead that was originally allocated to Project A prior to its suspension. This fixed overhead for the now suspended Project A is *unabsorbed* during the suspension period; there is no Project A revenue to absorb these overhead costs. In other words, the unabsorbed overhead of Project A must now be absorbed by the remaining projects. However, as the actual allocation base during suspension is less than that of the as-planned base, the remaining projects must support more overhead than was originally planned, or is "normal." For example, let us say that the normal overhead rate is 10 percent for all projects. Without Project A, it increases to 15 percent for the remaining projects. The suspension causes the overhead rate to increase. In this example, the overhead rate "differential" is a 5 percent increase.

It should be pointed out that overabsorption can also occur, such as when the base increases. For example, let us say that the suspended work of Project A is performed during a later period. Fixed overhead remains as planned, but the billings *increase.* Thus, the allocation base is higher than the as-planned base. Then, while the overhead rate may have been 10 percent, with a larger base it may actually drop to 5 percent. The overhead rate differential is then a 5 percent *decrease.* This is called *overabsorption* of fixed overhead.

Because the damages computed through the Eichleay method bear no relationship to actual absorption rates, contractors with home office overhead claims may be forced to rely upon other methods, such as two examples of construction delayed projects can illustrate. The first method is known as the *Comparative Absorption Rate* method (Figure 20.4). Under this method the underabsorbed overhead is determined by finding the difference between overhead actually incurred and overhead that would have been incurred if the contractor had been able to maintain a reasonable absorption rate. The second method is known as the *Burden Fluctuation* method (Figure 20.5). It has been used by Boards of Contract Appeals in calculating manufacturers' underabsorbed overhead claims. This method determines the underabsorbed overhead by finding the increase in absorption rate and allocating that increase to work on other projects that were forced to bear more than their fair share

Example No 1: A $400,000, 4-month contract with a 1-month suspension and no change orders.

	MO. 1	MO. 2	MO. 3	MO. 4	MO. 5
POTENTIAL PERFORMANCE					
Home Office Overhead	$ 40,000	$ 40,000	$ 40,000	$ 40,000	$ 40,000
Contract Billings	$100,000	$100,000	$100,000	$100,000	—
Other Contract Billings	$400,000	$400,000	$400,000	$400,000	$500,000
Total Billings	$500,000	$500,000	$500,000	$500,000	$500,000
ACTUAL PERFORMANCE					
Home Office Overhead	$ 40,000	$ 40,000	$ 40,000	$ 40,000	$ 40,000
Contract Billings	$100,000	$100,000	—	$100,000	$100,000
Other Contract Billings	$400,000	$400,000	$400,000	$400,000	$400,000
Total Billings	$500,000	$500,000	$400,000	$500,000	$500,000

METHOD #1—EICHLEAY FORMULA

$$\frac{\$400{,}000 \text{ (Contract Billings)}}{\$2{,}400{,}000 \text{ (Total Billings)}} \times \$200{,}000 \text{ (Total Overhead)} = \$33{,}333 \text{ (Allocable Overhead)}$$

$$\frac{\$33{,}333 \text{ (Allocable Overhead)}}{5 \text{ (Months of Performance)}} \times \$6667 \text{ (Monthly Contract Overhead)}$$

$$\times\ 1 \text{ Month Delay} = \$6667 \text{ Claim}$$

METHOD #2—COMPARATIVE ABSORPTION RATES

$$\frac{\$200{,}000 \text{ (Potential Total Overhead)}}{\$2{,}500{,}000 \text{ (Potential Total Billings)}} = 8\% \text{ Reasonable Overhead Ratio)}$$

$$8\% \text{ (Reasonable Overhead Ratio)} \times \$2{,}400{,}000 \text{ (Actual Total Billings)} = \$192{,}000 \text{ (Reasonable Total Overhead)}$$

$$\$200{,}000 \text{ (Actual Total Overhead)} - \$192{,}000 \text{ (Reasonable Total Overhead)} = \$8000 \text{ Underabsorbed Overhead Claim}$$

METHOD #3—BURDEN FLUCTUATION

$$\$2{,}400{,}000 \text{ (Total Billings)} - \$400{,}000 \text{ (Contract Billings)} = \$2{,}000{,}000 \text{ (Other Billings)}$$

$$\frac{\$200{,}000}{\$2{,}400{,}000} \text{ (Actual Overhead Rate)} - \frac{\$200{,}000}{\$2{,}500{,}000} \text{ (Potential Overhead Rate)} =$$

$$8.33\% - 8.00\% = 0.33\% \text{ (Burden Fluctuation)}$$

$$.33\% \times \$2{,}000{,}000 = \$6600 \text{ Underabsorbed Overhead Claim}$$

FIGURE 20.4 Comparative Absorption Rate Method of Computation. *(Reprinted from* Construction Claims Monthly *with permission from Business Publishers, Inc., 8737 Colesville Road, 10th Floor, Silver Spring, MD 20910-3928.)*

Example No 2: A $400,000, 4-month contract with no suspension of work and change orders worth $110,000 including $10,000 overhead mark-up. Potential performance is the same as Example No. 1.

	MO. 1	MO. 2	MO. 3	MO. 4	MO. 5
ACTUAL PERFORMANCE					
Home Office Overhead	$ 40,000	$ 40,000	$ 40,000	$ 40,000	$ 40,000
Contract Billings	$ 75,000	$ 75,000	$125,000	$135,000	$100,000
Other Contract Billings	$400,000	$400,000	$400,000	$400,000	$400,000
Total Billings	$475,000	$475,000	$525,000	$535,000	$500,000

METHOD #1—EICHLEAY FORMULA

$$\frac{\$510{,}000 \text{ (Contract Billings)}}{\$2{,}510{,}000 \text{ (Total Billings)}} \times \$200{,}000 \text{ (Total Overhead)} = \$40{,}637 \text{ (Allocable Overhead)}$$

$$\frac{\$40{,}637 \text{ (Allocable Overhead)}}{5 \text{ (Months of Performance)}} \times \$8127 \text{ (Monthly Contract Overhead)}$$

$$\times\ 1 \text{ Month Delay} = \$8127 \text{ Claim}$$

METHOD #2—COMPARATIVE ABSORPTION RATES

$$\frac{\$200{,}000 \text{ (Potential Total Overhead)}}{\$2{,}500{,}000 \text{ (Potential Total Billings)}} = 8\% \text{ (Reasonable Overhead Ratio)}$$

$$8\% \text{ (Reasonable Overhead Ratio)} \times \$2{,}510{,}000 \text{ (Actual Total Billings)} = \$200{,}800 \text{ (Reasonable Total Overhead)}$$

$$\$200{,}000 \text{ (Actual Total Overhead)} - \$200{,}800 \text{ (Reasonable Total Overhead)} = (\$800) \text{ Underabsorbed or } \$800 \text{ Overabsorbed Overhead}$$

METHOD #3—BURDEN FLUCTUATION

$$\$2{,}510{,}000 \text{ (Total Billings)} - \$510{,}000 \text{ (Contract Billings)} = \$2{,}000{,}000 \text{ (Other Billings)}$$

$$\frac{\$200{,}000}{\$2{,}510{,}000} \text{ (Actual Overhead Rate)} - \frac{\$200{,}000}{\$2{,}500{,}000} \text{ (Potential Overhead Rate)} =$$

$$7.97\% - 8.00\% = (0.03\%)\text{(Burden Fluctuation)}$$

$$(.03\%) \times \$2{,}000{,}000 = (\$600) \text{ Underabsorbed or } \$600 \text{ Overabsorbed Overhead}$$

FIGURE 20.5 Burden Fluctuation Method of Computation. *(Reprinted from* Construction Claims Monthly *with permission from Business Publishers, Inc., 8737 Colesville Road, 10th Floor, Silver Spring, MD 20910–3928.)*

of overhead expenses. An Eichleay calculation for each of the two examples is provided for further comparison.

Example 1 in Figure 20.4 illustrates a project that could have been performed by the contractor for $400,000 over a four-month duration, assuming no change orders were issued and no suspensions of work or other delays were encountered. The contractor in this example has a fixed home office overhead rate of $40,000 and regularly does $500,000 worth of total business per month, including the subject project. These data appear under the heading "Potential Performance" and reflect what would have happened if there had been no changes or delays caused by the owner on this project.

In the actual performance of the work, the contractor experienced no changes; however, there was a one-month suspension of work during the third month of the project. This resulted in a one-month delay to the contractor in question and the inability to take on $100,000 in new work after the planned completion date had passed. Under the Eichleay theory, the contractor would have a claim for only $6667 of extended overhead, based upon the one-month delay.

A comparison of absorption rates reveals that the contractor, in fact, suffered $8000 in underabsorbed overhead during the delay. This figure was arrived at by calculating a reasonable overhead rate, based upon what the contractor's overhead absorption rate would have been, had it not been for the delay. In this case, the rate equals 8 percent. By applying that rate to the actual billings for the actual period, the contractor can arrive at a reasonable total overhead of $192,000, which is $8000 less than the amount of overhead that the contractor had to absorb. In other words, the contractor was unable to maintain the 8 percent absorption rate during the delay. Had the contractor been able to maintain that rate, only $32,000 would have been expended on home office overhead during the delay.

Under the Burden Fluctuation method shown in Example 1 in Figure 20.4, the contractor would claim a 0.33 percent increase in its overhead rate. By applying that percentage increase to the work that had to bear the extra overhead cost, the contractor could claim $6600. Regardless of the method used, this example presents a classic case of underabsorption of home office overhead. Recovery, however, would be clearly dependent upon proof that additional work could have been obtained in other contracts had it not been for the delay on this project.

Example 2 in Figure 20.5 is based upon the same contract and contractor described in Example 1, but on a different type of delay. The actual performance included no suspension of work and resulted in the increase of the contract price by $110,000 for change orders, including $10,000 of overhead markup. Under the Eichleay formula, the contractor would have an extended home office overhead claim of $8127 despite the fact that change orders resulted in additional direct charges that could absorb the contractor's overhead.

Calculating the contractor's damages by comparing absorption rates shows that the contractor actually did not suffer from underabsorption of home office overhead. Instead of absorbing the overhead at the normal rate of 8 percent, the contractor, with the help of changes, was able to absorb overhead at the rate of 7.97 percent. This resulted in $800 of overabsorbed overhead. Moreover, if the $10,000 markup on the changes was required to be set off against the contractor's claim, the contractor would actually have absorbed its overhead at the rate of 6.8 percent. The Burden Fluctuation method would similarly reveal an overabsorption of overhead in the amount of $600, based upon the 0.03 percent decrease in the contractor's overhead rate.

The results obtained for the two examples through the Absorption Rate or Burden Fluctuation methods vary greatly, but the results obtained by using the Eichleay formula do not. This illustrates some of the objections to using Eichleay. The contractor in Example 1 who experiences a one-month suspension without receiving any extra work clearly suffers more damages than the contractor in Example 2, who receives extra work extending its contract through changes; yet the Eichleay formula would give the second contractor a larger recovery.

Flaws in Use of the Various Computation Methods

Several criticisms can be raised about the Burden Fluctuation method. Some of these criticisms also apply to the basic Eichleay approach. First, the use of a differential is not a cost accounting method but an approximation method that is based upon certain assumptions.

Burden Fluctuation assumes that all of the overhead rate differential is due to the one impacted project. It does not consider other factors that may impact a contractor's overhead costs, such as other sour jobs that a contractor may have. It may, therefore, overcompensate the contractor.

The normal rate is an *averaged* rate for all jobs. Different projects may, in fact, have overhead rates lower or higher than the average rate. How, then, is the normal rate for a particular job calculated? Even knowing the contractor's cost records and the nature of the indirect costs, it is difficult to find a way to know *directly* which part of fixed overhead differential is due to impacts, delays, suspensions, extensions, and so on, of one particular project.

Second, the unabsorbed overhead determined by the Burden Fluctuation method would be expected to be reduced by the amount of money received by the contractor on change orders.

Third, the Burden Fluctuation method does not account for the growth-decay bell curve of costs. This also applies to Eichleay.

Fourth, the method does not account for variable overhead costs. Burden Fluctuation treats all overhead costs as fixed, introducing some distortion.

Fifth, some home office overhead costs may be attributed directly to specific jobs or distributed on some percentage-of-effort basis instead of by allocation formula.

Sixth, a more appropriate base than total costs may be found to allocate overhead. One suggestion might be to use labor costs as a base. This would have the effect of overcompensation of a contractor on a labor-intensive job, or undercompensation of a contractor on a material-intensive job.

Thus, there are problems with the Burden Fluctuation method. The Eichleay method has some of the same problems. However, Burden Fluctuation does answer in part those complaints that the daily rate method of Eichleay is not related to actual costs. Through the use of actual overhead rates, the Burden Fluctuation method is more closely tied into actual costs.

While under certain circumstances, either the Burden Fluctuation method or the Comparative Absorption Rate formulas might be preferred to the Eichleay approach, in the sense of producing a more convincing claim, each of the alternative approaches is capable of producing claims that are inconsistent with the facts regarding a particular contractor's business, and thus may be unreasonable. The distortion is most likely when dealing with a large-volume contractor.

As an example, consider a large contractor ($300 million per year) whose $11 million, 20-month contract (which represents only a small fraction of the company's business) in affected by differing site conditions and takes 30 months for completion while generating $2 million through change orders. Plotted against these data, the expected performance on such a contract might be compared to actual performance as shown in Figure 20.6.

By comparing the expected and actual performance charts, it can be seen that (1) the contractor earned $116,667 per month less than projected from the extended (delayed) contract; (2) home office overhead increased each year (no direct correlation with the extended contract earnings); and (3) other billings were higher than projected for one year, and lower than projected for the last 10 months (no direct correlation with extended contract earnings).

Thus, at some time during the delay, additional work was available to absorb overhead costs (overabsorption), but at other times the delayed contract had to support

EXPECTED PERFORMANCE: 20 MONTHS, $11 MILLION	First Year	Second Year
Potential Performance	**Mo. 1–12**	**Mo. 13–20**
Home Office Overhead	$1 M/month	$1.3 M/month
Contract Billings	$550,000/month	$550,000/month
Other Billings	$25 M/month	$27 M/month
Total Billings	$25.55 M/month	$27.43 M/month

ACTUAL PERFORMANCE: 30 MONTHS, CHANGE ORDERS OF $2 MILLION	First Year	Second Year	Third Year
Actual Performance	**Mo. 1–12**	**Mo. 13–24**	**Mo. 25–30**
Home Office Overhead	$1 M/month	$1.3 M/month	$1.35 M/month
Contract Billings	$433,333/month	$433,333/month	$435,333/month
Other Billings	$25 M/month	$27 M/month	$22 M/month
Total Billings	$25.43 M/month	$27.43 M/month	$22.43 M/month

FIGURE 20.6 Comparison of Home Office Overhead Expected and Actual Performance. *(Reprinted from* Construction Claims Monthly *with permission from Business Publishers, Inc., 8737 Colesville Road, 10th Floor, Silver Spring, MD 20910-3928.)*

more than its projected portion of the overhead pool (underabsorption). Based upon these facts, the owner might argue that home office costs were covered despite the extended performance.

If the delay occurred during the first year of performance, when other contract earnings were higher than projected, the owner's argument might be meaningful, but still might not constitute an equitable basis for denying the payment of home office overhead to the contractor. Any overabsorption in the first year is caused not by the delayed contract, which is still earning less per month than reasonably projected, but by other factors, such as the contractor's managerial expertise or good fortune. For the owner to deny an equitable adjustment on that basis would be to let the owner take advantage of the contractor's good fortune or good business judgment to avoid paying for the logical consequences of the delay.

In the second year, the owner forced the contractor into a time of high overhead and low earnings that could not have been anticipated. Overabsorption in the first year, which was not caused by the owner, is balanced by underabsorption in the second year, which was a direct consequence of the owner-caused delay. The Eichleay formula ignores these considerations and bases its computation of the contractor's award on the actual allocated overhead expense and the number of compensable delays during the delay period. In comparison with the Burden Fluctuation method and the Comparative Absorption Rate, the Eichleay formula not only gives a smaller recovery, but is more realistic.

Eichleay	$201,196	($671/day × 300 calendar days)
Comparative Absorption Rate	$3.4 M	($11,333/day)
Burden Fluctuation	$2.94 M	($9,827/day)

While admittedly crude in terms of the number of variables it considers, the Eichleay formula per diem approach gives a daily value for home office support, such as management and clerical salaries, rent, computers, and so on, that bears a reasonable relationship to the value of the affected work (3.65 percent).

Scheduling Changes

Any scheduling change can have an important effect on project operations. When scheduling changes are implemented as a result of constraints created by the owner or architect/engineer, the basis for a contractor claim may exist. Whenever the owner or the architect/engineer issues any change order, field order, or work directive, or whenever a constructive change condition exists in any of the following areas, the contractor will begin to consider the impact of such action by the owner or architect/engineer upon its project costs and profitability.

The areas of primary sensitivity include owner- or architect/engineer-caused delays, requirements to deviate from the schedule, orders to expedite the work, job interference by the owner or architect/engineer, owner constraints on scheduling interfaces, late availability of owner-furnished material, owner-imposed acceleration or deceleration, impractical or impossible milestones, extra work, schedule impacts, sequence of construction by owner or architect/engineer, changes in completion dates, time extensions, utilization of scheduling float time, and schedule approvals. If a CPM schedule is not available during the job, any consistent scheduling method that illustrates the delay will help support the claim.

Any change in the schedule imposed by either the owner or the architect/engineer can become a basis for potential claim by the contractor. The best way to assure a sound, defensible claim is to utilize CPM scheduling techniques. This form of scheduling documentation offers the greatest protection.

The routine approval of contractor-prepared schedules can be the most costly mistake an owner can make. In the absence of a thorough review by an experienced, competent individual, contractually required schedules can become the primary source of documentation for successful contractor claims. The contractor can simply unveil its prebid, preconstruction, and progress schedules and compare them with the as-built or adjusted schedules. The comparison can be very graphic and may have a significant impact on a panel of arbitrators or a jury. As the owner had approved the contractor's schedules, the owner is left in a rather weak position.

The owner *should* ensure that updated contractor schedules are immediately reviewed to promptly resolve any problems: where is the critical path; what are the impacts of delayed/changed activities; and who is responsible. There will be few activities to consider and the recent reporting period's events will be fresh in everyone's minds.

As a job progresses, time extensions may be granted by the owner or architect/engineer. These would result in changes in the schedule and could result in revised interface dates or changed completion dates. The contractor will probably keep a close watch on the status of the schedules that it submits to the owner or architect/engineer for approval. Departures in the contractor's schedule from the schedule originally intended by the owner may be considered as a contract amendment after approval by the owner. If a contractor submits a shorter schedule than that originally called for under the contract, and it is approved by the owner or its architect/engineer, any liquidated damages provided under the contract may now be applied to the early finish date shown in the shortened schedule.

There is no question that the only way for a contractor to make a profit on a project is to get in and get out in the shortest possible time, thus reducing the home office overhead allocated to that project.

Constructive Changes

A *constructive* change is an informal act authorizing or directing a modification to the contract caused by the owner or architect/engineer through an act or failure to act. In contrast to the mutually recognized need for a change, certain acts or a failure to act by the owner that increases the contractor's cost and/or time of performance may be considered as grounds for a change order. This is termed as a constructive change; however, it must be claimed in writing by the contractor within the time specified in the contract documents; otherwise, the contractor may waive its rights to collect. Types of typical constructive changes may include:

- Defective plans and specifications
- Architect/engineer's interpretations of documents
- Higher standard of performance than specified
- Improper inspection and rejection
- Change in the method of performance by owner
- Change in construction sequence by owner
- Owner nondisclosure of pertinent facts
- Impossibility or impracticability of performance

An example of a constructive change resulting more *from failure to act* is the *Appeal of Continental Heller Corp.* [GSBCA 7140 (March 23, 1984)], where the government's failure to grant a legitimate request for a time extension has been held to be a constructive acceleration of the work schedule.

Continental Heller Corp. was awarded a contract for construction of a federal building in San Jose, California. Heavy rains at the start of the project made it impossible for the excavation subcontractor to proceed.

Nevertheless, the contracting officer insisted that the contractor stay on schedule and refused to grant a time extension until the contractor documented both the site conditions and the status of the activity on the critical path. An extension was finally granted 16 months after completion of the excavation. In the meantime, however, the subcontractor had switched to a more expensive method of excavation in order to remove the saturated soil in accordance with the original schedule.

The Board of Contract Appeals of the General Services Administration found classic elements of a construction acceleration case. The delay was excusable, yet the government forced the contractor to adhere to the original performance period, thereby causing the contractor to incur additional costs. Continental Heller was awarded $113,165 plus interest.

The Board of Contract Appeals, referring to the belated time extension granted to the contractor 16 months later, said: "As a defense to a claim of constructive acceleration, a belated time extension is worthless. . . . It had to have been clear to anyone who did not sleep through the entire two days that the soil at the site was saturated with moisture and could not be compacted as required. The government could not, by continually insisting on documentation of what was already known, justify its refusal to grant a time extension."

Other Causes of Claims and Disputes

Differing Site Conditions

Sometimes referred to as "changed conditions" or "unforeseen conditions," the term *differing site conditions* is typically used in all federal contracts, and there is a growing trend by many public agencies and a few private owners to adopt similar wording. Failure of an owner to provide payment for differing site conditions places the contractor in a difficult position. If the owner takes a hard-line position on this issue, the contractor may find it necessary to seek relief from the court, a process that is both lengthy and costly to both parties.

The federal policy is to make adjustments in time and/or price where unknown subsurface or latent conditions at the site are encountered by the contractor. The purpose is to have the owner accept certain risks and thus reduce the large contingency amounts in bids to cover such unknown conditions. The federal government and many local agencies include provisions in their construction contracts that will grant a price increase and/or time extension to a contractor who has encountered subsurface or latent conditions.

Under the definitions for differing site conditions given in Chapter 7, an existing underground pipeline either that was not shown on the drawings at all or that was incorrectly located on the contract drawings would be a Type 1 differing site condition. Unusually severe weather conditions for the time of year and location of a given project may well fall into a Type 2 differing site condition. The discovery of expansive clays in the excavation area, if not normally encountered in the location where the project was planned, and if not detected during soil investigations, may be either a Type 1 or 2 condition, depending upon circumstances. However, the discovery by the contractor of a permafrost condition in the tropics, for example, would most certainly be a Type 2 differing site condition.

Unusually Severe Weather Conditions

Severe rains or similar weather that prevents work from being done, or which in any way delays the project, may not always be excusable delays, and in some cases may be ruled excusable only and not compensable.

In the *Appeal of Inland Construction, Inc.* [ENG BCA 5033 (October 11, 1984)], a contractor was denied a differing site condition claim because the board found that the increased costs were caused by unusually severe rainfall, an event "which only entitles the contractor to an extension of time." In another case, although the contractor was delayed by rain, its claim of excusable delay was denied because the severity of the storm could have been foreseen by the contractor [*Appeal of B.D. Click Co., Inc.*, ASBCA 24586 (July 27, 1984)]. In that case, the Board of Contract Appeals ruled that weather delays are excusable only when the weather is abnormal. In the words of the board, "No matter how severe or destructive, if the weather is not unusual for the particular time and place, or if the contractor should have reasonably anticipated it, the contractor is not entitled to relief." Nevertheless, the owner or architect/engineer should document all weather delays as they occur. Determination of compensability can be made at a later date.

Although it is widely recognized that severe inclement weather is not necessarily a cause for excusable delay unless it is considered to be an abnormal weather condition that could not have been anticipated by the contractor, the problem is "What constitutes abnormal weather?"

Under a rule-of-thumb evaluation, any weather that results in an inability to perform work at the site, and which is in excess of the average annual rainfall over a historic period of, say, 5 or 10 years, might be considered excusable, but not compensable. Some specifiers attempt to develop a formula approach to resolving the difficulty of interpretation.

In documents prepared by HUD (Housing and Urban Development), a formula is used that seems quite generous to the contractor. Under the HUD documents it states, "Normal seasonal rainfall shall not be considered reason for a time extension. Normal seasonal rainfall shall be determined from the rainfall record of the U.S. Weather Bureau, taken at the recording station nearest the site of construction, and shall be monthly averages of the past thirty years, of the number of days in which 0.01 inch or more of rain was recorded." Thus, under the HUD formula, 0.01 inch of rainfall is generously considered sufficient to be defined as a "rain day."

Although the U.S. Weather Bureau has this information readily available, the HUD definition can result in a greatly erroneous conclusion as it is unduly generous to the contractor; it does not actually consider the weather at the project site, but rather the weather at the nearest U.S. Weather Bureau recording station. On a delay claim based on the HUD formula, in which the author was involved, the Contracting Officer had to rely on weather at the nearest weather recording station, which was at an airport at least 25 miles away.

Acceleration of the Work

Acceleration of the work is usually the result of an attempt by the contractor to take whatever means and measures are necessary to complete the work sooner than would normally be expected for a given project under stated conditions, or an attempt by the contractor to take extra measures to make up for delays, whatever the cause, by utilizing whatever means are at its disposal to accomplish the objective.

There are two types of acceleration:

1. Directed acceleration
2. Constructive acceleration

They are largely self-defining, but briefly, a "directed acceleration" occurs when the owner or architect/engineer specifically orders a contractor to speed up the work. The U.S. District Court outlined the necessary elements of a claim for constructive acceleration: "Constructive acceleration is present when (1) the contractor encountered an excusable delay entitling it to a time extension; (2) the contractor requested an extension; (3) the request was refused; (4) the contractor was ordered to accelerate its work, that is, to finish the project as scheduled despite the excused delays; and (5) the contractor actually accelerated the work."

In considering acceleration, the first consideration must be whether the so-called "acceleration" is the result of an order from the owner to a contractor who is behind schedule to get back onto schedule, or an order from the owner or architect/engineer to the contractor requiring a contractor either directly or constructively to complete the work prior to the scheduled completion date.

If a contractor is behind schedule and it is the desire of the owner or architect/engineer to require the contractor to get back on schedule, it is sometimes necessary to direct the contractor to take whatever means are necessary to assure completion by the originally scheduled completion date. Herein lies an administrative risk for the Resident Project Representative or contract administrator. If an order is issued to the contractor directing it to accelerate the work so as to catch up or make up for delays or lost time, the risk is that the contractor will perform as directed, then submit a claim for directed acceleration, arguing that although it appeared from the original schedule that it would not complete on time, actually the contractor may claim that there was no risk of failing to complete on time at all. The contractor might state that in the absence of the acceleration order it would have finished on time anyway. The proper handling of a situation such as this is not to issue an acceleration order (even though acceleration is justified or needed), but to send a letter to the contractor calling attention to the fact that completion by the scheduled date will be required, and that "according to the schedule it appears that the contractor will be unable to complete the project by the completion date indicated in the contract. Please resubmit a revised schedule showing how the contractor plans to complete the project by the scheduled date." This leaves the means and methods to the contractor, as it should be.

Often the problem starts in the design office. Project completion dates or time available to complete is frequently established by a designer who has little field experience, and thus is hardly qualified to properly establish the time needed to build a project. Typically, the time allotted to complete a project, as established by a designer, is either too short or too long.

If insufficient time is allotted to the completion of a project, all of the bidders will be forced to bid the job as accelerated work, thus increasing the costs materially. One indication that this may be happening is evident when all of the bids are in and, although they are grouped close together (a good sign of competent documents), they are considerably in excess of the amount of the engineer's estimate of anticipated project cost. This can be evidence that all of the bidders are bidding the job as accelerated work, thus increasing the cost of construction above that for a job completed within a normal schedule.

Another risk surfaces here as well. After figuring a bid for a job as being all accelerated work, a bidder may look in the contract documents to determine how much liquidated damages are being assessed in case of a completion delay. Upon finding daily liquidated damages to be assessed as somewhat less than $500 or $600 per day, the bidder now refigures the bid.

In the foregoing example, let us assume that we plan to build a project that on a normal schedule should take 16 months to complete, but the contract documents actually indicate that completion must be within 14 months. The bidder will first figure the bid for completion in 14 months, including the 2 months' acceleration of work that is necessary to complete by the required 14 months. Then the bidder may check the amount of liquidated damages called for in the documents. Let us say that the amount of liquidated damages was only $300 per day. An enterprising bidder will now go back to its 14-month bid and refigure the costs for completion in 16 months instead of the required 14 months. Then the bidder will determine that the difference in time between the time that was allowed by the owner for completion versus the time that a job such as that should normally take was two months, or 66 working days. The bidder may now multiply 66 times $300 per day and add the amount of liquidated damages to its total bid price for doing the work in 16 months, and submit it as its bid price.

The problem is, of course, that the cost of acceleration at $300 per day is considerably less than the daily cost to the contractor of acceleration; thus, the bidder submits a bid claiming that it will complete on time, when in fact it planned from the very beginning to complete the work two months late and pay liquidated damages costs. The owner and the designer, unfortunately, are the only persons who were unaware that the project was destined to be late before it was even started. Generally speaking, any job that has a Resident Project Representative operating out of an on-site field office can justify in excess of $1000 per day of liquidated damages without difficulty. Only then will liquidated damages serve as a deterrent to the contractor for finishing late.

A related condition called "deceleration" can also be experienced on a project. This occurs when a contractor is directed in writing or constructively to slow down its job progress. Many of the same considerations that apply to acceleration also apply to deceleration.

In preparing a claim for acceleration or deceleration, it should be borne in mind that the costs to the contractor for going into premium time, such as working an extended workweek, cannot be computed simply as including the added hourly costs multiplied by the additional hours. Studies have shown that as the workweek is extended, there is an accompanying loss in worker productivity. Furthermore, as the extended overtime is continued, the productivity rate continues to drop.

Productivity Losses

Scheduled overtime is not often seen on competitively bid lump-sum contracts, as most contractors are well aware of the negative effects of overtime on cost and productivity. Simple arithmetic shows that premium pay for double time or time and one-half makes overtime work much more expensive. However, people who insist on overtime seldom realize that other costs associated with overtime may be even more significant than premium pay. Premium pay affects only overtime hours, but continuing of scheduled overtime drastically affects costs of all hours. All available research findings indicate a serious inverse ratio between the amount and duration of scheduled construction overtime and the labor productivity achieved during both regular and overtime hours.

In the first few weeks of scheduled overtime, total productivity per person is normally greater than in a standard 40-hour week, but not as much more than the number of additional work hours. After seven to nine consecutive 50- or 60-hour weeks, the total weekly productivity is likely to be no more than that attainable by the same workforce in a 40-hour week.

Productivity will continue to diminish as the overtime schedule continues. After another eight weeks or so of scheduled overtime, the substandard productivity of later weeks can be expected to cancel out the costly gains in early weeks of the overtime schedule, so that the total work accomplished during the entire period over which weekly overtime was worked will be no greater, or possibly even less, than if no overtime had been worked at all.

When the loss of productivity is added to the higher wage cost (including premium pay), productive value per wage dollar paid after several weeks of scheduled overtime drops to less than 75 percent for five 10-hour days, less than 62 percent for six 10-hour days, and less than 40 percent for seven 12-hour days. When an overtime schedule is discontinued, it has been found that there is a dramatic jump in productivity per hour after return to a 40-hour week.

Construction delay claims involving acceleration of the work usually include claims for loss of productivity, which often exceed all other claimed amounts. The following breakdown of a claim on a treatment plant project serves to point out the relative magnitude of the claim for loss of productivity as compared with the other issues shown:

1. Extended project overhead	$1,019,099
2. Unabsorbed home office overhead	227,620
3. Labor escalation	142,430
4. Material escalation	148,329
5. Labor loss of productivity	2,442,409
6. Subcontractor claims	920,407
SUBTOTAL	4,900,294
7. Profit and overhead on items 1 to 5	1,298,575
8. Unresolved changes	157,993
9. Interest on money	1,073,897
10. Additional bond premium	11,844
TOTAL CLAIM	$7,442,603

Suspension of the Work; Termination

The work on the project can usually be suspended by the owner for any one or more of several reasons. In each case, the owner or architect/engineer should keep detailed cost isolation records of all activities affected by the suspension. It should be kept in mind that suspension of the work for any amount of time such that the completion date is extended may impact the contractor's costs through unabsorbed home office overhead and the real possibility of missing other projects due to the delay.

The contractor may also claim the effect upon its organization of the costs related to dismantling operations, mobilization or demobilization, direct costs, settlement expenses, escalation costs, prior commitments, post-termination continuing costs, unabsorbed overhead, unexpired leases, severance pay, implied agreements, restoration work, utility cutoff, inventory, replacement costs, and all other allocable costs. On suspensions of the work, be certain that all such orders are in writing and that a careful record is kept of its total effect on the contractor's time and costs (Figure 20.7).

Failure to Agree on Change Order Pricing

One of the most common causes of contractor claims occurs during attempts to price change orders. All too often, owner change orders contain a waiver clause that

FIGURE 20.7 Suspension of Work Necessary Because High Water Threatened to Flood the Lock Construction Site.

requires the contractor to guarantee that the price and time named in each individual change order represents the total cost to the owner for that change, and the contractor waives any rights to impact costs.

This, unfortunately, leaves the contractor with only one recourse—the claims process. The owner and architect/engineer should take note of the settlement ordered in at least one contract dispute. In the *Appeal of Centex Construction Co., Inc.* [ASBCA 26830 (April 29, 1983)], the Armed Services Board of Contract Appeals ruled that the government may not force a contractor to "forward-price" its impact costs when pricing a government-directed change.

In that case, Centex Construction Co., Inc., was a prime contractor on a military construction project. The government ordered a number of changes and asked Centex to submit cost proposals under the terms of its contract. Centex listed the *direct* costs of performing the changed work but did not list delay damages and other impact costs. When a change order was issued, Centex refused to release its right to submit a claim for impact costs.

Centex stated that it was impossible to forecast such costs accurately, particularly when an ongoing series of changes is involved. The Appeals Board agreed with Centex, stating:

> While it might be good contract administration on the part of the government to attempt to resolve all matters relating to a contract modification [change order] during the negotiation of the modification, use of a clause which imposes an obligation on the contractor to submit a price breakdown required to "cover all work involved in the modification" cannot be used to deprive the contractor from its right to file claims.

Errors and Omissions in Plans and Specifications

Errors and omissions usually are design or drafting deficiencies in the plans and specifications. Errors on plans and specifications are items that are shown incorrectly, while omissions are items that are not shown at all.

Errors and omissions in plans and specifications may be expected on any large, complex project and are usually brought to the attention of the owner through the Resident Project Representative or construction manager, the engineer, the construction contractor, or a supplier. Many errors are caught and corrected by alert inspectors before incurring unnecessary additional costs. However, some errors may lead to disputes between the engineer and construction contractor. The owner is responsible for resolving the dispute and may have difficulty in assessing the situation, since the engineer may have also been the designer and may strongly feel that the design was adequate. Moreover, there may be no clearly discernible line between design errors and faulty materials or installation by the construction contractor. The owner may seek opinions from independent third parties experienced in modern design practices to aid in the more complex decision making, but on federally funded projects the owner should request prior approval from the funding agency before contracting with a third party.

The terms of most federally funded projects require that the correction of errors/ omissions in the drawings or specifications be done without additional compensation. Such agencies do not accept time spent negotiating an engineer/construction contractor dispute as grounds for an extension of the time of completion. However, a finding of design deficiency may be grounds for time extension equal to that delay caused by the error itself.

The following questions must be examined before a decision can be made as to the action, if any, that is required to resolve the issue involved:

- What is the error or omission?
- Where are all the pertinent references in plans and specifications?
- Is the intention of the pertinent references obvious and will they result in meeting project objectives?
- Did the error/omission lead to increased construction contractor costs?

To the extent that was intended if the error or omission had not occurred, a change order may be necessary when some part of the project will not operate properly without making a change.

The owner should evaluate these requests carefully, as not all errors or omissions necessitate a change to the contract price and/or contract time. While the construction contractor may be entitled to an equitable adjustment, it may have incurred no additional expense and may not have a right to a claim even though an error or omission has occurred. Furthermore, if expense was decreased, a credit could be due the owner. The owner should pursue the available remedies against all parties who are responsible for the added costs of a change to protect itself.

The National Society of Professional Engineers/Professional Engineers in Private Practice (NSPE/PEPP), Professional Liability Committee reported[1] that in 1994 the National Research Council (NRC) published the results of a study it conducted on errors and omissions on federal projects in which it suggested that construction changes due to architectural and engineering errors and omissions should not increase the cost of construction more than 5 percent. A similar study was conducted by the Construction Industry Institute (CII) in the late 1980s in which it was stated that the correction of design errors and omissions might reasonably be expected to impact the cost of construction for a typical project in the range of 2 percent to 3 percent. From these studies it is clear that absolute perfection is not anticipated in construction, but the reported level of acceptability seems to vary significantly from 2 percent to 5 percent.

All too often there is a tendency to charge professional negligence, when the real problem is simply human error. The dividing line between negligence and ordinary errors and omission can best be explained by the following quotation from the standard civil jury instructions published by the Committee on Standard Jury Instructions, Civil, of the Superior Court of Los Angeles, California, and commonly referred to as BAJI (Book of Approved Jury Instructions—Civil).[2] Article 6.37 states:

PROFESSIONAL NEGLIGENCE—DUTY OF PROFESSIONAL

In performing professional services for a client, a [Engineer or Architect] has the duty to have that degree of learning and skill ordinarily possessed by reputable engineers or architects practicing in the same or a similar locality and under similar circumstances.

It is his or her further duty to use the care and skill ordinarily used in like cases by reputable members of his or her profession practicing in the same or a similar locality under similar circumstances, and to use reasonable diligence and his or her best judgment in the exercise of his professional skill and in the application of his learning, in an effort to accomplish the purpose for which he or she was employed.

A failure to fulfill any such duty is negligence.

Further, under BAJI Article 6.37.2 it goes on to address the subject of judgmental errors by a professional person:

PROFESSIONAL PERFECTION NOT REQUIRED

A [Engineer or Architect] is not necessarily negligent because he errs in judgment or because his efforts prove unsuccessful. Such a person is negligent if the error in judgment or lack of success is due to a failure to perform any of the duties as defined in these instructions.

[1]From *Construction Contingency: Standard of Care vs. Cost of Errors and Omissions,* by L.G. Lewis Jr., PE, NSPE/PEPP Professional Liability Committee. Reported in *Engineering Times,* published by the National Society of Professional Engineers.

[2]*California Jury Instructions—Civil* (7th edition), © 1986 by West Group, St Paul, MN 55164-0526. Reproduced with permission.

Conflicts in Plans and Specifications

This is an often misunderstood area in contractor claims. However, the probability of recovery by the contractor as the direct result of such conflicts is good insofar as the settlement is limited to the cost difference between the project cost as the plans and specifications are interpreted by the owner or architect/engineer and the contractor.

Public works contracts are called *contracts of adhesion,* which is a term applied to contract documents that are drawn by one party and offered to the other party on a take-it-or-leave-it basis, where there can be no discussion of terms nor contract modifications by the other party. The contractor, however, does have one advantage. In case of ambiguity, the court will interpret the contract in the contractor's favor. This does not relieve the contractor from the obligation of building the work in accordance with the interpretation of the architect or engineer, but only assures that the contractor will get paid for its trouble.

Frequently, the contractor will find in the specifications that outdated standards are specified, or that products are named that no longer exist. Often, the specifications will contain references stating that wherever codes or commercial standards are specified, the contractor is obligated to use the latest issue of that standard existing at the time that the project went to bid. Unfortunately, in many cases the designers failed to consider the fact that the design was based upon an old standard that exists in their files, or a standard that was current during the design phase, but which later may have been updated by the sponsoring agency without the designer's knowledge. Occasionally, serious difficulties arise from such practices, and the contractor certainly should have the right to the project cost difference resulting from the error.

Perfect specifications are hard to find. In fact, there has never been one known to the author, so the contractor must make a reasonable interpretation at the time of bidding the job and will then be in a good position for recovery if a variance exists. The contractor's interpretation must be based upon what a reasonable person would interpret the documents to mean; then the contractor will have the court on its side.

A contractor should never attempt to construct any questionable area without first submitting to the owner or architect/engineer for clarification or notifying them of an error. In many contract forms, failure to do this may serve as a bar to full recovery of contractor costs.

Miscellaneous Problems

Problems may arise that are directly related to the conduct of the owner's representative on the job, such as issuing changes in the work that are of such magnitude as to constitute a cardinal change (creating a breach of contract) or indirectly related, where one contractor may negligently delay another, which may result in the owner seeking recovery from the negligent contractor in order to pay the contractor that was harmed.

Some of the types of problems that may fall into this category include the following:

Damage to work by other prime contractors

Breach of contract

Cardinal changes

Work beyond contract scope
Beneficial use of the entire project before completion
Partial utilization of the project before completion
Owner nondisclosure of site-related information
Owner's failure to make payment when due

Resolving Differences

Even if a perfect set of contract documents were to be devised, there would still be disagreements between parties. Disagreements may arise between the contractor and the architect/engineer or owner concerning the interpretation of the contract; what constitutes extra work on the contract; payment for changes; extensions of time; damages for owner-directed acceleration or slowdown; costs occasioned by owner-caused delay; defective drawings or specifications; and changed conditions (sometimes referred to as unforeseen conditions, based upon the logic that the conditions have not actually "changed" but were that way all the time; discovery of the difference between the actual conditions and the way they were represented in the specifications or drawings is referred to under this concept as "unforeseen" differences). Similar matters may affect contract cost or time required to complete the work. Contract documents routinely include procedures to be followed in the settlement of such claims and disputes. The greatest difficulty arises from the fact that there is no uniform method of handling such unforeseen conditions, and numerous books and technical articles have been written about it; many cases are brought to suit over such differences; and the matter is not much closer to the development of a fair and equitable contract provision that will apply to all such cases than it was at the beginning.

Although their provisions may vary, construction contracts typically require the contractor to notify the architect/engineer immediately upon recognizing a situation that can lead to a claim or dispute. In cases of conflicts between the drawings and the specifications (a frequent cause of difficulty), it is the obligation of the contractor to notify the architect/engineer immediately without going further with what it knows to be improper work.

A word of caution at this point. The Resident Project Representative should always be aware of the possibility that a contractor who is alert for extras may recognize a defect in the plans or specifications, yet hastily construct the work exactly as shown on the defective documents, knowing all the time that it will be directed to remove such work and reconstruct it later as soon as the architect/engineer or owner discovers the same defect. A situation such as this could be a race between the contractor and the inspector to see if the contractor can get the incorrectly specified or detailed work constructed before the inspector notices it and reports it to the architect/engineer or owner. The logic of the plan is simple. If a contractor can feign ignorance of the problem and simply claim that it was just following the plans and specifications (which, undoubtedly, it would be), then, when the architect/engineer or owner discovers the defect in the plans and specifications and orders the newly constructed work to be removed and reconstructed, the contractor will willingly rebuild the work in any manner subsequently ordered by the architect/engineer or owner—followed later by bills for extra work, which, of course, the owner then becomes obligated to pay.

If a dispute cannot be settled with the architect/engineer or owner, and they order the contractor to continue with the work as directed by the architect/engineer or owner, the contractor cannot usually refuse without becoming potentially liable for breach of contract. Although the contractor may perform such work under written protest, it must continue to do the work and keep the operation on schedule, relying on the contractual remedies in the contract documents to settle the question of compensation or additional time.

Resolution by Negotiation

Although the individual provisions may vary, most construction contracts for public works contain language regarding the method of resolution of claims, either by litigation or arbitration. In the construction industry, there appears to be far more emphasis on the resolution of disputes by arbitration. Unfortunately, in a contract of adhesion (which includes any public works contract) the contractor is not offered a choice, except in those states where arbitration statutes prohibit contract provisions that agree at the time of signing the contract to submit to binding arbitration on future disputes.

Yet, in many cases, neither arbitration nor litigation is the sensible answer. Keep in mind that if the owner or architect/engineer goes to court or to arbitration, the costs on smaller claims may exceed the settlement, even if they win: Whenever the value of a claim is under \$50,000 or even \$100,000, it is doubtful whether either party can ever win financially. The only true winners will be the lawyers and claims consultants (Figure 20.8). Fighting for principles may be a fine hobby for another

IF YOUR CLAIM GETS TO COURT
THERE WILL BE ONLY ONE WINNER...
...THE LAWYERS

FIGURE 20.8 If Your Claim Gets to Court, There Will Be Only One Winner—The Lawyers.

Howard Hughes, but to most people and organizations it is financially unsound. The only economically sound solution is to negotiate.

Negotiation involves compromise, so it should be entered into with that in mind. Remember that claims in the dollar range mentioned previously cannot ever be won by anyone except the lawyers, so even if a party collected only 10 cents on the dollar, it would actually be ahead financially. Put up a good fight; do some hard bargaining, but be prepared to compromise. It should also not be forgotten that unless a little is left on the table for the other party to save face, it may be forced into arbitration or litigation—a position that neither party can afford.

Preparations for Claims Defense[3]

Documentation

One of the most important things to be considered in presenting any claim is documentation. In cases other than construction, very often the facts can be reconstructed from public records, police documents, or witnesses. The data required to support construction claims cannot generally be reconstructed after the fact without careful documentation. One of the most important functions of the Resident Project Representative is the establishment and implementation of procedures for documenting what may eventually become a claim. Figure 20.9 is a checklist that can be used to check present documentation policies for their potential usefulness in the defense of claims.

An owner would be wise to have its legal counsel investigate any engineering problem that might be likely to lead to a claim. Although there are instances where the architect/engineer can function effectively without the aid of legal counsel by making engineering determinations and documenting them for future use by legal counsel, there are very few instances where the attorneys should be involved in a potential construction claim without the advantage of the objective input of the architect/engineer. If negotiation fails, the documented data may be of considerable later value in litigation or in arbitration. The author recalls a few instances where failure of the firm's inspector to keep a diary or other daily record placed the employer in an indefensible position, and the firm was persuaded by its legal counsel to accept whatever out-of-court settlement could be reached, as they had no case at all if they went to court.

Organized documentation of the facts surrounding the alleged wrongdoing is the single most important weapon in a case. There is nothing that can replace regular, detailed record keeping during construction. The tremendous volume of documentation is common to all projects, good and bad; therefore, the key is in the organization of the various documents. It is an important duty of the Resident Project Representative and the project manager to provide or assist the attorney in producing a chronologically

[3]Adapted from I. M. Fogel, "The Claims Engineer," *The Military Engineer,* 457, September–October 1978, pp. 341–342. Copyright © 1978, The Society of American Military Engineers, Alexandria, VA. Reproduced by permission.

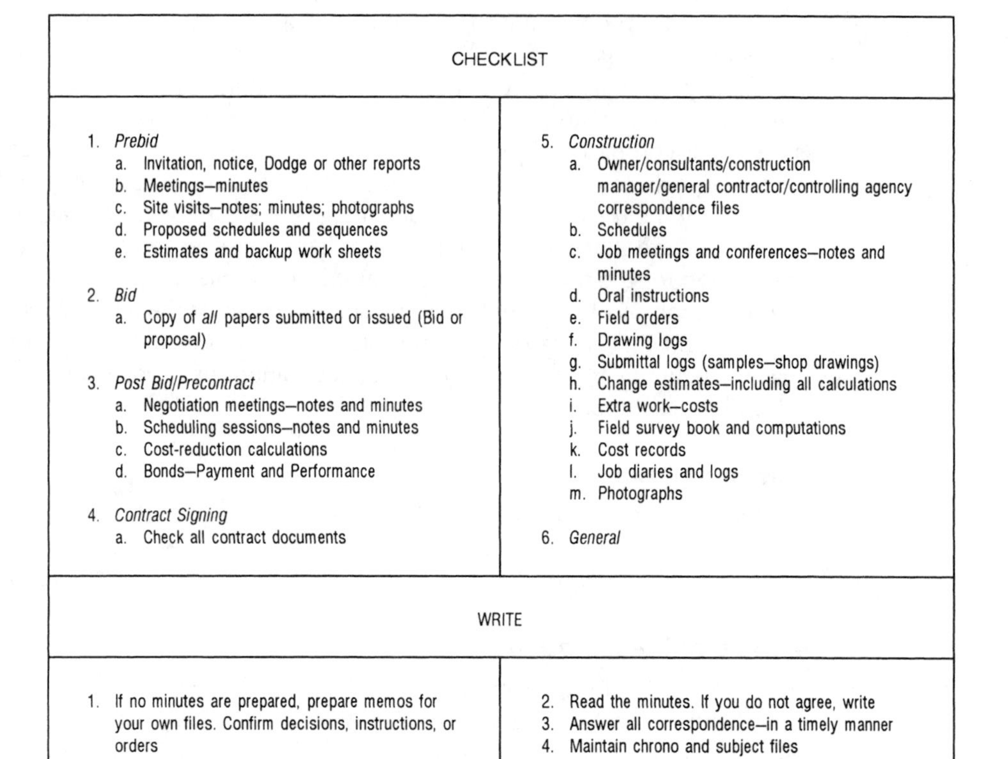

FIGURE 20.9 Checklist of Documentation for Claims Protection.

organized summary of facts of the dispute. At an early stage, the attorney should obtain copies of documents and other information held by the contractor through a pretrial process of discovery before they can be disturbed.

Another important source of evidence is the live witness, including the expert witness. The owner and architect/engineer should provide their attorneys with a list of all potential witnesses, including those who may testify against the claim, since it is equally important that the opposition's strengths are known early in the case. A live witness is especially important where the dispute concerns an oral agreement, such as an oral change order, for which there is no subsequent written memorandum. The exact spoken words can also provide valuable evidence where the dispute involves some subject matter that was left out of the written memorandum, either by error or because it was not directly relevant to the agreement. Given a solid reputation, the most important qualification of a person who will serve as an expert witness is integrity. The soundness of the integrity of the expert witness will be tested by the opposition in many ways, but most importantly by questioning the witness as to whether or not he or she is too closely connected with the case to give an impartial opinion.

In addition to proving the defendant's liability, the plaintiff is faced with the task of establishing the validity of the damages claimed. The general rule is that the prevailing party may recover whatever damages that may reasonably be supposed within the contemplation of the parties at the time that the contract was made. This is the most complex and time-consuming phase of construction litigation. Often, the case is split into two separate trials, one for determining liability and the other for determining the damages. The advantages of separating the trial into two separate actions are that the determination of liability issues may narrow and simplify preparation for the damages phase; that it may enable the parties to get an earlier determination of the entire dispute by settlement after the liability issue has been decided; and that it may reduce the number of parties involved in the damages phase. The greatest disadvantage is that separation into two separate actions may delay final resolution.

Burden of Proof

Generally, it is the plaintiff who has the burden to prove its claim. Where the contractor succeeds in proving substantial compliance against the owner's claim of nonperformance, the burden of proof reverts to the owner to prove its claim for residual defects.

Method of Presentation

The ordinary fact finder, whether it be a jury panel or a judge, is not likely to be familiar with the technical and administrative details of construction. Therefore, the factual issues must be presented in a simple, concise manner. Photographic evidence and charts are very effective in conveying the facts. Actual site visits may save time and energy consumed in otherwise describing the general physical conditions. An expert witness who can translate technical concepts into simple language is an invaluable evidentiary aid.

Evidence

In all disputes, regardless of the method of resolution, the prevailing party will be the one that can better support its burden of proof of its allegation. In preparing to bring a dispute to some kind of resolution, the contractor must assure that it is equipped with all of the necessary evidence in support of its claim or defense, and anticipate what the opponent will produce as evidence of their position on the claim. The following is a checklist of the possible sources of evidence that should be used in the course of dispute settlement:

Documentation

1. Bid documents
2. Boring logs and soils reports
3. Drawings: as-planned
4. Drawings: as-built

5. Specifications
6. General and supplemental general conditions
7. Schedule: as-planned (CPM, etc.)
8. Schedule: as-built (progress charts)
9. Addenda
10. Change orders
11. Instructions and directives
12. Inspection records
13. Contractor's logs
14. Architect/engineer's or owner's diaries (through deposition)
15. Correspondence files
16. Check registers
17. Purchase orders
18. Shipping and delivery tickets
19. Time cards
20. Person-loading charts
21. Memoranda (including memos to file)
22. Site photographs
23. Testing results
24. Public records
25. RFIs (Requests for Information) and responses

Witnesses

1. Contractor's project personnel
2. Owner's project personnel
3. Architect/engineer personnel
4. Construction manager personnel
5. Subcontractors
6. Inspectors
7. Suppliers
8. Testing lab personnel
9. Consultants (expert witnesses)
10. Adverse witnesses

There is no claims protection or support without adequate documentation. It is the single most important thing that an owner and architect/engineer or a contractor can do prior to the existence of a claim to provide the attorney with the tools needed for an owner, an architect/engineer, or a contractor to defend or prosecute its case, whether it be in arbitration or litigation.

Estimates

Invariably, every claim seems to include some form of estimate. An objective review of the source material and methods upon which the estimate was based will often show weaknesses or deviations from accepted industry practice. In any case, it is extremely difficult to determine the amount of a claim or counterclaim without preparing careful estimates of the work involved.

Schedules

Another facet of construction that seems to contribute to the most controversy is scheduling. Scheduling controversies arise not only from poor planning, on occasion, but also from disagreement over construction methods or sequences of operations. Often, these disputes occur where a project is constructed under multiple prime contracts, such as those commonly experienced in a professional construction management (PCM) contract. The absence of any contractual relationship between the individual prime contractors or the architect/engineer has been the root cause of numerous disputes.

Scheduling techniques such as CPM or other network scheduling systems and their associated computer-generated printouts have only served to compound the problem. Often the owner has stacks of computer printouts that it either does not want or does not know how to read. Yet somewhere in that stack of paper may well be the key to success in the defense of claims being made against it. The question of scheduling methods of operation and the impact of deviating from the anticipated schedule requires a careful analysis to determine the reasonableness of the originally anticipated schedule in conjunction with the planned methods of operation. The analysis of every schedule-related problem is unique, and every such analysis should include a review of the anticipated sequence and schedule together with a review of the actual progress of the work. The review must include an analysis of any delay and impact caused by all parties to the progress of the project.

Costs

It is in the presentation of cost data and their supporting documentation that most construction claims are the weakest. A carefully detailed reconstruction of all construction activities and their related costs must be made and presented so that they can be clearly identified for use in negotiating affirmatively or defensively. Construction-related organizations are reported to be poor in their cost record-keeping systems (or lack of systems) and to lack the ability to relate costs to possible claims areas. An accounting firm is not the answer either, as often those cost items relating to potential claims are not identified by an accounting firm because most accounting firms are not familiar with the cost control procedures for construction contracts.

Claims Administration

There are several common forms of settlement of contractor claims costs. Principally, these include lump-sum settlements with or without qualification, and time and materials settlements. The type of settlement generally favored under Clean

Water Grant Projects is the lump-sum approach. With that type of settlement the owner and the contractor reach an agreement on the total amount of money and time required to complete the work. This agreement may be reached either prior to or after the work has been completed, with the advantage that it places a price ceiling on the entire dispute, and on Clean Water Grant Projects facilitates the grantees' application for funding from the administrators of the funding.

The most desirable form of a lump-sum settlement is the so-called "no fault" settlement, which allows resolution of the dispute without determination of fault. With this type of settlement, in exchange for the issuance of a change order, the contractor gives a full accord and satisfaction and a waiver of all claims arising out of the incident.

A variation of the no-fault settlement is one wherein the owner may obtain a unilateral reservation of legal rights that will enable the owner to sue for a portion of the claim settlement after full payment has been made to the contractor upon completion of the project. This type of solution is of special value on Clean Water Grant Projects by helping to solve the contractor's cash flow problems related to the claim and gives the owner a chance to recover funds that may later be determined to be ineligible for payment by the administrators of the grant funds.

With the time and material settlements, the contractor performs the work and submits documented charges for all materials, labor, and equipment used on the affected work. This type of settlement has one advantage, namely, the opportunity for the contractor to begin work without prior negotiation of costs involved. It has a disadvantage as well. By this method a contractor can artificially increase the cost and time required to complete the work and inflate the costs submitted to the owner for payment.

A similar approach is sometimes referred to as *force account*. This is based upon a contractual agreement whose terms are generally spelled out in the General Conditions of the contract for payment of extra work. It is generally applied in conditions involving an unexpected condition that leads to extra work and involves a prearranged formula for reimbursing the contractor for the costs of labor, materials, and equipment plus a fixed percentage for overhead expenses and profit, and the allowable surcharge that will be permitted to be applied to the billings of its subcontractors and suppliers. A contractor does not waive its rights to claim extra costs over and above the amount allowed by force account, if it can be justified and verified; however, it is contractually bound to proceed with the extra work as long as the owner continues payments to the contractor under the formula. The advantages and disadvantages are similar to those for time and materials settlements as described earlier. It should be kept in mind that a force account settlement is open-ended, as opposed to a lump-sum settlement that may cover all costs of an item of extra work.

The Use of Project Records in Litigation

Planning Strategy

Before a dispute even reaches the courtroom, the value of project records becomes quite evident. During the early stages of case preparation they can be used in helping define the issues of a complaint. They are essential to the establishment of the facts in a case and to the planning of litigation strategy. In addition, project records are valuable as an aid to weighing the credibility of an additional source.

Prelitigation Use of Records

During the prelitigation phase of case preparation, the project records are invaluable in the preparation of interrogatories and depositions, for without competent and thorough documentation, key issues could easily be overlooked. They are also of considerable value during the preparation of the trial outline itself.

The Litigation Process

Initially, the project records may be used to define contractual parameters within which the parties should function. Then, during litigation, these records may be used to define relevance and establish probative value of facts under dispute.

The value of the records does not end here. They are also used to establish credibility or, conversely, for impeachment. Project records are also of considerable value in ascertaining the true impact (damages) from the harm that was alleged or suffered.

Records Are Your First Line of Defense

The construction environment of today is far too complex and vulnerable to claims to leave matters to be settled orally in the field. All matters of substance should always be in writing, and it is better to document too much than too little.

If sufficient documentation is maintained, many potential claims problems can be resolved equitably without resorting to the courts. The same records that can assist in the resolution of difficulties in the field can be used to support your position if a claim later goes to litigation or arbitration. Many of the records that are kept in the normal course of business are admissible as evidence in litigation or arbitration, so it is important that sufficient data be recorded and that the documents meet certain standards to be meaningful. Diaries can normally be used by a witness during testimony as an aid to his or her memory if such diaries meet the prescribed requirements for credibility. In some cases, such diaries are admissible as evidence in their own right as a record kept in the normal course of business.

Properly kept records are the first line of defense. Don't lose a case before it even starts by failure to develop and implement an effective record-keeping system. It is equally important that all parties working on a project be thoroughly indoctrinated in the proper use of the record-keeping system. The best system in the world is of no value if not faithfully used.

Order of Precedence of Contract Documents

To assist in the resolution of field disputes in cases where a discrepancy exists between the various parts of the contract documents, it is desirable to establish some system of order to indicate the relative order of importance of each. Although through legal precedent some degree of order has already been established, many

General Conditions cite the specific order so that a dispute may be settled without the need for interpretation or arbitration.

Although a specific listing of order, if specified in the contract documents, will govern over any policy listed here, the following guide is based upon the NSPE General Conditions, with some modifications.

> The contract documents are complementary; what is called for in one is as binding as if called for in all. If the contractor finds a conflict, error, or discrepancy in the contract documents, it should be called to the attention of the engineer in writing before proceeding with the work affected thereby. In resolving such conflicts, errors, and discrepancies, the documents should be given preference in the following order:
>
> 1. Agreement
> 2. Specifications
> 3. Drawings

Within the preceding elements, further preference can be established as outlined in the following breakdown of a typical set of public works contract documents (on private contracts it is uncommon to have the Notice Inviting Bids, the Instructions to Bidders, and the Bid or Proposal as a part of the contract documents).

Within the specifications, the order of precedence is as follows:

1. Change Orders or Work Change Directives
2. Agreement
3. Addenda
4. Contractor's Bid or Proposal
5. Supplementary General Conditions (Special Conditions)
6. Notice Inviting Bids
7. Instructions to Bidders
8. General Conditions of the Contract
9. Technical Specifications
10. Standard Specifications
11. Contract Drawings

With reference to the drawings, the order of precedence is as follows:

1. Figures govern over scaled dimensions.
2. Detail drawings govern over general drawings.
3. Change order drawings govern over contract drawings.
4. Contract drawings govern over standard drawings.
5. Contract drawings govern over shop drawings.

The acceptance of shop drawings that deviate substantially from the requirements of the contract documents must be accompanied by a written change order.

Obligations of the Contractor

As might be expected from a document prepared by the owner, the contractor's rights under the terms of the construction contract documents are often few and its obligations many. However, certain standard forms of General Conditions of the Construction Contract, such as those of the AIA, the EJCDC, and many public agencies, provide fairly equitable conditions for all parties to the contract, which is one reason that builders generally favor such standard forms over those typewritten, do-it-yourself varieties produced by the usually well-meaning but often poorly informed specifier or even attorney who is unfamiliar with construction contract requirements.

The contractor is expected to give its personal attention to the work, and either the contractor or its authorized representative must be on the construction site at all times that work is being performed there. Some contractors will leave the job site without adequate supervision in an attempt to entice the resident engineer or inspector into directing some of the construction work (actual direct supervision). Such action by the resident engineer or inspector would then shift a portion of the job responsibility to the owner or the architect/engineer. In cases of defects in construction in which the inspector has exercised such supervision, the contractor has grounds for being legally relieved of the financial responsibility of correcting such defects.

In addition to the requirements for providing continuous, adequate supervision, the contractor is also required to conform to all laws and ordinances concerning job safety, licensing, employment of labor, sanitation, insurance, zoning, building codes, environmental conditions, and similar constraints.

The general contractor, as the only party with a direct contractual relationship with the owner for construction, is responsible for and must guarantee all materials and workmanship, whether constructed by its own forces or by its subcontractors. The resident engineer or inspector should check the specifications thoroughly, as they may contain a provision that the contractor must arrange for an extension of the performance bond through the entire guarantee period, usually one year from the date of completion of the work and its acceptance by the owner. The requirement may call for the bond in the full amount of the original contract price or may allow a reduced amount such as 5 or 10 percent for the term of the guarantee period.

Even though the contractor normally has no direct responsibility for the plans and specifications, it can incur a contingent liability for proceeding with faulty work whose defects should be obvious to one in its business. Thus, if the contractor is asked to do something that it feels is improper or not in accordance with good construction practice, the contractor is entirely within its rights to submit a letter of protest to the architect/engineer or owner stating its position before proceeding with the matter in dispute. If after the contractor has submitted its written protest, the architect/engineer or owner prefers to require that the contractor proceed anyway, the contractor has protected itself in case of later flaws that may develop. It may even be possible that the insuring firm could get out of paying a claim if a flaw developed later as a direct result of not heeding the contractor's advice—a matter for the courts to decide.

Insurance coverage is an important contractual responsibility of the contractor, both as to type of insurance and policy limits. The contractor is required to provide insurance not only for its own direct and contingent liability, but also frequently for the owner's protection. The contractor is expected to exercise every reasonable safeguard for the protection of persons and property in, on, and adjacent to the construction site.

Some of the most important of the contractor's rights include progress payments, recourse by the contractor if the owner fails to make payments, termination of the contract for cause, rights to extensions of time and extra payment as provided, and appeals from the decisions of the owner and the architect/engineer. Subject to contractual requirements and contractual limitations and, in the case of public works contracts, subject to the local laws concerning the use of subcontractors, the general contractor is normally free to purchase materials from any source it wishes, and to schedule work in any manner that it sees fit in the absence of any contractual provisions to the contrary.

Alternative Methods for Dispute Resolution

Figure 20.10 shows a simplified summary of the principal methods available to the contractor for resolution of construction disputes.

Arbitration or Litigation?[4]

Many standard forms of construction contracts include provisions for binding arbitration of any future disputes. It is also the declared public policy of the federal government and many states to require arbitration in public contracts. Yet we find that most insurance carriers try to avoid arbitration wherever possible.

Parties to the construction process often gain a false sense of security from having an arbitration clause in their contracts. There is no perfect method for dispute resolution. Whether it is arbitration or litigation, the choice between the two is much like the choice of whether to be put to death in a gas chamber or in an electric chair. Under both arbitration and litigation the parties are asking an outside party to resolve their disputes. Common sense dictates that the quickest and most economical way of resolving a dispute is for the parties to resolve the dispute themselves, without the intervention of others.

Is arbitration the best method for resolving construction disputes? The answer is a definite "maybe." The proponents of arbitration hail it as being speedy and economical. Experience has shown, however, that this is not always true, especially in the larger cases. Arbitration can be extremely lengthy, expensive, and by its very nature can lead to inconsistent results. One arbitration case in which the author was involved lasted over five years, mostly because of hearing delays and the difficulty of scheduling hearing dates that were mutually acceptable to the arbitrators, the parties to the arbitration, and their attorneys, who also had to weave their arbitration

[4]Postner and Rubin, Attorneys at Law, New York, NY. Reproduced by permission.

Time	Settlement Cost	Binding Nature	Appeal
Negotiation			
1. Dependent on the parties' negotiators' objectives, attitude, and other factors. 2. Can be very fast.	1. Minimal. 2. Cost of compromised settlement.	1. Take it or leave it. 2. May lead to an agreement.	1. Waived if agreement is reached. 2. Arbitration or litigation if no agreement.
Mediation			
1. Same as negotiation. 2. There may be some limitations imposed by the mediator's schedule. 3. Usually fast.	1. Mediator's compensation, if any.	1. Take it or leave it. 2. May lead to an agreement. 3. Moral pressure to reach an agreement.	1. (same as above)
Mediation-Arbitration			
1. Speed depends on the procedure used. 2. If formalities can be waived, resolution is fast.	1. Mediator's compensation, if any.	1. May be agreed in advance in most states that parties will be bound to the decision.	1. (same as above)
Arbitration			
1. Faster than litigation. 2. Rules may impose some limitations. 3. Availability and schedule of arbitrators is a problem. 4. Preparation may take several months.	1. Filing fee. 2. Arbitrator's compensation after 2nd day. 3. Attorney fees, if any.	1. May be nonbinding or binding according to contract.	1. No review of merits in court. 2. Arbitrator not required to explain award.
Litigation			
1. May take up to 5 years or more to reach a trial. 2. Preparation itself may take years.	1. Prohibitively expensive, both in terms of attorney's fee and time costs.	1. Binding	1. Full appeal.
Drop Claim/Concede			
1. None	1. Value of claim.	1. Contractual agreement by mutual accord. 2. Waivers.	1. None, right waived in most cases.

FIGURE 20.10 Construction Claims Resolution Alternatives.

schedule in and out of equally demanding court appearance dates—many of which occurred on short notice, thus forcing the arbitration to be rescheduled.

Inconsistent Results

One difficulty is that the arbitrators are not bound by rules of evidence, or for that matter any previously rendered decisions, whereas under English common law principles, a court is obligated to be bound by previous decisions rendered in its state for similar circumstances.[5]

Consider also the case where an owner demands a general contractor to arbitrate a claim that the work performed by one of its subcontractors is defective. If the subcontract does not provide for arbitration, or if both arbitration agreements do not provide that third parties that have an interest in the issue to be arbitrated can be joined as parties to the arbitration, the general contractor may be unable to join the subcontractor in the arbitration with the owner over the subcontractor's work.

If this occurs, the arbitrators in the proceeding between the owner and the general contractor could find that the subcontractor's work was defective. This decision would not be binding on the subcontractor, however, who might convince a different panel of arbitrators that he or she was not responsible for the defects claimed. The general contractor would then be liable to the owner for the cost of correcting the defective work, although it will be unable to recover these costs from the subcontractor.

A parallel situation is one where an owner finds a defect in the work. The contractor claims that the defect was due to faulty design; the architect/engineer claims that it was due to defective work. In this connection it should be noted that standard AIA documents have been worded specifically to prevent consolidation or joinder of the architect/engineer along with the contractor in a single arbitration proceeding.

Cost of Arbitration

Arbitration is not always economical. The filing fees for initiating arbitration with the American Arbitration Association (AAA) greatly exceed the fees to file a complaint in most state courts. Under the rules of the AAA, the filing fee for a $100,000 claim is $1500. By contrast the fee for filing a complaint in most state courts is under $100. An additional expense of arbitration is the fee paid to the arbitrators, which can range from $300 to $1500 or more per day, depending upon the number and experience of the arbitrators. Other costs may include charges for a hearing room and a written transcript.

Speedy Results?

Another problem with arbitration is that the hearings are not normally held on a day-to-day basis. This is because arbitrators have their own business affairs to conduct, and it is rare to be able to schedule more than two hearings on consecutive days. This means that an arbitration that involves four days of hearings may actually take two

[5]Excluding the state of Louisiana, which does not operate under English common law, due to its French origins.

or three months before the hearings are completed. Similarly, a large, complex arbitration requiring 30 days of hearings (not unusual) can take upward of three years to resolve. It should be noted, too, that with hearings so far apart, the attorneys for the parties have the opportunity to prepare more, thus further increasing the cost of arbitration.

Discovery

Proponents of arbitration cite the fact that there is no discovery in arbitration, an expensive and time-consuming process involved in almost every lawsuit that is frequently the subject of abuse or overkill by one or both sets of lawyers. As a practical matter, however, discovery actually occurs in an arbitration during the hearings. Thus the time that is saved in arbitration by having no discovery beforehand is more often than not lost in the extension of the arbitration hearings themselves. It should also be noted that discovery in a lawsuit often discloses valid claims or defenses that can lead to settlement discussions. This opportunity is often lost in arbitration.

Arbitrary Arbitrators

Arbitrators are normally not bound by rules of law. For example, as an owner you may have insisted that your construction contracts include a clause providing that the contractor shall not be entitled to any damages for delay. Though the courts have held such clauses as enforceable in the face of contemplated delays, there is nothing to prevent an arbitrator from ignoring the clause and awarding the contractor damages for such delays.

THE MEDIATION PROCESS

Mediation is a process in which a trained third-party neutral attempts to assist the parties to a dispute in reaching an agreement that resolves the dispute. Another way of saying this is that mediation is a particular form of settlement negotiation in which a trained third-party neutral intervenes by agreement of the parties in order to guide and facilitate the parties' negotiations toward an agreement that resolves the dispute.

Why Is a Mediator Needed?

A mediator is needed when one or more of the following circumstances or desires exist:

1. Parties are unable to narrow the gap between the expectations of one group and the inflexibility of the other.
2. Too many issues are open and the parties are unable to get movement going.
3. The parties wish the mediator to explore and narrow the differences between them.
4. There is a desire to resolve a problem mutually and end a dispute amicably.
5. It is desired to furnish the parties with a realistic look at the demands and possibility of obtaining them.

6. There is a wish to give the parties some idea of how their positions look to an impartial person.
7. The parties wish a mediator to be used as a conduit through which private, confidential disclosures can be made without jeopardizing their original positions.
8. There is a wish to avoid negative consequences, such as a lawsuit.

Mediation as Distinguished from Arbitration

Like arbitration, mediation is a private, unofficial, confidential means of dispute resolution. Like arbitration, mediation is generally favored by modern law. And like arbitration, it generally promises savings in time and cost as compared with civil litigation. Yet mediation differs from arbitration in a number of important ways, including:

1. An arbitrator has final power of decision, subject only to very limited judicial review; a mediator has absolutely *no* power beyond the power of illumination and persuasion.
2. An arbitrator normally takes a rather passive role, leaving it largely to the parties to present and press their cases; a mediator is normally an active intervenor in, and principal shaper of, the process.
3. Although arbitration is more informal than civil litigation, and the usual order of proceedings is generally predictable and tracks the usual order of proceeding in litigation, the order of proceeding in a mediation is much less structured than either arbitration or litigation.
4. The arbitrator avoids all ex-parte communication with the parties on the merits of the dispute or the parties' positions; the mediator will virtually always caucus privately with each party for the express purpose of learning as much as they are willing to disclose about their private views and interests with respect to the dispute.
5. Although an arbitration is normally a private proceeding, information about it may become public knowledge in some circumstances: for example, if the losing party seeks to have a court vacate the award. Mediations are settlement discussions, and, as such, some information about them is inadmissible.
6. Perhaps most important, the arbitrator imposes upon the parties a decision that he or she has made and that may please neither party; the mediator, if successful, guides the parties to a decision that they can have a major role in shaping and with which, by definition, they both agree.

Use of Mediation as a Dispute Resolution Tool

Mediation is generally used first. Its use may precede either arbitration or litigation. It may be used, however, even after an arbitration or a litigation has been initiated but prior to an award. Because a mediation is treated as a settlement negotiation, evidence concerning it is not admissible in any subsequent arbitration or litigation.

Settlement of Disputes by Arbitration

In overturning a decision rendered by the California Supreme Court [*Southland Corporation v. Keating,* 52 U.S.L.W. 4131 (January 23, 1984)], the U.S. Supreme Court ruled that the Federal Arbitration Act [9 USC Sec 1–4] takes precedence over state statutes prohibiting arbitration of certain disputes. Although the case did not involve a construction dispute, it has significant implications for the construction industry.

Business Disputes

In the world of business, disputes are inevitable. One person may understand rights and obligations differently from another no matter how carefully a contract is written. This could lead to delayed shipments, complaints about the quality of the work, claims of nonperformance of the terms of the contract, and similar misunderstandings. Even with the best of intentions, parties often perform less than they promise.

Such controversies seldom involve great legal issues. On the contrary, they generally deal with the same facts and interpretation of contract terms that owners and contractors are used to dealing with every day. Therefore, when disputes arise out of day-to-day activities, the parties frequently like to settle them privately and informally. This is what commercial arbitration is for.

What Is Arbitration?

Arbitration is the voluntary submission of a dispute to one or more impartial persons for final and binding determination. It is private and informal and is designed for quick, practical, inexpensive settlements. But at the same time, arbitration is an *orderly* proceeding, governed by the rules of procedure and standards of conduct that are prescribed by law. The most commonly used arbitration procedures are those administered by the American Arbitration Association. Both the AIA and the EJCDC General Conditions call for arbitration under the Construction Industry Arbitration Rules of the American Arbitration Association. The association does not act as arbitrator. Its function is to *administer* arbitrations in accordance with the agreement of the parties and to maintain panels from which arbitrators may be chosen.

The arbitrators are quasi-judicial officers. Their decisions represent their judgments of the rights of the parties to a dispute. One significant difference between the rules that govern civil action in court and an arbitration proceeding is that, in arbitration, the strict rules of evidence do not apply. The arbitrator's guiding principle is to hear all the evidence that may be material and to hear no arguments or evidence from one side that the other has no opportunity to comment upon or to rebut.

The arbitrator is the final judge of matters considered. His or her decision will not be reviewed on its merits where procedures were fair and impartial. More than a hundred years ago, in 1854, the U.S. Supreme Court said:

> If an award is within the submission, and contains the honest decision of the arbitrators, after a full and fair hearing of the parties, a court of equity will not set it aside for error, either in law or in fact. A contrary course would be a substitute of the judgment of the chancellor in place of the judges chosen by the parties, and would make an award the commencement, not the end, of litigation.

Most arbitration cases are heard by a panel of three arbitrators, each usually representing a different field of specialization. Unless the agreement of the parties requires a unanimous decision, the arbitrators are governed by majority rule, both in procedural decisions and in rendering the award. In some cases, usually those in which the amount in question is relatively small, a single arbitrator may serve if desired by the parties. When three-person boards are used, it is customary for one of the arbitrators to serve as a chairperson, whose powers, however, are exactly the same as those of the other two.

Authority of the Arbitrator

The arbitrator has broad powers to determine matters of fact, law, and procedure. This decision-making authority must be exercised by the arbitrator alone to the best of its ability, and it may not be delegated to others. It would render his or her award subject to attack in court if, for example, the arbitrator sought clarification of a point of law by outside consultation. Under the rules of arbitration, the award must be in writing, and it must represent the judgment of at least the majority of the board unless the contract of the parties requires a unanimous decision. The arbitrator must word the award clearly and definitely and must answer all the questions but may not deal with any matter not submitted to arbitration. The relief granted in the judgment must be consistent with the contract and may include specific performance as well as monetary damages. Arbitrators are not required to write opinions explaining the reasons for their decisions. As a general rule, their awards consist of a brief direction to the parties on a single sheet of paper. There are a number of reasons for this. One is that written opinions might open avenues for attack on the award by the losing party.

As stated earlier, arbitrators are the final judges of all matters of both fact and law before them. Courts will not review their decisions on the merits of the case even when the arbitrators have come to a conclusion that is different from that which the court might have reached. In particular, courts are concerned only with the face of the award itself, not any additional explanatory matter.

The members of the National Panel of Arbitrators of the American Arbitration Association volunteer their time and talent without any fee, although on hearings that last for more than one day, some payment is made. There is an administrative fee that must be paid to the American Arbitration Association for handling claims; however, the amount is relatively small. The American Arbitration Association was founded in 1926 as a private, nonprofit organization "to foster the study of arbitration, to perfect its techniques and procedures under arbitration law, and to advance generally the science of arbitration."

Arbitration Agreements Regulated by Law

Commercial arbitration agreements today are recognized by statute by the U.S. Arbitration Act and in all but one state. However, the extent to which these agreements are enforceable differs widely. In overruling common law, many states distinguish between agreements to submit present disputes to arbitration and agreements to arbitrate unknown disputes that might arise in the future. The most modern arbitration laws provide that all agreements to arbitrate, whether of present or future controversies, are valid, irrevocable, and enforceable. In some other states, the laws merely provide that present and known disputes are arbitrable when the parties agreed by contract to arbitrate them. In cases of present disputes, however, many statutes require that the arbitration agreement be made a rule of court before it becomes irrevocable and enforceable.

Some confusion has been caused by some writers who state that in some of these older laws, agreements to arbitrate future disputes are void. In most states such agreements are valid in the sense that it is permissible to enter into such agreements. The only problem that remains is the one of enforcing it.

For the basis of comparison, the various arbitration laws of the different states will be divided into two groups. The breakdown will be based upon those states that have, by statute or judicial law, changed the common law rule of revocability that permitted either party to an arbitration agreement to terminate it at his or her will at any time prior to the rendition of an award.

Group I: Statutes Allowing Arbitration of Present and Future Disputes

The first group of states will be those in which there is specific authority by statute or judicial ruling that holds that an arbitration clause in a commercial contract for *construction* is valid, irrevocable, and enforceable. The 44 states listed, as well as Puerto Rico, the District of Columbia, and the federal government, have statutes that conform to or embody features that are essentially similar to "modern arbitration law" in which all written agreements to arbitrate, whether present controversies or future controversies between the contracting parties, are valid, irrevocable, and enforceable.

The 44 states, Puerto Rico, the District of Columbia, and federal law in Group I are as follows:[6]

1. ***Alaska*** Alaska Stats. Ann. §09.43.010 et seq. (1968)* (4).
2. ***Arizona*** Ariz. Rev. Stats. §12–1501 et seq. (1962)* (4).
3. ***Arkansas*** Ark. Stats. Ann. §34–511 et seq. (1971)* (2, 4, 7).
4. ***California*** Cal. Code of Civil Procedure §1280 et seq. (1961).
5. ***Colorado*** Colo. Rev. Stats. §13–22–201 (1975).*
6. ***Connecticut*** Conn. Gen. Stats. Ann. §52–408 et seq. (1958).
7. ***Delaware*** Del. Code Ann., Title 10, §5701 et seq. (1973)* (4).
8. ***Florida*** Fla. Stats. Ann. §682.01 et seq. (1969).
9. ***Hawaii*** Hawaii Rev. Stats. §658–1 et seq. (1955).
10. ***Idaho*** Idaho Code, Chap. 9, §7–901 et seq. (1975)* (4).
11. ***Illinois*** Ill. Rev. Stats. Chap. 10, §101 et seq. (1961).*

[6]Asterisks indicate laws that are referred to as the Uniform Arbitration Act. Numbers following * indicate statute exclusions as to (1) Insurance, (2) Leases, (3) Labor Contracts, (4) Loans, (5) Sales, (6) Torts, (7) Uninsured Motorists, and (8) Doctors and Lawyers.

12. ***Indiana*** Ind. Stats. Ann. §344–2–1 et seq. (1968)* (3, 5, 6).
13. ***Iowa*** Iowa Code §679A.1 et seq.
14. ***Kansas*** Kansas Stats. Ann., Chap. 5, §401 et seq. (1973)* (2, 4, 7).
15. ***Kentucky*** Ky. Rev. Stat. Ann.; [1984 Ky. Acts ch. 278].
16. ***Louisiana*** La. Rev. Stats. §9:4201 et seq. (1948)* (4).
17. ***Maine*** Me. Rev. Stats. Ann., Title 14, §5927 et seq. (1967)* (8).
18. ***Maryland*** Md. Courts and Judicial Proceedings §3–201 et seq. (1965)* (4).
19. ***Massachusetts*** Mass. Ann. Laws, Chap. 251 §1 et seq. (1960)* (4).
20. ***Michigan*** Mich. Compiled Laws Ann. §600.5001 et seq. (1963).
21. ***Minnesota*** Minn. Stats. Ann. §572.08 et seq. (1957).*
22. ***Missouri*** Mo. Ann. Stats. §§435.350 to 435.470 (1980).
23. ***Montana*** Mont. Code Ann. Chap. 27 [S.B. 110(1985)].
24. ***Nevada*** Nev. Rev. Stats., Chap. 38, §38.015 et seq. (1967).*
25. ***New Hampshire*** N.H. Rev. Stats. Ann. §542:1 et seq. (1955).
26. ***New Jersey*** N.J. Stats. Ann. §2A:24–1 et seq. (1952).
27. ***New Mexico*** N.M. Stats. Ann. 44–7–1 et seq.
28. ***New York*** N.Y. C.P.L.R. §7501 et seq. (1920).
29. ***North Carolina*** No. Carolina Gen. Stats., §1–567.1 et seq. (1973)* (4).
30. ***Ohio*** Ohio Rev. Code Ann. §271 1.01 et seq. (1955).
31. ***Oklahoma*** Okla. Supp. 1978, §801 et seq., Title 15 (1978)* (2, 4).
32. ***Oregon*** Ore. Rev. Stats. §33.210 et seq. (1955).
33. ***Pennsylvania*** Pa. Stats. Ann., Title 22, §7301 et seq.
34. ***Rhode Island*** R.I. Gen. Laws §10–3–1 et seq. (1956).
35. ***South Carolina*** S.C. Code of Laws §15–48–10 et seq. (1978)* (2, 4, 7, 9).
36. ***South Dakota*** S.D. Comp. Laws §21–25A–1 et seq. (1971)* (2).
37. ***Tennessee*** Tenn. Code. Ann. §29–5–301 et seq.
38. ***Texas*** Tex. Rev. Civ. Stat. Ann. art. 224 et seq.
39. ***Utah*** Utah Code Ann. [tit. 78, ch. 31(a) (1985)].
40. ***Vermont*** Vt. Stat. Ann. tit. 12, §5651 et seq.
41. ***Virginia*** Va. Code, Vol. 2 §8–503 et seq. (1986).
42. ***Washington*** Wash. Rev. Code §7.04.010 et seq. (1943).
43. ***Wisconsin*** Wisc. Stats. Ann. §788.01 et seq.
44. ***Wyoming*** Wyo. Stats. §1–36–101 et seq.

District of Columbia D.C. Code Ann. §16–4301 et seq.

Puerto Rico Puerto Rico, Title 32, §3201 et seq. (1951).

United States Code United States Arbitration Act, 43 Stat. 883, 9 U.S.C.A. §1 et seq. (1925).

Group II: Statutes Allowing Arbitration of Present Disputes Only

The six states listed below have statutes that provide essentially that agreements to arbitrate *present* controversies are valid. The following list indicates the states in Group II in which arbitration statutes exist, but they contain varying restrictions that could affect the application of a future arbitration clause in the General Conditions of a *construction contract:*

1. ***Alabama*** Code of Alabama, §6–6–1 et seq.
2. ***Georgia*** Ga. Code §9–9–1 et seq.
3. ***Mississippi*** Miss. Code Ann., Chap. 15, ¶ 11–15–1.
4. ***Nebraska*** Nebr. Rev. Stats. ¶ 25–2103.
5. ***North Dakota*** N. Dak. Rev. Code, Chap. 32–29, ¶ 32–29–01.
6. ***West Virginia*** W.V. Code, Chap. 55, Art. 10, ¶ 55–10–1.

Preliminary Notice of Potential Claim

Most states require subcontractors and material suppliers to notify the prime contractor, owner, or surety of a potential lien or bond claim. There are, however, a limited number of states that have preliminary notice requirements. In these states, notice of any potential claim must be given before or within 60 days of first beginning the performance of the subcontractor's work.

Problems

1. What are the four principal methods available for the resolution of construction claims?
2. Of the four claims-resolution alternatives discussed in the book, which two are considered as binding?
3. What are the chances of a successful appeal of an award under American Arbitration Association rules?
4. Where a contract document lists an order of precedence, what effect does that provision have on disputes related to conflicts between the different documents?
5. In a construction claim, is the burden of proof on the plaintiff or the defendant?
6. Can recovery of contractor costs be made under a "Type 2 Differing Site Conditions" clause for weather-related delays?
7. Show how extended home office overhead is computed under the Eichleay formula.
8. What is the difference between compensable and noncompensable delays?
9. Name the five principles of contract administration.
10. A contractor is working on a $500,000 construction contract that was scheduled to be completed within 12 months; however, the owner delayed the contractor 1 month. Using the Eichleay formula, compute the amount of reimbursement that the contractor can claim for home office overhead if the contractor's total billing for the contract period was $10,000,000 and the total overhead for the contract period was $300,000.

21

PROJECT CLOSEOUT

ACCEPTANCE OF THE WORK

At first it would seem that all that would be necessary to close out a project would be to inspect it, accept it for the owner, and see that the contractor receives the final payment. But what about all the guarantees, operating instructions for equipment, keying schedule, record drawings, bonds, and similar items that must be accounted for first? What about liens that may have been filed against the property by subcontractors? Each of these items will require careful handling to assure the owner of a quality product that is free of encumbrances and that will be backed up by the guarantees that were called for in the original documents.

Acceptance of the work and final payment to the contractor must proceed in accordance with the terms of the construction contract documents. Although the methods may vary somewhat from job to job, basically they all begin with a request from the contractor to make a final inspection of the work. Generally, there may be at least two inspections required to close out the project. The first will establish those areas still requiring correction or other remedial work, and the final inspection will be a checkoff to assure that all work is substantially complete and that all corrections have been made.

The checkoff list, or "punch list" as it is normally called, is a detailed list made near the end of the project, showing all items still requiring completion or correction before the work can be accepted and a *Certificate of Completion* issued. Before acceptance, all workmanship must meet specified standards, all work must be installed and complete, and all equipment must be tested and operational.

In some cases it is possible to accept a project as being "substantially complete" if only minor items remain to be finished. This simply means that the project is close enough to being completed that it can be put to use for the purpose it was intended, and that all remaining incomplete work is comprised of relatively minor items that the contractor agrees to correct while the structure is occupied: for example, maybe the wrong wall switch plates were installed. In this case, the owner could use the building while waiting for the contractor to receive the proper wall plates and replace them.

If a *Certificate of Completion* or *Substantial Completion* is filed, and written on the certificate is a complete list of all work remaining to be done to complete the project, anything that is not indicated on the list of deficiencies requiring correction on the Certificate of Substantial Completion is considered as being satisfactory as is. Often, retention funds are held for 35 to 45 days after completion and are not released until correction of all remaining deficiencies and waiver of liens. Final payment, as it is often defined, does not include the release of retainage held during the project, which generally amounts to about 5 to 10 percent of the total cost of the project. The retainage is usually held for an additional 30 to 40 days to cover the lien filing period and to assure the completion of remaining punch list work.

Guarantee Period

Generally, the work covered in a construction contract includes a stated guarantee period, which is frequently one or two years. In some cases, the overall project may be guaranteed for only one year, although certain portions of the work may be covered by supplementary guarantees for longer periods. Normally, there is no need to withhold payment from the contractor for the purpose of assuring performance during the stated guarantee period, as the performance bond may be written to cover this period. Although some contracts call for 100 percent of the performance bond to be continued in force during the entire guarantee period (a costly arrangement for the contractor), many contracts allow for a reduced portion of the performance bond to cover any defects noted during the guarantee period. After all, if the project is 100 percent complete, there is little reason to believe that it will *all* fail. Thus, many such bonds are reduced to 10 percent or some other reduced percentage of the performance bond during the guarantee period. This seems quite reasonable, because frankly, if a significant percentage of failures were noted in the first year following completion, it would certainly appear to cast doubt over the quality of the inspection that was provided. As an alternative to the surety bond, some owners require the contractor to post a cash bond to cover the guarantee period.

The contractor's warranty does not imply liability, however, for the adequacy of the plans and specifications unless they were prepared by the contractor personally, or unless the contractor agreed to guarantee their adequacy. A contractor is only required to construct in accordance with the terms of the contract documents, and when this is done, the contractor cannot ordinarily be held to guarantee that the architect's or engineer's design will be free from defects or that the completed job will accomplish the purpose intended. The contractor is responsible only for the quality of the workmanship, the quality of the materials used, and for performance of the contract. It should be kept in mind that in the evaluation of the adequacy of the contractor's work, the standard of comparison should not be based on the previous experience of the inspector as to what is considered to be inferior work, but rather upon specific or substantial conformance with the terms set forth in the contract documents.

FIGURE 21.1 Failure of Improperly Designed Breakwater That Was Built in Accordance with Plans and Specifications.

The contractor can only be obligated to perform that which it specifically agreed to do in the written contract; anything required of the contractor that is beyond this would be valid grounds for a claim for additional compensation (Figure 21.1).

In Wisconsin, a court ruled that a contractor does not absolutely guarantee its work against defects or losses prior to the owner's final acceptance [*E.D. Wesley Co. v. City of New Berlin,* 215 N.W.2d 657 (1974)]. The case involved a problem with a booster pump and the owner withheld the contractor's final payment. The city maintained that because *it had not yet accepted the work,* the contractor was absolutely liable for all repair costs. The court found that the damage was not caused either by a defect in the pump or by the contractor in its installation. Therefore, the court ruled that the damage was not within the scope of the contractor's liability either before or after acceptance of the work.

CONTRACT TIME

Most contracts are quite specific regarding the amount of construction time allowed to complete the work, and many provide for the payment of "liquidated damages" by the contractor to the owner for failure to complete on time or, in some cases, to complete portions of the work that interface with other contract schedules where multiple prime contracts have been executed. It should be noted, however, that in the absence of a provision establishing "time as the essence of the contract," neither the beginning date nor the date of completion can be considered to create an absolute schedule for the purpose of imposing the provisions of a liquidated damages clause for exceeding the specified completion dates. According to a ruling by the Supreme Court of Nebraska, time is not of the essence in a construction contract unless specifically stated in the contract documents [*Kingery Construction Co. v. Scherbarth Welding, Inc.*, 185 N.W.2d 857 (1971)]. If a contractor's failure to complete a project on schedule was the result of delays that occurred because of the owner's action, however, a North Carolina Court of Appeals has ruled that the contractor would not be liable for liquidated damages [*Dickerson, Inc. v. Bd. of Transportation*, Court of Appeals of North Carolina (June 18, 1975)].

When computing contract time, particular attention should be paid to the contract wording—is it "calendar days" or "working days"? If it is working days, particular care will have to be taken to determine the definition of a working day. Generally, this will have to be resolved by checking the master labor agreement for the area involved to see what is considered a holiday and what is not. The easiest time to compute is *calendar* days, as this method includes all days, including Saturdays, Sundays, and holidays.

Normally, construction time is computed from the date on the written Notice to Proceed given to the contractor at the beginning of the job (Figure 12.17). If no such notice was issued, the determination of the actual contract period may be indefinite. Preferably, a Notice to Proceed should be issued at the beginning of each project to assure a complete understanding as to the actual date construction work was authorized to begin (see "The 5-Step Process of Initiating a Project" in Chapter 1). The Notice to Proceed should not be confused with the Notice of Award (Figure 12.15)—the latter document simply establishes the identity of the contractor who will be given the contract to do the work and obligates the owner to sign the contract. It does not establish the date of starting construction, as no contract exists at that time. Once a contract has been signed, however, most specifications require that the contractor begin work within 10 or 15 days after the signing of the agreement. If this takes the place of a formal Notice to Proceed, then to compute the completion date, you will automatically have to allow the contractor an additional 10 or 15 days added to the contract term to allow for the latest date it could begin the work. As noted before, unless time is stated as the "essence of the contract" in the contract documents, there is always the possibility that a contractor may be reasonably secure from charges of liquidated damages for relatively minor time overruns.

In the absence of a formal document that specifically identifies the date upon which all work was completed, the establishment of the date of completion of a project may be more difficult. It is in the owner's best interests to file a formal certificate of completion upon completing the project. This should be filed as early as possible before the release of the contractor's final payment and retention money. The filing and recording of such a notice generally sets the lien law "time" running in the owner's favor. Sufficient holding time should be provided in the specifications to enable the owner to retain the contractor's final payment until a waiver of claims or final payment of all subcontractors and material suppliers has been made; otherwise, the contractor will have received the final payment and all of the retainage money, and the owner will be encumbered with stop notices or liens on its property. Of course, the owner has legal recourse to recover from the general contractor, but that could be a long and costly process. Meanwhile, the owner would be obligated to pay the liens filed by the subcontractors and suppliers to clear title to its own property (see "Final Payment to the Contractor" in Chapter 17).

Liquidated Damages for Delay

On many projects, where time is the essence of construction, the owner and the contractor agree under the contract terms that if the contractor fails to complete the project by the stipulated date, it is financially liable to the owner for a preagreed sum for each day beyond the specified completion date that it takes the contractor to finish the work. This amount of money represents the financial losses to the owner for such delays, and because it is difficult to determine the real values of the owner's losses, the preagreed sum is used in lieu of a determination of the actual damages suffered. This assessment is referred to as *liquidated damages,* and it is common practice throughout the construction industry to require that the contractor pay the owner this fixed sum of money for each calendar day that it exceeds the specified date of completion.

Liquidated damages, when provided for by contract, are enforceable at law provided that they represent an estimated forecast based upon knowledge possessed at the time of entering into contract of the anticipated damages that the parties agree to at the time of executing the agreement.[1] To collect, the owner simply deducts the amount of accumulated liquidated damages from the sum due to the contractor at the time of final payment. In recent years, some organizations also have successfully assessed liquidated damages for failure of the contractor to meet specified key dates for completion of certain specified portions of the work that affect the ability of other prime contractors to deliver their portions of the work on time (see "Liquidated Damages During Construction" in Chapter 17). Once established under the contract, it is enforceable as long as the date was missed, even if the owner suffered no loss at all.

It should be emphasized that the courts enforce liquidated damages only when they appear to represent reasonably the actual damages expected to be suffered by the owner based on the information available at the time of bidding. When it has been established that the amount was excessive and unreasonable, the courts have ruled that such payment by the contractor to the owner constituted a *penalty* and was not enforceable. Another thing that is considered by the court is whether *time* was the essence of the contract. In the absence of anything in the terms of the contract to specify this, the owner is placed in a weak position to sustain its rights to liquidated damages for time overruns [cf. *Kingery Construction Co. v. Scherbarth Welding, Inc.,* 185 N.W.2d 857 (1971)].

[1]California Government Code §53069.85 states, "The sum so specified shall be valid unless manifestly unreasonable under the circumstances existing at the time the contract was made."

The estimated amount of the liquidated damages per day may be a function of many things. It can represent the loss of rental fees in an apartment, utility fees for a public utility company, or any other losses to the owner in connection with a revenue-producing project. Similarly, it can be keyed to daily costs to the owner of interest on loans or investment as a direct result of the failure of the contractor to finish the project on time.

The basic rule is that a liquidated damages provision is enforceable if the amount represents a reasonable forecast, at the time of signing the contract, of the actual damages the owner might incur if the project is not completed by the contractual deadline. It is recognized that a precise determination of the owner's delay damages is not possible. This is why it is desirable to "liquidate" the damages, that is, to establish them in advance as a specified sum. But a project owner may be called upon to show that it made a good faith effort to estimate its actual delay damages at the time the amount was inserted in the contract.

If the owner does not make a reasonable attempt to forecast its actual delay damages, the provision may be considered an unenforceable penalty, or an attempt to provide a negative incentive for timely contractor performance [*San OreGardner v. Missouri Pacific Railroad Co.*, 658 F.2d 562 (8th Cir. 1981); *Appeal of Great Western Utility Corp.*, ENG BCA No. 4934 (April 5, 1985)].

In contracts, "liquidated" damages take the place of "actual" damages. You cannot have both in a contract. Once you have established liquidated damages in the contract, it will normally bar recovery of actual damages. Furthermore, under a liquidated damages provision you need not even prove loss; in fact, you need not suffer any loss at all, and you are still entitled to collect the full amount of the specified liquidated damages.

Computation of Liquidated Damages

A common situation occurs when an owner uses a stock formula for determining the amount of liquidated damages and is unable to justify the validity of its formula. These formulas are particularly prevalent in public contracting. One example is that contained in the Washington Department of Transportation Standard Specifications, Section 1-08.9, where it specifies the following formula for determination of liquidated damages:

$$LD = \frac{0.15C}{T}$$

where LD = liquidated damages per working day (rounded to the nearest dollar)
C = original contract amount
T = original time for physical completion

The weakness in this approach is that the amount of liquidated damages is based upon the cost of the project and not on the actual anticipated losses suffered by the owner. Another approach popular with some public agencies, also based upon the cost of the project instead of the owner's anticipated losses, is the use of a table such as that illustrated in Figure 21.2 that is quoted from the Federal Highway

Table 108-1 Charge for Liquidated Damages for Each Day Work Is Not Substantially Completed

Original Contract Price		Daily Charge
For More Than----	To and Including----	
$0	$250,000	$300
250,000	1,000,000	500
1,000,000	2,000,000	800
2,000,000	5,000,000	1000
5,000,000	10,000,000	1400
10,000,000	and more	2100

FIGURE 21.2 Charge for Liquidated Damages for Each Day Work Is Not Substantially Completed. From Federal Highway Administration Standard Specifications, Table 108-1.

Administration Standard Specifications FP-96. A similar table is used by a number of other public agencies.

The amounts shown in the table in Figure 21.2 for projects under $250,000 are somewhat low, and thus provide little or no incentive to finish on time, as the cost of paying liquidated damages will be less than the contractor's cost of acceleration of the work to make up for falling behind schedule. As you consider the smaller projects, there are certain fixed costs associated with field inspection and administration that can easily justify a higher rate for smaller projects.

As with the formula approach described earlier, the chart is still based solely on a percentage of the project cost. Nevertheless, if, as in the previous example, the owner has adopted a policy or guidelines such as a formula or chart governing the amount of liquidated damages, any attempts to vary from that policy without explanation may result in the liquidated damages clause being ruled as an unenforceable penalty [*Appeal of Dave's Excavation,* ASBCA No. 36161 (June 8, 1988)].

Another way that project owners may render their liquidated damages clauses unenforceable is to reduce the original daily rate after contract formation. In the *Appeal of Coliseum Construction, Inc.,* ASBCA No. 36642 (December 6, 1988), the contract called for liquidated damages of $1820 per day. The government's contracting officer decided this was excessive and assessed the tardy contractor at a rate of $220 per day. The Armed Services Board of Contract Appeals said this was a tacit acknowledgment that no reasonable effort had been made to predict damages at the time of contract formation. Even if $220 per day was a fair approximation of the government's actual delay damages, the entire clause was unenforceable, and the government could not assess any liquidated damages.

It is important to remember that the validity of a liquidated damages clause is determined from knowledge possessed at the time that the contract was formed. If a reasonable effort was made to estimate damages, the clause is enforceable regardless of the actual delay damages the owner ultimately does or does not incur.

An example of the approach to the computation of liquidated damages preferred by the author is illustrated in Figure 21.3.

COMPUTATION OF LIQUIDATED DAMAGES

IMPORTANT: TO BE RETAINED IN THE PROJECT FILE

An estimate of daily liquidated damages based upon circumstances existing at the time that the Contract was executed

PROJECT NAME *INTERSECTION IMPROVEMENTS*	PROJECT NO. *00-01*
DATE OF ESTIMATE *12 JAN 2000*	DATE OF CONTRACT
LIQUIDATED DAMAGES AT COMPLETION OF THE WORK - Yes ☒ No ☐	
LIQUIDATED DAMAGES AT PROJECT MILESTONES - Yes ☐ No ☒	
If at milestones, identify all milestones: *NONE*	

COMPUTATION	EST. DAILY COSTS
INSPECTION COSTS *1* City Inspectors at *$55* hr for *8* hrs per (day) ~~week~~ *1* Technicians at *$75* hr for *1* hrs per (day) ~~week~~ Inspection vehicle (Daily rate) *$10* per day for *5* days per week	*440.00* *75.00* $ *50.00*
TEST LABORATORY SERVICES *(2 EA @ $35 ÷ 5 DAYS)* Sampling *1* items at *$35* each Pro-rata cost per day $ *14* Testing ___ items at ___ each Pro-rata cost per day $ ___ Pick up test specimens ___ trips at ___ trips/week Average daily total	$ *14.00*
ENGINEER/ARCHITECT SITE OBSERVATION *($100/HR x 8 HR ÷ 5 DAYS)* Cost per day *$800* for *1* days per week divided by number of working days per week *5*	$ *160.00*
DAILY REVENUES LOST (Tangible losses only--No compensation for "inconvenience" Nature of revenues lost *NONE* (rentals, leases, utility charges, Green Fees, Tolls, Parking Fees, etc.)	$ *—*
INTEREST ON CONSTRUCTION LOAN Identify loan *NONE* Daily interest rates *N.A.*	$ *—*
OTHER FIELD COSTS Increased bond costs (prorated per day) *NONE* Other *NONE*	$ $ *—*
TOTAL DAILY LIQUIDATED DAMAGES	$ *739.00*

FIGURE 21.3 Example of a Liquidated Damages Estimate for a Street Improvement Project.

You are urged to use caution in setting the amount of liquidated damages. If the amount set as liquidated damages is too low, it has a reverse effect . . . it is like offering the contractor a bonus for finishing late, as the cost of liquidated damages may be lower than the cost of accelerating the work to finish on time.

The author's preference is to adopt a policy of estimating liquidated damages by itemizing anticipated costs (see Figure 21.3). My rule of thumb is this: If you cannot justify liquidated damages over $500 per day, my suggestion is that you do not specify any liquidated damages at all. In that way you still have the option to file for actual damages, whereas if you have a liquidated damages clause in the contract, it acts as a bar to later claims of actual damages.

Ideally, the contract terms governing liquidated damages are best set out in the General Conditions of the contract, and the dollar amount specified in the Agreement itself.

Cleanup

In addition to the requirement that the job site be kept clean during the progress of the work, the contractor is similarly obligated to clean up the construction site thoroughly at the end of the job before the work can be accepted. The final cleanup is of significantly greater proportions than previous cleanup work, as all of the various items of demobilization rightfully are included under the cleanup category. This includes removal of temporary utilities, haul roads, temporary fences, field offices, detours, stockpiles, surplus materials, scrap, replacement of landscaping where it had been temporarily removed, street cleaning, and the obtaining of releases from the various city, county, or other governmental authorities having jurisdiction.

The contractor is obligated to clean up the site of its own operations as well as all areas under the control of the owner that may have been used by the contractor in connection with the work on the project, and should be required to remove all temporary construction, equipment, waste, and surplus material of every nature unless the owner has approved otherwise in writing. Final acceptance of the work should be withheld until the contractor has satisfactorily complied with all of the requirements for final cleanup of the project site. This includes cleanup of city streets as well, where dirt or other deposits have accumulated as a result of the contractor's operations.

Disposal of all waste and refuse should be at the contractor's expense. No waste or rubbish of any nature should be allowed to be buried or otherwise disposed of at the site except upon receipt of written approval of the owner.

The Punch List

There is probably no period during construction that is troubled with more time-consuming delays and the resulting exasperation than the period involving the corrective work prior to final acceptance. Theoretically, if every trade performed its work in strict compliance with the contract requirements and the best of craftsmanship, what is known as a *punch list* might never have come to exist. Thus, it should be the objective of everyone connected with a project to minimize the number of punch lists required.

Electronic Punch List

Instead of the traditional handwritten punch list, another alternative is available if the owner or architect/engineer and their Resident Project Representatives have access to a computerized project management system, such as illustrated in Figure 5.11. Projects that lend themselves to a prototype or model being constructed as a definition of the expected quality standard could benefit by that approach.

As the Work nears its completion in various phases, a prepunch-list activity could be encountered that contains activities that would be likely to show up on a punch list. Such activities can be identified and corrected while the majority of the craftspersons and mechanics are still on the project.

When the time arrives for preparing the punch list, the use of a spreadsheet is recommended as an input device to be used as a template by the users of the program. In the spreadsheet, experienced persons should be asked to list the typical punch-list items, which they know from experience would be likely to be encountered. These items can then be coded to the responsible contractors, along with the amount of time that would be allowed for correction. The spreadsheet can then be used to print out a list to be used as a collection tool in the field.

Once the punch list is in the database, a PDA (personal digital assistant) may be used for checking off the items that have been completed and accepted. A hard copy of the information can be printed at any time. No action is required in the database until the item is actually accepted.

Punch-List Obligations of the Contractor and Subcontractors

It is the contractor and its subcontractors who must assume the greatest responsibility for the existence of work that must be corrected. More critical, exacting, and progressive supervision is required of the contractor so that all the trades will perform their work in accordance with the highest standards of quality workmanship. To that end, the following procedures are recommended:

1. The contractor should carefully check its own work and that of the subcontractors while the work is being performed.
2. From the very beginning of a project, it is suggested that the contractor's superintendent prepare and maintain a written record of deficiencies observed as the job progresses to preclude their being overlooked or forgotten.
3. Unsatisfactory work should be corrected immediately and not permitted to remain and become a part of the punch list.
4. Corrections should be made before any particular subtrade leaves the project. Unless this is done, the door is left open for later evasion and disclaiming of responsibility for extended delays.
5. During the finishing stages of the project, the contractor should make frequent and periodic inspections with the subcontractors and the Resident Project Representative to check progressively for and correct any faulty work.

6. When the contractor has decided that the project has been completed satisfactorily in accordance with the terms of the contract, the architect/engineer should be notified, through the Resident Project Representative, for the purpose of obtaining acceptance of the work.

Punch-List Obligations of the Architect/Engineer

The architect/engineer can also make a positive contribution toward the efficient handling of the final inspection process by following a few simple procedures of its own.

1. During the progress of the work, the Resident Project Representative or the inspectors should make frequent and careful inspections of all work and should point out any deficiencies as they are discovered instead of waiting to place the items on the punch list.
2. During the finishing stages of the work, the contractor and the Resident Project Representative, accompanied by any affected subcontractors, should make frequent and careful inspections of the work to check progressively for and assure the correction of any faulty or deficient work.
3. When the contractor has determined that the work has been completed satisfactorily in accordance with the terms of the contract, he or she should promptly notify the Resident Project Representative (Figure 21.4).
4. Upon receiving such notification from the contractor, the Resident Project Representative should notify the project manager and promptly make arrangements for the *prefinal inspection* of the work. The representatives of the contractor and the subcontractors should participate in the inspection tour to respond to any questions that may be raised by the representatives of the architect/engineer.
5. Preferably prior to, but in no case later than during, the *prefinal inspection* period, dates should be established for equipment testing, systems validation, acceptance periods, warranty dates, and instructional requirements that may not have been agreed upon previously.
6. Following the prefinal inspection of the work, the Resident Project Representative should prepare a punch list setting forth in accurate detail any items of work that have been found not to be in accordance with the requirements of the contract documents (Figure 21.5). Following preparation of the punch list, the contractor, the subcontractors, the Resident Project Representative, and the architect/engineer's project manager shall meet to make a tour of the entire project and identify and explain all of the items on the punch list. At that time the architect/engineer representatives should be ready to answer any questions that might arise so that there will be no misunderstanding of what is required before the project can be accepted as complete.
7. If the contractor gives notice that a major subcontractor has completed its punch-list items, the Resident Project Representative should inspect that portion of the work, and if those items are found to be satisfactory, the contractor and subcontractor should be advised accordingly. If some items of

CONTRACTOR'S CERTIFICATION OF COMPLETION

DATE 9 August 1990

TO: Mr. L.B. Farrington
General Manager
AIC Mutual Water District
123 Central Avenue
Winchester, Indiana 00000

PROJECT Pyramid Pumping Plant

JOB NO. C1234-87

CONTRACT NO. W1234-5678

ATTN: Resident Project Rep.

OWNER ABC Mutual Water District

FROM: L&M Construction Company, Inc.
(Firm or Corporation)

This is to certify that I, Delbert C. Martin am an authorized official of L&M Construction Company, Inc. working in the capacity of Vice President of Operations and have been properly authorized by said firm or corporation to sign the following statements pertaining to the subject contract:

I know of my own personal knowledge, and do hereby certify, that the work of the contract described above has been performed, and materials used and installed in every particular, in accordance with, and in conformity to, the contract drawings and specifications.

The contract work is now complete in all parts and requirements, and ready for your final inspection.

I understand that neither the determination by the Engineer-Architect that the work is complete, nor the acceptance thereof by the Owner, shall operate as a bar to claim against the Contractor under the terms of the guarantee provisions of the contract documents.

BY Delbert C. Martin
Delbert C. Martin

TITLE Vice President, Operations

FOR L&M Construction Company, Inc.

DISTRIBUTION: 1. Proj. Mgr.
2. Field Ofc.
3. File

Wiley-Fisk Form 17-2

FIGURE 21.4 Contractor's Notification to the Owner and Architect/Engineer That the Work Is Certified to Be Complete. *(Fisk, Edward R.,* Construction Engineer's Complete Handbook of Forms, *1st edition, © 1993. Reprinted by permission of Pearson Education, Inc., Upper Saddle River, NJ.)*

E. R. FISK & ASSOCIATES
PO Box 6448 • Orange, CA 92613-6448

PUNCH LIST

Project: Zone 3 Booster Pumping Station No. 532-52

Location: Mission Viejo, Orange County, CA Date: 11 Dec 1989

Inspection was conducted at above project by R. M. Bendix at 10:00am o'clock this date.

REPRESENTATION

CONTRACTOR-OWNER		ENGINEER-ARCHITECT	
Contractor:	Farris & Booth, Inc	Design:	E. R. Fisk & Associates
Owner:	Santa Ana Water District	Staff Spec.:	J. F. Beardsley
	Improvement District No. 1	Job Supervisor:	R. M. Bendix

The following items are to be corrected or completed to comply with the contract documents:

Type of Inspection	Check	Final X	1 Yr. Guar.	Guar.

NO.	ITEM
9	Cracked globe on vapor-proof fixture in lower gallery
13	Field Coating req'd for damaged shop coat on LPG tank
17	Replace defective stair tread
18	Replace door No. 2 stop w/331ES or 431ES
19	Finish paint access panels after factory prime
20	Readjust swing of Door No. 5
21	MasterKeying not in accordance w/Spec
28	Readjust air systems to within 5% of design requirements
31	Provide handwheel for emergency manual operation of plug
	valve motor operator as per spec.
37	Adjust access MH cover to finish grade
39	Replace existing welded gate post with spec. one-piece hot-
	dip galv.-after-fab post
40	Cleanup premises Remove & dispose of all debris
41	Remove temp. power facilities
43	Retest standby engine-generator set on auto mode
44	Pump station system control check & validation test
	Data used are fictitious for illustration only

DISTRIBUTION:
1. Project Manager
2. Contractor's Representative
3. Resident Project Representative
4. File

Wiley-Fisk Form 17-1

FIGURE 21.5 Example of a Final Punch List Showing Use of the Same Item Numbers That Appeared on the First List. *(Fisk, Edward R.,* Construction Engineer's Complete Handbook of Forms, *1st edition, © 1993. Reprinted by permission of Pearson Education, Inc., Upper Saddle River, NJ.)*

work still remain to be picked up, the cycle should be repeated with a new punch list until all of the items on the list have been satisfactorily corrected.

8. Each punch list should be dated and signed by the person who prepared it, and all items on the original list should be numbered consecutively. Upon issuance of any subsequent punch lists containing only the remaining uncorrected items, the original item numbers should be retained to assure proper identification. If additional items are later discovered, they should be added to the end of the list and assigned item numbers in sequence following the last number used on the *original list*. All punch lists should be dated and signed. Failure to do so has sometimes resulted in unfavorable actions when a case went to arbitration to resolve a dispute.
9. When advised by the contractor that all punch-list items have been completed, the architect/engineer representatives, accompanied by the contractor and subcontractors, should conduct the *final inspection* of the work. Then, if all punch-list items have been completed satisfactorily, the Certificate of Completion should be issued.
10. If, following the *final inspection* of any portion of the work, there remains a question as to whether one or more punch-list items have not been properly completed, but otherwise the overall project is substantially complete, the owner, through its architect/engineer, may issue a *Certificate of Substantial Completion* noting the uncompleted punch-list items. Final payment of any retainage for that portion of the work should be paid after deducting an amount that the architect/engineer reasonably estimates will more than cover the cost of the completion of the remaining punch-list items. Any such amounts withheld should be retained only until completion of such items to the satisfaction of the architect/engineer.
11. If the owner or installer of the owner's equipment and furnishings should damage any work that was completed and accepted previously, the owner should be advised accordingly and made aware of its obligation to bear the responsibility of repair costs for any such damaged work.
12. When preparing the punch list or any subsequent updates of the punch list, items of maintenance or any work damaged by the owner after he or she has taken occupancy or beneficial use should not be included. Should the owner want the contractor to repair or replace any such damaged work, he or she should be separately reimbursed for such costs through the issuance of a formal change order.

If these procedures are followed, it is reasonable to assume that the initial punch list will be minimal and that there should be no more than one additional punch list between the period of initial occupancy or use and final acceptance. The issuance of multiple punch lists in series is considered by many to be a sign of improper project control, and it is considered as unnecessary under good field management by both the contractor and the staff of the architect/engineer. However, there is no fixed limit to the number of punch lists that can be—and often are—used.

Preparations for Closeout

Contrary to the layperson's belief, the resident engineer's or inspector's job is not finished when the contractor completes the work at the site and a Certificate of Completion is executed. It is only then that the work of closing out the project *really* begins. The field staff must be reduced to the minimum number of persons necessary to complete the closeout activities; all office equipment must be returned to the office that supplied it; telephone and utility services must be terminated; radio paging devices or similar communications equipment must be returned and their contracts terminated; project records must be transferred into the office of the resident engineer's or inspector's employer; and a copy of a complete field office inventory should precede the return of all supplies and equipment and records of the field office.

Notices of address change should be sent to all parties who were previously addressing correspondence to the field office, and a closing schedule should be sent to the design firm and the owner so that appropriate plans can be made for the smooth transition of field personnel being released from the closing project into another assignment. Generally, the closeout will be accomplished by one or two persons: the Resident Project Representative and, if the work load warrants, a field clerk-typist.

The closeout period may actually begin several weeks to a month before the contractor completes the work on the project and can often extend for a month after completion of the work. In many cases the Resident Project Representative will be required to assure that all construction data have been posted on a copy of the plans showing the actual manner and location of all work as actually completed. These are referred to as "record drawings," and they are normally in the form of red pencil marks on a set of prints of the contractor's. These are often required to be turned over to the architect or engineer or owner when completed, and from these the architect or engineer or the owner may be required to have their drafters revise the original tracings to reflect recorded field information.

A summary of the principal closeout activities for a medium-sized to large project is shown in the flow diagram illustrated in Figure 21.6 and in the following list. Although some of the items may vary from job to job and the order may vary somewhat, the tasks do not vary significantly.

1. Perform closeout inspections as outlined under "The Punch List" in the preceding section of this chapter.
2. During the closeout inspection phase of a project, careful notes should be kept by each inspector in the field diary of all corrective, remedial, or extra work required to meet acceptance standards, and these data should be used to develop a preliminary list of all items still requiring completion or correction before acceptance of the work. The contractor must make corrections before the final inspection date.
3. Begin a partial reduction of field office inspection staff if the project is large enough to require several full-time inspectors in addition to the Resident Project Representative.

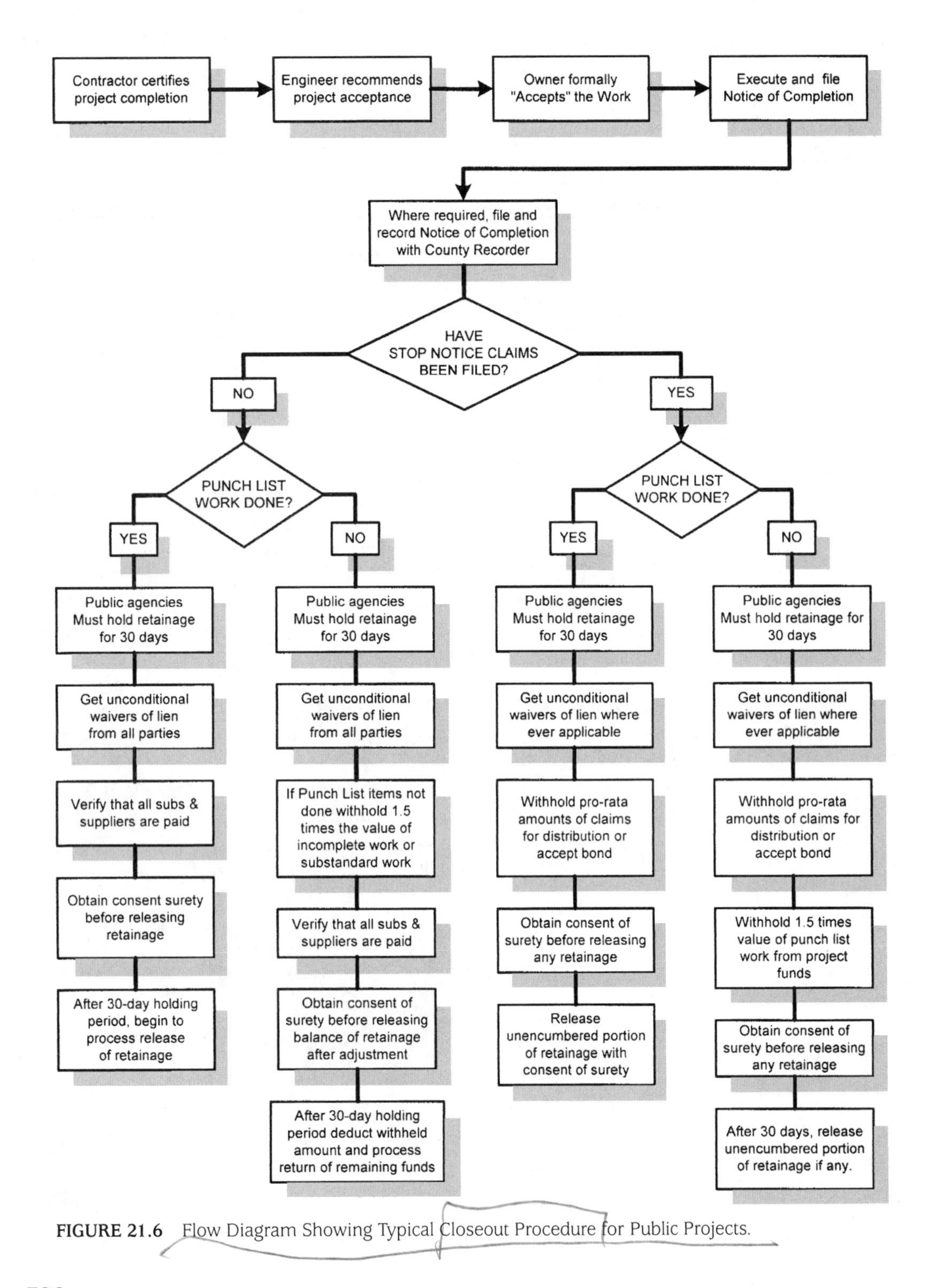

FIGURE 21.6 Flow Diagram Showing Typical Closeout Procedure for Public Projects.

4. Begin an inventory of all architect/engineer or owner property in the field office. List separately all office equipment, records and reports, supplies, field inspection and testing equipment, vehicles, cameras and photographic supplies, office furniture, and other property.
5. Complete final reduction of the field office inspection staff to the minimum number of persons necessary to complete the closeout administrative activities.
6. Prepare for final inspection. The final inspection team should include the same persons who participated in the prefinal inspection. All items indicated as requiring correction on the preliminary punch list should be reinspected, and all tests that were originally unsatisfactory should be conducted again. Tests not only should include run-up of motors, pumps, air-conditioning systems, fire protection systems, communications systems, and similar installed work, but also should include the testing of fail-safe devices, switches, controls, and other emergency devices. In addition, all doors should be operated, all locksets operated, and any other moving parts operated to assure that they will function properly before acceptance of the work. A final punch list should be developed for any outstanding deficiencies still requiring correction.
7. The contractor's record drawing set should be checked to see that all changes and variations from the original contract drawings have been marked on the set of prints or tracings that it may be required to submit under the terms of the contract. If properly completed, these should be turned over to the Resident Project Representative for transmittal to the architect/engineer or owner.
8. Prior to acceptance of the work, the Resident Project Representative should obtain the following items from the contractor if required under the contract:

   Guarantees
   Certificates of inspection
   Operating manuals and instructions for equipment items
   Keying schedule
   Maintenance stock items; spare parts; special tools
   Record drawings
   Bonds (roof bonds, maintenance bonds, guarantee bonds, etc.)
   Certificates of inspection and compliance by local agencies
   Waivers of liens (varies from state to state)
   Consent of Surety for Final Payments

9. If all work has been substantially completed and all punch-list items accomplished to the satisfaction of the inspecting team, a *Certificate of Completion* or *Substantial Completion* (Figures 21.7 and 21.8) should be prepared. If some items remain to be corrected and the owner elects to move into the facilities prior to their completion, an attachment may be

CERTIFICATE OF SUBSTANTIAL COMPLETION

DATE OF ISSUANCE ______________

OWNER ______________
CONTRACTOR ______________
Contract: ______________
Project: ______________

OWNER's Contract No. ______________ ENGINEER's Project No. ______________

This Certificate of Substantial Completion applies to all Work under the Contract Documents or to the following specified parts thereof:

To ______________
OWNER

And To ______________
CONTRACTOR

The Work to which this Certificate applies has been inspected by authorized representatives of OWNER, CONTRACTOR and ENGINEER, and that Work is hereby declared to be substantially complete in accordance with the Contract Documents on

____
DATE OF SUBSTANTIAL COMPLETION

A tentative list of items to be completed or corrected is attached hereto. This list may not be all-inclusive, and the failure to include an item in it does not alter the responsibility of CONTRACTOR to complete all the Work in accordance with the Contract Documents. The items in the tentative list shall be completed or corrected by CONTRACTOR within _______ days of the above date of Substantial Completion.

EJCDC No. 1910-8-D (1996 Edition)
Prepared by the Engineers' Joint Contract Documents Committee and endorsed by The Associated General Contractors of America and the Construction Specifications Institute.

FIGURE 21.7a EJCDC Certificate of Substantial Completion. *(Copyright © 1996 by National Society of Professional Engineers.)*

The responsibilities between OWNER and CONTRACTOR for security, operation, safety, maintenance, heat, utilities, insurance and warranties and guarantees shall be as follows:

OWNER: ______________________________

____

____

CONTRACTOR: ______________________________

____

____

The following documents are attached to and made a part of this Certificate:

[For items to be attached see definition of Substantial Completion as supplemented and other specifically noted conditions precedent to achieving Substantial Completion as required by Contract Documents.]

This certificate does not constitute an acceptance of Work not in accordance with the Contract Documents nor is it a release of CONTRACTOR's obligation to complete the Work in accordance with the Contract Documents.

Executed by ENGINEER on ____________________
Date

____
ENGINEER

By: ______________________________
(Authorized Signature)

CONTRACTOR accepts this Certificate of Substantial Completion on ____________________
Date

____
CONTRACTOR

By: ______________________________
(Authorized Signature)

OWNER accepts this Certificate of Substantial Completion on ____________________
Date

____
OWNER

By: ______________________________
(Authorized Signature)

FIGURE 21.7b (continued)

City of Thousand Oaks

NOTICE OF COMPLETION

City's Project No. 89/90-00 Engineer's Project No. ____________

Project Thousand Oaks Utilities Department Treatment Plant Addition

Contractor XYZ Constructors, Inc., Thousand Oaks, CA

Contract For Construction of 5 MGD Treatment Plant Addition

Project or Specified Part Shall Include 5 MGD Treatment Plant Addition and all Appurtenant Work.

Contract Date 7-10-89

The work performed under this contract has been inspected by authorized representatives of the City, Contractor, and Engineer, and the Project (or specified part of the Project, as indicated above) is hereby accepted by the City and declared to be substantially completed on the above date.

Completion of the Work shall be the date of such acceptance of the Work by the City, as provided under California Civil Code Section 3086. Completion shall mean substantial performance of the contract as such is defined in Black's Law Dictionary, Revised Fourth Edition, West Publishing Company.

A list of all items remaining to be completed or corrected is appended hereto. All such work shall be completed or corrected to the satisfaction of the City within 30 calendar days after the above Contract Date, otherwise the Contractor does hereby waive any and all claims to all monies withheld by the City under the Contract to cover the value of such uncompleted or uncorrected items.

City Engineer, City of Thousand Oaks By L. M. Barker 3-12-90
Engineer Authorized Representative/Date
L. M. Barker, City Engineer

The Contractor hereby accepts the above Notice of Completion and agrees to complete and correct all of the items on the appended list within 30 calendar days or waives all rights to any monies withheld therefor.

XYZ Constructors, Inc. By C. P. Warren 3-12-90
Contractor Authorized Representative/Date
C. P. Warren, President

The City accepts the project or specified area of the project as substantially completed and will assume full possession of the Project or specified area of the Project at 10:00 am (time), on 3-15-90 (date). The responsibility for heat, utilities, security, and insurance under the Contract Documents will be assumed by the City after that date.

Data used are fictitious for illustration only

FOR THE CITY OF THOUSAND OAKS By J. Erdam
Authorized Representative/Date
J. Erdam, City Clerk

REMARKS

The following or supplementary sheets listing such items remaining to be completed or corrected are hereby made a part of this document by this reference thereto.

Punch List No. 3, dated 9 March 1990, comprising one sheet and 7 line items.
No other attachments

FIGURE 21.8 Notice of Completion (Substantial Completion) Suitable for Either Buildings or Engineering Works. *(Fisk, Edward R.,* Construction Engineer's Complete Handbook of Forms, *1st edition, © 1993. Reprinted by permission of Pearson Education, Inc., Upper Saddle River, NJ.)*

made to the *Certificate* listing all remaining remedial work required to be done, as noted on the final punch list. Either a *Certificate of Completion* or a *Certificate of Substantial Completion* may serve to certify fulfillment of the contract by the contractor. (See "Completion versus Substantial Completion" later in this chapter.) Execution of a *completion* or *substantial completion* notice constitutes acceptance by the owner of the work *as is*—the presumption being made that it is contractually complete. (See definition of the terms *Completion* and *Substantial Performance* as quoted from *Black's Law Dictionary* later in this chapter.) Under such conditions, in many cases the contractor may have no further obligations under the contract except possibly to satisfy any claims made under the provisions of the guarantee or any exceptions taken by supplemental agreement.

An owner who unqualifiedly accepts the work and makes final payment on a construction contract without taking exception to any part of the work has been held by the Montana Supreme Court to waive the owner's right to demand correction or get damages for defects in the work that are known at the time. Furthermore, such flaws do not fall under the coverage of the guarantee clause [*Grass Range High School District No. 27 v. Wallace Diteman, Inc.*, Supreme Court of Montana, 465 P. 2d 814 (1970)].

10. Receive the contractor's request for its final progress payment. This does *not* include a release of its retainage money at this time, however. The flow diagram illustrated in Figure 21.6 shows the typical closeout steps involved in public agency contracts.
11. Check all work quantities and the value of the work completed from the punch list, retaining funds for those portions of the work named as still required to be done on the Certificate of Substantial Completion.
12. Submit contractor's payment request to the owner through the design or construction management firm with recommendation to pay, if warranted, less the retainage specified in the contract plus an amount sufficient to cover double the cost of remaining punch-list items.
13. Obtain signatures of the architect/engineer, the contractor, and the owner or their authorized representatives on the Certificate of Completion or Certificate of Substantial Completion; then file the certificate for recording in the office of the county recorder, where it serves as a public record and puts all potential lien holders on notice that their lien filing period has begun.
14. Notify owner, through the architect/engineer, that the building or other project is ready for occupancy or beneficial use, subject to completion of pickup work during occupancy.
15. If using a contractor-furnished field office, terminate any architect/engineer or owner obligations for telephone or other utility services to the field office. Transfer all records, supplies, equipment, and all other items on the inventory to the architect/engineer's or owner's office. Move into other quarters during the final administrative functions involved in the termination of all owner/contractor obligations.

16. If final completion of all pickup work noted on the Certificate of Substantial Completion has been accomplished, and if no liens have been filed during the holding period specified in the contract for making final payment, the architect/engineer may recommend to the owner that all remaining retainage funds be released. Prior to releasing retainage, however, the following certificates should be executed by the contractor and its surety, by all subcontractors and all suppliers: final waiver of lien and a properly signed Consent of Surety for Final Payment.
17. The making of the final payment and release of retainage by the owner will normally constitute a waiver of further claims by the owner, except those arising from:
    (a) Unsettled liens or stop notice claims.
    (b) Faulty or defective work appearing after substantial completion.
    (c) Failure of the work to comply with the requirements of the contract documents.
    (d) Terms of special guarantees required by the contract documents.

The making of final payment may also constitute a waiver of all claims by the contractor *except* those previously made and still unsettled.

Completion versus Substantial Completion

Substantial completion is based on the owner being able to use the project for the purposes intended and that the remaining activity of the contractor will not interfere with such use. The definitions of AIA and EJCDC vary slightly in their definition as to what substantial completion actually is; the AIA definition in Document A201 (1987) states that substantial completion is the stage in the progress of the Work when the Work or designated portion thereof is sufficiently complete in accordance with the Contract Documents so that the owner can occupy or utilize the work for its intended use.

Then AIA has another "stage" of completion, called Final Completion, which it defines as that stage of the Work where the Architect finds the Work acceptable under the Contract Documents and the Contract fully performed. At that point the Architect issues a final Certificate for payment and a statement that the Work has been completed in accordance with the terms and conditions of the Contract Documents.

Particular note should be made of the AIA references to both "Substantial Completion" and "Final Completion" as though they represented two separate stages of completion of a project, notwithstanding the legal definitions to the contrary of "completion" and "substantial performance" in *Black's Law Dictionary.* In this area, the AIA documents create some confusion with the wording used in contract law, in particular the definitions established under the lien laws of the various states.

It is possible, of course, to have a different definition for substantial completion and substantial performance if they are carefully defined in the contract; however, this leads to confusion and ultimately into litigation. It is much more economical to avoid such confusion of terms in the contract documents.

The EJCDC definition of substantial completion is more consistent with the legal definition of substantial performance of a contract. In EJCDC Document 1910-8 (1996) the language has been improved since the 1983 edition. Article 14.04 is reproduced in Figure 21.9.

There is no attempt by EJCDC to suggest two stages of completion. Substantial Completion is recognized as contractual fulfillment of the obligation of the contractor to the owner, and that the existence of small defects or omissions will not jeopardize that. It has been held in law that once substantial completion has been achieved and acknowledged, a contractor cannot be defaulted for failure to pick up the remaining punch-list items, quite inconsistent with the AIA definition. The EJCDC addresses the subject of "Final Payment and Acceptance" as opposed to AIA's "Final Completion and Final Payment," thus not necessarily agreeing that the work is acceptable merely

14.04. Substantial Completion

A. When CONTRACTOR considers the entire Work ready for its intended use CONTRACTOR shall notify OWNER and ENGINEER in writing that the entire Work is substantially complete (except for items specifically listed by CONTRACTOR as incomplete) and request that ENGINEER issue a certificate of Substantial Completion. Within a reasonable time thereafter, OWNER, CONTRACTOR and ENGINEER shall make an inspection of the Work to determine the status of completion. If ENGINEER does not consider the Work substantially complete, ENGINEER will notify CONTRACTOR in writing giving the reasons therefor. If ENGINEER considers the Work substantially complete, ENGINEER will prepare and deliver to OWNER a tentative certificate of Substantial Completion which shall fix the date of Substantial Completion. There shall be attached to the certificate a tentative list of items to be completed or corrected before final payment. OWNER shall have seven days after receipt of the tentative certificate during which to make written objection to ENGINEER as to any provisions of the certificate or attached list. If, after considering such objections, ENGINEER concludes that the Work is not substantially complete, ENGINEER will within fourteen days after submission of the tentative certificate to OWNER notify CONTRACTOR in writing, stating the reasons therefor. If, after consideration of OWNER's objections, ENGINEER considers the Work substantially complete, ENGINEER will within said fourteen days execute and deliver to OWNER and CONTRACTOR a definitive certificate of Substantial Completion (with a revised tentative list of items to be completed or corrected) reflecting such changes from the tentative certificate as ENGINEER believes justified after consideration of any objections from OWNER. At the time of delivery of the tentative certificate of Substantial Completion ENGINEER will deliver to OWNER and CONTRACTOR a written recommendation as to division of responsibilities pending final payment between OWNER and CONTRACTOR with respect to security, operation, safety and protection of the Work, maintenance, heat, utilities, insurance and warranties and guarantees. Unless OWNER and CONTRACTOR agree otherwise in writing and so inform ENGINEER in writing prior to ENGINEER's issuing the definitive certificate of Substantial Completion, ENGINEER's aforesaid recommendation will be binding on OWNER and CONTRACTOR until final payment.

B. OWNER shall have the right to exclude CONTRACTOR from the Site after the date of Substantial Completion, but OWNER shall allow CONTRACTOR reasonable access to complete or correct items on the tentative list.

FIGURE 21.9 Provisions for Substantial Completion from the EJCDC Standard General Conditions of the Construction Contract 1910-8 (1996). *(Copyright © 1996 by National Society of Professional Engineers.)*

because "substantial completion" has been achieved. This difference is normally resolved either by the contractor's completion of the outstanding punch-list items, or by a *quanti minoris* settlement wherein the owner withholds an amount of money from the retainage equal to substantially more than the value of the remaining punch-list items as a cash settlement for failing to complete these items.

Both the AIA and EJCDC publish a document entitled *Certificate of Substantial Completion* but no document called *Notice of Completion* or *Notice of Final Completion.* Under the EJCDC terms stated earlier, this document is adequate, as EJCDC recognizes substantial completion as "completion." However, the AIA documents leave some doubt. In fact, in California, where such documents are required by law to be recorded with the office of the county recorder within 10 days of actual completion, the recorders have generally refused to recognize and record a "substantial" completion document, stating simply, "Come back when it is finished." The layperson does not generally recognize the significance of the legal terminology involved in certifying project completion and does not realize that substantial performance of a contract is defined in law as completion of a contract. (See *Black's Law Dictionary,* 4th edition, pages 357 and 1597 definitions of "completion" and "substantial performance" that follow.)

The provisions of the EJCDC Standard General Conditions of the Construction Contract for "substantial completion" are reasonable and accurate, but the AIA provisions leave the actual interpretation of the term *substantial completion* to the architect/engineer using the documents.

As if to further complicate matters, the courts have their own definition of substantial completion, which appears to differ materially from the interpretations of the term by the architect/engineers using the AIA and the EJCDC. Some documents tend to confuse the unwary by treating the term *substantial completion* as if it meant "nearly finished" as opposed to *final completion,* a term that seems predicated on the concept of absolute performance of the contract in every detail. Unfortunately, this leads many owners and architect/engineers astray, as the courts take a different and considerably more liberal view as to when a construction contract is done. The term *substantial completion* as applied to a construction contract means "substantial performance" of the terms of the contract. *Black's Law Dictionary,* 4th edition, defines "completion" and "substantial performance" as follows:

> COMPLETION. The finishing or accomplishing in full of something theretofore begun; *substantial performance of what one has agreed to do;* state in which no essential element is lacking. *Flad v. Murphysboro & S.I.R. Co.,* C.C.A.Ill., 283 F. 386, 390.

> SUBSTANTIAL PERFORMANCE. Exists where there has been no willful departure from the terms of the contract, and no omission in essential points, and the contract has been honestly and faithfully performed in its material and substantial particulars, and the only variance from the strict and literal performance consists of technical or unimportant omissions or defects. Cotherman v. Oriental Oil Co., Tex.Civ.App., 272 S.W. 616, 619; Brown v. Aguilar, 202 Cal. 143, 259 P. 735, 737; Cramer v. Esswein, 220 App.Div. 10, 220 N.Y.S. 634; Connell v. Higgins, 170 Cal. 541, 150 P. 769, 774. Performance except as to unsubstantial omissions with compensation therefor, Cassino v. Yacevish, 261 App.Div. 685, 27 N.Y.S. 2d 95, 97, 99. (Equitable doctrine of "substantial performance," protects against forfeiture, for technical inadvertence or trivial variations or omissions in performance. Sgarlat v. Griffith, 349 Pa. 42, 36 A.2d 330, 332.)

The word *completion*, then, properly means that a project is finished when all of the work of any substantial nature has been done, regardless of whether a few minor pickup or call-back items remain to be done. It is the interpretation of the term *minor* that seems to be causing problems for many architects or engineers. The work done in the following cases offers an example of the courts' interpretation of the term *minor.* In each of these cases, the court held that the remaining punch-list item was sufficient to *prevent* "completion" of the work, because it was a specified item and the contractor failed to make an honest attempt to meet the terms of the contract.

> Application of a second coat of paint on the porch floors and steps, where the second coat was required by the specifications. [*Rockwell v. Light* (1907) 6 Cal.App.563, 92 Pac 649] Installation of soap dispensers which were specified in the contract. [*Lewis v. Hopper* (1956) 140 Cal.App.2d 365, 295 Pac. 2d 93.]

Some architect/engineers attempt to issue a Notice of Substantial Completion prematurely, such as in the case of a wastewater treatment plant where the owner wanted to issue the Notice of Substantial Completion as soon as the plant was operable but before completion of fencing, paving, and landscaping. This, however, did not meet the legal criteria for substantial completion, and the owner incurred the risk of being vulnerable to liens filed later than the normal filing time if a claimant could succeed in having a court invalidate a prematurely issued Notice of Substantial Completion.

It should also be kept in mind that the contractor's failure to complete remaining punch-list items after the issuance of a Notice of Substantial Completion may not be used as grounds for declaring the contractor in default. Substantial completion in the eyes of the court may be viewed as "completion" of the contract, and as such the contractor can no longer be declared in default for failure to perform subsequent to the date of substantial completion [*Appeal of Wolfe Construction Co.,* ENG BCA No. 3610 (June 29, 1984)]. See Figure 21.10. In a conflicting opinion, the Veterans Administration Board of Contract Appeals ruled that a contractor *may* be terminated for default by failing to meet a deadline for completion of punch-list items [*Appeal of Dimarco Corp.* VABCA No. 1953 (June 22, 1984)].

Certain other facts are worthy of consideration before executing a Notice of Substantial Completion. If the contractor is to be excluded from the area involved except for access necessary to correct or complete the pickup work items on the punch list, it will be necessary for the architect/engineer or owner to indicate who will have the responsibility for maintenance, heat, and utilities for the area involved. The extent to which the issuance of a Certificate of Substantial Completion will affect the insurable interest of the owner and the contractor in the project should be discussed with an insurance counselor.

The execution of a Certificate of Substantial Completion releases the contractor of all responsibility and obligation for further maintenance of the work, and ownership of the project passes to the owner. It is important to see that the owner is fully aware of the significance of this document in terms of its added responsibility. Normally, a Certificate of Substantial Completion will relieve the contractor of the hazards of liquidated damages for any work performed subsequent to its execution. Thus it certainly seems to be in the contractor's best interest to obtain such

FIGURE 21.10 75MGD (ult.) Water Treatment Plant for the City of Escondido, Vista Irrigation District, California, "Substantially Completed." *(Photo courtesy of MWH Global, Broomfield, CO.)*

a certificate as early as possible. As for a deficiency noted on the punch list that required the replacement of a piece of defective equipment, it would seem within reason that the guarantee period for the affected equipment should begin after the receipt of the replacement equipment and conclusion of a satisfactory operational test.

Substantial Completion versus Beneficial Occupancy or Use

The meaning of *substantial completion* as it applies to a construction project has been found to be somewhat ambiguous. Although the AIA and EJCDC General Conditions both provide a good definition and are quite acceptable as far as they go, they do not determine the responsibilities of the owner, architect/engineer, and contractor, nor do they recommend how certain matters should be resolved.

By a general definition, the owner's occupancy of a project prior to its being 100 percent complete may be defined as "beneficial occupancy." Thus the terms *substantial completion* and *beneficial occupancy* (or "use") are complementary.

The date of *substantial completion* or *beneficial occupancy* will normally establish the beginning of the specified period on the guarantees, unless a prior commitment has been made for acceptance of a portion of the total project (or *partial occupancy* or use). This concept is not uncommon in large residential tract housing and apartment complexes, where completed units are generally accepted and placed for sale while other portions of the project are in various stages of completion. It is *important* on tract work to identify the specific units that are accepted and cover in the specifications who is responsible for security, maintenance, heat, and general care during the period between the substantial completion of the early units and the date of sale and eventual move-in by the new owners. It should clearly be established that after acceptance of a single structure as being substantially complete, the owner will accept total responsibility for the building. Otherwise, the difficulty may arise, as it did on a case where the author was the arbitrator, where certain apartment units were substantially completed but no separate certificates had been issued, yet a final punch list had been satisfactorily completed. Both the contractor and the sales agency denied responsibility for weather damage occurring to some of the units because of lack of heat and for doors left open during the rainy season. The units were completed by the contractor, so no further maintenance was being provided. Furthermore, the sales agency had been leaving the doors open, causing some of the units to suffer severe weather damage. This condition existed for over six months with some of the completed units. The judgment was in favor of the contractor in this case, as there is no reason, contractually or morally, why the contractor had to play nursemaid to the owner's buildings while the owner was not only trying to sell them, but also was not even showing normal consideration and care when an apartment was shown to a prospective buyer (see Figure 21.11).

FIGURE 21.11 Apartment Complex.

One solution, or an approach to one at least, is to execute a separate Certificate of Substantial Completion for each apartment unit as it is completed. Then the responsibility clearly passes to the owner, where it belongs.

The laws appear to provide generally that when a work of improvement consists of separate residential units, the owner may record a Notice of Completion as each unit is finished. Similarly, a subdivider may record separate Notices of Completion covering each house in the subdivision as it is finished rather than waiting until the entire subdivision is completed.

In one sense, a condominium project also consists of separate residential units. Each condominium is a residential unit that may be owned separately from the other condominiums in the project. However, in a California decision, it has been held that an apartment-type condominium, at least, is a single project requiring one Notice of Completion at the end of the entire job [*McGillicuddy Construction Co. v. Noll Recreation Association, Inc.*, 31 Cal. App. 3d 891, 107 Cal. Rptr. 899 (1973)].

Immediately following *substantial completion* or *beneficial occupancy*, the owner must assume the complete responsibility for the maintenance and operation of all fuel and service utilities. The owner will normally also become responsible for all maintenance and damage and/or wear and tear and, with the exception of items that are specifically under guarantee or warranty, the cost of repairs or restoration during the period between *substantial completion* and *final completion*. The owner should also have the responsibility to have in effect all necessary insurance for protection against any losses not directly attributable to the contractor's negligence.

The contractor must be required to arrange a schedule so that punch-list items will be completed within the designated time by working during regular working hours. If the architect/engineer determines that the work interferes with the beneficial use of the project, and the owner is unable to adjust the operations to permit the contractor to perform punch-list work during regular working hours, the architect/engineer must certify to the owner that this work must be performed on an overtime basis, and the owner should compensate the contractor for the additional expense.

The purpose of retention is both as a hedge against lien claims and a guarantee to the owner that sufficient funds will remain to pay another party to complete the work or correct unsatisfactory items if the contractor refuses to make or delays making the corrections for an unreasonable length of time. If all lien waivers have been received, the amount of the retention should be adjusted so that the sum has a direct relation to the value of the work included on the punch list. It is recommended that the proper amount of retention to cover punch-list items be equal to approximately twice the value of the punch-list items, as determined by the architect/engineer.

Beneficial Use/Partial Utilization

Generally, during the administration of a construction contract for a project such as a water or wastewater treatment plant addition, it will be necessary for the owner to begin to use completed portions of the new facilities well before total project completion.

This creates a very sensitive, high-risk relationship between the parties to the contract, as several serious problems can and often do occur as a result of such utilization.

1. Identification of latent defects in equipment are almost impossible, and malfunctions are generally claimed to be the result of improper maintenance of equipment being utilized by the owner.
2. Equipment warranty dates are in question.
3. Maintenance responsibility frequently becomes a controversial issue between the contractor and the owner.
4. Security of the site and the on-site safety responsibility are no longer clear-cut issues during beneficial use. Generally, the owner will inherit these risks.
5. One of the principal risks involved in items 3 and 4 is that each party will assume that the other is responsible for maintenance, security, and safety, and then neither party will perform. This can result in increasing the severity of the problem and increasing the volume and magnitude of disputes and claims.

It is recommended that wherever practical the owner first be advised against assuming the risk of taking beneficial use or partial utilization unless the owner is fully prepared to accept all attendant risks from the contractor.

If the owner does elect to utilize any portion or portions of a project prior to total project completion, the following guidelines are suggested:

1. Issue a letter to the contractor advising that the owner will take beneficial use of a particular described portion of the work as of a specified date. This will serve to begin the product warranty date if it has not already begun upon purchase by the contractor. This should not affect the contractor's normal one-year guarantee period, dated from the *completion* date of the total project. This letter will place all parties on notice that the owner is taking partial utilization and has agreed to accept certain responsibilities (Figure 21.12).
2. A copy of this letter should be sent to the owner's operations and maintenance department.
3. The letter to the contractor should indicate that a particular portion of the Work is designated as being "sufficiently complete so as to allow beneficial use by the owner."
4. Be careful of the word *complete* in any notice to the contractor. The term *operationally complete* should be avoided in favor of terms utilizing the phrase *beneficial use* or one of its variants.
5. Do not refer to the act of taking "beneficial use" as constituting "preliminary acceptance." The term *acceptance* should not be used in any context at this stage of the project, as it has a strict legal meaning on public works projects.
6. Advise the owner to take positive steps to take over maintenance of any affected equipment.

City of Thousand Oaks

UTILITIES DEPARTMENT
DONALD H. NELSON, DIRECTOR

10 August 1990

XYZ Constructors, Inc.
1000 Magnolia Street
Camarillo, CA 90000

Attention: Mr. Conrad W. Belker
Project manager

Dear Mr. Belker:

The City plans to take beneficial use of a portion of the Lakeside Treatment Plant Addition prior to completion of the remaining portion of the project by utilizing or placing into service the following items during the period when your firm will be working elsewhere on the site.

> Pump Nos. 203 and 205 and appurtenance piping from valve 17 to valve 27 and from the pumps to, but not including, the circuit breaker controlling said pumps.

In consideration of such beneficial use, the City hereby accepts responsibility for the maintenance and protection of the specific portion of the project so used. It is further understood that the manufacturer's warranties on any piece of equipment placed into beneficial use by the City will commence as of the date of this Notice of Beneficial Use.

The Contractor shall retain full responsibility for satisfactory operation of the total project, however, and the Contractor's one-year maintenance and repair period and/or guarantee provided under the Contract Documents shall commence only after acceptance by the City and issuance of a final Notice of Completion for the entire project. Such guarantee of total systems operation shall include that portion or portions of the Work previously placed into beneficial use by the City.

Very truly yours,

WENDELL T. RIGBY
SENIOR CIVIL ENGINEER

WTR:db

CC: O&M Department

2150 WEST HILLCREST DRIVE THOUSAND OAKS, CALIFORNIA 91320 (805) 497-8611

Purpose: Notice to Contractor that the City will take Beneficial Use/Partial Utilization of a portion of the project prior to completion of the remaining portion of the project.

Prepared by: Project Manager/Engineer, Construction Manager, or Resident Project Representative.

Directed to: Primarily to the Contractor, but in reality also intended to place the City's Operations and Maintenance personnel on ''Notice'' as well.

Copies to: Contractor, City, Project Manager/Engineer, Resident Project Representative.

Comments: Notice to Contractor that the City will accept certain responsibilities for items placed into beneficial use; that it is understood that warranties will commence on those items; and, that the City will accept responsibility for providing all maintenance, security, and safety related to the affected portions of the Work after the date of this letter. Contractor is reminded that his or her project guarantee or maintenance and repair agreement still covers the affected items during the normal one-year postconstruction period.

FIGURE 21.12 Notice of Taking Beneficial Use of a Part of a Project (Partial Utilization).

7. Advise the owner to take positive steps to institute security measures, as necessary, to protect the equipment in use.
8. The owner should be advised not to occupy or use any portion of the Work until after official notification of beneficial use has been made to the contractor.

Such notification serves a dual purpose. It not only alerts the owner's operational and maintenance staff that they are under obligation to maintain the equipment now being utilized by the owner, but it also affords the owner the opportunity to establish in writing the limits of responsibility that the owner will accept, without making a big issue of it.

In the current edition of the Standard General Conditions of the Construction Contract published by the EJCDC, the subject is recognized and an orderly procedure for administering a contract involving partial utilization is specified (Figure 21.13).

14.05. Partial Utilization

A. Use by OWNER at OWNER's option of any substantially completed part of the Work which has specifically been identified in the Contract Documents, or which OWNER, ENGINEER and CONTRACTOR agree constitutes a separately functioning and usable part of the Work that can be used by OWNER for its intended purpose without significant interference with CONTRACTOR's performance of the remainder of the Work, may be accomplished prior to Substantial Completion of all the Work subject to the following conditions.

1. OWNER at any time may request CONTRACTOR in writing to permit OWNER to use any such part of the Work which OWNER believes to be ready for its intended use and substantially complete. If CONTRACTOR agrees that such part of the Work is substantially complete, CONTRACTOR will certify to OWNER and ENGINEER that such part of the Work is substantially complete and request ENGINEER to issue a certificate of Substantial Completion for that part of the Work. CONTRACTOR at any time may notify OWNER and ENGINEER in writing that CONTRACTOR considers any such part of the Work ready for its intended use and substantially complete and request ENGINEER to issue a certificate of Substantial Completion for that part of the Work. Within a reasonable time after either such request, OWNER, CONTRACTOR and ENGINEER shall make an inspection of that part of the Work to determine its status of completion. If ENGINEER does not consider that part of the Work to be substantially complete, ENGINEER will notify OWNER and CONTRACTOR in writing giving the reasons therefor. If ENGINEER considers that part of the Work to be substantially complete, the provisions of paragraph 14.04 will apply with respect to certification of Substantial Completion of that part of the Work and the division of responsibility in respect thereof and access thereto.
2. No occupancy or separate operation of part of the Work may occur prior to compliance with the requirements of paragraph 5.10 regarding property insurance.

FIGURE 21.13 Provisions for Partial Utilization from the EJCDC Standard General Conditions of the Construction Contract 1910-8 (1996). *(Copyright © 1996 by National Society of Professional Engineers.)*

LIENS AND STOP ORDERS

Mechanics' lien laws applicable to construction are designed to protect subcontractors, material suppliers, laborers, and in some cases architects, engineers, and other design professionals who contribute to a work of improvement. These individuals and entities are all potential lien claimants on the projects they worked on. Due to the wide variation in state laws, it is impossible to do more than discuss lien rights generally.

Although public property is not subject to liens, some states have lien laws that entitle an unpaid claimant to place a lien on public funds that may be due a contractor. Under such schemes, the unpaid claimant advises the agency of its claim in accordance with the statutory notice requirements, and then the agency must stop further disbursements to the contractor of the affected funds until payment has been received by the claimant. Generally, once a claim has been made, the lien claimant must foreclose on its lien within a set time period in order to collect; otherwise, the rights will be lost. The agency must hold the liened funds until the lien has been satisfied through foreclosure, at which time the funds are paid over to the claimant. California, Louisiana, New York, and Texas are examples of four states that have provisions allowing liens to be filed against construction funds. In California, a lien against construction funds is accomplished by the use of a *Stop Notice,* which, when served, gives the agency notice to stop payment to the contractor.

Lien Waivers

A contractor may be asked to submit lien waivers as a condition of receiving payment. Although this is not at all uncommon at the end of the job, it is becoming increasingly popular as partial waivers of lien on a monthly basis, each covering only the value of the currently completed month's work.

There are basically two types of waivers of lien and two versions of each.

1. Conditional waiver and release upon receipt of progress payment upon final payment (Figure 21.14).
2. Unconditional waiver and release upon receipt of progress payment upon final payment (Figure 21.15).

A contractor will generally be asked to submit lien waivers for itself and each of its subcontractors before the owner will issue a check. Some owners don't seem to have the least bit of sympathy for the fact that it demands an unconditional release—they could refuse to pay at all, and the contractor would have no recourse for collection. The answer, of course, is either use the unconditional lien waivers when receiving payment in person, trading the contractor's lien release for the owner's check, or if being paid by mail, submit only a conditional waiver and release. A contractor is not at risk as long as conditional waivers are used, but should not consent to submitting unconditional releases unless the transaction is being handled in person and the check is tendered at the same time that the release is submitted.

CONDITIONAL WAIVER
AND RELEASE UPON FINAL PAYMENT
[California Civil Code §3262(d)(3)]

Upon receipt by the undersigned of a check from XYZ Constructors, Inc. (Maker of Check) in the sum of $ 5,278.43 (Amount of Check) payable to ABC Paving Contractors, Inc (Payee or Payees of Check) and when the check has been properly endorsed and has been paid by the bank upon which it is drawn, this document shall become effective to release any mechanic's lien, stop notice, or bond right the undersigned has on the job of City of Thousand Oaks (Owner) located at Janss Road at Drake (Job Description)

This release covers the final payment to the undersigned for all labor, services, equipment or material furnished on the job, except for disputed claims for additional work in the amount of $ 243.75. Before any recipient of this document relies on it, the party should verify evidence of payment to the undersigned.

Dated: 7 Aug 1990

ABC Paving Contractors, Inc. (Company Name)

By [signature] (Title)

Thousand Oaks
Form 17-7a

Purpose: Protect the City from lien (stop notice) claims.

Prepared by: Prime and Subcontractors; Suppliers.

Directed to: City's Representative.

Copies to: City; Project Manager or Construction Manager; Surety Company; Contractor.

Comments: Application for final payment should be accompanied by final waivers of lien from the Prime Contractor, all Subcontractors, and Suppliers who have not previously furnished such final waivers. If partial lien waivers have been used, the amount of such final lien waiver would not be required to exceed the last month's payment. *NOTE: That a conditional waiver does not become effective unless or until AFTER the funds have been credited to the lienholder's account.*

FIGURE 21.14 Conditional Waiver of Lien for Final Payment.

UNCONDITIONAL WAIVER
AND RELEASE UPON FINAL PAYMENT

[California Civil Code §3262(d)4]

The undersigned has been paid in full for all labor, services, equipment or material

furnished to XYZ CONSTRUCTORS, INC.

on the job of CITY OF THOUSAND OAKS STREET WIDENING

located at JANSS ROAD AT DRAKE

and does hereby waive and release any right to a mechnaic's lien, stop notice, or any right against a labor or material bond on the job, except for disputed claims for extra work in the amount of:

$ 5,278.43

Dated 5 AUG 1999

ABC PAVING CONTRACTORS, INC.
(Company Name)

By F Cochrane.
(Title)

NOTICE TO PERSONS SIGNING THIS WAIVER: THIS DOCUMENT WAIVES RIGHTS UNCONDITIONALLY AND STATES THAT YOU HAVE BEEN PAID FOR GIVING UP THOSE RIGHTS. THIS DOCUMENT IS ENFORCEABLE AGAINST YOU IF YOU SIGN IT, EVEN IF YOU HAVE NOT BEEN PAID. IF YOU HAVE NOT BEEN PAID, USE A CONDITIONAL RELEASE FORM.

Data used are fictitious
for illustration only

FIGURE 21.15 Unconditional Waiver of Lien for Final Payment.

Final Payment and Waiver of Liens

On public projects the closeout procedure is often more complex. A typical closeout procedure on a public works project is illustrated in Figure 21.6. Although mechanics' liens cannot be applied to public property, the lien laws of a number of states do provide the means to allow access to, or freezing of, public funds (Chapter 20). The lien laws throughout the country are by no means uniform. Lien rights, however, are based on the contract and provide for a lien on the property improved. Stop notices or "freeze orders" may also provide for a lien on funds payable by virtue of the improvement. A lien is only an additional remedy for securing payment of labor and materials furnished on a project. A final waiver of lien (Figures 21.14 and 21.15) is a receipt for an exact sum of money paid as of a certain date for certain services, labor, and material that were supplied under the contract for a specific improvement and property (see "Measurement for Payment" in Chapter 17).

If such a receipt is made for funds received by the contractor in payment for its work, any claim of lien to the extent of such payment against the money owed is waived. To implement this effectively in each state, a local attorney should be consulted, as the laws vary from state to state.

Every time a payment is made to a contractor, you should insist on a corresponding waiver of lien rights. Progress payments should result in partial lien waivers, and final payment should result in a full waiver and release.

Along with each lien waiver, you should insist on an affidavit from the contractor swearing that all subcontractors and suppliers who furnished labor or materials for the work covered by the owner's payment have been paid in full. This should include an indemnification provision whereby the contractor agrees to reimburse the owner if any of the subs or suppliers later assert a lien. It is important for each administrator of a construction contract to become familiar with the lien laws of his or her own state. It should be kept in mind that even individuals may be able to assert lien rights over the owner's property due to failure of their contractor employer to pay them wages due them.

It is important to note, however, that on projects for which the contractor has provided surety bonds, it is essential to obtain a written release or *consent of surety* prior to release of the final payment to the contractor (Figure 21.16).

Stop Notice Release Bond

Most states that have statutory provisions enabling lien claimants to lien construction funds usually have provisions to help the contractor protect itself against fraudulent, improper, or disputed claims. In some states, if the contractor posts a bond in a specified amount to cover the amount of the lien, the agency is then free to release the construction funds that have been held up in accordance with the original lien. However, it should be noted that agencies will often demand that all liens and lien-related issues be resolved by the contractor before the project is accepted and final payment is made.

City of Palm Springs

CONSENT OF SURETY
For Final Payment

Project Name 1991 Expansion of Palm Springs Convention Center

Location 277 North Avenida Caballeros

Project No. CP90-38 Contract No. ____

Type of Contract Construction

Amount of Contract $5,678,000.00

In accordance with the provisions of the above-named contract between the Owner and the Contractor, the following named surety:

Guaranteed Indemnity Corporation, a legal entity organized and existing in the City of Los Angeles, State of California.

on the Payment Bond of the following named Contractor:

XYZ Constructors, Inc., a California Corporation, organized and existing in The City of Santa Ana, California

hereby approves of final payment to the Contractor, and further agrees that said final payment to the Contractor shall not relieve the Surety Company named herein of any of its obligations to the following named Owner: as set forth in said Surety company's bond:

City of Palm Springs, a legal entity, organized and existing in the County of Riverside, State of California.

IN WITNESS WHEREOF, the Surety Company has hereunto set its hand and seal this 20th day of August 1990 19____

Guaranteed Indemnity Corporation
(Name of Surety Company)

(Signature of Authorized Representative)

(Affix corporate seal here)

TITLE Vice President

Wiley-Fisk Form 17-5

Purpose: The surety's financial interest could be jeopardized by premature final payment by the owner.

Prepared by: Surety company upon specific request of the contractor.

Directed to: Owner's Project Manager.

Copies to: Surety company; owner; Project Manager; Resident Project Representative.

Comments: Final payment by owner, prior to a consent of the surety, could place owner in a vulnerable position in case contractor fails to pay his subcontractors or suppliers.

FIGURE 21.16 Consent of Surety for Final Payment to the Contractor. *(Fisk, Edward R.,* Construction Engineer's Complete Handbook of Forms, *1st edition, © 1993. Reprinted by permission of Pearson Education, Inc., Upper Saddle River, NJ.)*

One method open to a public agency to allow the release to the contractor of retainage funds being held to cover stop notice claims is through the acceptance of a *Stop Notice Release Bond* from a reputable surety company in favor of the general contractor. A public agency that honors a Stop Notice Release Bond by releasing retainage funds to the contractor may be relieved of any liability resulting from the stop notice. The public agency is no longer a stakeholder when it accepts a release bond. Its duty has been discharged by release of the withheld funds, and the claimant must then look to the surety company for payment. Under such conditions, the claimant has no right of action against the public agency [*Cal-Pacific Materials Co. v. Redondo Beach City School Dist.*, 94 Cal. App. 3d 652, 156 Cal. Rptr. 590 (1979)].

Post Completion

As soon as the contractor's final payment and release of retainage have been made, another, and final, phase of the project begins. At this point, assuming that all contractual obligations of both parties have been met, the project manager takes over all of the activities on the project, and the Resident Project Representative is completely phased out in most organizations.

The visual satisfactions of watching the project develop into reality are now gone, and in its place is the rather mundane task of bookkeeping and report writing. Of the several tasks that must be accomplished during this period, the most prominent ones are:

1. Financial accounting of the entire project, summarizing all costs, expenditures, overhead, profit, and other cost-related items for the entire life of the project.
2. Preparation of the final project report to the management team of the architect/engineer's or owner's organization.
3. Assembly, organization, analysis, and filing of complete project records for the master file. These are quite important, as they may yet be needed by the legal department in case of later claims.

Problems

1. What types of coverage are provided under a guarantee? What types under a warranty?
2. Under a liquidated damages clause, is it necessary to prove that the owner suffered a loss to be entitled to withhold liquidated damages?
3. Is it necessary to include a provision for a bonus clause in order to have a valid liquidated damages clause?
4. Define *substantial completion.*

5. In a construction contract, list the effects of achieving substantial completion upon the contract requirements for project insurance, liquidated damages, time for filing lien claims, processing of payments, and retainage.
6. What is the effect of "beneficial use" on the foregoing list of contract requirements?
7. Define the difference between a conditional and an unconditional waiver of lien.
8. Define "punch list" and indicate the principal elements that a punch list must contain.
9. What phrase regarding time must be included in the contract to support a liquidated damages claim?
10. Liquidated damages are intended to represent anticipated losses to the owner based upon circumstances existing at the time the contract was made. List at least five types of potential losses to the owner that would qualify for determination of such potential losses.

BIBLIOGRAPHY

ACRET, J. *California Construction Law Manual,* 3rd ed. Colorado Springs, CO: Shepard's/McGraw-Hill, 1982.

ANTILL, J. M., AND WOODHEAD, R. W. *Critical Path Methods in Construction Practice,* 3rd ed., New York, NY: Wiley, 1982.

ASCE Committee on Specifications, "Summary Report of Questionnaire on Specifications (Contractor Returns)," ASCE Vol. 104, No. C03, Proceedings Paper No. 14001, September 1978, pp. 353–359.

ASCE Committee on Specifications, "Summary Report of Questionnaire on Specifications (Owner and Owner Representative Returns)," *Journal of the Construction Division,* ASCE Vol. 105, No. C03, Proceedings Paper No. 14799, September 1979, pp. 163–186.

ASKEN, G. "Resolving Construction Contract Disputes through Arbitration." *Delays and Disputes in Building Construction,* New York, The American Arbitration Association, 1981.

Associated General Contractors of America. "Standard Form of Agreement between Owner and Construction Manager (Guaranteed Maximum Price Option)," AGC Document No. 8. Washington, D.C.

BARRIE, D. S., AND PAULSON, B. C. JR., "Professional Construction Management," *Journal of the Construction Division,* Vol. 102, No. C03, Proceedings of the American Society of Civil Engineers, September 1976, pp. 427, 428.

BONNY AND FREIN, *Handbook of Construction Management and Organization,* Van Nostrand Reinhold Company, New York, 1973.

A Businessman's Guide to Commercial Arbitration and a Manual for Commercial Arbitrators and Construction Industry Arbitration Rules. New York, The American Arbitration Association, 1981.

California Jury Instructions Civil, Book of Approved Jury Instructions (BAJI), 7th ed. by Charles A. Loring, Judge of the Superior Court (Retired), West Publishing Company, St Paul, MN, 1986.

California Public Contract Law Conference, published papers, Sacramento, CA, April 1970, pp. 3, 4.

CASEY, J. J. (President, Gordon H. Ball, Inc., New York, NY). "Identification and Nature of Risks in Construction Projects: A Contractor's Perspective." *Construction Risks and Liability Sharing,* ASCE Conference (January 1979), published papers, Vol. 1, pp. 17–23.

CIAC Construction Industry Affairs Committee of Chicago, Recommendations No. 3, *Management Control of Construction Operations.*

CIAC Construction Industry Affairs Committee of Chicago, Recommendations No. 7, *Construction Contract Change Orders.*

CIAC Construction Industry Affairs Committee of Chicago, Recommendations No. 8, *Punch List.*

CIAC Construction Industry Affairs Committee of Chicago, Recommendations No. 9, *Substitutions.*

CIAC Construction Industry Affairs Committee of Chicago, Recommendations No. 12, *Construction Completion Schedules Related to Building Costs.*

CIAC Construction Industry Affairs Committee of Chicago, Recommendations No. 13, *Waiver of Lien Procedure.*

CIAC Construction Industry Affairs Committee of Chicago, Recommendations No. 16, *Bonds.*

City of Palm Springs, CA, Procurement Department, 3200 E. Tahquitz Canyon Drive, Palm Springs, CA 92263-2743.

City of Thousand Oaks, CA, Public Works Department, Engineering Division, 2100 Thousand Oaks Blvd., Thousand Oaks, CA 91362.

Clarke, J. R., Esq. *Commentary on Contract Documents.* 1990 ed. Alexandria, VA 22314, Professional Engineers in Private Practice, a practice division of the National Society of Professional Engineers, Alexandria, VA 22314.

Clarke, J. R., Esq. *Focus on Shop Drawings,* 1990 ed., prepared for Engineer's Joint Contract Documents Committee, published by the National Society of Professional Engineers, Alexandria, VA 22314.

Clough, R. H. *Construction Contracting,* 5th ed., New York: Wiley, 1986.

Clough, R. H. *Construction Project Management,* New York: Wiley, 1972.

Coffin, R. A. *The Negotiator: A Manual for Winners.* New York, AMACM, a Division of American Management Associations, 1973.

Collins, F. T. *Network Planning and Critical Path Scheduling,* Berkeley, CA: Know-How Publications, 1964.

Constructability Task Force. "Guidelines for Implementing a Constructability Program," Publication 3-2, July 1987, Construction Industry Institute, The University of Texas at Austin, Austin, Texas 78705.

Construction Claims Monthly, April 1983, "Differing Site Conditions."

Construction Claims Monthly, August 1983, "Recent Decisions Affecting Recovery of Home Office Overhead," by Robert G. Watt, Esq., and David C. Romm, Esq.

Construction Claims Monthly, December 1983, "Recovery of Home Office Overhead—A Different Point of View," by Phillip R. McDonald, Esq.

Construction Claims Monthly, July 1984, "The Return of Eichleay: Is it Here to Stay?" Part II, by W. John Irwin II, P. E., and "Denial of Time Extension Is a Constructive Acceleration."

Construction Claims Monthly, 8737 Colesville Road, Suite 1100, Silver Spring, MD 20910, Phone (301) 587-6300.

Contractor's State Licensing Board, Sacramento, CA. *California Contractor's License Law and Reference Book.*

CSI Manual of Practice, Vol. 2, "Formats: Specifications and Manuals." Washington, D.C., The Construction Specifications Institute, Inc.

Department of the Navy. "Quality Level and Quality Control" NAVFAC P-455, *Construction Engineering Handbook,* July 1974 ed., Book 1—General Requirements, Naval Facilities Engineering Command, Alexandria, VA.

Dixon, S. A. and Crowell, R. D. *The Contract Guide.* Design Professionals Insurance Company (DPIC), Monterey, CA, 1993.

EDGEWATER Services, ProjectEDGE, 225 Greenfield Parkway, Suite 102, Liverpool, NY 13088 (ProjectEDGE.com).

EJCDC Standard General Conditions of the Construction Contract, prepared by the Engineer's Joint Contract Documents Committee, as issued and published by Professional Engineers in Private Practice, a practice division of the National Society of Professional Engineers;

American Consulting Engineer's Council; American Society of Civil Engineers; approved and endorsed by the Associated General Contractors of America (1996 edition).

Engineers Joint Contract Documents Committee (EJCDC), *Standard Forms of Agreement,* Document No. 1910-1-A (1996 edition).

FIDIC Conditions of Contract for Works of Civil Engineering Construction, Parts I and II, Federation Internationale des Ingenieurs-Conseil Fourth Edition 1987, Lausanne, Switzerland.

FISK, E. R. *Construction Contracts and Specifications for California Public Works Projects,* a technical book, Brea, CA, 1990.

FISK, E. R. *Construction Engineer's Complete Handbook of Forms.* New York: Prentice Hall, 1981, 1992.

FISK, E. R. "Designer Evaluation of Contractor Comments on Specifications," *Journal of the Construction Division,* ASCE Vol. 104, No. C01, Proceedings Paper No. 13585, March 1978, pp. 77–83; presented at the October 17–21, 1977, ASCE National Convention and Exposition, held at San Francisco, CA (Preprint No. 2930).

FISK, E. R. "Evaluation of Owner Comments on Specifications," *Journal of the Construction Division,* ASCE Vol. 106, No. C04, Proceedings Paper 15875, December 1980, pp. 469–476; presented at the April 2–6, 1979, ASCE Spring Convention, held at Boston, MA.

FISK, E. R. *Inspector Responsibility and Authority,* a video course prepared for the University of California, Berkeley, Center for Media and Independent Learning, 1997.

FISK, E. R. "Inspector Training and Knowledge of Specifications," a technical paper presented at the Feb. 16–18, 1981, ASCE Construction Division Committees on Inspection and Specifications Specialty Conference on *Reducing Risk and Liability through Better Specifications and Inspection,* held in San Diego, CA.

FISK, E. R. "Management Systems for Claims Protection," a technical paper presented at the March 17–19, 1982, ASCE Construction Division Committee on Professional Construction Management Specialty Conference on *Engineering and Construction Projects—The Emerging Management Roles,* New Orleans, LA.

FISK, E. R. "Project Administration," a technical paper presented at the April 15–17, 1985, ASCE Construction Division Committee on Contract Administration Specialty Conference on *The Resident Engineer,* Tampa, FL.

FISK, E. R. "Risk Management and Liability Sharing," a technical paper presented at the May 1–3, 1985, Annual State Conference of the Arizona Water & Pollution Control Association (AWPCA), held at Lake Havasu, AZ.

FISK, E. R. "Specifications-Design Professionals Forgotten Challenge," *Journal of the Construction Division,* ASCE Vol. 102, No. C02, Proceedings Paper 12171, June 1976, pp. 303–306; presented at the November 3–7, 1976, ASCE National Convention and Expo, held at Denver, CO (Preprint No. 2522).

FISK, E. R. "The Use of Project Records for Litigation," a technical paper presented at the ASCE Construction Division Committee on Professional Construction Management Symposium on *Liability and the Construction Manager,* held at the October 17–21, 1983, ASCE Convention and Exposition held in Houston, TX.

FISK, E. R. AND CALHOUN, J. C. *Contracts and Specifications for Public Works Projects.* New York, Wiley, 1991.

FISK, E. R., AND NEGELE, J. R. *Contractor's Project Guide to Public Works Contracts.* New York: Prentice Hall, 1987.

General Services Administration, Public Buildings Service, Project Management Division, Washington, D.C. *Guide Specifications,* U.S. Army, Corps of Engineers.

GORDON, C., P.E. "Accident Prevention." Unpublished paper.

HAMMOND, D. G. (Vice President, Daniel, Mann, Johnson, & Mendenhall, Baltimore, MD). "Minimizing Risks and Mitigating Losses," *Construction Risks and Liability Sharing* (ASCE Conference, January 1979) published papers Vol. 1, pp. 133–143.

HUIE, W. S. (Partner, Kutak, Rock, & Huie, Atlanta, GA) "Identification and Nature of Risks," *Construction Risks and Liability Sharing* (ASCE Conference, January 1979) published papers Vol. 1, pp. 15–27.

JACOBY, H. J. (Chairman, Grow Tunneling Corporation, New York, NY). "Summary Session Commentary," Vol. 2, pp. 178–183.

KUESEL, T. R. (Senior Vice President, Parsons, Brinckerhoff, Quade, & Douglas, Inc., New York, NY). "Allocation of Risks," *Construction Risks and Liability Sharing* (ASCE Conference, January 1979) published papers Vol. 1, pp. 51–60.

Managing Construction of Clean Water Grant Projects, a clean water grant program guideline by Division of Water Quality, California State Water Resources Control Board, March 1979.

MCLEAN, R. C. "Construction Planning and Scheduling" from Havers and Stubbs (eds.), *Handbook of Heavy Construction,* 2nd ed. NY: McGraw-Hill, 1971.

NADEL, N. A. (President, MacLean Grove & Company, Inc., New York, NY). "Allocation of Risks—A Contractor's View," *Construction Risks and Liability Sharing* (ASCE Conference, January 1979) published papers Vol. 1, pp. 61–67.

National Joint Guideline. "Overtime, Construction Costs and Productivity." American Subcontractor's Association, Associated General Contractors of America, and Associated Specialty Contractors, Inc., 1979.

The National Society of Professional Engineers. *Professional Index of Private Practice.* Washington, D.C.

NAVFAC Construction Quality Control Manual, NAVFAC P-445. Department of the Navy, Naval Facilities Engineering Command, Alexandria, VA.

NEAL, RICHARD H., AND DAVID E. *Construction Planning.* London: Thomas Telford.

Network Float—What Is It and Who Owns It? and Some Comments on the Precedence Diagraming Technique of Network Analysis Systems, 1976. Forward Associates, Ltd., Novato, CA, 1991.

NSPE/PEPP Guidelines for Development of Architect/Engineer Quality Control Manual. Washington, D.C., Professional Engineers in Private Practice, a Division of National Society of Professional Engineers, NSPE Pub. No. 1957, January 1977.

O'BRIEN, J. J. *CPM in Construction Management,* 2nd ed. New York: McGraw-Hill, 1971.

PEURIFOY, R. L. *Construction Planning, Equipment, and Methods.* New York: McGraw-Hill, 1956.

Prevention, Management and Resolution of Construction Contractor Claims on Clean Water Grant Projects, Vol. I—Discussion. California State Water Resources Control Board, Division of Water Quality, June 1980.

SEARLES, W. H., IVES, H. C., AND KISSAM, P. *Field Engineering,* 22nd ed. New York: Wiley.

Smart Computing, "How It Works: Digital Cameras." Lincoln, Nebraska, April 2000.

State of California, Department of Transportation, *Construction Manual,* Section 1-60 Safety; issued by the Division of Construction and Research, Office of Construction, Sacramento, CA, 1990.

State of California, Department of Transportation, *Construction Manual,* Section 2-08 Prosecution and Progress; issued by the Division of Construction and Research, Office of Construction, Sacramento, CA, 1990.

State of California, Department of Transportation, *Construction Manual,* Section 2-70 Protests, Potential Claims, and Claims; issued by the Division of Construction and Research, Office of Construction, Sacramento, CA, 1990.

State of Washington, Department of Transportation, *Construction Manual M 41-01,* Olympia, WA, http://www.wsdot.wa.gov/fossc/cons/const.pdf

SWEET, JUSTIN. *Sweet on Construction Industry Contracts: Major AIA Documents.* Wiley Law Publications, John Wiley & Sons, New York, 1987.

The Uniform Arbitration Act. New York: The American Arbitration Association, 1990.

U.S. Army Corps of Engineers. *Resident Engineer's Management Guide,* pamphlet EP 415-1-260, dated 6 December 1990.

U.S. Army Management Engineering Training Agency. *Principles and Applications of Value Engineering,* Department of Defense Joint Course Book, Vol. I. Rock Island, IL.

U.S. Government Printing Office. *Value Engineering,* Department of Defense Handbook 5010.8-H, 12 September 1968, Superintendent of Documents, Washington, D.C.

U.S. Department of the Interior, Bureau of Reclamation. *Metric Manual,* by L. D. Pedde, W. E. Foote, L. F. Scott, D. L. King, and D. L. McGalliard. Bureau of Reclamation, Engineering and Research Center, Denver Federal Center, P.O. Box 25007, Denver, CO 80225, Attn: 922, dated 1978.

INDEX

FORMS INDEX

ABOUT THE AUTHORS

Edward R. Fisk, PE, LS, is a construction consultant in Orange, California. He is a licensed civil and structural engineer, land surveyor, and licensed general contractor and holds licenses in 13 states where he has been involved on projects. Before becoming an independent consultant, he was president of Gleason, Peacock, & Fisk, Inc., of Brea, CA, construction consultants; vice president of Lawrence, Fisk, & McFarland, Inc., engineers, of Santa Barbara, CA; and vice president of Construction Services for Wilsey & Ham, engineers, Foster City, CA. Prior to that he served as Corporate Director of Construction Management for J.M. Montgomery Engineers (now MWH Global), and VTN Consolidated, Inc., and was an Engineer and a Field Engineer for Bechtel Corp., Power Division. He has had extensive experience in both the public and private sectors. He is a Life Fellow of the American Society of Civil Engineers and former chairman of its Construction Division, and is a Fellow of the National Academy of Forensic Engineers. He taught short courses statewide for the University of California, Berkeley, Institute of Transportation Studies, and for the University of Washington, Engineering Professional Programs, Seattle for many years. He lectured nationally and internationally for the American Society of Civil Engineers until succeded by co-author Wayne Reynolds.

Wayne D. Reynolds, PE, is an Associate Professor in the Department of Technology at Eastern Kentucky University. He teaches introduction to construction, project organization and supervision, scheduling and cost control, contracts and bidding, quantity and cost estimating, engineering economy, and statics in the Construction Management program. He also lectures nationally and internationally for the American Society of Civil Engineers. He has a B.S. from the U.S. Military Academy and an M.S. in civil engineering from the Ohio State University. Mr. Reynolds has experience in design and construction projects including transportation facilities, navigation structures, flood protection, and institutional buildings, and has served as Inspector, Project Engineer, Assistant Resident Engineer, Project Manager, Deputy District Engineer, and Contracting Officer. He was an Assistant Professor of Civil Engineering at the U.S. Air Force Academy and completed service in the U.S. Army Corps of Engineers as a Lieutenant Colonel. He is a member of American Society of Civil Engineers, the Society of American Military Engineers, and the Associated Schools of Construction.